Springer **M***onographs in* **M***athematics*

Heinz-Dieter Ebbinghaus
Jörg Flum

Finite Model Theory

Second Revised and Enlarged
Edition 1999

Heinz-Dieter Ebbinghaus
Jörg Flum
Mathematisches Institut
Abteilung für Mathematische Logik
Universität Freiburg
Eckerstraße 1
79104 Freiburg, Germany
e-mail: hde@math.uni-freiburg.de
joerg.flum@math.uni-freiburg.de

The softcover edition was published in 1999 by Springer-Verlag under the same title in the series *Perspectives in Mathematical Logic.*

Library of Congress Control Number: 2005932862

Mathematics Subject Classification (2000): 03C13, 03C80, 03D15, 68P15

ISSN 1439-7382
ISBN-10 3-540-28787-6 Springer Berlin Heidelberg New York
ISBN-13 978-3-540-28787-2 Springer Berlin Heidelberg New York
ISBN 3-540-65758-4 2nd Edition Springer-Verlag Berlin Heidelberg New York

Springer is a part of Springer Science+Business Media
springeronline.com

Printed in Germany

Typeset by the authors using a Springer LaTeX macro package
Production: LE-TeX Jelonek, Schmidt & Vöckler GbR, Leipzig
Cover design: Erich Kirchner, Heidelberg

Printed on acid-free paper 46/3142YL - 5 4 3 2 1 0

Preface

Finite model theory, the model theory of finite structures, has roots in classical model theory; however, its systematic development was strongly influenced by research and questions of complexity theory and of database theory.

Model theory or the theory of models, as it was first named by Tarski in 1954, may be considered as the part of the semantics of formalized languages that is concerned with the interplay between the syntactic structure of an axiom system on the one hand and (algebraic, settheoretic, ...) properties of its models on the other hand. As it turned out, first-order language (we mostly speak of first-order logic) became the most prominent language in this respect, the reason being that it obeys some fundamental principles such as the compactness theorem and the completeness theorem. These principles are valuable modeltheoretic tools and, at the same time, reflect the expressive weakness of first-order logic. This weakness is the breeding ground for the freedom which modeltheoretic methods rest upon.

By compactness, any first-order axiom system either has only finite models of limited cardinality or has infinite models. The first case is trivial because finitely many finite structures can explicitly be described by a first-order sentence. As model theory usually considers *all* models of an axiom system, modeltheorists were thus led to the second case, that is, to infinite structures. In fact, classical model theory of first-order logic and its generalizations to stronger languages live in the realm of the infinite. Basic methods such as the methods of constructing models, and basic aims such as a structure theory for first-order axiomatizable classes of structures are essentially concerned with infinite structures and transfinite or settheoretic combinatorics.

Nevertheless, there are natural reasons to consider finite structures. Historically, the most important one is the finite model property for certain classes of first-order formulas, that is, the equivalence of satisfiability and satisfiability in the finite. It was this property that settled the positive cases of the decision problem for prefix classes of first-order logic.

However, it took some twenty years until the middle of the twentieth century to really ask questions of a modeltheoretic flavour within the world of finite structures. The first landmarks here are Trahtenbrot's Theorem (1950) on the failure of the completeness theorem in the finite and the formulation of the spectrum problem by Scholz (1952) that asks for a characterization of spectra in the finite of first-order sentences.

In both cases, computational aspects play their part: Trahtenbrot's proof rests on the undecidability of the halting problem for machines, and the spectrum problem turned out to be intimately linked to the question whether deterministic and nondeterministic polynomial time complexity coincide. The influence of computational aspects can be explained in various ways. First of all, finite structures can be coded as words and hence, can be objects of computations. Moreover, finite structures can serve to describe finite runs of machines. Finally, formulas of a logical language often can be interpreted as programs that, given a structure as input, perform the corresponding evaluation. This viewpoint is of importance, for example, in database theory, where relational databases are considered as finite structures.

We now give a short description of the contents of the book, at the same time motivating the choice of material and the emphasis it has been given.

Chapter 1 provides basic material concerning first-order logic. Chapters 2 and 3 are of a purely modeltheoretic character. When restricting oneself to finite structures, the essential theorems of first-order logic fail (this is documented at the end of Chapter 3) and important methods get trivial or useless. However, the gametheoretic methods of Ehrenfeucht and Fraïssé survive or even gain a special power. They are developed for first-order logic in Chapter 2 and for second-order and infinitary extensions in Chapter 3. Our representation is strongly based on isomorphism types (or, Hintikka formulas) as a unifying feature.

When turning to the finite, settheoretic combinatorics get replaced by finite combinatorics. As a consequence, there are new questions, for instance, questions of the kind: What is the relative frequency of graphs versus structures or of connected graphs versus graphs? Chapter 4 is dedicated to results aiming in this direction, so-called 0-1 laws, for first-order logic and some extensions. They say that for relational formulas either almost all or almost no finite structure is a model.

Chapter 5 treats the finite model property for some fragments of first-order logic, namely for the two-variable logic that consists of first-order formulas with only two variables, and for the class of $\forall\exists$-formulas. The choice of the first logic is motivated by the methodological role of logics with a restricted number of variables, the choice of the second one by its relationship to concepts of the preceding chapter.

As has already been remarked, many questions in finite model theory are related to or even arose from questions in complexity theory. Chapter 6 gives an account of these connections on the computational level of finite automata.

When considering automata or even more powerful machines, it quickly becomes clear that first-order logic does not provide an adequate framework on the logical side because first-order logic lacks the ability of adequately expressing recursive procedures. This weakness of expressive power corresponds to a similar phenomenon in classical model theory, and in a similar way as there, one can try to overcome this deficiency by introducing more powerful languages. In finite model theory, various so-called fixed-point logics have turned out to be promising candidates. They allow to speak about fixed-points of definable iteration procedures; the weakest one just allows to speak about the transitive closure of a definable binary relation.

Chapter 7 (together with Chapter 8) forms the core of the book. After a short introduction to fixed-point logics it develops what is known as the theory of descriptive complexity. Given a resource-bounded machine model, say the model of polynomially time-bounded Turing machines, and a suitable logic $\mathcal{L}$, say least fixed-point logic, then for every machine M of this kind there is a sentence φ of $\mathcal{L}$ whose models are just the structures accepted by M, and for every sentence φ of $\mathcal{L}$ there is a machine M of this kind that just accepts the models of φ. Hence, the classes of structures that are acceptable by a polynomially time-bounded machine, correspond to the classes that are axiomatizable in $\mathcal{L}$. In this way, one obtains logical descriptions of complexity classes and, therefore, logical analogues of major problems in complexity theory. For example, the PTIME = PSPACE-problem now amounts to the question whether two fixed-point logics have the same expressive power in finite structures. The chapter gives an account of the most important results of this kind. As it turns out, they mostly presuppose that the structures considered are ordered. This assumption seems to be natural, since we induce an ordering on the universe of a structure, when encoding it as an input string. On the other hand, some major open problems of descriptive complexity theory are concerned with the question to what extent the descriptive results can be generalized to structures without an ordering.

The proof indicated above, for the fact that least fixed-point logic corresponds to PTIME, yields that any sentence φ of this logic is equivalent to a sentence describing the behaviour of a machine and hence, has a special syntactical structure. So, as a byproduct of the logical characterizations of complexity classes, one gets certain normal forms for the logics involved.

Chapter 8 (with parts of Chapter 9) presents the model theory of fixed-point logics for finite structures. The material presented includes a thorough discussion of transitive closure logics, as well as connections to second-order logic and infinitary logics.

Chapter 9 takes up the viewpoint of formulas as programs or queries, introducing a bunch of programming languages of the DATALOG family in database theory. The main aim here is to show the equivalence of these languages with certain fixed-point languages, thus opening another methodological gateway to the latter ones. The way pays, as can be seen by new and far reaching possibilities of obtaining normal forms. The chapter concludes with the investigation of the fine structure of fixed-point languages.

Chapter 10 lies outside the mainstream of the book. It is concerned with a logically-oriented representation of optimization problems, but is restricted to basic material.

Chapter 11 is concerned with one of the most prominent open problems of finite model theory, namely whether there is a logic that captures PTIME also on unordered structures. As it turns out, this question is linked to the problem of feasibily characterizing finite structures by invariants. By coding structures in graphs we reduce the problem to the case of graphs. The chapter ends by providing positive answers for some classes of graphs, thus illustrating an at the present time active direction of research.

Finally, Chapter 12 discusses a concept that is well-known from classical model theory, the concept of a quantifier. The idea here is the following: In order to find a logic $\mathcal{L}$ that corresponds to a complexity class $\mathcal{C}$ in the sense described above, one can try to start with a simple logic, say first-order logic, and add a quantifier that incorporates a $\mathcal{C}$-complete problem. When pursuing this idea, one has to find analogues of notions from complexity theory such as reductions, hardness, completeness, etc. The chapter concludes with an example of the analogy between quantifiers in logic and oracles in complexity theory.

As the preceding description shows, the core of the book is centered around modeltheoretic issues related to descriptive complexity. Such a concentration seemed to be necessary in order to come along with a book of a reasonable size. Among the major gaps the reader will encounter we mention two ones lying at opposite ends of the spectrum of possible topics: the theory of circuits on the more computational side and the work related to a structure theory for the finite on the more modeltheoretic side. Concerning topics of the first group we refer the reader to Neil Immerman's recent monograph "Computational Complexity".

Some chapters are as independent from each other as their contents allows. In particular, a reader interested in descriptive complexity theory should not have problems to start with Chapter 7. And a modeltheorist interested in the finite model theory of fixed-point logics can start with Chapter 8.

This second edition is a thoroughly revised and enlarged version of the original text, the most relevant addition being the new Chapter 11. Moreover, we have tried to take into consideration criticism and suggestions of readers of the first edition. Again, with gratitude we mention Martin Grohe: his proposals and his results have influenced and enriched considerable parts of the text.

Freiburg, April 1999

Heinz-Dieter Ebbinghaus
Jörg Flum

Table of Contents

Preface V

1. Preliminaries 1

2. The Ehrenfeucht-Fraïssé Method 13
- 2.1 Elementary Classes 13
- 2.2 Ehrenfeucht's Theorem 15
- 2.3 Examples and Fraïssé's Theorem 20
- 2.4 Hanf's Theorem 26
- 2.5 Gaifman's Theorem 30

3. More on Games 37
- 3.1 Second-Order Logic 37
- 3.2 Infinitary Logic: The Logics $L_{\infty\omega}$ and $L_{\omega_1\omega}$ 40
- 3.3 The Logics FO^s and $L^s_{\infty\omega}$ 46
  - 3.3.1 Pebble Games 49
  - 3.3.2 The s-Invariant of a Structure 54
  - 3.3.3 Scott Formulas 56
- 3.4 Logics with Counting Quantifiers 58
- 3.5 Failure of Classical Theorems in the Finite 62

4. 0–1 Laws 71
- 4.1 0–1 Laws for FO and $L^\omega_{\infty\omega}$ 71
- 4.2 Parametric Classes 74
- 4.3 Unlabeled 0–1 Laws 77
  - 4.3.1 Appendix 82
- 4.4 Examples and Consequences 84
- 4.5 Probabilities of Monadic Second Order Properties 88

5. Satisfiability in the Finite 95
- 5.1 Finite Model Property of FO^2 95
- 5.2 Finite Model Property of $\forall^2\exists^*$-Sentences 99

6. Finite Automata and Logic: A Microcosm of Finite Model Theory ... 105
6.1 Languages Accepted by Automata ... 105
6.2 Word Models ... 108
6.3 Examples and Applications ... 111
6.4 First-Order Definability ... 114

7. Descriptive Complexity Theory ... 119
7.1 Some Extensions of First-Order Logic ... 120
7.2 Turing Machines and Complexity Classes ... 124
7.2.1 Digression: Trahtenbrot's Theorem ... 127
7.2.2 Structures as Inputs ... 129
7.3 Logical Descriptions of Computations ... 133
7.4 The Complexity of the Satisfaction Relation ... 147
7.5 The Main Theorem and Some Consequences ... 151
7.5.1 Appendix ... 162

8. Logics with Fixed-Point Operators ... 165
8.1 Inflationary and Least Fixed-Points ... 165
8.2 Simultaneous Induction and Transitivity ... 177
8.3 Partial Fixed-Point Logic ... 191
8.4 Fixed-Point Logics and $L^{\omega}_{\infty\omega}$ ... 198
8.4.1 The Logic $FO(PFP_{PTIME})$ ... 205
8.4.2 Fixed-Point Logic with Counting ... 207
8.5 Fixed-Point Logics and Second-Order Logic ... 210
8.5.1 Digression: Implicit Definability ... 217
8.6 Transitive Closure Logic ... 220
8.6.1 FO(DTC) < FO(TC) ... 221
8.6.2 FO(posTC) and Normal Forms ... 224
8.6.3 FO(TC) < FO(LFP) ... 229
8.7 Bounded Fixed-Point Logic ... 235

9. Logic Programs ... 239
9.1 DATALOG ... 239
9.2 I-DATALOG and P-DATALOG ... 245
9.3 A Preservation Theorem ... 250
9.4 Normal Forms for Fixed-Point Logics ... 253
9.5 An Application of Negative Fixed-Point Logic ... 263
9.6 Hierarchies of Fixed-Point Logics ... 268

10. Optimization Problems ... 275
10.1 Polynomially Bounded Optimization Problems ... 275
10.2 Approximable Optimization Problems ... 280

11. Logics for PTIME 287
11.1 Logics and Invariants 288
11.2 PTIME on Classes of Structures 295

12. Quantifiers and Logical Reductions 307
12.1 Lindström Quantifiers 308
12.2 PTIME and Quantifiers 314
12.3 Logical Reductions 320
12.4 Quantifiers and Oracles 330

References 339

Index 349

1. Preliminaries

The purpose of this section is to fix notations and terminology for the basic notions related to first-order logic. We assume that the reader has already some familiarity with these concepts, as it is obtained by an introductory course in mathematical logic. For more detailed information we refer to textbooks such as [32, 35]. Parts of our exposition in part B follow [9].

A Structures

Vocabularies are finite sets that consist of *relation symbols* P, Q, R, ... and *constant symbols* (for short: *constants*) c, d, ...[1] Every relation symbol is equipped with a natural number ≥ 1, its *arity*. We denote vocabularies by $\tau, \sigma, \ldots$. A vocabulary is *relational*, if it does not contain constants.

A *structure* $\mathcal{A}$ of vocabulary τ (by short: a *τ-structure*) consists of a nonempty set A, the *universe* or *domain* of $\mathcal{A}$, of an n-ary relation $R^{\mathcal{A}}$ on A for every n-ary relation symbol R in τ, and of an element $c^{\mathcal{A}}$ of A for every constant c in τ. (Mostly we use the notations R^A for $R^{\mathcal{A}}$ and c^A for $c^{\mathcal{A}}$.) An n-ary *relation* R^A on A is a subset of A^n, the set of n-tuples of elements of A. We mostly write $R^A a_1 \ldots a_n$ instead of $(a_1, \ldots, a_n) \in R^A$. A structure $\mathcal{A}$ is *finite*, if its universe A is a finite set.

We give some examples of structures that will play a prominent role in the book.

A1 *Graphs*

Let $\tau = \{E\}$ with a binary relation symbol E. A *graph* (or, *undirected graph*) is a τ-structure $\mathcal{G} = (G, E^G)$ satisfying

(1) for all $a \in G$: not $E^G aa$
(2) for all a, $b \in G$: if $E^G ab$ then $E^G ba$.

By GRAPH we denote the class of *finite* graphs. If only (1) is required, we speak of a *digraph* (or, *directed graph*). The elements of G are sometimes called *points* or *vertices*, the elements of E^G *edges*. The following figures represent

[1] Usually also function symbols are allowed in vocabularies. For the purposes of this book our definition does not represent an essential restriction. We explain in part D how the results can be extended to vocabularies with function symbols.

the graph $(\{a,b,c,d\},\{(a,b),(b,a),(b,c),(c,b),(b,d),(d,b),(c,d),(d,c)\})$ and the digraph $(\{a,b,c,d\},\{(a,b),(b,a),(b,c),(b,d),(d,c)\})$.

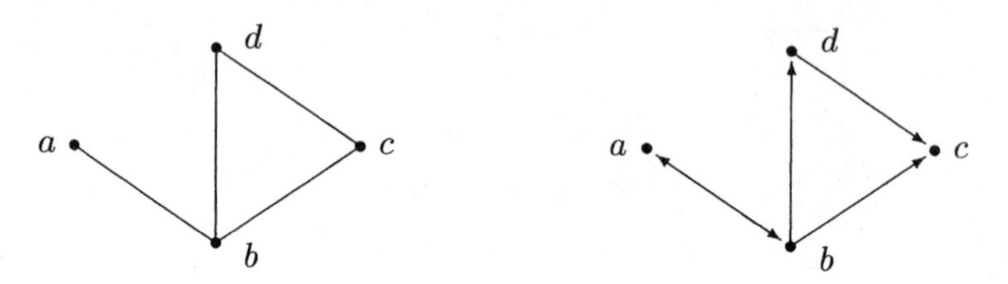

A subset X of the universe of a graph $\mathcal{G}$ is a *clique*, if $E^G ab$ for all $a, b \in X$, $a \neq b$.

Let $\mathcal{G}$ be a digraph. If $n \geq 1$ and

$$E^G a_0 a_1, E^G a_1 a_2, \dots, E^G a_{n-1} a_n$$

then $a_0, \dots, a_n$ is a *path* from a_0 to a_n of *length* n. If $a_0 = a_n$ then $a_0, \dots, a_n$ is a *cycle*. $\mathcal{G}$ is *acyclic* if it has no cycle. A path $a_0, \dots, a_n$ is *Hamiltonian* if $G = \{a_0, \dots, a_n\}$ and $a_i \neq a_j$ for $i \neq j$. If, in addition, $E^G a_n a_0$ we speak of a *Hamiltonian circuit*.

Let $\mathcal{G}$ be a graph. Write $a \sim b$ if $a = b$ or if there is a path from a to b. Clearly, $\sim$ is an equivalence relation. The equivalence class of a is called the *(connected) component* of a. $\mathcal{G}$ is *connected* if $a \sim b$ for all $a, b \in G$, that is, if there is only one connected component. Let CONN be the class of finite connected graphs.

Denote by $d(a,b)$ the length of a shortest path from a to b; more precisely, define the *distance function* $d : G \times G \to \mathbb{N} \cup \{\infty\}$ by [2]

$$d(a,b) = \infty \text{ iff } a \not\sim b; \qquad d(a,b) = 0 \text{ iff } a = b;$$

and otherwise,

$$d(a,b) = \min\{n \geq 1 \mid \text{there is a path from } a \text{ to } b \text{ of length } n\}.$$

Obviously,

$$d(a,c) \leq d(a,b) + d(b,c),$$

where we use the natural conventions for ∞.

For simplicity, we give the following definitions only for *finite* digraphs. A vertex b is a *successor* of a vertex a (and a a *predecessor* of b) if $E^G ab$. The *in-degree* of a vertex is the number of its predecessors, the *out-degree* the number of its successors. In graphs the in-degree and the out-degree of a vertex a coincide and are called the *degree* of a.

[2] $\mathbb{N}$ denotes the set of natural numbers, $\mathbb{N} = \{0, 1, 2, \dots\}$.

A *root* of a digraph is a vertex with in-degree 0 and a *leaf* a vertex with out-degree 0.

A *forest* is an acyclic digraph where each vertex has in-degree at most 1. A *tree* is a forest with connected underlying graph, i.e., a forest (G, E^G) such that $(G, \{(a,b) \mid E^G ab \text{ or } E^G ba\})$ is connected. Note that a finite tree has exactly one root. Let TREE be the class of finite trees.

A2 *Orderings*

Let $\tau = \{<\}$ with a binary relation symbol $<$. A τ-structure $\mathcal{A} = (A, <^A)$ is called an *ordering* if for all a, b, $c \in A$:

(1) not $a <^A a$
(2) $a <^A b$ or $b <^A a$ or $a = b$
(3) if $a <^A b$ and $b <^A c$ then $a <^A c$.

Sometimes we consider finite orderings also as $\{<, S, \min, \max\}$-structures. Here S is a binary relation symbol representing the successor relation and min and max are constants for the first and the last element of the ordering. (When considering the natural ordering on $\{0, \ldots, n\}$ we often refer to min as the zero-th element.) Thus, a finite $\{<, S, \min, \max\}$-structure $\mathcal{A}$ is an ordering if, in addition to (1), (2), (3), for all $a, b \in A$:

(4) $S^A ab$ iff ($a <^A b$ and for all c, if $a <^A c$ then $b <^A c$ or $b = c$)
(5) $\min^A <^A a$ or $\min^A = a$
(6) $a <^A \max^A$ or $a = \max^A$.

For other purposes it might be advantageous to consider finite orderings as $\{<, \min, \max\}$-structures. Suppose that τ_0 is a vocabulary with $\{<\} \subseteq \tau_0 \subseteq \{<, S, \min, \max\}$ and let σ be an arbitrary vocabulary with $\tau_0 \subseteq \sigma$. Later in this book the class $\mathcal{O}[\sigma]$ of finite *ordered* σ-structures will play a prominent role. Here a finite σ-structure $\mathcal{A}$ is said to be *ordered*, if the *reduct* $\mathcal{A}|\tau_0$ (i.e., the τ_0-structure obtained from $\mathcal{A}$ by forgetting the interpretations of the symbols in $\sigma \setminus \tau_0$) is an ordering.

A3 *Operations on Structures*

Two τ-structures $\mathcal{A}$ and $\mathcal{B}$ are *isomorphic*, written $\mathcal{A} \cong \mathcal{B}$, if there is an *isomorphism* from $\mathcal{A}$ to $\mathcal{B}$, i.e., a bijection π: $A \to B$ preserving relations and constants, that is,

– for n-ary $R \in \tau$ and $a_1, \ldots, a_n \in A$,

$$R^A a_1 \ldots a_n \quad \text{iff} \quad R^B \pi(a_1) \ldots \pi(a_n)$$

– for $c \in \tau$, $\pi(c^A) = c^B$.

For τ-structures $\mathcal{A}$ and $\mathcal{B}$, the *product* $\mathcal{A} \times \mathcal{B}$ of $\mathcal{A}$ and $\mathcal{B}$ is the τ-structure with domain $A \times B := \{(a,b) \mid a \in A, b \in B\}$, which is given by

– for n-ary R in τ and $(a_1, b_1), \dots, (a_n, b_n) \in A \times B$,

$$R^{A\times B}(a_1,b_1)\dots(a_n,b_n) \quad \text{iff} \quad R^A a_1 \dots a_n \text{ and } R^B b_1 \dots b_n$$

– for c in τ, $c^{A\times B} := (c^A, c^B)$.

For relational τ we introduce the *union* (or, *disjoint union*) of structures. Assume that $\mathcal{A}$ and $\mathcal{B}$ are τ-structures with $A \cap B = \emptyset$. Then $\mathcal{A}\dot{\cup}\mathcal{B}$, the *union* of $\mathcal{A}$ and $\mathcal{B}$, is the τ-structure with domain $A \cup B$ and

$$R^{\mathcal{A}\dot{\cup}\mathcal{B}} \quad := \quad R^{\mathcal{A}} \cup R^{\mathcal{B}}$$

for any R in τ. In case $\mathcal{A}$ and $\mathcal{B}$ are structures with $A \cap B \neq \emptyset$, we take isomorphic copies $\mathcal{A}'$ of $\mathcal{A}$ and $\mathcal{B}'$ of $\mathcal{B}$ with disjoint universes (e.g., with universes $A \times \{1\}$ and $B \times \{2\}$) and set $\mathcal{A}\dot{\cup}\mathcal{B} := \mathcal{A}'\dot{\cup}\mathcal{B}'$.

Note that the union of ordered structures is not an ordered structure. The situation is different for the so-called *ordered sum*: Let τ with $<\in \tau$ be relational and let $\mathcal{A}$ and $\mathcal{B}$ be ordered τ-structures. Assume that $A \cap B = \emptyset$. Define $\mathcal{A} \lhd \mathcal{B}$, the *ordered sum* of $\mathcal{A}$ and $\mathcal{B}$, as $\mathcal{A}\dot{\cup}\mathcal{B}$ but setting

$$<^{\mathcal{A}\lhd\mathcal{B}} \quad := \quad <^{\mathcal{A}} \cup <^{\mathcal{B}} \cup \{(a,b) \mid a \in A, b \in B\},$$

that is, in $\mathcal{A} \lhd \mathcal{B}$ all elements of A precede all elements of B.

Note that $\mathcal{A} \times \mathcal{B} \cong \mathcal{B} \times \mathcal{A}$, $\mathcal{A}\dot{\cup}\mathcal{B} \cong \mathcal{B}\dot{\cup}\mathcal{A}$ but, in general, $\mathcal{A} \lhd \mathcal{B} \ncong \mathcal{B} \lhd \mathcal{A}$. The definition of product, union, and ordered sum can easily be extended to more than two structures setting, for example, $\mathcal{A} \lhd \mathcal{B} \lhd \mathcal{C} := ((\mathcal{A} \lhd \mathcal{B}) \lhd \mathcal{C})$. For a finite nonempty set I we denote by $\mathcal{A}^I$, $\dot{\cup}_I\mathcal{A}$, and $\lhd_I\mathcal{A}$ the product, the union, and the ordered sum, respectively, of $\|I\|$ copies of $\mathcal{A}$. Here $\|I\|$ denotes the cardinality of I.

B Syntax and Semantics of First-Order Logic

We now turn to the syntactic notions of first-order logic FO. Fix a vocabulary τ. Each formula of first-order logic will be a string of symbols taken from the alphabet consisting of

- $v_1, v_2, v_3, \dots$ (the *variables*)
- $\neg, \vee$ (the *connectives not, or*)
- $\exists$ (the *existential quantifier*)
- $=$ (the *equality symbol*)
-), (
- the symbols in τ.

A *term* of vocabulary τ is a variable or a constant in τ. Henceforth, we often shall use the letters x, y, z, ... for variables and t, t_1, ... for terms.

The *formulas* of first-order logic of vocabulary τ are those strings which are obtained by finitely many applications of the following rules:

(F1) If t_0 and t_1 are terms then $t_0 = t_1$ is a formula.
(F2) If R in τ is n-ary and $t_1, \ldots, t_n$ are terms then $Rt_1 \ldots t_n$ is a formula.
(F3) If φ is a formula then $\neg\varphi$ is a formula.
(F4) If φ and ψ are formulas then $(\varphi \vee \psi)$ is a formula.
(F5) If φ is a formula and x a variable then $\exists x \varphi$ is a formula.

Denote by FO[τ] the set of formulas of first-order logic of vocabulary τ. Formulas obtained by (F1) or (F2) are called *atomic* formulas. For formulas φ and ψ we use $(\varphi \wedge \psi)$, $(\varphi \rightarrow \psi)$, $(\varphi \leftrightarrow \psi)$, and $\forall x \varphi$ as abbreviations for the formulas $\neg(\neg\varphi \vee \neg\psi)$, $(\neg\varphi \vee \psi)$, $((\neg\varphi \vee \psi) \wedge (\neg\psi \vee \varphi))$, and $\neg\exists x \neg\varphi$, respectively.

We shall often omit parentheses in formulas if they are not essential such as the outermost parentheses in disjunctions $(\varphi \vee \psi)$. In examples, different letters x, y, $z \ldots$ will always stand for different variables.

The axioms for graphs stated above have the following formalizations in FO[$\{E\}$]:

$$\forall x \neg Exx$$
$$\forall x \forall y (Exy \rightarrow Eyx),$$

those for orderings the following formalizations in FO[$\{<\}$]:

$$\forall x \neg x < x$$
$$\forall x \forall y (x < y \vee y < x \vee x = y)$$
$$\forall x \forall y \forall z ((x < y \wedge y < z) \rightarrow x < z).$$

For orderings as $\{<, S, \min, \max\}$-structures we need in addition

$$\forall x \forall y (Sxy \leftrightarrow (x < y \wedge \forall z (x < z \rightarrow (y < z \vee y = z))))$$
$$\forall x (\min < x \vee \min = x)$$
$$\forall x (x < \max \vee x = \max).$$

All these formulas are *sentences*, i.e., formulas in which every variable in an atomic subformula is in the scope of a corresponding quantifier. Such occurrences are called *bound* occurrences. The last occurrence of x in

$$(\forall x \neg Exx \wedge \exists y Exy)$$

is not in the scope of a quantifier binding it. Such occurrences are called *free*. The notion of a *free* variable of a formula φ is made precise by the following definition, a definition by *induction on* (the length of) φ: The set free(φ) of *free variables* of a formula φ is defined by:

- If φ is atomic then the set free(φ) of free variables of φ is the set of variables occurring in φ
- $\text{free}(\neg\varphi) := \text{free}(\varphi)$
- $\text{free}(\varphi \vee \psi) := \text{free}(\varphi) \cup \text{free}(\psi)$
- $\text{free}(\exists x \varphi) := \text{free}(\varphi) \setminus \{x\}$.

It is common practice to use the notation $\varphi(x_1, \ldots, x_n)$ to indicate that $x_1, \ldots, x_n$ are distinct and $\text{free}(\varphi) \subseteq \{x_1, \ldots, x_n\}$ without implying that all x_i are actually free in φ (in some chapters of the book we shall give another meaning to this notation). Often we abbreviate an n-tuple $x_1, \ldots, x_n$ of variables by $\overline{x}$, for example, writing $\varphi(\overline{x})$ for $\varphi(x_1, \ldots, x_n)$. Usually we do not make explicit the length of $\overline{x}$ (here n), its size either being inessential or clear from the context. Moreover, we often omit commas writing, for example, $\overline{x} = x_1 \ldots x_n$.

So far, terms and formulas of FO are simply finite strings of symbols. We now assign the intended meanings to the logical symbols so that, in particular, the sentences above really formalize the axioms of graphs and orderings. We do this by defining the *satisfaction relation* $\mathcal{A} \models \varphi$ between structures on the one hand and sentences on the other hand.

Let $\mathcal{A}$ be a τ-structure. An *assignment* in $\mathcal{A}$ is a function α with domain the set of variables and with values in A, $\alpha : \{v_n \mid n \geq 1\} \to A$. Think of α as assigning the meaning $\alpha(x)$ to the variable x. Extend α to a function defined for all terms by setting $\alpha(c) := c^{\mathcal{A}}$ for c in τ. Denote by $\alpha\frac{a}{x}$ the assignment that agrees with α except that $\alpha\frac{a}{x}(x) = a$.

We define the relation

$$\mathcal{A} \models \varphi[\alpha]$$

("the assignment α *satisfies* the formula φ in $\mathcal{A}$" or "φ is *true* in $\mathcal{A}$ under α") as follows:

$\mathcal{A} \models t_1 = t_2[\alpha]$	iff	$\alpha(t_1) = \alpha(t_2)$
$\mathcal{A} \models Rt_1 \ldots t_n[\alpha]$	iff	$R^{\mathcal{A}}\alpha(t_1) \ldots \alpha(t_n)$
$\mathcal{A} \models \neg\varphi[\alpha]$	iff	not $\mathcal{A} \models \varphi[\alpha]$
$\mathcal{A} \models (\varphi \vee \psi)[\alpha]$	iff	$\mathcal{A} \models \varphi[\alpha]$ or $\mathcal{A} \models \psi[\alpha]$
$\mathcal{A} \models \exists x\varphi[\alpha]$	iff	there is an $a \in A$ such that $\mathcal{A} \models \varphi[\alpha\frac{a}{x}]$.

Note that the truth or falsity of $\mathcal{A} \models \varphi[\alpha]$ depends only on the values of α for those variables x which are free in φ. That is, if $\alpha_1(x) = \alpha_2(x)$ for all $x \in \text{free}(\varphi)$, then $\mathcal{A} \models \varphi[\alpha_1]$ iff $\mathcal{A} \models \varphi[\alpha_2]$. Thus, if $\varphi = \varphi(x_1, \ldots, x_n)$ and $a_1 = \alpha(x_1), \ldots, a_n = \alpha(x_n)$, then we may write $\mathcal{A} \models \varphi[a_1, \ldots, a_n]$ for $\mathcal{A} \models \varphi[\alpha]$. In particular, if φ is a sentence, then the truth or falsity of $\mathcal{A} \models \varphi[\alpha]$ is completely independent of α. Thus we may write $\mathcal{A} \models \varphi$ (read: $\mathcal{A}$ is a *model* of φ, or $\mathcal{A}$ *satisfies* φ), if for some (hence every) assignment α, $\mathcal{A} \models \varphi[\alpha]$. For a set Φ of formulas, $\mathcal{A} \models \Phi[\alpha]$ means that $\mathcal{A} \models \varphi[\alpha]$ for all $\varphi \in \Phi$. Φ is *satisfiable* if there is a structure $\mathcal{A}$ and an assignment α in $\mathcal{A}$ such that $\mathcal{A} \models \Phi[\alpha]$.

A formula ψ is a *consequence* of Φ, written $\Phi \models \psi$, if $\mathcal{A} \models \psi[\alpha]$ whenever $\mathcal{A} \models \Phi[\alpha]$. The formula ψ is *logically valid*, written $\models \psi$, if $\emptyset \models \psi$, that is, if ψ is true in all structures under all assignments. And formulas φ and ψ are *logically equivalent* if $\models \varphi \leftrightarrow \psi$. When only taking into consideration finite structures, we use the notations $\Phi \models_{\text{fin}} \psi$ and $\models_{\text{fin}} \psi$, and speak of *equivalent* formulas; hence, φ and ψ are equivalent if $\models_{\text{fin}} \varphi \leftrightarrow \psi$, that is, if $\varphi \leftrightarrow \psi$ holds in all *finite* structures under all assignments.

At some places it will be convenient to assume that first-order logic contains two zero-ary relation symbols T, F. In every structure, T and F are interpreted as TRUE (i.e., as being true) and FALSE, respectively. Hence, the atomic formula T is logically equivalent to $\exists x x = x$ and F to $\neg\exists x x = x$. If $\Phi = \{\varphi_1, \ldots, \varphi_n\}$ we sometimes write $\bigwedge \Phi$ for $\varphi_1 \wedge \ldots \wedge \varphi_n$ and $\bigvee \Phi$ for $\varphi_1 \vee \ldots \vee \varphi_n$.[3] In case $\Phi = \emptyset$ we set $\bigwedge \Phi = \text{T}$ and $\bigvee \Phi = \text{F}$. Then, for arbitrary finite Φ,

$$\mathcal{A} \models \bigwedge \Phi \quad \text{iff} \quad \text{for all } \varphi \in \Phi,\ \mathcal{A} \models \varphi.$$

We introduce some further syntactic notions and notations. The *quantifier rank* $\text{qr}(\varphi)$ of a formula φ is the maximum number of nested quantifiers occurring in it:

$$\begin{array}{rclcrcl} \text{qr}(\varphi) & := & 0, \quad \text{if } \varphi \text{ is atomic;} & & \text{qr}(\neg\varphi) & := & \text{qr}(\varphi); \\ \text{qr}(\varphi \vee \psi) & := & \max\{\text{qr}(\varphi), \text{qr}(\psi)\}; & & \text{qr}(\exists x \varphi) & := & \text{qr}(\varphi) + 1. \end{array}$$

It can be shown that every first-order formula is logically equivalent to a formula in prenex normal form, that is, to a formula of the form $Q_1 x_1 \ldots Q_s x_s \psi$, where $Q_1, \ldots, Q_s \in \{\forall, \exists\}$ and where ψ is quantifier-free. Such a formula is called Σ_n, if the string of quantifiers consists of n consecutive blocks, where in each block all quantifiers are of the same type (i.e., all universal or all existential), adjacent blocks contain quantifiers of different type, and the first block is existential. Π_n formulas are defined in the same way, but now we require that the first block consists of universal quantifiers. A Δ_n-formula is a formula logically equivalent to both a Σ_n-formula and a Π_n-formula.

If $\varphi(x_1, \ldots, x_n)$ is a formula and $t_1, \ldots, t_n$ are terms then $\varphi\frac{t_1 \ldots t_n}{x_1 \ldots x_n}$ or, more simply, $\varphi(t_1, \ldots, t_n)$ denotes the result of simultaneously replacing all free occurrences of $x_1, \ldots, x_n$ by $t_1, \ldots, t_n$, respectively. This presupposes that none of the variables in $t_1, \ldots, t_n$ gets into the scope of a corresponding quantifier; otherwise, the bound variables in φ must be renamed in some canonical fashion before replacing.

Given a formula $\varphi(x, \overline{z})$ and $n \geq 1$,

$$\exists^{\geq n} x \varphi(x, \overline{z})$$

is an abbreviation for the formula

$$\exists x_1 \ldots \exists x_n \Big(\bigwedge_{1 \leq i \leq n} \varphi(x_i, \overline{z}) \wedge \bigwedge_{1 \leq i < j \leq n} \neg x_i = x_j \Big)$$

[3] For definiteness, given the vocabulary τ, fix an ordering on the alphabet of first-order logic and interpret a conjunction $\bigwedge \Phi$ for a finite set Φ as the iterative conjunction of the elements of Φ in the induced lexicographic ordering, say.

expressing that there are at least n elements x with $\varphi(x,\overline{z})$. $\exists^{=n}x\varphi(x,\overline{z})$ and $\exists^{\leq n}x\varphi(x,\overline{z})$ are defined similarly. Moreover, we set

$$\varphi_{\geq n} := \exists^{\geq n}x x = x; \quad \varphi_{=n} := \exists^{=n}x x = x; \quad \varphi_{\leq n} := \exists^{\leq n}x x = x.$$

Clearly,

$$\mathcal{A} \models \varphi_{\geq n} \quad \text{iff} \quad \|A\| \geq n$$

and, similarly, for $\varphi_{=n}$ and $\varphi_{\leq n}$.

C Some Classical Results of First-Order Logic

First-order logic has been used as a framework to analyze the notion of mathematical proof. A result of this analysis is Gödel's Completeness Theorem: A mathematical proof of ψ from Φ shows that $\Phi \models \psi$, i.e. that ψ is a consequence of Φ. A natural question is whether the converse holds, too, that is, whether for any consequence there is a proof. To give an answer, Gödel used a notion of *formal proof* that is based on a finite system of formal rules. A *formal proof* of ψ from Φ consists of a sequence of applications of these rules leading from the formulas in Φ to ψ.[4] In a natural way every formal proof corresponds to a mathematical proof. Thus, if ψ is formally provable from Φ then ψ is a consequence of Φ. Moreover, Gödel showed:

Theorem 1.0.1 (Completeness Theorem) *ψ is a consequence of Φ iff ψ is formally provable from Φ.*

Two immediate consequences are:

Theorem 1.0.2 *The set of logically valid sentences of first-order logic is recursively enumerable.*

Theorem 1.0.3 (Compactness Theorem) (a) *If ψ is a consequence of Φ then ψ is already a consequence of a finite subset of Φ.*
(b) *If every finite subset of Φ is satisfiable then Φ is satisfiable.*

The proof of the Completeness Theorem often leads to a proof of

Theorem 1.0.4 (Löwenheim-Skolem Theorem) *If Φ has a model then Φ has an at most countable model.*

Neither 1.0.2 nor 1.0.3 remain valid if one only considers finite structures. A counterexample for the Compactness Theorem is given by the set $\Phi_\infty := \{\varphi_{\geq n} \mid n \geq 1\}$: Each finite subset of Φ_∞ has a finite model, but Φ_∞ has no finite model.

The failure of 1.0.2 is documented by

Theorem 1.0.5 (Trahtenbrot's Theorem) *The set of sentences of first-order logic valid in all finite structures is not recursively enumerable.*

[4] We do not give detailed definitions, since we do not need them later.

A proof will be given in section 7.2.

We derive a consequence of the Compactness Theorem, which will be used in section 2.5.

Lemma 1.0.6 *Let $\varphi \in \mathrm{FO}[\tau]$ and for $i \in I$, let $\Phi^i \subseteq \mathrm{FO}[\tau]$. Assume that*

$$(*) \qquad \models \varphi \leftrightarrow \bigvee_{i \in I} \bigwedge \Phi^i.$$[5]

Then there is a finite $I_0 \subseteq I$ and, for every $i \in I_0$, a finite $\Phi^i_0 \subseteq \Phi^i$ such that

$$\models \varphi \leftrightarrow \bigvee_{i \in I_0} \bigwedge \Phi^i_0.$$

Proof. For simplicity we assume that φ is a sentence and that every Φ^i is a set of sentences. By hypothesis, for $i \in I$ we have $\Phi^i \models \varphi$; hence, by the Compactness Theorem, $\Phi^i_0 \models \varphi$ for some finite $\Phi^i_0 \subseteq \Phi^i$, and therefore, $\models \bigvee_{i \in I_0} \bigwedge \Phi^i_0 \rightarrow \varphi$ for each finite subset $I_0 \subseteq I$. If there is no such I_0 with $\models \varphi \rightarrow \bigvee_{i \in I_0} \bigwedge \Phi^i_0$, then each finite subset of $\{\varphi\} \cup \{\neg \bigwedge \Phi^i_0 \mid i \in I\}$ has a model. Hence, by the Compactness Theorem, there is a model of φ which for all $i \in I$ satisfies $\neg \bigwedge \Phi^i_0$. This contradicts $(*)$. □

Structures $\mathcal{A}$ and $\mathcal{B}$ (of the same vocabulary) are said to be *elementarily equivalent*, written $\mathcal{A} \equiv \mathcal{B}$, if they satisfy the same first-order sentences. The preceding lemma has the following

Corollary 1.0.7 *Let Φ be a set of first-order sentences. Assume that any two structures that satisfy the same sentences of Φ are elementarily equivalent. Then any first-order sentence is equivalent to a boolean combination of sentences of Φ (that is, is equivalent to a sentence obtainable by closing Φ under $\neg$ and $\vee$).*

Proof. For any structure $\mathcal{A}$ set

$$\Phi(\mathcal{A}) \quad := \quad \{\psi \mid \psi \in \Phi, \mathcal{A} \models \psi\} \cup \{\neg\psi \mid \psi \in \Phi, \mathcal{A} \models \neg\psi\}.$$

Let φ be any first-order sentence. By the preceding lemma it suffices to show that

$$\models \varphi \leftrightarrow \bigvee_{\mathcal{A} \models \varphi} \bigwedge \Phi(\mathcal{A}).$$

Clearly, if $\mathcal{B} \models \varphi$ then $\mathcal{B} \models \bigvee_{\mathcal{A} \models \varphi} \bigwedge \Phi(\mathcal{A})$, since $\mathcal{B} \models \Phi(\mathcal{B})$. For the converse, suppose $\mathcal{B} \models \bigvee_{\mathcal{A} \models \varphi} \bigwedge \Phi(\mathcal{A})$. Then for some model $\mathcal{A}$ of φ, $\mathcal{B} \models \Phi(\mathcal{A})$. By definition of $\Phi(\mathcal{A})$, $\mathcal{A}$ and $\mathcal{B}$ satisfy the same sentences of Φ and hence, by hypothesis, are elementarily equivalent. Therefore, $\mathcal{B} \models \varphi$. □

---

[5] This means that under any assignment in any structure, φ is true iff for some $i \in I$ all formulas in Φ^i are true.

D Model Classes and Global Relations

Fix a vocabulary τ. For a sentence φ of $\mathrm{FO}[\tau]$ we denote by $\mathrm{Mod}(\varphi)$ the class of *finite* models of φ. If π is an isomorphism from $\mathcal{A}$ to $\mathcal{B}$, $\varphi(x_1,\ldots,x_n) \in \mathrm{FO}[\tau]$, and $a_1,\ldots,a_n \in A$, then an easy induction on formulas shows

$$(1) \qquad \mathcal{A} \models \varphi[a_1,\ldots,a_n] \quad \text{iff} \quad \mathcal{B} \models \varphi[\pi(a_1),\ldots,\pi(a_n)].$$

In particular, if φ is a sentence then

$$\mathcal{A} \models \varphi \quad \text{iff} \quad \mathcal{B} \models \varphi.$$

Hence, $\mathrm{Mod}(\varphi)$ is closed under isomorphisms, that is,

$$(2) \qquad \mathcal{A} \in \mathrm{Mod}(\varphi) \text{ and } \mathcal{A} \cong \mathcal{B} \text{ imply } \mathcal{B} \in \mathrm{Mod}(\varphi).$$

For $\varphi(x_1,\ldots,x_n) \in \mathrm{FO}[\tau]$ and a structure $\mathcal{A}$ let

$$\varphi^{\mathcal{A}}(\_) \quad := \quad \{(a_1,\ldots,a_n) \mid \mathcal{A} \models \varphi[a_1,\ldots,a_n]\}$$

be the set of n-tuples *defined by φ in $\mathcal{A}$*. For $n = 0$ this should be read as[6]

$$(3) \qquad \varphi^{\mathcal{A}} := \begin{cases} \text{TRUE} & \text{if } \mathcal{A} \models \varphi \\ \text{FALSE} & \text{if } \mathcal{A} \not\models \varphi. \end{cases}$$

Using this notation we can rewrite (1) as

$$(4) \qquad \text{if } \pi : \mathcal{A} \cong \mathcal{B} \text{ then } \pi(\varphi^{\mathcal{A}}(\_)) = \varphi^{\mathcal{B}}(\_)$$

where for $X \subseteq A^n$ we set $\pi(X) := \{(\pi(a_1),\ldots,\pi(a_n)) \mid (a_1,\ldots,a_n) \in X\}$.

Later we are going to study various logics that extend first-order logic. In all of these logics only structural properties, that is, properties invariant under isomorphisms, will be expressible. So the analogues of (2) and (4) will be true. In particular, (2) says that only classes of structures closed under isomorphisms can be axiomatizable in these extensions. Only such classes will be of interest. We therefore agree upon the following convention:

Throughout the book all classes K of structures considered will tacitly be assumed to be closed under isomorphisms, *i.e.*,

$$\mathcal{A} \in K \text{ and } \mathcal{A} \cong \mathcal{B} \text{ imply } \mathcal{B} \in K.$$

(4) shows that properties expressible in logics correspond to so-called global relations:

[6] TRUE corresponds to the set $\{\emptyset\}$ (consisting of the "empty sequence" $\emptyset$) and FALSE to the empty set $\emptyset$.

Definition 1.0.8 Let K be a class of τ-structures. An n-ary *global relation* Γ *on* K is a mapping assigning to each $\mathcal{A} \in K$ an n-ary relation $\Gamma(\mathcal{A})$ on $\mathcal{A}$ satisfying

$$\Gamma(\mathcal{A})a_1 \ldots a_n \qquad \text{iff} \qquad \Gamma(\mathcal{B})\pi(a_1) \ldots \pi(a_n)$$

for every isomorphism $\pi : \mathcal{A} \cong \mathcal{B}$ and every $a_1, \ldots, a_n \in A$. If K is the class of all finite τ-structures, then we just speak of an n-ary *global relation.* □

Examples 1.0.9 (a) Any formula $\varphi(x_1, \ldots, x_n)$ of FO[τ] defines the global relation $\mathcal{A} \mapsto \varphi^{\mathcal{A}}(\_)$.

(b) The "transitive closure relation" TC is the binary global relation on GRAPH with

$$\text{TC}(\mathcal{G}) \quad := \quad \{(a, b) \mid a, b \in G, \text{ there is a path from } a \text{ to } b\}.$$

(c) For $m \geq 0$, Γ_m is a unary global relation on GRAPH, where

$$\Gamma_m(\mathcal{G}) := \{a \mid \|\{b \in G \mid E^G ab\}\| = m\}$$

is the set of elements of $\mathcal{G}$ of degree m. □

In the definition of a global relation we also allow the case $n = 0$. There are only two 0-ary relations on a structure, TRUE and FALSE. Often one identifies a 0-ary global relation Γ on K with the class

$$\{\mathcal{A} \in K \mid \Gamma(\mathcal{A}) = \text{TRUE}\}.$$

By this identification, the global relation associated with a first-order sentence φ is the class Mod(φ) of finite models of φ (compare (3)).

An important issue in model theory is the study of properties of classes of structures that are axiomatizable in a given logic $\mathcal{L}$ and, in particular, to determine what classes of structures are axiomatizable and what global relations are definable in $\mathcal{L}$. However, since we only consider "function-free" vocabularies, how can we examine such problems, say, for the class of groups?

Nearly all the methods and results presented in this book can directly be extended to vocabularies containing function symbols. Moreover, by replacing functions by their graphs one can always pass to function-free vocabularies. We sketch how to get rid of the function symbols in a vocabulary τ. The price is the introduction of a new $(n+1)$-ary relation symbol F for every n-ary $f \in \tau$. Let the vocabulary τ^r consist of the relation symbols and constants from τ together with the new relation symbols. Thus τ^r is function-free. For a τ-structure $\mathcal{A}$, let $\mathcal{A}^r$ be the τ^r-structure obtained from $\mathcal{A}$ by replacing every n-ary function f^A by its graph F^A,

$$F^A \quad := \quad \{(a_1, \ldots, a_n, f(a_1, \ldots, a_n)) \mid a_1, \ldots, a_n \in A\}.$$

(So, for example, we look at a group $(G, \circ^G, e^G)$ as the $\{R, e\}$-structure (G, R^G, e^G), where the ternary relation R is interpreted as the graph of $\circ^G$, i.e., $R^G = \{(a, b, a \circ^G b) \mid a, b \in G\}$ of $\circ^G$.) The class of τ^r-structures of the form $\mathcal{A}^r$ is the class of models of the conjunction of the formulas

$$\forall x_1 \ldots \forall x_n \exists^{=1} y F x_1 \ldots x_n y$$

where $f \in \tau$. For every τ-sentence φ there is a τ^r-sentence φ^r and for every τ^r-sentence ψ there is a τ-sentence ψ^{-r} such that for every τ-structure $\mathcal{A}$ we have

$$\begin{array}{ccc} \mathcal{A} \models \varphi & \text{iff} & \mathcal{A}^r \models \varphi^r \\ \mathcal{A} \models \psi^{-r} & \text{iff} & \mathcal{A}^r \models \psi. \end{array}$$

(For example, if $\varphi := \exists x \forall y f(g(y)) = x$ then $\varphi^r = \exists x \forall y \exists u (Gyu \wedge Fux)$, and if $\psi := \forall x \exists y (Fxc \wedge \neg Gcy)$ then $\psi^{-r} = \forall x \exists y (f(x) = c \wedge \neg g(c) = y)$. Note that, in general, $\mathrm{qr}(\varphi^r) > \mathrm{qr}(\varphi)$.) Hence, a class K of τ-structures is the class of models of a first-order sentence iff $K^r := \{\mathcal{A}^r \mid \mathcal{A} \in K\}$ is the class of models of a first-order sentence.

E Relational Databases and Query Languages

Suppose that a database contains the names of the main cities in the world and the pairs (a, b) of such cities such that a given airline offers service from a to b without stopover. We can view the database as a first-order structure, namely as a digraph $\mathcal{G} = (G, E^G)$, where G is the set of cities and $E^G ab$ means that there is a flight without stopover from a to b. Now, first-order logic can be considered as a query language. For example, let

$$\varphi(x, y) \quad := \quad Exy \vee \exists z (Exz \wedge Ezy).$$

If φ is thought of as a query to the database, then the response is the set of pairs (a, b) of cities such that b can be reached from a with at most one stop. (We obtain a global relation if we assign to any database (digraph) the response corresponding to the query φ.)

First-order logic provides a rich class of database queries. However, some plausible queries are not first-order expressible. For example, it is impossible to express the query "Can one fly from x to y" by a first-order formula such that we get the right answer in all possible databases (digraphs). We thus are led to ask for stronger logics (or, query languages).

The last two sections have revealed a close relationship between classes of structures, global relations, and queries. Depending on the type of problem we are studying and the methods that are involved, we shall use one or the other terminology, even though we mostly use the terminology related to classes of structures.

2. The Ehrenfeucht-Fraïssé Method

The Ehrenfeucht-Fraïssé method is among the few tools of model theory that survive when we restrict our attention to finite structures. We present the method in its gametheoretic, its algebraic, and its logical form (due to Ehrenfeucht, Fraïssé, and Hintikka, respectively). Later we shall see that generalizations are also available for some extensions of first-order logic. The detailed presentation for the case of first-order logic will help to understand these extensions, where in each case we only will indicate the changes that are necessary.

We always refer to a fixed vocabulary τ. As already mentioned in the preliminaries, τ contains only relation symbols and constants.

2.1 Elementary Classes

The Ehrenfeucht-Fraïssé method is a useful tool for showing that a given class of structures or a given global relation is (not) definable in first-order logic. We start with some easy remarks concerning the expressive power of first-order logic in the finite.

Proposition 2.1.1 *Every finite structure can be characterized in first-order logic up to isomorphism, i.e., for every finite structure $\mathcal{A}$ there is a sentence $\varphi_\mathcal{A}$ of first-order logic such that for all structures $\mathcal{B}$ we have*

$$\mathcal{B} \models \varphi_\mathcal{A} \qquad \textit{iff} \qquad \mathcal{A} \cong \mathcal{B}.$$

Proof. Suppose $A = \{a_1, \ldots, a_n\}$. Set $\overline{a} = a_1 \ldots a_n$. Let

$$\begin{aligned}\Theta_n := \{\psi \mid \psi \text{ has the form } Rx_1 \ldots x_k,\ x = y,\ \text{ or } c = x,\\ \text{and variables among } v_1, \ldots, v_n\}\end{aligned}$$

and

$$\begin{aligned}\varphi_\mathcal{A} \ := \ & \exists v_1 \ldots \exists v_n (\bigwedge\{\psi \mid \psi \in \Theta_n, \mathcal{A} \models \psi[\overline{a}]\} \wedge \\ & \bigwedge\{\neg\psi \mid \psi \in \Theta_n, \mathcal{A} \models \neg\psi[\overline{a}]\} \wedge \forall v_{n+1}(v_{n+1} = v_1 \vee \ldots \vee v_{n+1} = v_n)).\end{aligned}$$

$\square$

Corollary 2.1.2 *Let K be a class of finite structures. Then there is a set Φ of first-order sentences such that*

$$K = \mathrm{Mod}(\Phi),$$

that is, K is the class of finite models of Φ.

Proof. Let K be a class of finite structures. For each n there is only a finite number of pairwise nonisomorphic structures of cardinality n. Let $\{\mathcal{A}_1, \ldots, \mathcal{A}_k\}$ be a maximal subset of K of pairwise nonisomorphic structures of cardinality n. Set

$$\psi_n := (\varphi_{=n} \rightarrow (\varphi_{\mathcal{A}_1} \vee \ldots \vee \varphi_{\mathcal{A}_k})),$$ [1]

where $\varphi_{=n}$ is a first-order sentence expressing "there are exactly n elements" (see 1.B)). Then $K = \mathrm{Mod}(\{\psi_n \mid n \geq 1\})$. □

In many situations we want to know more, namely whether a class K of finite structures or, equivalently, a property of finite structures, is axiomatizable by a *single* first-order sentence, i.e., whether K is elementary in the sense of the following definition.

Definition 2.1.3 Let K be a class of finite structures. K is called *axiomatizable in first-order logic* or *elementary*, if there is a sentence φ of first-order logic such that $K = \mathrm{Mod}(\varphi)$.[2] □

For structures $\mathcal{A}$ and $\mathcal{B}$ and $m \in \mathbb{N}$ we write $\mathcal{A} \equiv_m \mathcal{B}$ and say that $\mathcal{A}$ and $\mathcal{B}$ are *m-equivalent*, if $\mathcal{A}$ and $\mathcal{B}$ satisfy the same first-order sentences of quantifier rank $\leq m$. The following theorem contains necessary and – as we shall see in 2.2.12 – sufficient conditions for a class K to be elementary. Since, in general, it is used to prove nonaxiomatizability results (see 2.3.5–2.3.9), we formulate it in the corresponding way.

Theorem 2.1.4 *Let K be a class of finite structures. Suppose that for every m there are finite structures $\mathcal{A}$ and $\mathcal{B}$ such that*

$$\mathcal{A} \in K, \ \mathcal{B} \notin K, \ \text{and} \ \mathcal{A} \equiv_m \mathcal{B}.$$

Then K is not axiomatizable in first-order logic.

Proof. Let φ be any first-order sentence. Set $m := \mathrm{qr}(\varphi)$. By our assumption there are $\mathcal{A}$ and $\mathcal{B}$ such that $\mathcal{A} \in K$, $\mathcal{B} \notin K$, and $\mathcal{A} \equiv_m \mathcal{B}$; hence, $K \neq \mathrm{Mod}(\varphi)$. □

---

[1] In case $k = 0$ recall our convention that the empty disjunction is F; one also could set $\psi_n := (\varphi_{=n} \rightarrow \neg\exists x\, x = x)$ in this case.

[2] In the literature, instead of *axiomatizable* one often uses the term *finitely axiomatizable.* In (general) model theory a class of *arbitrary* structures K is called *elementary* if, for some φ, K is the class of arbitrary models of φ. And given classes K_0 and K with $K_0 \supseteq K$, it is said that K *is elementary relative to* K_0 if, for some φ, K is the class of models of φ in K_0. In this terminology our notion of *elementary* corresponds to *elementary relative to the class of finite structures.*

2.2 Ehrenfeucht's Theorem

In this section we present a purely gametheoretic characterization of the relation $\equiv_m$. It will be a useful tool, in particular, for applying Theorem 2.1.4 to concrete classes, at the same time helping us to understand the expressive power of first-order logic. One of the central ingredients of the characterization are partial isomorphisms.

Definition 2.2.1 Assume $\mathcal{A}$ and $\mathcal{B}$ are structures. Let p be a map with $\mathrm{do}(p) \subseteq A$ and $\mathrm{rg}(p) \subseteq B$, where $\mathrm{do}(p)$ and $\mathrm{rg}(p)$ denote the domain and the range of p, respectively. Then p is said to be a *partial isomorphism* from $\mathcal{A}$ to $\mathcal{B}$ if

- p is injective
- for every $c \in \tau$: $c^A \in \mathrm{do}(p)$ and $p(c^A) = c^B$
- for every n-ary $R \in \tau$ and all $a_1, \ldots, a_n \in \mathrm{do}(p)$,

$$R^A a_1 \ldots a_n \quad \text{iff} \quad R^B p(a_1) \ldots p(a_n).$$

We write $\mathrm{Part}(\mathcal{A}, \mathcal{B})$ for the set of partial isomorphisms from $\mathcal{A}$ to $\mathcal{B}$. □

In the following we identify a map p with its graph $\{(a, p(a)) \mid a \in \mathrm{do}(p)\}$. Then $p \subseteq q$ means that q is an extension of p.

Remarks 2.2.2 (a) The empty map, $p = \emptyset$, is a partial isomorphism from $\mathcal{A}$ to $\mathcal{B}$ just in case the vocabulary contains no constants.

(b) If $p \neq \emptyset$ is a map with $\mathrm{do}(p) \subseteq A$ and $\mathrm{rg}(p) \subseteq B$, then p is a partial isomorphism from $\mathcal{A}$ to $\mathcal{B}$ iff $\mathrm{do}(p)$ contains c^A for all constants $c \in \tau$ and $p : \mathrm{do}(p)^{\mathcal{A}} \cong \mathrm{rg}(p)^{\mathcal{B}}$ (where $\mathrm{do}(p)^{\mathcal{A}}$ and $\mathrm{rg}(p)^{\mathcal{B}}$ denote the substructures of $\mathcal{A}$ and $\mathcal{B}$ with universes $\mathrm{do}(p)$ and $\mathrm{rg}(p)$, respectively).

(c) For $\overline{a} = a_1 \ldots a_s \in A$ and $\overline{b} = b_1 \ldots b_s \in B$ the following statements are equivalent:

(i) The clauses

$$p(a_i) = b_i \text{ for } i = 1, \ldots, s$$

and

$$p(c^A) = c^B \text{ for } c \text{ in } \tau$$

define a map, which is a partial isomorphism from $\mathcal{A}$ to $\mathcal{B}$ (henceforth denoted by $\overline{a} \mapsto \overline{b}$, a notation that suppresses the constants).

(ii) For all quantifier-free $\varphi(v_1, \ldots, v_s)$: $\mathcal{A} \models \varphi[\overline{a}]$ iff $\mathcal{B} \models \varphi[\overline{b}]$.

(iii) For all atomic $\varphi(v_1, \ldots, v_s)$: $\mathcal{A} \models \varphi[\overline{a}]$ iff $\mathcal{B} \models \varphi[\overline{b}]$.

Proof. Note that for an arbitrary structure $\mathcal{D}$ and $\overline{d}$ in D,

$$\begin{array}{lll} d_i = d_j & \text{iff} & \mathcal{D} \models v_i = v_j[\overline{d}] \\ c^D = d_j & \text{iff} & \mathcal{D} \models c = v_j[\overline{d}] \\ R^D c^D d_i d_j & \text{iff} & \mathcal{D} \models Rcv_iv_j[\overline{d}] \end{array}$$

($c, R \in \tau, R$ ternary). Using such equivalences, it is easy to show that (i) and (iii) are equivalent. Clearly, (ii) implies (iii), and (ii) follows from (iii), since every quantifier-free formula is a boolean combination of atomic formulas. □

In general, a partial isomorphism does not preserve the validity of formulas with quantifiers: Let $\tau = \{<\}$, $\mathcal{A} = (\{0,1,2\},<)$, $\mathcal{B} = (\{0,1,2,3\},<)$, where in both cases $<$ denotes the natural ordering. Then $p_0 := 02 \mapsto 01$ is a partial isomorphism from $\mathcal{A}$ to $\mathcal{B}$ such that

$$\mathcal{A} \models \exists v_3(v_1 < v_3 \wedge v_3 < v_2)\,[0,2]$$

but

$$\mathcal{B} \not\models \exists v_3(v_1 < v_3 \wedge v_3 < v_2)\,[p_0(0), p_0(2)].$$

Since

$$\mathcal{A} \models (v_1 < v_3 \wedge v_3 < v_2)\,[0,2,1]$$

we see that, for any $p \in \mathrm{Part}(\mathcal{A},\mathcal{B})$ with $\mathrm{do}(p) = \{0,2\}$, the validity of

$$\mathcal{B} \models \exists v_3(v_1 < v_3 \wedge v_3 < v_2)\,[p(0), p(2)]$$

is equivalent to the existence of some $q \in \mathrm{Part}(\mathcal{A},\mathcal{B})$ which extends p and has 1 in its domain.

This example indicates that the truth of formulas with quantifiers is preserved under partial isomorphisms provided they admit certain extensions. It embodies the basic idea behind the algebraic characterization of $\equiv_m$ we have in mind: The m-equivalence of structures amounts to the existence of partial isomorphisms that can be extended m times. In the gametheoretic terms introduced by the next definition, $\overline{a} \mapsto \overline{b}$ is such a partial isomorphism if the duplicator has a winning strategy for the Ehrenfeucht game $\mathrm{G}_m(\mathcal{A}, \overline{a}, \mathcal{B}, \overline{b})$.

Let $\mathcal{A}$ and $\mathcal{B}$ be τ-structures, $\overline{a} \in A^s$, $\overline{b} \in B^s$, and $m \in \mathbb{N}$. The *Ehrenfeucht game* $\mathrm{G}_m(\mathcal{A}, \overline{a}, \mathcal{B}, \overline{b})$ is played by two players called the *spoiler* and the *duplicator*. Each player has to make m moves in the course of a play. The players take turns. In his i-th move the spoiler first selects a structure, $\mathcal{A}$ or $\mathcal{B}$, and an element in this structure. If the spoiler chooses e_i in $\mathcal{A}$ then the duplicator in his i-th move must choose an element f_i in $\mathcal{B}$. If the spoiler chooses f_i in $\mathcal{B}$ then the duplicator must choose an element e_i in $\mathcal{A}$.

	$\mathcal{A}, \overline{a}$	$\mathcal{B}, \overline{b}$
first move	e_1	f_1
second move	e_2	f_2
$\vdots$	$\vdots$	$\vdots$
m-th move	e_m	f_m

As illustrated by this table, at the end elements $e_1, \ldots, e_m$ in $\mathcal{A}$ and $f_1, \ldots, f_m$ in $\mathcal{B}$ have been chosen. The duplicator *wins* iff $\overline{a}\,\overline{e} \mapsto \overline{b}\,\overline{f} \in \mathrm{Part}(\mathcal{A},\mathcal{B})$ (in case $m = 0$ we just require that $\overline{a} \mapsto \overline{b} \in \mathrm{Part}(\mathcal{A},\mathcal{B})$). Otherwise, the spoiler wins.

Equivalently, the spoiler wins if, after some $i \leq m$, $\overline{a}e_1 \ldots e_i \mapsto \overline{b}f_1 \ldots f_i$ is not a partial isomorphism. We say that a player, the spoiler or the duplicator, has a *winning strategy* in $\mathrm{G}_m(\mathcal{A}, \overline{a}, \mathcal{B}, \overline{b})$, or shortly, that he *wins* $\mathrm{G}_m(\mathcal{A}, \overline{a}, \mathcal{B}, \overline{b})$, if it is possible for him to win each play whatever choices are made by the opponent. We omit a formal definition of the notion of a winning strategy; it is implicit in the algebraic characterization given below (cf. the proof of Theorem 2.3.3). If $s = 0$ (and hence $\overline{a}$ and $\overline{b}$ are empty), we denote the game by $\mathrm{G}_m(\mathcal{A}, \mathcal{B})$.

Lemma 2.2.3 (a) *If $\mathcal{A} \cong \mathcal{B}$ then the duplicator wins $\mathrm{G}_m(\mathcal{A}, \mathcal{B})$.*
(b) *If the duplicator wins $\mathrm{G}_{m+1}(\mathcal{A}, \mathcal{B})$ and $\|A\| \leq m$ then $\mathcal{A} \cong \mathcal{B}$.*

Proof. (a) Suppose $\pi : \mathcal{A} \cong \mathcal{B}$. A winning strategy for the duplicator consists in always choosing the image or preimage under π of the spoiler's selection; that is, if the spoiler chooses $a \in A$ then the duplicator chooses $\pi(a)$; and if the spoiler chooses $b \in B$ then the duplicator answers with $\pi^{-1}(b)$.

(b) Suppose that the duplicator wins $\mathrm{G}_{m+1}(\mathcal{A}, \mathcal{B})$, and assume that $A = \{a_1, \ldots, a_m\}$. Consider a play where the spoiler, in his first m moves, chooses $a_1, \ldots, a_m$. Let $b_1, \ldots, b_m$ be the responses of the duplicator according to his winning strategy. Then $p : \overline{a} \mapsto \overline{b} \in \mathrm{Part}(\mathcal{A}, \mathcal{B})$ with $\mathrm{do}(p) = A$. It even is an isomorphism from $\mathcal{A}$ onto $\mathcal{B}$. Otherwise, we have $\mathrm{rg}(p) \neq B$. Then the spoiler, in the last move of the play, chooses some element $b \in B \setminus \mathrm{rg}(p)$. As there is no answer for the duplicator leading to a win, we get a contradiction. □

The following lemma collects some facts about the Ehrenfeucht game. Their proofs are immediate from the definition.

Lemma 2.2.4 *Let $\mathcal{A}$ and $\mathcal{B}$ be structures, $\overline{a} \in A^s$, $\overline{b} \in B^s$, and $m \geq 0$.*

(a) *The duplicator wins $\mathrm{G}_0(\mathcal{A}, \overline{a}, \mathcal{B}, \overline{b})$ iff $\overline{a} \mapsto \overline{b}$ is a partial isomorphism.*
(b) *For $m > 0$ the following are equivalent:*
  (i) *The duplicator wins $\mathrm{G}_m(\mathcal{A}, \overline{a}, \mathcal{B}, \overline{b})$.*
  (ii) *For all $a \in A$ there is $b \in B$ such that the duplicator wins the game $\mathrm{G}_{m-1}(\mathcal{A}, \overline{a}a, \mathcal{B}, \overline{b}b)$ and for all $b \in B$ there is $a \in A$ such that the duplicator wins $\mathrm{G}_{m-1}(\mathcal{A}, \overline{a}a, \mathcal{B}, \overline{b}b)$.*
(c) *If the duplicator wins $\mathrm{G}_m(\mathcal{A}, \overline{a}, \mathcal{B}, \overline{b})$ and if $m' < m$, the duplicator wins $\mathrm{G}_{m'}(\mathcal{A}, \overline{a}, \mathcal{B}, \overline{b})$.* □

Parts (a) and (b) give us a hint how to relate the game to the validity of formulas.

Let $\mathcal{A}$ be given. For $\overline{a} = a_1 \ldots a_s \in A$ and $m \geq 0$ we introduce a formula $\varphi^m_{\overline{a}}(v_1, \ldots, v_s)$ that describes the gametheoretic properties of $\overline{a}$ in any game $\mathrm{G}_m(\mathcal{A}, \overline{a}, \ldots)$; more precisely, we want to define $\varphi^m_{\overline{a}}$ in such a way that for any $\mathcal{B}$ and $\overline{b} = b_1 \ldots b_s \in B$,

$$\mathcal{B} \models \varphi^m_{\overline{a}}\,[\overline{b}] \quad \text{iff} \quad \text{the duplicator wins } \mathrm{G}_m(\mathcal{A}, \overline{a}, \mathcal{B}, \overline{b}).$$

If the structure $\mathcal{A}$ is not clear from the context, we use the notation $\varphi^m_{\mathcal{A},\overline{a}}$ for $\varphi^m_{\overline{a}}$. We also allow $s = 0$, the case of the empty sequence $\emptyset$ of elements in A, and write $\varphi^m_{\mathcal{A}}$ for the sentence $\varphi^m_{\mathcal{A},\emptyset}$.

Definition 2.2.5 Let $\overline{v}$ be $v_1, \ldots, v_s$.

$$\varphi^0_{\overline{a}}(\overline{v}) \;:=\; \bigwedge\{\varphi(\overline{v}) \mid \varphi \text{ atomic or negated atomic, } \mathcal{A} \models \varphi[\overline{a}]\}$$

and for $m > 0$,

$$\varphi^m_{\overline{a}}(\overline{v}) \;:=\; \bigwedge_{a\in A} \exists v_{s+1} \varphi^{m-1}_{\overline{a}a}(\overline{v}, v_{s+1}) \;\wedge\; \forall v_{s+1} \bigvee_{a\in A} \varphi^{m-1}_{\overline{a}a}(\overline{v}, v_{s+1}). \qquad \square$$

$\varphi^0_{\overline{a}}$ describes the isomorphism type of the substructure generated by $\overline{a}$ in $\mathcal{A}$; and for $m > 0$ the formula $\varphi^m_{\overline{a}}$ tells us to which isomorphism types the tuple $\overline{a}$ can be extended in m steps adding one element in each step. $\varphi^m_{\overline{a}}$ is called the *m-isomorphism type* (or *m-Hintikka formula*) of $\overline{a}$ in $\mathcal{A}$.

Since $\{\varphi(v_1, \ldots, v_s) \mid \varphi \text{ atomic or negated atomic}\}$ is finite, a simple induction on m shows:

Lemma 2.2.6 *For $s, m \geq 0$, the set $\{\varphi^m_{\mathcal{A},\overline{a}} \mid \mathcal{A}$ a structure and $\overline{a} \in A^s\}$ is finite.* $\square$

In particular, the conjunctions and disjunctions in the above definition are finite.

Lemma 2.2.7 (a) $\mathrm{qr}(\varphi^m_{\overline{a}}) = m$.

(b) $\mathcal{A} \models \varphi^m_{\overline{a}}[\overline{a}]$.

(c) *For any $\mathcal{B}$ and $\overline{b}$ in B,*

$$\mathcal{B} \models \varphi^0_{\overline{a}}[\overline{b}] \quad \textit{iff} \quad \overline{a} \mapsto \overline{b} \in \mathrm{Part}(\mathcal{A}, \mathcal{B}).$$

Proof. The proofs of (a) and (b) are straightforward. (c) holds by part (c) of 2.2.2. $\square$

Theorem 2.2.8 (Ehrenfeucht's Theorem) *Given $\mathcal{A}$ and $\mathcal{B}$, $\overline{a} \in A^s$ and $\overline{b} \in B^s$, and $m \geq 0$, the following are equivalent:*

(i) *The duplicator wins $\mathrm{G}_m(\mathcal{A}, \overline{a}, \mathcal{B}, \overline{b})$.*

(ii) $\mathcal{B} \models \varphi^m_{\overline{a}}[\overline{b}]$.

(iii) *$\overline{a}$ and $\overline{b}$ satisfy the same formulas of quantifier rank $\leq m$, that is, if $\varphi(x_1, \ldots, x_s)$ is of quantifier rank $\leq m$, then*

$$(*) \qquad \mathcal{A} \models \varphi[\overline{a}] \quad \textit{iff} \quad \mathcal{B} \models \varphi[\overline{b}].$$

Proof. (iii) implies (ii) since $\mathrm{qr}(\varphi^m_{\overline{a}}) = m$ and $\mathcal{A} \models \varphi^m_{\overline{a}}[\overline{a}]$. We prove the equivalence of (i) and (ii) by induction on m. For $m = 0$

the duplicator wins $\mathrm{G}_0(\mathcal{A}, \overline{a}, \mathcal{B}, \overline{b})$ iff $\overline{a} \mapsto \overline{b} \in \mathrm{Part}(\mathcal{A}, \mathcal{B})$ (cf. 2.2.4(a))
iff $\mathcal{B} \models \varphi^0_{\overline{a}}[\overline{b}]$ (by 2.2.7(c)).

For $m > 0$,

the duplicator wins $\mathrm{G}_m(\mathcal{A}, \overline{a}, \mathcal{B}, \overline{b})$

iff for all $a \in A$ there is $b \in B$ such that the duplicator wins $\mathrm{G}_{m-1}(\mathcal{A}, \overline{a}a, \mathcal{B}, \overline{b}b)$, and for all $b \in B$ there is $a \in A$ such that the duplicator wins $\mathrm{G}_{m-1}(\mathcal{A}, \overline{a}a, \mathcal{B}, \overline{b}b)$ (cf. 2.2.4(b))

iff for all $a \in A$ there is $b \in B$ with $\mathcal{B} \models \varphi^{m-1}_{\overline{a}a}[\overline{b}b]$, and for all $b \in B$ there is $a \in A$ with $\mathcal{B} \models \varphi^{m-1}_{\overline{a}a}[\overline{b}b]$ (ind. hyp.)

iff $\mathcal{B} \models \bigwedge_{a \in A} \exists v_{s+1} \varphi^{m-1}_{\overline{a}a}(\overline{v}, v_{s+1}) \wedge \forall v_{s+1} \bigvee_{a \in A} \varphi^{m-1}_{\overline{a}a}(\overline{v}, v_{s+1})[\overline{b}]$

iff $\mathcal{B} \models \varphi^m_{\overline{a}}[\overline{b}]$.

(i) $\Rightarrow$ (iii): The proof proceeds by induction on m. The case $m = 0$ is handled as above. Let $m > 0$ and suppose that the duplicator wins $\mathrm{G}_m(\mathcal{A}, \overline{a}, \mathcal{B}, \overline{b})$. Clearly, the set of formulas $\varphi(x_1, \ldots, x_s)$ satisfying $(*)$ contains the atomic formulas and is closed under $\neg$ and $\vee$. Suppose that $\varphi(\overline{x}) = \exists y \psi$ and $\mathrm{qr}(\varphi) \leq m$. Since $y \notin \mathrm{free}(\varphi)$, we can assume that y is distinct from the variables in $\overline{x}$. Hence, $\psi = \psi(\overline{x}, y)$. Assume, for instance, $\mathcal{A} \models \varphi[\overline{a}]$. Then there is $a \in A$ such that $\mathcal{A} \models \psi[\overline{a}, a]$. As, by (i), the duplicator wins $\mathrm{G}_m(\mathcal{A}, \overline{a}, \mathcal{B}, \overline{b})$, there is $b \in B$ such that the duplicator wins $\mathrm{G}_{m-1}(\mathcal{A}, \overline{a}a, \mathcal{B}, \overline{b}b)$. Since $\mathrm{qr}(\psi) \leq m-1$, the induction hypothesis yields $\mathcal{B} \models \psi[\overline{b}, b]$, hence $\mathcal{B} \models \varphi[\overline{b}]$. □

Corollary 2.2.9 *For structures $\mathcal{A}$, $\mathcal{B}$ and $m \geq 0$ the following are equivalent:*
(i) *The duplicator wins $G_m(\mathcal{A}, \mathcal{B})$.* (ii) $\mathcal{B} \models \varphi^m_{\mathcal{A}}$.
(iii) $\mathcal{A} \equiv_m \mathcal{B}$. □

By 2.2.3(b) we get

Corollary 2.2.10 *Let $\mathcal{A}$ be a structure with $\|A\| \leq m$. Then for all $\mathcal{B}$,*

$$\mathcal{B} \models \varphi^{m+1}_{\mathcal{A}} \quad \textit{iff} \quad \mathcal{A} \cong \mathcal{B}.$$ □

The next result shows that the formulas $\varphi^m_{\overline{a}}$ give a clear picture of the expressive power of first-order logic.

Theorem 2.2.11 *Let $\varphi(v_1, \ldots, v_s)$ be a formula of quantifier rank $\leq m$. Then*

$$\models \varphi \leftrightarrow \bigvee \{\varphi^m_{\mathcal{A},\overline{a}} \mid \mathcal{A} \textit{ a structure}, \ \overline{a} \in \mathcal{A}, \textit{ and } \mathcal{A} \models \varphi[\overline{a}]\}.$$[3]

Proof. Suppose first that $\mathcal{B} \models \varphi[\overline{b}]$. Then the formula $\varphi^m_{\mathcal{B},\overline{b}}$ is a member of the disjunction on the right side of the equivalence, which therefore is satisfied by $\overline{b}$. Conversely, suppose $\mathcal{B} \models \bigvee\{\varphi^m_{\mathcal{A},\overline{a}} \mid \mathcal{A} \models \varphi[\overline{a}]\}[\overline{b}]$. Then, for some $\mathcal{A}$ and $\overline{a}$ such that $\mathcal{A} \models \varphi[\overline{a}]$, we have $\mathcal{B} \models \varphi^m_{\mathcal{A},\overline{a}}[\overline{b}]$. By 2.2.8, $\overline{a}$ and $\overline{b}$ satisfy the same formulas of quantifier rank $\leq m$ and therefore, $\mathcal{B} \models \varphi[\overline{b}]$. □

[3] By 2.2.6 the disjunction is taken over a finite set.

The following result contains the desired characterization of classes axiomatizable in first-order logic. As 2.1.4, we formulate it in a "negative" form, since the implication (ii) $\Rightarrow$ (i) turns out to be one of the main tools to obtain nonaxiomatizability results for first-order logic.

Theorem 2.2.12 *For a class K of finite structures the following are equivalent:*

(i) *K is not axiomatizable in first-order logic.*

(ii) *For each m there are finite structures $\mathcal{A}$ and $\mathcal{B}$ such that*

$$\mathcal{A} \in K, \ \mathcal{B} \notin K \text{ and } \mathcal{A} \equiv_m \mathcal{B}.$$

Proof. (ii) $\Rightarrow$ (i) was proven in 2.1.4. For the converse, suppose that (ii) does not hold, i.e., that for some m and all finite $\mathcal{A}$ and $\mathcal{B}$,

$$\mathcal{A} \in K \text{ and } \mathcal{A} \equiv_m \mathcal{B} \text{ imply } \mathcal{B} \in K.$$

Then $K = \text{Mod}(\bigvee\{\varphi^m_{\mathcal{A}} \mid \mathcal{A} \in K\})$, and thus K is axiomatizable. □

2.3 Examples and Fraïssé's Theorem

To apply Ehrenfeucht's characterization to concrete examples, it is more convenient to use an algebraic version due to Fraïssé, even though it lacks the intuitive appeal of the gametheoretical approach.

Given structures $\mathcal{A}$, $\mathcal{B}$ and $m \in \mathbb{N}$, let $W_m(\mathcal{A}, \mathcal{B}) :=$

$$\{\overline{a} \mapsto \overline{b} \mid s \geq 0, \overline{a} \in A^s, \overline{b} \in B^s, \text{ the duplicator wins } \mathrm{G}_m(\mathcal{A}, \overline{a}, \mathcal{B}, \overline{b})\}$$

be the set of winning positions for the duplicator. The sequence of the $W_m(\mathcal{A}, \mathcal{B})$ has the back and forth properties as introduced in the following definition.

Definition 2.3.1 Structures $\mathcal{A}$ and $\mathcal{B}$ are said to be *m-isomorphic*, written $\mathcal{A} \cong_m \mathcal{B}$, if there is a sequence $(I_j)_{j \leq m}$ with the following properties:

(a) Every I_j is a nonempty set of partial isomorphisms from $\mathcal{A}$ to $\mathcal{B}$.

(b) (*Forth property*) For every $j < m$, $p \in I_{j+1}$, and $a \in A$ there is $q \in I_j$ such that $q \supseteq p$ and $a \in \text{do}(q)$.

(c) (*Back property*) For every $j < m$, $p \in I_{j+1}$, and $b \in B$ there is $q \in I_j$ such that $q \supseteq p$ and $b \in \text{rg}(q)$.

If $(I_j)_{j \leq m}$ has the properties (a), (b), and (c), we write $(I_j)_{j \leq m} : \mathcal{A} \cong_m \mathcal{B}$ and say that $\mathcal{A}$ and $\mathcal{B}$ are *m-isomorphic via* $(I_j)_{j \leq m}$. □

Exercise 2.3.2 Suppose $(I_j)_{j \leq m} : \mathcal{A} \cong_m \mathcal{B}$. Then $(\tilde{I}_j)_{j \leq m} : \mathcal{A} \cong_m \mathcal{B}$ with $\tilde{I}_j := \{q \in \text{Part}(\mathcal{A}, \mathcal{B}) \mid q \subseteq p \text{ for some } p \in I_j\}$. In particular, $\emptyset \mapsto \emptyset \in I_j$ for all $j \leq m$. Moreover, $\widetilde{W_j(\mathcal{A}, \mathcal{B})} = W_j(\mathcal{A}, \mathcal{B})$.

Using the results of the preceding section we obtain:

Theorem 2.3.3 *For structures $\mathcal{A}$ and $\mathcal{B}$, $\overline{a} \in A^s$, $\overline{b} \in B^s$, and $m \geq 0$ the following are equivalent:*

(i) *The duplicator wins* $\mathrm{G}_m(\mathcal{A}, \overline{a}, \mathcal{B}, \overline{b})$.

(ii) $\overline{a} \mapsto \overline{b} \in W_m(\mathcal{A}, \mathcal{B})$ *and* $(W_j(\mathcal{A}, \mathcal{B}))_{j \leq m} : \mathcal{A} \cong_m \mathcal{B}$.

(iii) *There is* $(I_j)_{j \leq m}$ *with* $\overline{a} \mapsto \overline{b} \in I_m$ *such that* $(I_j)_{j \leq m} : \mathcal{A} \cong_m \mathcal{B}$.

(iv) $\mathcal{B} \models \varphi^m_{\overline{a}}[\overline{b}]$.

(v) $\overline{a}$ *satisfies in* $\mathcal{A}$ *the same formulas of quantifier rank* $\leq m$ *as* $\overline{b}$ *in* $\mathcal{B}$.

Proof. By the definition of $W_m(\mathcal{A}, \mathcal{B})$ and 2.2.4, (i) implies (ii). Obviously, (ii) implies (iii). Therefore it suffices to show the implication (iii) $\Rightarrow$ (i), the remaining equivalences being clear from 2.2.8. For (iii) $\Rightarrow$ (i) suppose that $(I_j)_{j \leq m} : \mathcal{A} \cong_m \mathcal{B}$ and $\overline{a} \mapsto \overline{b} \in I_m$. We describe a winning strategy in $\mathrm{G}_m(\mathcal{A}, \overline{a}, \mathcal{B}, \overline{b})$ for the duplicator: In his i-th move he should choose the element e_i (or f_i, respectively) such that for $p_i : \overline{a}e_1 \ldots e_i \mapsto \overline{b}f_1 \ldots f_i$ it is true that $p_i \subseteq q$ for some $q \in I_{m-i}$; this is always possible because of the back and forth properties of $(I_j)_{j \leq m}$. Looking at $i := m$ we see that the duplicator wins. □

For $s = 0$ in view of Exercise 2.3.2 the preceding theorem yields the following extension of 2.2.9.

Corollary 2.3.4 *For structures $\mathcal{A}$, $\mathcal{B}$ and $m \geq 0$ the following are equivalent:*

(i) *The duplicator wins* $\mathrm{G}_m(\mathcal{A}, \mathcal{B})$. (ii) $(W_j(\mathcal{A}, \mathcal{B}))_{j \leq m} : \mathcal{A} \cong_m \mathcal{B}$.

(iii) $\mathcal{A} \cong_m \mathcal{B}$. (iv) $\mathcal{B} \models \varphi^m_{\mathcal{A}}$.

(v) $\mathcal{A} \equiv_m \mathcal{B}$. □

The equivalence of (iii) and (v) is known as *Fraïssé's Theorem.* The proof of the preceding theorem, and especially that of the equivalence of (i) and (iii) shows, that Ehrenfeucht's Theorem and Fraïssé's Theorem are different formulations of the same fact. In particular, the proof exhibits the close relationship between sequences $(I_j)_{j \leq m}$ and winning strategies for the duplicator in $\mathrm{G}_m(\mathcal{A}, \overline{a}, \mathcal{B}, \overline{b})$. Therefore, one often speaks of the *Ehrenfeucht-Fraïssé game* or the *Ehrenfeucht-Fraïssé method.*

Example 2.3.5 Let τ be the empty vocabulary and $\mathcal{A}$ and $\mathcal{B}$ be τ-structures (i.e., nonempty sets). Suppose $\|A\| \geq m$ and $\|B\| \geq m$. Then $\mathcal{A} \cong_m \mathcal{B}$. In fact, $(I_j)_{j \leq m} : \mathcal{A} \cong_m \mathcal{B}$ with $I_j := \{p \in \mathrm{Part}(\mathcal{A}, \mathcal{B}) \mid \|\mathrm{do}(p)\| \leq m - j\}$.

As a consequence the class $\mathrm{EVEN}[\tau]$ of finite τ-structures of even cardinality is not axiomatizable in first-order logic. In fact, for each $m > 0$, let $\mathcal{A}_m$ be a structure of cardinality m. Then, $\mathcal{A}_m \in \mathrm{EVEN}[\tau]$ iff $\mathcal{A}_{m+1} \notin \mathrm{EVEN}[\tau]$, but $\mathcal{A}_m \cong_m \mathcal{A}_{m+1}$. Now apply 2.2.12. The reader is encouraged to show for arbitrary τ that $\mathrm{EVEN}[\tau]$ is not axiomatizable. □

Example 2.3.6 Let $\tau = \{<, \min, \max\}$ be a vocabulary for finite orderings as introduced in 1.A2 and $m \geq 1$. Suppose that $\mathcal{A}$ and $\mathcal{B}$ are finite orderings, $\|A\| > 2^m$ and $\|B\| > 2^m$. Then $\mathcal{A} \cong_m \mathcal{B}$. Hence, the class of finite orderings of even cardinality is not axiomatizable in first-order logic. Clearly, the last statement remains true, if we consider orderings as $\{<, S, \min, \max\}$-structures.

For a *proof*, given any ordering $\mathcal{C}$, we define its distance function d by

$$d(a, a') \quad := \quad \|\{b \in C \mid (a < b \leq a') \text{ or } (a' < b \leq a)\}\|.$$

And, for $j \geq 0$, we introduce the "truncated" j-distance function d_j on $C \times C$ by

$$d_j(a, a') \quad := \quad \begin{cases} d(a, a') & \text{if } d(a, a') < 2^j \\ \infty & \text{else.} \end{cases}$$

Now, suppose that $\mathcal{A}$ and $\mathcal{B}$ are finite orderings with $\|A\|, \|B\| > 2^m$. For $j \leq m$ set

$$I_j \quad := \quad \{p \in \text{Part}(\mathcal{A}, \mathcal{B}) \mid d_j(a, a') = d_j(p(a), p(a')) \text{ for } a, a' \in \text{ do}(p)\}.$$

Then $(I_j)_{j \leq m} : \mathcal{A} \cong_m \mathcal{B}$: By assumption on the cardinalities of $\mathcal{A}$ and $\mathcal{B}$ we have $\{(\min^A, \min^B), (\max^A, \max^B)\} \in I_j$ for every $j \leq m$. To give a proof of the forth property of $(I_j)_{j \leq m}$ (the back property can be proven analogously), suppose $j < m$, $p \in I_{j+1}$, and $a \in A$. We distinguish two cases, depending on whether or not the following condition

$$(*) \qquad \text{there is an } a' \in \text{do}(p) \text{ such that } d_j(a, a') < 2^j$$

is satisfied. If $(*)$ holds then there is exactly one $b \in B$ for which $p \cup \{(a, b)\}$ is a partial isomorphism preserving d_j-distances. Now assume that $(*)$ does not hold and let $\text{do}(p) = \{a_1, \ldots, a_r\}$ with $a_1 < \ldots < a_r$. We restrict ourselves to the case $a_i < a < a_{i+1}$ for some i. Then, $d_j(a_i, a) = \infty$ and $d_j(a, a_{i+1}) = \infty$; hence, $d_{j+1}(a_i, a_{i+1}) = \infty$ and therefore, $d_{j+1}(p(a_i), p(a_{i+1})) = \infty$. Thus there is a b such that $p(a_i) < b < p(a_{i+1})$, $d_j(p(a_i), b) = \infty$, and $d_j(b, p(a_{i+1})) = \infty$. One easily verifies that $q := p \cup \{(a, b)\}$ is a partial isomorphism in I_j. □

Example 2.3.7 Let $\tau = \{<, \min, \max\}$ be as in the preceding example and $\sigma = \tau \cup \{E\}$ with a binary relation symbol E. For $n \geq 3$ let $\mathcal{A}_n$ be the ordered τ-structure with $A_n = \{0, \ldots, n\}$, $\min^{A_n} = 0$, $\max^{A_n} = n$, where $<^{A_n}$ is the natural ordering on A_n, and

$$E^{A_n} = \{(i, j) \mid |i - j| = 2\} \cup \{(0, n), (n, 0), (1, n-1), (n-1, 1)\}.$$

(A_n, E^{A_n}) is a graph that is connected iff n is odd. Now, let $m \geq 2$ and $l, k \geq 2^m$.

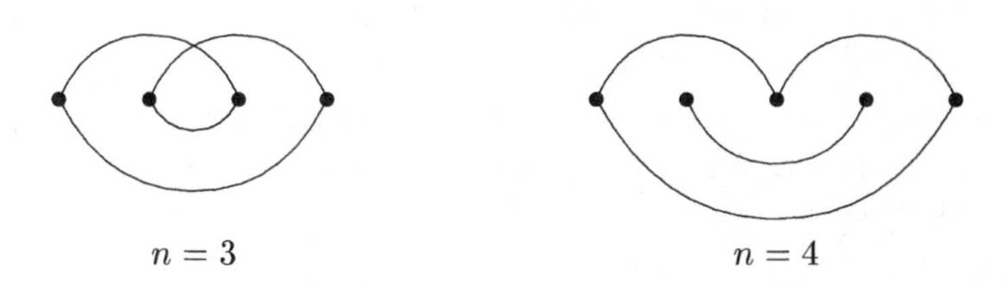

Let I_j be the set of partial isomorphisms from $\mathcal{A}_l|\tau$ to $\mathcal{A}_k|\tau$ as introduced in the preceding example. For $j \geq 2$ any $p \in I_j$ preserves E, too, that is, $I_j \subseteq \text{Part}(\mathcal{A}_l, \mathcal{A}_k)$. Hence, $(I_{j+2})_{j\leq m-2} : \mathcal{A}_l \cong_{m-2} \mathcal{A}_k$, and, by 2.2.12, we obtain:

– The class of finite connected ordered graphs is not first-order axiomatizable. □

The following example is related to the preceding one, but does not involve orderings.

Example 2.3.8 For $l \geq 1$, let $\mathcal{G}_l$ be the graph given by a cycle of length $l+1$. To be precise, set

$$G_l := \{0, \ldots, l\}, \;\; E^{G_l} := \{(i, i+1) \mid i < l\} \cup \{(i+1, i) \mid i < l\} \cup \{(0, l), (l, 0)\}.$$

Thus, for $l, k \in \mathbb{N}$, the disjoint union $\mathcal{G}_l \dot{\cup} \mathcal{G}_k$ consists of a cycle of length $l+1$ and of a cycle of length $k+1$ (for $\mathcal{A}_l$ as defined in the preceding example we have $\mathcal{A}_l|\{E\} \cong \mathcal{G}_l$ for l odd, and $\mathcal{A}_l|\{E\} \cong \mathcal{G}_{\frac{l}{2}-1} \dot{\cup} \mathcal{G}_{\frac{l}{2}}$ for l even). We show:

$$\text{If } l, k \geq 2^m \text{ then } \mathcal{G}_l \cong_m \mathcal{G}_k \text{ and } \mathcal{G}_l \cong_m \mathcal{G}_l \dot{\cup} \mathcal{G}_l.$$

In fact, for $j \in \mathbb{N}$, define the distance function d_j on a graph $\mathcal{G}$ by

$$d_j(a, a') \;\; := \;\; \begin{cases} d(a, b) & \text{if } d(a, b) < 2^{j+1} \\ \infty & \text{else} \end{cases}$$

(where d denotes the distance function on $\mathcal{G}$ introduced in 1.A1). To show, say, that $\mathcal{G}_l$ and $\mathcal{G}_l \dot{\cup} \mathcal{G}_l$ are m-isomorphic, one verifies $(I_j)_{j\leq m} : \mathcal{G}_l \cong_m \mathcal{G}_l \dot{\cup} \mathcal{G}_l$ where I_j is the set of $p \in \text{Part}(\mathcal{G}_l, \mathcal{G}_l \dot{\cup} \mathcal{G}_l)$ with

$$\|\text{do}(p)\| \leq m - j \quad \text{and} \quad d_j(a, b) = d_j(p(a), p(b)) \text{ for } a, b \in \text{do}(p)$$

(the proof is similar to that of Example 2.3.6). We note two consequences.

– The class CONN of connected finite graphs is not axiomatizable in first-order logic.

In fact, by 2.2.12, CONN is not axiomatizable, since for each m we have

$$\mathcal{G}_{2^m} \in \text{CONN}, \;\; \mathcal{G}_{2^m} \dot{\cup} \mathcal{G}_{2^m} \notin \text{CONN}, \;\; \mathcal{G}_{2^m} \equiv_m \mathcal{G}_{2^m} \dot{\cup} \mathcal{G}_{2^m}.$$

– The global relation TC (cf. 1.0.9), the relation of transitive closure on the class GRAPH of finite graphs, is not first-order definable.

In fact, suppose $\psi(x, y)$ is a first-order formula defining TC on GRAPH. Then CONN would be the class of finite models of $\forall x \forall y\, (\neg x = y \rightarrow \psi(x, y))$ (and the graph axioms). □

Exercise 2.3.9 Set $\tau = \{E\}$. For $l \geq 1$ let $\mathcal{B}_l$ and $\mathcal{D}_l$ be the τ-structures given by

$$B_l := \{0, \dots, l\}, \quad E^{B_l} := \{(i, i+1) \mid i < l\},$$
$$D_l := \{0, \dots, l\}, \quad E^{D_l} := \{(i, i+1) \mid i < l\} \cup \{(l, 0)\}.$$

Given $m \geq 0$, show that $\mathcal{B}_l \cong_m \mathcal{B}_l \dot{\cup} \mathcal{D}_l$ for sufficiently large l. Conclude that the class of finite acyclic digraphs (cf. 1.A1) is not axiomatizable in first-order logic. □

Proposition 2.3.10 *The product, the disjoint union, and the ordered sum (cf. 1.A3) preserve* $\equiv_m$, *i.e.,*

(a) *If* $\mathcal{A}_1 \equiv_m \mathcal{B}_1$ *and* $\mathcal{A}_2 \equiv_m \mathcal{B}_2$ *then* $\mathcal{A}_1 \times \mathcal{A}_2 \equiv_m \mathcal{B}_1 \times \mathcal{B}_2$.
(b) *If* $\mathcal{A}_1 \equiv_m \mathcal{B}_1$ *and* $\mathcal{A}_2 \equiv_m \mathcal{B}_2$ *then* $\mathcal{A}_1 \dot{\cup} \mathcal{A}_2 \equiv_m \mathcal{B}_1 \dot{\cup} \mathcal{B}_2$.
(c) *If* $\mathcal{A}_1 \equiv_m \mathcal{B}_1$ *and* $\mathcal{A}_2 \equiv_m \mathcal{B}_2$ *then* $\mathcal{A}_1 \lhd \mathcal{A}_2 \equiv_m \mathcal{B}_1 \lhd \mathcal{B}_2$.

Proof. Suppose $\mathcal{A}_1 \equiv_m \mathcal{B}_1$ and $\mathcal{A}_2 \equiv_m \mathcal{B}_2$. By Ehrenfeucht's Theorem there are winning strategies for the duplicator in the games $G_m(\mathcal{A}_1, \mathcal{B}_1)$ and $G_m(\mathcal{A}_2, \mathcal{B}_2)$. We refer to these strategies as S_1 and S_2.

(a) The following gives a winning strategy for the duplicator in the game $G_m(\mathcal{A}_1 \times \mathcal{A}_2, \mathcal{B}_1 \times \mathcal{B}_2)$: We simultaneously play games in $G_m(\mathcal{A}_1, \mathcal{B}_1)$ and $G_m(\mathcal{A}_2, \mathcal{B}_2)$. Suppose that in his i-th move the spoiler chooses, say, $(a_1, a_2) \in A_1 \times A_2$. Let $b_1 \in B_1$ and $b_2 \in B_2$ be answers to a_1 and a_2 according to S_1 and S_2, respectively. Then the duplicator chooses (b_1, b_2).

(b),(c) The proofs for parts (b) and (c) proceed in the same way. So let $* \in \{\dot{\cup}, \lhd\}$. The following represents a winning strategy for the duplicator in $G_m(\mathcal{A}_1 * \mathcal{A}_2, \mathcal{B}_1 * \mathcal{B}_2)$ (when describing it we use moves of plays in $G_m(\mathcal{A}_1, \mathcal{B}_1)$ and $G_m(\mathcal{A}_2, \mathcal{B}_2)$). Suppose that in his i-th move the spoiler selects, say, $a \in A_1 * A_2$. Then the duplicator gets his answer by applying S_1 if $a \in A_1$, and S_2 if $a \in A_2$. □

The last proof yields more:

Corollary 2.3.11 (a) *If* $(\mathcal{A}_1, \overline{a}_1) \equiv_m (\mathcal{B}_1, \overline{b}_1)$ *and* $(\mathcal{A}_2, \overline{a}_2) \equiv_m (\mathcal{B}_2, \overline{b}_2)$ *then* $(\mathcal{A}_1 \dot{\cup} \mathcal{A}_2, \overline{a}_1, \overline{a}_2) \equiv_m (\mathcal{B}_1 \dot{\cup} \mathcal{B}_2, \overline{b}_1, \overline{b}_2)$.

(b) *If* $(\mathcal{A}_1, \overline{a}_1) \equiv_m (\mathcal{B}_1, \overline{b}_1)$ *and* $(\mathcal{A}_2, \overline{a}_2) \equiv_m (\mathcal{B}_2, \overline{b}_2)$ *then* $(\mathcal{A}_1 \lhd \mathcal{A}_2, \overline{a}_1, \overline{a}_2)$ $\equiv_m (\mathcal{B}_1 \lhd \mathcal{B}_2, \overline{b}_1, \overline{b}_2)$.[4] □

---

[4] In this corollary, $\overline{a}_1$ and $\overline{a}_2$ can be sequences of different length.

Exercise 2.3.12 Suppose that $\tau = \{R_1, \ldots, R_k\}$ with unary relation symbols R_i. For $\alpha : \{1, \ldots, k\} \to \{0, 1\}$ and a τ-structure $\mathcal{A}$ denote by A_α the subset $X_1 \cap \ldots \cap X_k$, where $X_i := R_i^A$ if $\alpha(i) = 1$, and $X_i := A \setminus R_i^A$ otherwise. Show for any τ-structures $\mathcal{A}$, $\mathcal{B}$ and $m \geq 1$ that $\mathcal{A} \equiv_m \mathcal{B}$ iff $\min\{\|A_\alpha\|, m\} = \min\{\|B_\alpha\|, m\}$ holds for all α. Conclude that every sentence φ from FO[τ] is equivalent to a boolean combination of sentences of the form $\exists^{=l} x R^\alpha x$, where $R^\alpha x := \varphi_1 \wedge \ldots \wedge \varphi_k$ with $\varphi_i = R_i x$ if $\alpha(i) = 1$, and $\varphi_i = \neg R_i x$ if $\alpha(i) = 0$. Hint: First, consider sentences φ of the form $\varphi_{\mathcal{A}}^m$. Then apply 2.2.11. □

Exercise 2.3.13 Let $<$ be a binary symbol of the relational vocabulary τ. For $m \geq 1$ and $l, k > 2^m$ show that $\triangleleft^l \mathcal{A} \cong_m \triangleleft^k \mathcal{A}$ holds for ordered $\mathcal{A}$. Here $\triangleleft^l \mathcal{A}$ denotes the ordered sum of l copies of $\mathcal{A}$. (Hint: Use the idea of the proof of 2.3.6). □

Exercise 2.3.14 Show for $\overline{a} \mapsto \overline{b} \in \text{Part}(\mathcal{A}, \mathcal{B})$ and $m \geq 0$ that

$$\mathcal{B} \models \varphi_{\overline{a}}^m[\overline{b}] \qquad \text{iff} \qquad (\mathcal{A}, \overline{a}) \cong_m (\mathcal{B}, \overline{b}).$$

□

Exercise 2.3.15 Show that $(I_j)_{j \leq m} : \mathcal{A} \cong_m \mathcal{B}$ implies $I_j \subseteq W_j(\mathcal{A}, \mathcal{B})$ for $j \leq m$. □

Remark 2.3.16 Let $\mathcal{Z} := (\mathbb{Z}, <)$ and $\mathcal{Q} := (\mathbb{Q}, <)$ be the integers and the rationals with their orderings, respectively. For

$$\varphi \quad := \quad \exists x \exists y (x < y \wedge \forall z \neg (x < z \wedge z < y))$$

we have

$$(*) \qquad \mathcal{Z} \models \varphi \text{ and } \mathcal{Q} \not\models \varphi.$$

Hence, $\mathcal{Z} \not\equiv_3 \mathcal{Q}$ and therefore, $\mathcal{Z} \not\cong_3 \mathcal{Q}$. The spoiler can "transform" the information $(*)$ into a winning strategy for the game $G_3(\mathcal{Z}, \mathcal{Q})$ as given by the table

$\mathcal{Z}$	$\mathcal{Q}$
$\underline{5}$	a
$\underline{6}$	b
?	$\underline{\frac{a+b}{2}}$

We have underlined the selections of the spoiler. Note that no third move of the duplicator will lead to a partial isomorphism (since for $a < b$ we have $a < \frac{a+b}{2} < b$ and there is no integer between 5 and 6). In this strategy of the spoiler his selections in $\mathcal{Z}$ correspond to the existential quantifiers in φ and his selections in $\mathcal{Q}$ to the universal quantifiers. This connection can be made precise and is implicit in the proof of 2.2.8. It is the reason why moves in $G_m(\mathcal{A}, \mathcal{B})$ in which the spoiler chooses an element of $\mathcal{A}$ (of $\mathcal{B}$) are sometimes called $\exists$-*moves* ($\forall$-*moves*). □

As in the present section, the formulas $\varphi_{\mathcal{A},\overline{a}}^m$, the m-isomorphism type of $\overline{a}$ in $\mathcal{A}$, will also play a crucial role in subsequent considerations. In our opinion their methodological importance stems from the following two facts:

(1) They have a clear algebraic meaning.
(2) Every first-order formula is equivalent to a disjunction of such formulas (see 2.2.11).

Classical model theory has been characterized by the equation

$$\text{model theory} = \text{universal algebra} + \text{logic}$$

(see [20]). By (1) and (2) above, it is clear that the formulas $\varphi^j_{\mathcal{A},\overline{a}}$ provide a bridge between structures and first-order formulas, that is, between our main notions from (universal) algebra and from (first-order) logic, respectively. Therefore the experience that they are a valuable tool in model theory might not come as a surprise.

There is a more algebraic, sort of logic-free, way to define m-isomorphism types, say for $\overline{a} = a_1 \dots a_s$ in $\mathcal{A}$ by

$$\mathrm{IT}^0(\mathcal{A}, \overline{a}) \quad := \quad \{\varphi \mid \mathcal{A} \models \varphi[\overline{a}],\ \varphi(v_1, \dots, v_s) \text{ atomic}\}$$

and

$$\mathrm{IT}^{m+1}(\mathcal{A}, \overline{a}) \quad := \quad \{\mathrm{IT}^m(\mathcal{A}, \overline{a}a) \mid a \in A\}.$$

One easily verifies that for any $\mathcal{B}$ and $\overline{b} \in \mathcal{B}$,

$$\mathrm{IT}^m(\mathcal{A}, \overline{a}) = \mathrm{IT}^m(\mathcal{B}, \overline{b}) \qquad \text{iff} \qquad \varphi^m_{\mathcal{A},\overline{a}} = \varphi^m_{\mathcal{B},\overline{b}}\,.$$

2.4 Hanf's Theorem

All vocabularies in this and the next section will be relational unless stated otherwise. For a nonempty subset M of a structure $\mathcal{A}$ we denote by $\mathcal{M}$ the substructure of $\mathcal{A}$ with universe M.

Given a structure $\mathcal{A}$, we define the binary relation E^A on A by

$$E^A := \{(a, b) \mid a \neq b, \text{ and there are } R \text{ in } \tau \text{ and } \overline{c} \in A \text{ such that } R^A\overline{c} \text{ and } a \text{ and } b \text{ are components of the tuple } \overline{c}\}.$$

The structure $\mathcal{G}(\mathcal{A}) := (A, E^A)$ is called the *Gaifman graph of* $\mathcal{A}$. Obviously, if $\mathcal{A}$ itself is a graph then $\mathcal{G}(\mathcal{A}) = \mathcal{A}$. For a in A and $r \in \mathbb{N}$ we denote by $S(r, a)$ (or $S^A(r, a)$) the *r-ball* of a,

$$S(r, a) \quad := \quad \{b \in A \mid d(a, b) \leq r\}.^5$$

$\mathcal{S}(r, a)$ (or $\mathcal{S}^A(r, a)$) stands for the substructure of $\mathcal{A}$ with universe $S(r, a)$. Note that for $b, c \in S(r, a)$ we have $d(b, c) \leq 2r$. For $\overline{a} = a_1 \dots a_s$ we set $S(r, \overline{a}) := S(r, a_1) \cup \dots \cup S(r, a_s)$.

[5] d denotes the distance function of $\mathcal{G}(\mathcal{A})$ as defined in 1.A1.

We define the *r-ball type* of a point a in $\mathcal{A}$ to be the isomorphism type of $(\mathcal{S}(r,a),a)$, i.e., points a in $\mathcal{A}$ and b in $\mathcal{B}$ have the same r-ball type iff $(\mathcal{S}^{\mathcal{A}}(r,a),a) \cong (\mathcal{S}^{\mathcal{B}}(r,b),b)$.

In 2.3.8 we showed for certain graphs that they are m-isomorphic and hence, m-equivalent, using a sequence $(I_j)_{j\leq m}$, where – in the terminology just introduced – for each $p \in I_j$ and $a \in \mathrm{do}(p)$ there was an isomorphism of $\mathcal{S}(2^j-1,a)$ onto $\mathcal{S}(2^j-1,p(a))$ compatible with p. We use generalizations of this idea to show two further theorems on the expressive power of first-order logic. The first one (Hanf's Theorem) is obtained by applying the idea just mentioned to graphs of structures. In the second one (Gaifman's Theorem) the requirement of isomorphism of corresponding balls is weakened to l-equivalence for a suitable l.

Theorem 2.4.1 (Hanf's Theorem) *Let $\mathcal{A}$ and $\mathcal{B}$ be τ-structures and let $m \in \mathbb{N}$. Suppose that for some $e \in \mathbb{N}$ the 3^m-balls in $\mathcal{A}$ and $\mathcal{B}$ have less than e elements*[6] *and that for each 3^m-ball type ι,* (i) *or* (ii) *holds where*

(i) *$\mathcal{A}$ and $\mathcal{B}$ have the same number of elements of 3^m-ball type ι,*
(ii) *both $\mathcal{A}$ and $\mathcal{B}$ have more than $m \cdot e$ elements of 3^m-ball type ι.*

Then $\mathcal{A} \equiv_m \mathcal{B}$.

Proof. Since for $n < l$ the l-ball type of an element determines its n-ball type, we see that for $n < 3^m$ and every n-ball type ι, $\mathcal{A}$ and $\mathcal{B}$ have the same number of elements of n-ball type ι or both, $\mathcal{A}$ and $\mathcal{B}$, have more than $m \cdot e$ elements of n-ball type ι.

We show that $(I_j)_{j\leq m} : \mathcal{A} \cong_m \mathcal{B}$, where I_j is the set

$$\{\overline{a} \mapsto \overline{b} \in \mathrm{Part}(\mathcal{A},\mathcal{B}) \mid (\mathcal{S}(3^j,\overline{a}),\overline{a}) \cong (\mathcal{S}(3^j,\overline{b}),\overline{b}) \text{ and } \mathrm{length}(\overline{a}) \leq m-j\},$$

and where, for $\mathrm{length}(\overline{a}) = 0$, we set $(\mathcal{S}(3^j,\overline{a}),\overline{a}) = \emptyset$ and agree that $\emptyset \cong \emptyset$. Therefore, we have $\emptyset \mapsto \emptyset \in I_m$. Concerning the back and forth properties it is enough, by symmetry, to prove the forth property. Thus suppose that $0 \leq j < m$, $a \in A$ and $\overline{a} \mapsto \overline{b} \in I_{j+1}$, say,

$$(*) \qquad \pi : (\mathcal{S}(3^{j+1},\overline{a}),\overline{a}) \cong (\mathcal{S}(3^{j+1},\overline{b}),\overline{b}).$$

Case 1: $a \in S(2\cdot 3^j,\overline{a})$.

Then $S(3^j,\overline{a}a) \subseteq S(3^{j+1},\overline{a})$. Setting $b := \pi(a)$, we have $\pi : (\mathcal{S}(3^j,\overline{a}a),\overline{a}a) \cong (\mathcal{S}(3^j,\overline{b}b),\overline{b}b)$, hence $\overline{a}a \mapsto \overline{b}b \in I_j$.

Case 2: $a \notin S(2\cdot 3^j,\overline{a})$ (and hence, $S(3^j,a) \cap S(3^j,\overline{a}) = \emptyset$).

Let ι be the 3^j-ball type of a. By $(*)$, $\mathcal{S}(2\cdot 3^j,\overline{a})$ and $\mathcal{S}(2\cdot 3^j,\overline{b})$ contain the same number of elements of 3^j-ball type ι which, by our assumption on the cardinality of balls, is $\leq \mathrm{length}(\overline{a}) \cdot e \leq m \cdot e$. Therefore, by (i) or

[6] Note that for finite $\mathcal{A}$ and $\mathcal{B}$ there is always such an e.

(ii), there must be an element $b \notin S(2 \cdot 3^j, \bar{b})$ with 3^j-ball type ι. Choose $\pi' : (\mathcal{S}(3^j, a), a) \cong (\mathcal{S}(3^j, b), b)$. Then the corresponding restriction of $\pi \cup \pi'$ is an isomorphism of $(\mathcal{S}(3^j, \bar{a}a), \bar{a}a)$ onto $(\mathcal{S}(3^j, \bar{b}b), \bar{b}b)$. □

We give an application. Note that a graph $\mathcal{G}$ is connected if each nonempty subset of G closed under the graph relation E^G contains already all elements of G, i.e., if $\mathcal{G}$ is a model of the "second-order sentence"

$$\forall P((\exists x Px \wedge \forall x \forall y((Px \wedge Exy) \rightarrow Py)) \rightarrow \forall z Pz).$$

We are going to show that the class of connected graphs is not axiomatizable by a second-order sentence of the form $\exists P_1 \ldots \exists P_r \psi$, where $P_1, \ldots, P_r$ are unary and ψ is first-order (see section 3.1 for a precise definition of second-order logic).

For $l \geq 1$, let $\mathcal{D}_l = (D_l, E_l)$ be a digraph consisting of a cycle of length $l + 1$, e.g.

$$D_l := \{0, \ldots, l\}, \quad E_l := \{(i, i+1) \mid i < l\} \cup \{(l, 0)\}.$$

Lemma 2.4.2 *Suppose $\tau = \{E, P_1, \ldots, P_r\}$ where $P_1, \ldots, P_r$ are unary, and let $m \geq 0$. Then there is an $l_0 \geq 1$ such that for any $l \geq l_0$ and any τ-structure of the form $(\mathcal{D}_l, P_1, \ldots, P_r)$ there are $a, b \in D_l$ with disjoint and isomorphic 3^m-balls.*

Proof. For the structures under consideration any 3^m-ball contains exactly $2 \cdot 3^m + 1$ elements (note that $P_1, \ldots, P_r$ are unary and therefore do not influence the distances induced by the underlying digraphs). Let i be the number of possible isomorphism types of 3^m-balls. Then in a structure of cardinality $\geq l_0 := (i + 1)(2 \cdot 3^m + 1)$ there must be two points with disjoint 3^m-balls of the same isomorphism type. □

Lemma 2.4.3 *Suppose $(\mathcal{D}_l, P_1, \ldots, P_r)$ is a τ-structure (τ as in the preceding lemma) containing elements a and b with disjoint and isomorphic 3^m-balls. Denote by a_- and b_- the elements of D_l with $E_l a_- a$ and $E_l b_- b$, respectively (see the figure below). Let $(D_l, E_l', P_1, \ldots P_r)$ be the structure obtained by splitting the cycle $(\mathcal{D}_l, P_1, \ldots, P_r)$ into two cycles by removing the edges $(a_-, a), (b_-, b)$ and adding edges (b_-, a), (a_-, b) instead; more formally:*

$$E_l' \;:=\; (E_l \setminus \{(a_-, a), (b_-, b)\}) \cup \{(b_-, a), (a_-, b)\}.$$

Then $(\mathcal{D}_l, P_1, \ldots, P_r) \cong_m (D_l, E_l', P_1, \ldots, P_r)$.

Proof. Immediate by Hanf's Theorem, since both structures have the same number of 3^m-balls of any given isomorphism type. □

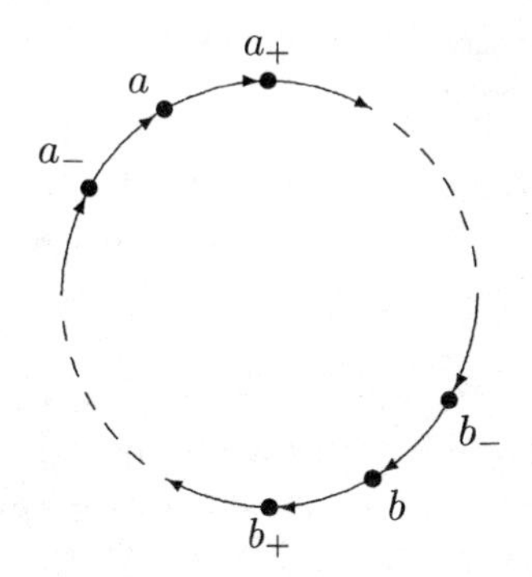

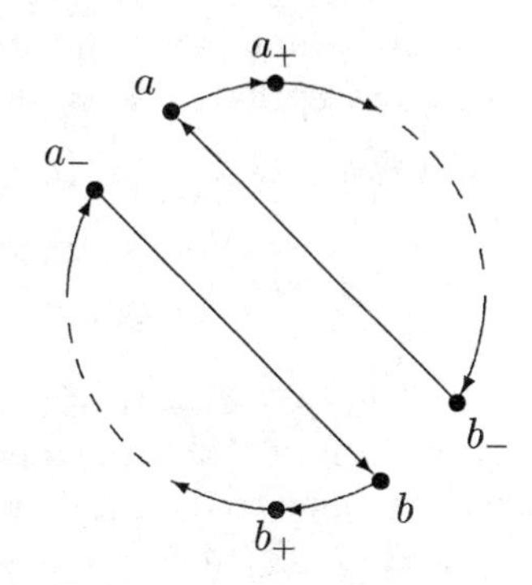

Since a partial isomorphism between digraphs is a partial isomorphism of the associated graphs, we get from the two preceding lemmas:

Lemma 2.4.4 *For $\tau = \{E, P_1, \ldots, P_r\}$ and $m \geq 0$ choose l_0 according to 2.4.2. Let $l \geq l_0$ and $(\mathcal{G}_l, P_1, \ldots, P_r)$ be a τ-structure, where $\mathcal{G}_l$ is the Gaifman graph $\mathcal{G}(\mathcal{D}_l)$ of $\mathcal{D}_l$, that is, $\mathcal{G}_l$ is a cycle of length $l+1$. Let $\mathcal{G}'_l$ be the Gaifman graph $\mathcal{G}((D_l, E'_l))$, where (D_l, E'_l) is defined as in the preceding lemma. Then*

$$(\mathcal{G}_l, P_1, \ldots, P_r) \equiv_m (\mathcal{G}'_l, P_1, \ldots, P_r). \qquad \square$$

We are now in a position to show:

Proposition 2.4.5 *The class of finite and connected graphs cannot be axiomatized by a formula of the form*

$$(*) \qquad \exists P_1 \ldots \exists P_r \psi,$$

where $P_1, \ldots, P_r$ are unary relation symbols and ψ is a first-order sentence over the vocabulary $\{E, P_1, \ldots, P_r\}$.

Proof. Suppose that for the sentence $(*)$ and any finite graph $\mathcal{G}$, we have:

$$\mathcal{G} \text{ is connected iff for some } P_1, \ldots, P_r \subseteq G: \ (\mathcal{G}, P_1, \ldots, P_r) \models \psi.$$

For $m := \mathrm{qr}(\psi)$ choose l_0 as in 2.4.2. Since $\mathcal{G}_{l_0}$ is connected, there are $P_1, \ldots, P_r$ such that $(\mathcal{G}_{l_0}, P_1, \ldots, P_r) \models \psi$. Then, $(\mathcal{G}'_{l_0}, P_1, \ldots, P_r) \models \psi$ by 2.4.4, but $\mathcal{G}'_{l_0}$ is not connected, a contradiction. $\square$

On the other side we have

Proposition 2.4.6 *The class of finite and connected graphs can be axiomatized by a formula of the form $\exists R\, \psi$, where R is binary and ψ is a first-order sentence over the vocabulary $\{E, R\}$.*

Proof. Let ψ be a sentence expressing that R is an irreflexive and transitive relation with a minimal element, and that Exy holds for any immediate R-successor y of x; that is, ψ is the conjunction of

$$\forall x \neg Rxx \wedge \forall x \forall y \forall z((Rxy \wedge Ryz) \rightarrow Rxz),$$
$$\exists x \forall y(x = y \vee Rxy),$$
$$\forall x \forall y((Rxy \wedge \forall z \neg(Rxz \wedge Rzy)) \rightarrow Exy).$$

Let $\mathcal{G}$ be a graph. Clearly, if $\mathcal{G}$ is a model of $\exists R\psi$, say $(\mathcal{G}, R^A) \models \psi$, then for any element of $\mathcal{G}$ there is a path connecting it with the minimal element; hence, $\mathcal{G}$ is connected. Conversely, suppose $\mathcal{G}$ is connected. Choose an arbitrary $a \in G$. For $n \in \mathbb{N}$ set $L_n := \{b \mid d(a,b) = n\}$ and take as R the transitive closure of $\{(b,c) \mid E^G bc$ and for some n, $b \in L_n$ and $c \in L_{n+1}\}$. □

Exercise 2.4.7 Show the results corresponding to 2.4.5 and 2.4.6 for the class of finite acyclic digraphs (cf. 1.A1). (Hint: Denote by $\mathcal{H}_l$ the digraph $(\{0,\ldots,l\}, \{(i,i+1) \mid i < l\})$. For the nonaxiomatizability result use Hanf's Theorem similarly as above, cutting off in some $(\mathcal{H}_l, P_1, \ldots, P_r)$ an interval with endpoints a and b for suitable a and b and forming a cycle out of it.) Moreover show that the class of finite acyclic digraphs can be axiomatized by a sentence $\forall P\psi$, where P is unary and ψ is a first-order sentence over $\{E, P\}$. □

Exercise 2.4.8 Show that there is a formula $\psi(x,y)$ of the form $\exists P\varphi$, where P is unary and φ is a first-order formula over $\{E, P\}$, expressing in finite graphs that x and y are in the same connected component. In fact, as $\psi(x,y)$ one can take a formula expressing that $x = y$ or that there is a subset P (a "path from x to y") containing x and y such that both x and y have an edge to exactly one member of P and every other member of P has an edge to precisely two members of P. Conclude that $\forall x \forall y \exists P\varphi$ is not equivalent to a sentence of the form $\exists P_1 \ldots \exists P_r \chi$ with unary $P_1, \ldots, P_r$ and first-order χ. □

2.5 Gaifman's Theorem

Fix a relational τ. Let $\mathcal{A}$ be a τ-structure. A subset M of A is *l-scattered*, if the distance (in the Gaifman graph $\mathcal{G}(\mathcal{A})$) between any two elements of M exceeds l. Given $r, n \geq 1$ and a τ-formula $\psi(x)$, it is easy to write down (see below) a first-order sentence asserting that there is a $2r$-scattered subset M of cardinality at least n such that $\mathcal{S}(r,a) \models \psi[a]$ for all $a \in M$.[7] Gaifman's Theorem states that every first-order sentence is logically equivalent to a boolean combination of such sentences. It thus is a further formulation of

[7] Note that, due to $2r$-scatteredness, the balls $S(r,a)$ for $a \in M$ are pairwise disjoint.

the fact already present in Hanf's Theorem that first-order sentences only capture local properties of structures.

In order to give a more precise formulation we need some notations. First note that there is a τ-formula $\theta_n(x, y)$ such that for any τ-structure $\mathcal{A}$ and $a, b \in A$,

$$\mathcal{A} \models \theta_n(x, y)\ [a, b] \qquad \text{iff} \qquad d(a, b) \leq n.$$

In fact, set $\theta_0(x, y) \ := \ x = y$ and (denoting by $a(R)$ the arity of R)

$$\theta_{n+1}(x, y) \ := \ \theta_n(x, y) \vee \exists z(\theta_n(x, z) \wedge \bigvee_{R \in \tau} \exists u_1 \ldots \exists u_{a(R)}$$
$$(Ru_1 \ldots u_{a(R)} \wedge \bigvee_{1 \leq i,j \leq a(R)} (u_i = z \wedge u_j = y))).$$

From now on we write $d(x, y) \leq n$ for $\theta_n(x, y)$. For $\overline{x} = x_1 \ldots x_m$ let

$$d(\overline{x}, y) \leq n \ := \ (d(x_1, y) \leq n \vee \ldots \vee d(x_m, y) \leq n).$$

Let $k \in \mathbb{N}$. With every τ-formula $\varphi = \varphi(\overline{x}, \overline{y})$ we associate a formula $\varphi^{S(k,\overline{x})}(\overline{x}, \overline{y})$ such that for any τ-structure $\mathcal{A}$, $\overline{a} \in A$, and $\overline{b} \in S(k, \overline{a})$,

$$\mathcal{A} \models \varphi^{S(k,\overline{x})}[\overline{a}, \overline{b}] \qquad \text{iff} \qquad \mathcal{S}(k, \overline{a}) \models \varphi[\overline{a}, \overline{b}].$$

To define $\varphi^{S(k,\overline{x})}$, first replace any bound occurrence in φ of a variable in $\overline{x}$ by a new variable and then inductively relativize the quantifiers to $S(k, \overline{x})$, e.g.,

$$[\exists z \varphi]^{S(k,\overline{x})} \ := \ \exists z(d(\overline{x}, z) \leq k \wedge \varphi^{S(k,\overline{x})}).$$

Call a sentence *basic local* if it has the form

$$\exists x_1 \ldots \exists x_n \bigwedge_{1 \leq i < j \leq n} (d(x_i, x_j) > 2r \ \wedge \ \psi^{S(r,x_i)}(x_i)),$$

where $\psi = \psi(x)$ is a first-order formula. Note that for $l \leq k$,

$$\models \varphi^{S(l,\overline{x})} \leftrightarrow [\varphi^{S(k,\overline{x})}]^{S(l,\overline{x})} \quad \text{and} \quad \models \varphi^{S(l,\overline{x})} \leftrightarrow [\varphi^{S(l,\overline{x})}]^{S(k,\overline{x})};$$

in particular, any sentence of the form

$$\exists x_1 \ldots \exists x_n \bigwedge_{1 \leq i < j \leq n} (d(x_i, x_j) > 2r \ \wedge \ \psi^{S(l,x_i)}(x_i))$$

with $l \leq r$ is logically equivalent to a basic local sentence.

A *local* sentence is a boolean combination of basic local sentences.

Theorem 2.5.1 (Gaifman's Theorem) *Every first-order sentence is logically equivalent to a local sentence.*

By 1.0.7 it suffices to show

Lemma 2.5.2 *Suppose $\mathcal{A}$ and $\mathcal{B}$ satisfy the same basic local sentences. Then $\mathcal{A} \equiv \mathcal{B}$.*

Proof. We show that $\mathcal{A} \cong_m \mathcal{B}$ for $m \in \mathbb{N}$. The argument parallels that for Hanf's Theorem. There the sets I_j consisted of partial isomorphisms $\overline{a} \mapsto \overline{b}$ such that length$(\overline{a}) \leq m - j$ and

$$(\mathcal{S}(3^j, \overline{a}), \overline{a}) \cong (\mathcal{S}(3^j, \overline{b}), \overline{b}).$$

Here we replace $\cong$ by $\equiv_{g(j)}$ and take balls of radius 7^j; the values $g(0)$, $g(1), \ldots$ of the function g can be defined by induction: $g(j)$ only has to be greater than some values which one gets in the course of the proof.

So let I_j comprise all the partial isomorphisms $\overline{a} \mapsto \overline{b}$ from $\mathcal{A}$ to $\mathcal{B}$ such that length$(\overline{a}) \leq m - j$ and

$$(\mathcal{S}(7^j, \overline{a}), \overline{a}) \equiv_{g(j)} (\mathcal{S}(7^j, \overline{b}), \overline{b}).$$

Again, in case length$(\overline{a}) = 0$, we set $(\mathcal{S}(7^j, \overline{a}), \overline{a}) = \emptyset$ and agree that $\emptyset \equiv_k \emptyset$ for all k. In particular, $\emptyset \mapsto \emptyset \in I_m$. We show $(I_j)_{j \leq m} : \mathcal{A} \cong_m \mathcal{B}$. By symmetry, we can restrict ourselves to the forth property. Thus suppose $0 \leq j < m$, $a \in A$, $\overline{a} \mapsto \overline{b} \in I_{j+1}$; hence

$$(1) \qquad (\mathcal{S}(7^{j+1}, \overline{a}), \overline{a}) \equiv_{g(j+1)} (\mathcal{S}(7^{j+1}, \overline{b}), \overline{b}).$$

We introduce a useful abbreviation: For $\overline{d}$ in a structure $\mathcal{D}$ let

$$\psi^j_{\overline{d}}(\overline{x}) \quad := \quad \left[\varphi^{g(j)}_{\mathcal{S}(7^j, \overline{d}), \overline{d}}(\overline{x})\right]^{S(7^j, \overline{x})}$$

that is, $\psi^j_{\overline{d}}(\overline{x})$ expresses that $(\mathcal{S}(7^j, \overline{d}), \overline{d}) \equiv_{g(j)} (\mathcal{S}(7^j, \overline{x}), \overline{x})$ (recall that $\varphi^l_{\mathcal{D}, \overline{d}}$ denotes the l-isomorphism type of $\overline{d}$ in $\mathcal{D}$).

Case 1. $a \in S(2 \cdot 7^j, \overline{a})$.

Then

$$\mathcal{S}(7^{j+1}, \overline{a}) \models \exists z (d(\overline{a}, z) \leq 2 \cdot 7^j \wedge \psi^j_{\overline{a}a}(\overline{a}z)).$$

We assume that the quantifier rank of this formula is $\leq g(j+1)$ (this gives us a first condition on the value of $g(j+1)$). Hence by (1),

$$\mathcal{S}(7^{j+1}, \overline{b}) \models \exists z (d(\overline{b}, z) \leq 2 \cdot 7^j \wedge \psi^j_{\overline{a}a}(\overline{b}z)),$$

so that for some b we have

$$(\mathcal{S}(7^j, \overline{a}a), \overline{a}a) \equiv_{g(j)} (\mathcal{S}(7^j, \overline{b}b), \overline{b}b).$$

Therefore $\overline{a}a \mapsto \overline{b}b \in I_j$.

Case 2. $a \notin S(2 \cdot 7^j, \overline{a})$, i.e., $S(7^j, \overline{a}) \cap S(7^j, a) = \emptyset$.

For $s \geq 1$, the following formula $\delta_s(x_1, \ldots, x_s)$ expresses that $\{x_1, \ldots, x_s\}$ is a $4 \cdot 7^j$-scattered set of elements whose 7^j-ball has the same $g(j)$-isomorphism type as that of a:

$$\delta_s \quad := \quad \bigwedge_{1 \leq l < k \leq s} d(x_l, x_k) > 4 \cdot 7^j \wedge \bigwedge_{1 \leq l \leq s} \psi_a^j(x_l).$$

Compare the cardinalities (e and i below) of maximal $4 \cdot 7^j$-scattered sets in $\mathcal{S}(2 \cdot 7^j, \overline{a})$ and in $\mathcal{A}$, respectively, consisting of such elements. More precisely, let e and i be such that

(2) $$\mathcal{S}(7^{j+1}, \overline{a}) \models \exists x_1 \ldots \exists x_e (\bigwedge_{1 \leq k \leq e} d(\overline{a}, x_k) \leq 2 \cdot 7^j \wedge \delta_e)$$

(3) $$\mathcal{S}(7^{j+1}, \overline{a}) \not\models \exists x_1 \ldots \exists x_{e+1} (\bigwedge_{1 \leq k \leq e+1} d(\overline{a}, x_k) \leq 2 \cdot 7^j \wedge \delta_{e+1})$$

(4) $$\mathcal{A} \models \exists x_1 \ldots \exists x_i \delta_i, \quad \mathcal{A} \not\models \exists x_1 \ldots \exists x_{i+1} \delta_{i+1};$$

if no such i exists, set $i = \infty$ (note that e is bounded by the length of $\overline{a}$ (and hence by m), since any two elements of the same ball of radius $2 \cdot 7^j$ have a distance at most $4 \cdot 7^j$). Clearly, $e \leq i$. Moreover we claim that the corresponding numbers e and i determined in $\mathcal{S}(7^{j+1}, \overline{b})$ and $\mathcal{B}$, respectively, are the same: Concerning $\mathcal{B}$, this holds since the sentences in (4) are basic local up to logical equivalence. Concerning $\mathcal{S}(7^{j+1}, \overline{b})$ note that $(\mathcal{S}(7^{j+1}, \overline{a}), \overline{a}) \equiv_{g(j+1)} (\mathcal{S}(7^{j+1}, \overline{b}), \overline{b})$ and that $g(j+1)$ is greater than the quantifier rank of the sentences in (2) and (3) (this gives a second condition on the value of $g(j+1)$; recall that e is bounded by m).

Case 2.1. $e = i$.

Then all elements satisfying $\psi_a^j(x)$ have a distance

$$\leq 4 \cdot 7^j + 2 \cdot 7^j = 6 \cdot 7^j < 7^{j+1}$$

from $\overline{a}$ (in fact, if one such element a' satisfies $d(\overline{a}, a') > 6 \cdot 7^j$ then a' together with e many witnesses for (2) show that $i \geq e + 1$). In particular, this holds for a. Since $a \notin S(2 \cdot 7^{j+1}, \overline{a})$,

$$\mathcal{S}(7^{j+1}, \overline{a}) \models \exists z (2 \cdot 7^j < d(\overline{a}, z) \leq 6 \cdot 7^j \wedge \psi_a^j(z) \wedge \psi_{\overline{a}}^j(\overline{a})).$$

Then, by (1),

$$\mathcal{S}(7^{j+1}, \overline{b}) \models \exists z (2 \cdot 7^j < d(\overline{b}, z) \leq 6 \cdot 7^j \wedge \psi_a^j(z) \wedge \psi_{\overline{a}}^j(\overline{b}))$$

(this gives us a third condition on the value of $g(j+1)$). Thus there is b with $2 \cdot 7^j < d(\overline{b}, b) \leq 6 \cdot 7^j$, and

(5) $$(\mathcal{S}(7^j, a), a) \equiv_{g(j)} (\mathcal{S}(7^j, b), b).$$

Moreover,

(6) $$(\mathcal{S}(7^j, \bar{a}), \bar{a})) \equiv_{g(j)} (\mathcal{S}(7^j, \bar{b}), \bar{b}).$$

Since the universes of the structures on the left sides of (5) and (6) are disjoint, and the same applies to the right sides, we obtain from 2.3.10(b) that

$$(\mathcal{S}(7^j, \bar{a}a), \bar{a}a) \equiv_{g(j)} (\mathcal{S}(7^j, \bar{b}b), \bar{b}b),$$

thus $\bar{a}a \mapsto \bar{b}b \in I_j$.

Case 2.2. $e < i$.

Then

$$\mathcal{B} \models \exists x_1 \dots \exists x_{e+1} \delta_{e+1}.$$

Hence there must be an element b in B such that

$$S(7^j, \bar{b}) \cap S(7^j, b) = \emptyset \quad \text{and} \quad \mathcal{B} \models \psi^j_a(x)[b],$$

in particular, $(\mathcal{S}(7^j, a), a) \equiv_{g(j)} (\mathcal{S}(7^j, b), b)$. Now one can argue as at the end of the preceding case. □

We close with an application of Gaifman's Theorem which will be used in Chapter 9. In the rest of this section all structures are assumed to be finite. Recall that τ is relational. Given τ-structures $\mathcal{A}$ and $\mathcal{B}$, a mapping $h : A \to B$ is a *homomorphism* if for all $R \in \tau$ and $\bar{a} \in A$, $R\bar{a}$ implies $Rh(\bar{a})$. The homomorphism is said to be *strict* if, in addition, for all $R \in \tau$ and $\bar{a} \in A$ with $R^B h(\bar{a})$ there is $\bar{e} \in A$ such that $R^A \bar{e}$ and $h(\bar{e}) = h(\bar{a})$.

A sentence φ is *preserved under (strict) homomorphisms* if for all $\mathcal{A}$, $\mathcal{B}$ and any (strict) homomorphism $h : A \to B$,

$$\mathcal{A} \models \varphi \quad \text{implies} \quad \mathcal{B} \models \varphi.$$

Exercise 2.5.3 Every existential positive sentence, that is, every sentence built up from atomic formulas with the connectives $\wedge$ and $\vee$, and the quantifier $\exists$, is preserved under homomorphisms. □

A model $\mathcal{A}$ of a sentence φ is said to be *minimal* if no proper substructure is a model of φ, that is, if

$$\mathcal{B} \subseteq \mathcal{A} \text{ and } \mathcal{B} \models \varphi \quad \text{imply} \quad \mathcal{B} = \mathcal{A}.$$

Exercise 2.5.4 For any φ, every model of φ contains a minimal model, that is, if $\mathcal{B} \models \varphi$ then there is $\mathcal{A} \subseteq \mathcal{B}$ such that $\mathcal{A}$ is a minimal model of φ (recall that we restrict ourselves to finite structures). □

Theorem 2.5.5 *If φ is preserved under strict homomorphisms then there are l and m such that no minimal model of φ contains an l-scattered subset of cardinality m.*

Proof. By Gaifman's Theorem, φ is logically equivalent to a boolean combination of basic local sentences $\varphi_1, \dots, \varphi_k$. Suppose

$$\varphi_i = \exists x_1 \dots \exists x_{n_i} (\bigwedge_{1 \le s < t \le n_i} (d(x_s, x_t) > 2r_i \land \psi_i^{S(r_i, x_s)}(x_s))).$$

Set $r := \max\{r_i \mid 1 \le i \le k\}$, $l := 2r$ and $m := 2^k + 1$. Let $\mathcal{A}$ be a minimal model of φ. We show that $\mathcal{A}$ contains no l-scattered subset of cardinality m. By contradiction, suppose that M is l-scattered and $\|M\| \ge m$. For $i = 1, \dots, k$ let $\rho_i(u)$ express "there is v such that $d(u, v) \le r_i$ and $\psi_i^{S(r_i, v)}(v)$". By choice of m there are $a, a' \in M$ such that $a \ne a'$ and for $i = 1, \dots, k$,

$$\mathcal{A} \models \rho_i[a] \text{ iff } \mathcal{A} \models \rho_i[a'].$$

Let $\mathcal{B}$ be the substructure of $\mathcal{A}$ with universe $A \setminus \{a\}$. Then $\mathcal{B} \not\models \varphi$, since $\mathcal{A}$ is a minimal model of φ.

Set $n := \max\{n_i \mid i = 1, \dots, k\}$, $\mathcal{B}_n := \dot{\bigcup}_{j=1,\dots,n} \mathcal{B}$ (the disjoint union of n copies of $\mathcal{B}$), and $\mathcal{A}_n := \mathcal{A} \dot{\cup} \mathcal{B}_n$. The projection of $\mathcal{B}_n$ to $\mathcal{B}$ is a strict homomorphism, therefore, $\mathcal{B}_n \not\models \varphi$. Since the inclusion map of $\mathcal{A}$ to $\mathcal{A}_n$ is a strict homomorphism, too, and $\mathcal{A} \models \varphi$, we have $\mathcal{A}_n \models \varphi$. We obtain the desired contradiction, if we show that for $i = 1, \dots, k$,

$$\mathcal{A}_n \models \varphi_i \quad \text{iff} \quad \mathcal{B}_n \models \varphi_i.$$

Fix i. Suppose first that $\mathcal{B}_n \models \varphi_i$. Then, for some $b \in B_n$, $\mathcal{S}^{\mathcal{B}_n}(r_i, b) \models \psi_i[b]$. View b as an element of $\mathcal{B}$. Then the B_n-part of $\mathcal{A}_n$ contains n_i – even n – copies of the element b, which are pairwise at infinite distance. Since $\mathcal{S}^{\mathcal{B}_n}(r_i, b) \cong \mathcal{S}^{\mathcal{A}_n}(r_i, b)$, we obtain $\mathcal{A}_n \models \varphi_i$.

Assume now that $\mathcal{A}_n \models \varphi_i$. Choose $e \in A_n$ such that $\mathcal{S}^{\mathcal{A}_n}(r_i, e) \models \psi_i[e]$. If $a \notin S^A(r_i, e)$ then $\mathcal{S}^{\mathcal{A}_n}(r_i, e) \cong \mathcal{S}^{B}(r_i, e)$, and we argue as above. Finally, assume that $a \in S^A(r_i, e)$. Then, $\mathcal{A} \models \rho_i[a]$, and hence, $\mathcal{A} \models \rho_i[a']$. Thus there is $e' \in A$ such that $d(e', a') \le r$ and $\mathcal{S}^A(r_i, e') \models \psi_i[e']$. Now $d(e', a) \ge d(a', a) - d(a', e') > l - r = 2r - r \ge r_i$. Therefore, $a \notin S^A(r_i, e')$, and hence $\mathcal{S}^A(r_i, e') \cong \mathcal{S}^B(r_i, e')$. Therefore, $\mathcal{B}_n$ contains n_i copies of e', and thus, $\mathcal{B}_n \models \varphi_i$. □

Notes 2.5.6 As already mentioned in the introduction to this chapter, the algebraic characterization of m-equivalence is due to Fraïssé [43], its game-theoretic version to Ehrenfeucht [33]. The j-isomorphism types were introduced by Hintikka [85] in a different context. Theorem 2.4.1 is due to Hanf [79], Theorem 2.5.1 to Gaifman [45], and Theorem 2.5.5 to Ajtai and Gurevich [7]. Further references for the results in this chapter are [5, 78, 39]. The "local" character as it becomes apparent for first-order logic in the theorems of Hanf and Gaifman is studied in [83, 115, 116, 69].

3. More on Games

In this section we show that for some fragments and extensions of first-order logic there are corresponding variants of the Ehrenfeucht-Fraïssé method. First we deal with second-order logic, mainly with monadic second-order logic, then with infinitary languages, and finally with restrictions of these infinitary languages and of first-order language to fragments consisting of formulas which only contain a fixed finite number of variables.

3.1 Second-Order Logic

Second-order logic, SO, is an extension of first-order logic which allows to quantify over relations. In addition to the symbols of first-order logic, its alphabet contains, for each $n \geq 1$, countably many n-ary *relation* (or *predicate*) *variables* $V_1^n, V_2^n, \ldots$ To denote relation variables we use letters $X, Y, \ldots$. We define the set of second-order formulas of vocabulary τ to be the set generated by the rules for first-order formulas extended by:

- If X is n-ary and $t_1, \ldots, t_n$ are terms then $Xt_1 \ldots t_n$ is a formula.
- If φ is a formula and X is a relation variable then $\exists X \varphi$ is a formula.

The free occurrence of a variable or of a relation variable in a second-order formula is defined in the obvious way and the notion of satisfaction is extended canonically. Then, given $\varphi = \varphi(x_1, \ldots, x_n, Y_1, \ldots, Y_k)$ with free (individual and relation) variables among $x_1, \ldots, x_n, Y_1, \ldots, Y_k$, a τ-structure $\mathcal{A}$, elements $a_1, \ldots, a_n \in A$, and relations $R_1, \ldots, R_k$ over A of arities corresponding to $Y_1, \ldots, Y_k$, respectively,

$$\mathcal{A} \models \varphi[a_1, \ldots, a_n, R_1, \ldots, R_k]$$

means that $a_1, \ldots, a_n$ together with $R_1, \ldots, R_k$ satisfy φ in $\mathcal{A}$.

For any τ the class EVEN$[\tau]$ of finite τ-structures of even cardinality is axiomatizable in second-order logic (but not in first-order logic, as we saw in 2.3.5). In fact, EVEN$[\tau] = \text{Mod}(\varphi)$, where φ is a sentence expressing "there is a binary relation which is an equivalence relation having only equivalence classes with exactly two elements", e.g.,

$$\exists X(\forall x Xxx \wedge \forall x\forall y(Xxy \rightarrow Xyx) \wedge \forall x\forall y\forall z((Xxy \wedge Xyz) \rightarrow Xxz) \wedge \forall x\exists^{=1}y(Xxy \wedge y \neq x)).$$

We are mainly interested in the fragment MSO of second-order logic known as *monadic second-order logic.* In formulas of MSO only *unary* relation variables ("set variables") are allowed. We write $\mathcal{A} \equiv_m^{\text{MSO}} \mathcal{B}$ if $\mathcal{A}$ and $\mathcal{B}$ satisfy the same monadic second-order sentences of quantifier rank $\leq m$ (the quantifier rank is the maximal number of nested first-order and second-order quantifiers).

As in first-order logic, $\equiv_m^{\text{MSO}}$ can be characterized by an Ehrenfeucht-Fraïssé game, MSO-$\text{G}_m(\mathcal{A}, \mathcal{B})$. The rules are the same as in the first-order Ehrenfeucht-Fraïssé game, but now in every move the spoiler can decide whether to make a *point move* or a *set move.* The point moves are as the moves in the first-order case. In a set move the spoiler chooses a subset $P \subseteq A$ or $Q \subseteq B$, and then the duplicator answers by a subset $Q \subseteq B$ or $P \subseteq A$, respectively. After m moves, elements $a_1, \ldots, a_r$ and subsets $P_1, \ldots, P_s$ in A, and corresponding elements $b_1, \ldots, b_r$ and subsets $Q_1, \ldots, Q_s$ in B (with $m = r + s$) have been chosen. The duplicator wins if $\overline{a} \mapsto \overline{b} \in \text{Part}((\mathcal{A}, P_1, \ldots, P_s), (\mathcal{B}, Q_1, \ldots, Q_s))$.

Theorem 3.1.1 $\mathcal{A} \equiv_m^{\text{MSO}} \mathcal{B}$ *iff the duplicator wins* MSO-$\text{G}_m(\mathcal{A}, \mathcal{B})$.

The following exercise leads to a proof of this theorem (along the lines of the proof of the corresponding result 2.2.8). □

Exercise 3.1.2 Given $\mathcal{A}, \overline{a} \,(= a_1 \ldots a_r)$ in A, and $\overline{P} \,(= P_1 \ldots P_s)$ a sequence of subsets of A, define the formulas $\psi^j_{\overline{a},\overline{P}}$ similar to the j-isomorphism type $\varphi^j_{\overline{a}}$, but now taking into account also the second-order set quantifiers:

$$\psi^0_{\overline{a},\overline{P}} := \bigwedge\{\varphi(v_1, \ldots, v_r, V_1, \ldots, V_s) \mid \varphi \text{ atomic or negated atomic}, \mathcal{A} \models \varphi[\overline{a}, \overline{P}]\}$$

$$\psi^{j+1}_{\overline{a},\overline{P}} := \bigwedge_{a \in A} \exists v_{r+1} \psi^j_{\overline{a}a,\overline{P}} \wedge \forall v_{r+1} \bigvee_{a \in A} \psi^j_{\overline{a}a,\overline{P}} \wedge \bigwedge_{P \subseteq A} \exists V_{s+1} \psi^j_{\overline{a},\overline{P}P} \wedge \forall V_{s+1} \bigvee_{P \subseteq A} \psi^j_{\overline{a},\overline{P}P}.$$

Show the equivalence of

(i) The duplicator wins MSO-$\text{G}_m((\mathcal{A}, \overline{P}, \overline{a}), (\mathcal{B}, \overline{Q}, \overline{b}))$;

(ii) $\mathcal{B} \models \psi^m_{\overline{a},\overline{P}}[\overline{b}, \overline{Q}]$;

(iii) $\overline{a}, \overline{P}$ satisfies in $\mathcal{A}$ the same formulas of MSO of quantifier rank $\leq m$ as $\overline{b}, \overline{Q}$ in $\mathcal{B}$. □

As in first-order logic (compare 2.2.6) or by a direct proof, one easily gets the following result which we need later.

Proposition 3.1.3 *For a fixed vocabulary and* $m \in \mathbb{N}$, *the relation* $\equiv_m^{\text{MSO}}$ *is an equivalence relation with finitely many equivalence classes.* □

We give an application of the gametheoretic characterization.

Proposition 3.1.4 *The disjoint union and the ordered sum*[1] *preserve the relation* $\equiv_m^{\mathrm{MSO}}$, *i.e., for relational* τ *we have:*

(a) *If* $\mathcal{A}_1 \equiv_m^{\mathrm{MSO}} \mathcal{B}_1$ *and* $\mathcal{A}_2 \equiv_m^{\mathrm{MSO}} \mathcal{B}_2$ *then* $\mathcal{A}_1 \dot{\cup} \mathcal{A}_2 \equiv_m^{\mathrm{MSO}} \mathcal{B}_1 \dot{\cup} \mathcal{B}_2$.

(b) *If* $\mathcal{A}_1 \equiv_m^{\mathrm{MSO}} \mathcal{B}_1$ *and* $\mathcal{A}_2 \equiv_m^{\mathrm{MSO}} \mathcal{B}_2$ *then* $\mathcal{A}_1 \lhd \mathcal{A}_2 \equiv_m^{\mathrm{MSO}} \mathcal{B}_1 \lhd \mathcal{B}_2$.

Proof. The proofs for (a) and (b) proceed in the same way. So let $* \in \{\dot{\cup}, \lhd\}$. Assume

$$\mathcal{A}_1 \equiv_m^{\mathrm{MSO}} \mathcal{B}_1 \quad \text{and} \quad \mathcal{A}_2 \equiv_m^{\mathrm{MSO}} \mathcal{B}_2.$$

By hypothesis and the last theorem there are winning strategies S_1 and S_2 for the duplicator in the games MSO-$\mathrm{G}_m(\mathcal{A}_1, \mathcal{B}_1)$ and MSO-$\mathrm{G}_m(\mathcal{A}_2, \mathcal{B}_2)$, respectively. Then the following represents a winning strategy for the duplicator in MSO-$\mathrm{G}_m(\mathcal{A}_1 * \mathcal{A}_2, \mathcal{B}_1 * \mathcal{B}_2)$ (when describing it we use moves of plays in MSO-$\mathrm{G}_m(\mathcal{A}_1, \mathcal{B}_1)$ and MSO-$\mathrm{G}_m(\mathcal{A}_2, \mathcal{B}_2)$). Suppose first that the i-th move of the spoiler is a point move where he selects, say, $a \in A_1 * A_2$. Then the duplicator gets his answer by applying S_1 if $a \in A_1$, and S_2 if $a \in A_2$. Now assume that the spoiler selects, say, $P \subseteq A_1 \cup A_2$. Set $P_1 := P \cap A_1$ and $P_2 := P \cap A_2$. Let Q_1 and Q_2 be the selections of the duplicator according to S_1 and S_2, respectively. Then, in the game MSO-$\mathrm{G}_m(\mathcal{A}_1 * \mathcal{A}_2, \mathcal{B}_1 * \mathcal{B}_2)$, the duplicator chooses $Q_1 \cup Q_2$. □

An easy induction using equivalences such as

$$\models \neg\exists X\varphi \leftrightarrow \forall X\neg\varphi, \qquad \models (\varphi \vee \forall Y\psi) \leftrightarrow \forall Y(\varphi \vee \psi) \quad \text{if } Y \text{ is not free in } \varphi,$$

shows that each (M)SO-formula is logically equivalent to an (M)SO-formula in prenex normal form, that is, to a formula of the form

$$Q_1\alpha_1 \ldots Q_s\alpha_s\psi,$$

where $Q_1, \ldots, Q_s \in \{\forall, \exists\}$, and where $\alpha_1, \ldots, \alpha_s$ are first-order or second-order variables and ψ is quantifier-free. Moreover, since

$$\models \exists x Q_1\alpha_1 \ldots Q_s\alpha_s\psi \ \leftrightarrow\ \exists X Q_1\alpha_1 \ldots Q_s\alpha_s(\exists^{=1}xXx \wedge \forall x(Xx \to \psi)),$$
$$\models \forall x Q_1\alpha_1 \ldots Q_s\alpha_s\psi \ \leftrightarrow\ \forall X Q_1\alpha_1 \ldots Q_s\alpha_s(\exists^{=1}xXx \to \forall x(Xx \to \psi)),$$

every (M)SO-formula is logically equivalent to one in prenex normal form in which each second-order quantifier precedes all first-order quantifiers. Such a formula is called (M)Σ^1_n, if the string of second-order quantifiers consists of n consecutive blocks, where in each block all quantifiers are of the same type (i.e., all universal or all existential), adjacent blocks contain quantifiers of different type, and the first block is existential. (M)Π^1_n-formulas are defined in the same way, but now we require that the first block consists of universal quantifiers.

[1] Compare 1.A3 for the definition of disjoint union and ordered sum.

In particular, $\exists X \exists Y \forall Z \forall x \exists y \varphi$ with quantifier-free φ is a Σ^1_2-formula. Clearly, the negation of a Σ^1_n-formula is logically equivalent to a Π^1_n-formula, and the negation of a Π^1_n-formula to a Σ^1_n-formula. Moreover, denoting by Δ^1_n the set of formulas that are logically equivalent to both a Σ^1_n-formula and a Π^1_n-formula, we have up to logical equivalence

$$\begin{array}{ccccccccc} & & \Sigma^1_1 & & & & \Sigma^1_2 & & \\ & \subseteq & & \subseteq & & \subseteq & & & \\ \Delta^1_1 & & & & \Delta^1_2 & & & & \dots \\ & \subseteq & & \subseteq & & \subseteq & & & \\ & & \Pi^1_1 & & & & \Pi^1_2 & & \end{array}$$

which can easily be verified by adding dummy variables. The same inclusions hold for the monadic classes.

It can be shown that for arbitrary models all the inclusions above are proper (this also holds for MSO). The question to what extent the hierarchies are proper in the finite is related to important questions of complexity theory (cf. Chapter 7). We have seen in section 2.4 that the class of finite, connected graphs is $\text{M}\Pi^1_1$-axiomatizable but not $\text{M}\Sigma^1_1$-axiomatizable. Thus, in the finite, $\text{M}\Sigma^1_1 \neq \text{M}\Pi^1_1$.

3.2 Infinitary Logic: The Logics $\text{L}_{\infty\omega}$ and $\text{L}_{\omega_1\omega}$

In recent years even *infinitary* logics have turned out to be relevant in the context of finite model theory. In fact, they contain some of the logics important in descriptive complexity theory (see Chapter 7), but – in contrast to them – their expressive power allows a manageable characterization in terms of games, similar to that of first-order logic. In this and the next section we introduce these logics and deal with their gametheoretic aspects.

The infinitary logics $\text{L}_{\infty\omega}$ and $\text{L}_{\omega_1\omega}$ allow arbitrary and countable disjunctions (and hence conjunctions), respectively. More formally: Let τ be a vocabulary. The class of $\text{L}_{\infty\omega}$-formulas over τ is given by the following clauses:

- it contains all atomic first-order formulas over τ
- if φ is a formula then so is $\neg\varphi$
- if φ is a formula and x a variable then $\exists x \varphi$ is a formula
- if Ψ is a set of formulas then $\bigvee \Psi$ is a formula.

For $\text{L}_{\omega_1\omega}$ we replace the last clause by

- if Ψ is a *countable* set of formulas then $\bigvee \Psi$ is a formula.

The semantics is a direct extension of the semantics of first-order logic with $\bigvee \Psi$ being interpreted as the disjunction over all formulas in Ψ; hence, neglecting the interpretation of the free variables,

$$\mathcal{A} \models \bigvee \Psi \quad \text{iff} \quad \text{for some } \psi \in \Psi,\ \mathcal{A} \models \psi.$$

We set $\bigwedge\Psi := \neg\bigvee\{\neg\psi \mid \psi \in \Psi\}$. Then $\bigwedge\Psi$ is interpreted as the conjunction over all formulas in Ψ. By identifying $(\varphi \vee \psi)$ with $\bigvee\{\varphi, \psi\}$ we see that $L_{\infty\omega}$ and $L_{\omega_1\omega}$ are extensions of first-order logic.

Examples 3.2.1 (a) For any τ, the models of the $L_{\omega_1\omega}$-sentence

$$\bigvee\{\varphi_{=n} \mid n \geq 1\},$$

where $\varphi_{=n}$ is a first-order sentence expressing that the universe has cardinality n, are the finite τ-structures. The $L_{\omega_1\omega}$-sentence $\bigvee\{\varphi_{=2n} \mid n \geq 1\}$ axiomatizes the class EVEN$[\tau]$. Similarly, if M is any nonempty set of positive natural numbers, then the class of models of the $L_{\omega_1\omega}$-sentence $\bigvee\{\varphi_{=k} \mid k \in M\}$ corresponds to the query "$\|A\| \in M$?". In particular, we see that nonrecursive queries are $L_{\omega_1\omega}$-definable. We get even more:

(b) Any class of finite structures is axiomatizable in $L_{\omega_1\omega}$. In fact, let K be a class of finite structures. Choose, using 2.1.2, a set Φ of first-order sentences such that $K = \text{Mod}(\Phi)$. Then $K = \text{Mod}(\varphi)$ for the $L_{\omega_1\omega}$-sentence $\varphi := \bigwedge\Phi$.

(c) "Connectivity" is a property of graphs expressible in $L_{\omega_1\omega}$ by

$$\forall x \forall y(\neg x = y \rightarrow \bigvee\{\varphi_n(x,y) \mid n \geq 1\}),$$

where $\varphi_n(x,y)$ is a first-order formula saying that there is a path from x to y of length n,

$$\varphi_n(x,y) \ := \ \exists z_0 \ldots \exists z_n(z_0 = x \wedge z_n = y \wedge Ez_0z_1 \wedge \ldots \wedge Ez_{n-1}z_n). \quad \Box$$

$L_{\infty\omega}$-*sentences* are $L_{\infty\omega}$-formulas without free variables. While $L_{\infty\omega}$-formulas may have infinitely many free variables ($\bigvee\{\neg v_i = v_j \mid 1 \leq i < j\}$ is an example), it can easily be seen that subformulas of $L_{\infty\omega}$-sentences only have finitely many free variables. In the following we restrict ourselves to $L_{\infty\omega}$-formulas with only finitely many free variables.

The next two propositions – and in particular the proof of the first one – show that we need not more than $L_{\omega_1\omega}$ in the context of finite model theory, although it has become popular – and we follow this tradition – to mainly consider $L_{\infty\omega}$.

Proposition 3.2.2 (a) *In the finite, every* $L_{\infty\omega}$*-formula* $\varphi(\overline{x})$ *is equivalent to an* $L_{\omega_1\omega}$*-formula* $\psi(\overline{x})$.

(b) *Assume* $\mathcal{A}$ *and* $\mathcal{B}$ *are finite. For every* $L_{\infty\omega}$*-formula* $\varphi(\overline{x})$ *there is an* FO*-formula* $\psi(\overline{x})$ *such that*

$$\mathcal{A} \models \forall\overline{x}(\varphi(\overline{x}) \leftrightarrow \psi(\overline{x})) \quad \textit{and} \quad \mathcal{B} \models \forall\overline{x}(\varphi(\overline{x}) \leftrightarrow \psi(\overline{x})).$$

In both cases, (a) *and* (b), *the formula* ψ *can be chosen such that* $\text{free}(\psi) \subseteq \text{free}(\varphi)$ *and that every variable occurring in* ψ *(free or bound) occurs in* φ.

Proof. The proof of (a) and (b) is by induction over the rules for $\mathrm{L}_{\infty\omega}$-formulas. The translation procedure preserves the "structure" of formulas and only replaces infinitary disjunctions by countable ones (by finite ones in (b)). In the main step suppose that

$$\varphi(\overline{x}) = \bigvee\{\varphi_i(\overline{x}) \mid i \in I\}$$

is an $\mathrm{L}_{\infty\omega}$-formula. For each finite structure $\mathcal{C}$ with universe $\{1, 2, \ldots, \|C\|\}$ (for $\mathcal{C} \in \{\mathcal{A}, \mathcal{B}\}$ in (b)) and each $\overline{c} \in C$, if there exists an $i \in I$ such that $\mathcal{C} \models \varphi_i[\overline{c}]$, choose such an i. Let I_0 be the set of i's chosen in this way. Then I_0 is countable (finite in (b)) and $\bigvee\{\varphi_i(\overline{x}) \mid i \in I\}$ and $\bigvee\{\varphi_i(\overline{x}) \mid i \in I_0\}$ are equivalent in the finite (in $\mathcal{A}$ and $\mathcal{B}$). □

Since every finite structure can be characterized in first-order logic, we obtain the following improvement of part (a) of the preceding proposition (cf. 3.2.1(b)):

Proposition 3.2.3 *In the finite, every $\mathrm{L}_{\infty\omega}$-formula $\varphi(\overline{x})$ is equivalent to a countable disjunction – and hence to a countable conjunction – of first-order formulas. In fact, in the finite, $\varphi(\overline{x})$ is equivalent to*

$$\bigvee\{\varphi^{\|A\|+1}_{\mathcal{A},\overline{a}}(\overline{x}) \mid \mathcal{A} \text{ finite}, \overline{a} \in A, \mathcal{A} \models \varphi[\overline{a}]\}.$$

Proof. For simplicity we restrict ourselves to sentences. Let $\mathcal{B}$ be a finite structure. If $\mathcal{B} \models \varphi$ then $\varphi^{\|B\|+1}_{\mathcal{B}}$ is a member of the disjunction which, therefore, is satisfied by $\mathcal{B}$. Conversely, if $\mathcal{B}$ satisfies the disjunction, then for some finite $\mathcal{A}$ with $\mathcal{A} \models \varphi$ we have $\mathcal{B} \models \varphi^{\|A\|+1}_{\mathcal{A}}$. Therefore by 2.2.10, $\mathcal{A} \cong \mathcal{B}$; hence, $\mathcal{B} \models \varphi$. □

We say that $\mathcal{A}$ and $\mathcal{B}$ are *$\mathrm{L}_{\infty\omega}$-equivalent*, $\mathcal{A} \equiv^{\mathrm{L}_{\infty\omega}} \mathcal{B}$, if $\mathcal{A}$ and $\mathcal{B}$ satisfy the same $\mathrm{L}_{\infty\omega}$-sentences. In order to characterize $\equiv^{\mathrm{L}_{\infty\omega}}$, one needs Ehrenfeucht games with *infinitely* many moves.

Definition 3.2.4 Let $\mathcal{A}$ and $\mathcal{B}$ be structures, $\overline{a} \in A^s$, and $\overline{b} \in B^s$. The game $\mathrm{G}_\infty(\mathcal{A}, \overline{a}, \mathcal{B}, \overline{b})$ is the same as the game $\mathrm{G}_m(\mathcal{A}, \overline{a}, \mathcal{B}, \overline{b})$ up to the fact that now each player has to make infinitely many moves. Thus, in the course of a play of $\mathrm{G}_\infty(\mathcal{A}, \overline{a}, \mathcal{B}, \overline{b})$, elements $e_1, e_2, \ldots$ in $\mathcal{A}$ and $f_1, f_2, \ldots$ in $\mathcal{B}$ are chosen. The duplicator *wins the play* if $\overline{a}e_1 \ldots e_i \mapsto \overline{b}f_1 \ldots f_i \in \mathrm{Part}(\mathcal{A}, \mathcal{B})$ for *all* i, and the spoiler wins if $\overline{a}e_1 \ldots e_i \mapsto \overline{b}f_1 \ldots f_i \notin \mathrm{Part}(\mathcal{A}, \mathcal{B})$ for *some* i. The *duplicator wins* $\mathrm{G}_\infty(\mathcal{A}, \overline{a}, \mathcal{B}, \overline{b})$ if he has a winning strategy. □

Immediately from the definition we get:

Lemma 3.2.5 *Suppose that the duplicator wins $\mathrm{G}_\infty(\mathcal{A}, \overline{a}, \mathcal{B}, \overline{b})$. Then*

(a) $\overline{a} \mapsto \overline{b} \in \mathrm{Part}(\mathcal{A}, \mathcal{B})$.

(b) *For $a \in A$ there is $b \in B$ such that the duplicator wins $\mathrm{G}_\infty(\mathcal{A}, \overline{a}a, \mathcal{B}, \overline{b}b)$.*

(c) *For $b \in B$ there is $a \in A$ such that the duplicator wins $G_\infty(\mathcal{A}, \bar{a}a, \mathcal{B}, \bar{b}b)$.* □

Since we want to give the Ehrenfeucht-Fraïssé type characterization of $\equiv^{L_{\infty\omega}}$ in its different forms we define:

Definition 3.2.6 (a) $\mathcal{A}$ and $\mathcal{B}$ are said to be *partially isomorphic*, written $\mathcal{A} \cong_{\text{part}} \mathcal{B}$, if there is a nonempty set I of partial isomorphisms from $\mathcal{A}$ to $\mathcal{B}$ with the back and forth properties (that is, for every $p \in I$ and every $a \in A$ ($b \in B$) there is $q \in I$ with $q \supseteq p$ and $a \in \text{do}(q)$ ($b \in \text{rg}(q)$)). We then write $I : \mathcal{A} \cong_{\text{part}} \mathcal{B}$.

(b) $W_\infty(\mathcal{A}, \mathcal{B}) :=$

$$\{\bar{a} \mapsto \bar{b} \mid s \in \mathbb{N}, \bar{a} \in A^s, \bar{b} \in B^s, \text{ the duplicator wins } G_\infty(\mathcal{A}, \bar{a}, \mathcal{B}, \bar{b})\}$$

is the set of winning positions for the duplicator. □

$L_{\infty\omega}$-equivalence and the notions just introduced are intimately related:

Theorem 3.2.7 *For structures $\mathcal{A}$ and $\mathcal{B}$, $\bar{a} \in A^s$, and $\bar{b} \in B^s$ the following are equivalent:*

(i) *The duplicator wins $G_\infty(\mathcal{A}, \bar{a}, \mathcal{B}, \bar{b})$.*
(ii) *$\bar{a} \mapsto \bar{b} \in W_\infty(\mathcal{A}, \mathcal{B})$ and $W_\infty(\mathcal{A}, \mathcal{B}) : \mathcal{A} \cong_{part} \mathcal{B}$.*
(iii) *There is a set I with $\bar{a} \mapsto \bar{b} \in I$ such that $I : \mathcal{A} \cong_{part} \mathcal{B}$.*
(iv) *$\bar{a}$ and $\bar{b}$ satisfy the same formulas of $L_{\infty\omega}$ in $\mathcal{A}$ and $\mathcal{B}$, respectively, that is, if $\varphi(x_1, \ldots, x_s)$ is a formula of $L_{\infty\omega}$, then*

$$\mathcal{A} \models \varphi[\bar{a}] \quad \textit{iff} \quad \mathcal{B} \models \varphi[\bar{b}].$$

Proof. For (i) ⇒ (ii) see the preceding lemma; clearly (ii) implies (iii). For (iii) ⇒ (i) note that a set I with $\bar{a} \mapsto \bar{b} \in I$ and $I : \mathcal{A} \cong_{\text{part}} \mathcal{B}$ can be viewed as a winning strategy for the duplicator for the game $G_\infty(\mathcal{A}, \bar{a}, \mathcal{B}, \bar{b})$. Hence, it suffices to show the equivalence of (iii) and (iv). Let I be as in (iii). By (transfinite) induction on the quantifier rank of the $L_{\infty\omega}$-formula $\varphi(x_1, \ldots, x_s)$ we prove

$$(*) \qquad \mathcal{A} \models \varphi[\bar{e}] \quad \text{iff} \quad \mathcal{B} \models \varphi[\bar{f}]$$

for any $e_1 \ldots e_s \mapsto f_1 \ldots f_s \in I$. The case of quantifier rank 0 is handled by part (c) of 2.2.2. For arbitrary quantifier rank note that the class of formulas satisfying $(*)$ contains the atomic formulas and is closed under $\neg$ and $\bigvee$. Suppose that $\varphi(x_1, \ldots, x_s) = \exists y \psi(x_1, \ldots, x_s, y)$. Assume, for example, that $\mathcal{A} \models \varphi[e_1, \ldots, e_s]$. Then there is $a \in A$ such that $\mathcal{A} \models \psi[e_1, \ldots, e_s, a]$. By the forth property of I, there is $b \in B$ such that $e_1 \ldots e_s a \mapsto f_1 \ldots f_s b \in I$. Since $\text{qr}(\psi) < \text{qr}(\varphi)$, the induction hypothesis yields $\mathcal{B} \models \psi[f_1, \ldots, f_s, b]$; hence, $\mathcal{B} \models \varphi[f_1, \ldots, f_s]$.

Suppose now that (iv) holds, and let I be the set of all partial isomorphisms $e_1 \dots e_r \mapsto f_1 \dots f_r$ (with $r \geq 0$) such that for all $\mathrm{L}_{\infty\omega}$-formulas $\varphi(x_1, \dots, x_r)$,

$$\mathcal{A} \models \varphi[\overline{e}] \quad \text{iff} \quad \mathcal{B} \models \varphi[\overline{f}].$$

By (iv), $\overline{a} \mapsto \overline{b} \in I$. We show that I has the back and forth properties. So, let $e_1 \dots e_r \mapsto f_1 \dots f_r \in I$ and $a \in I$. For each $b \in B$, if there is a formula $\varphi(x_1, \dots, x_r, x)$ of $\mathrm{L}_{\infty\omega}$ such that

$$\mathcal{A} \models \varphi(\overline{x}, x)[\overline{e}a] \text{ and } \mathcal{B} \models \neg\varphi(\overline{x}, x)[\overline{f}b],$$

let $\varphi_b(\overline{x}, x)$ be such a formula; otherwise, set $\varphi_b(\overline{x}, x) := \ x = x$. Since $\mathcal{A} \models \exists x \bigwedge\{\varphi_b \mid b \in B\}[\overline{e}]$, we have $\mathcal{B} \models \exists x \bigwedge\{\varphi_b \mid b \in B\}[\overline{f}]$. Hence, there is $b' \in B$ such that $\mathcal{B} \models \bigwedge\{\varphi_b \mid b \in B\}[\overline{f}b']$. Using the definition of $\varphi_{b'}$, one easily sees that $\overline{e}a$ and $\overline{f}b'$ satisfy the same formulas of $\mathrm{L}_{\infty\omega}$ in $\mathcal{A}$ and $\mathcal{B}$, respectively, and hence, $\overline{e}a \mapsto \overline{f}b' \in I$. The back property is proven similarly. □

Since $W_\infty(\mathcal{A}, \mathcal{B}) \neq \emptyset$ iff $\emptyset \mapsto \emptyset \in W_\infty(\mathcal{A}, \mathcal{B})$ (cf. 3.2.7), we obtain the

Corollary 3.2.8 *For $\mathcal{A}$ and $\mathcal{B}$ the following are equivalent.*
(i) *The duplicator wins $\mathrm{G}_\infty(\mathcal{A}, \mathcal{B})$.* (ii) $W_\infty(\mathcal{A}, \mathcal{B}) : \mathcal{A} \cong_{part} \mathcal{B}$.
(iii) $\mathcal{A} \cong_{\text{part}} \mathcal{B}$. (iv) $\mathcal{A} \equiv^{\mathrm{L}_{\infty\omega}} \mathcal{B}$. □

To give some applications we first show:

Lemma 3.2.9 *Let A and B be countable.*

(a) *If $\mathcal{A} \cong_{\text{part}} \mathcal{B}$ then $\mathcal{A} \cong \mathcal{B}$.*
(b) *If $I : \mathcal{A} \cong_{\text{part}} \mathcal{B}$ and $p_0 \in I$ then p_0 can be extended to an isomorphism from $\mathcal{A}$ onto $\mathcal{B}$.*

Proof. Let $A = \{a_1, a_2, \dots\}$ and $B = \{b_1, b_2, \dots\}$. It suffices to show (b). Suppose $I : \mathcal{A} \cong_{\text{part}} \mathcal{B}$ and $p_0 \in I$. By repeated application of the back and forth properties, one obtains $p_1, p_2, \dots$ in I such that $p_0 \subseteq p_1 \subseteq \dots$ and such that $a_1 \in \mathrm{do}(p_1), b_1 \in \mathrm{rg}(p_2), a_2 \in \mathrm{do}(p_3), \dots$. Then $\bigcup_{n \geq 0} p_n$ is an isomorphism from $\mathcal{A}$ onto $\mathcal{B}$. □

Corollary 3.2.10 *If $\mathcal{A}$ and $\mathcal{B}$ are countable and $\mathrm{L}_{\infty\omega}$-equivalent then they are isomorphic.*

Proof. The claim follows immediately from 3.2.8 and 3.2.9(a). □

The following example will be of importance in the next chapter.

Example 3.2.11 Let τ be relational. For $r \geq 0$ let Δ_{r+1} be the set

$$\Delta_{r+1} := \{\varphi(v_1, \dots, v_r, v_{r+1}) \mid \varphi \text{ has the form } R\overline{x}, \text{where } R \in \tau \text{ and where } v_{r+1} \text{ occurs in } \overline{x}\}.$$

For a subset Φ of Δ_{r+1} the sentence $\chi_\Phi :=$ [2]

$$\forall v_1 \ldots \forall v_r (\bigwedge_{1\le i<j\le r} v_i \neq v_j \to \exists v_{r+1} (\bigwedge_{1\le i\le r} v_i \neq v_{r+1} \wedge \bigwedge_{\varphi\in\Phi} \varphi \wedge \bigwedge_{\varphi\in\Phi^c} \neg\varphi))$$

where $\Phi^c := \Delta_{r+1} \setminus \Phi$, is called an *extension axiom*, more precisely, an $(r+1)$-*extension axiom.* The set T_{rand} of all extension axioms gives the *random structure theory.*[3] Clearly,

– every model of T_{rand} is infinite.

The extension axioms in T_{rand} guarantee for any models $\mathcal{A}$ and $\mathcal{B}$ of T_{rand} that the set

$$I \quad := \quad \{\bar{a} \mapsto \bar{b} \mid \bar{a} \in A,\ \bar{b} \in B, \text{ and } \varphi^0_{\mathcal{A},\bar{a}} = \varphi^0_{\mathcal{B},\bar{b}}\}$$

has the back and forth properties. For a proof let $\bar{a} \mapsto \bar{b} \in I$, where $\bar{a} = a_1 \ldots a_r$ and $a_1, \ldots, a_r$ can be assumed to be distinct, and let, say, a_{r+1} be in $A \setminus \{a_1, \ldots, a_r\}$. Set $\Phi := \{\varphi(v_1, \ldots, v_{r+1}) \mid \varphi \in \Delta_{r+1}, \mathcal{A} \models \varphi[\bar{a}a_{r+1}]\}$. As $\mathcal{B} \models \chi_\Phi$, there is an element $b_{r+1} \in B$ such that $\varphi^0_{\mathcal{B},\bar{b}b_{r+1}} = \varphi^0_{\mathcal{A},\bar{a}a_{r+1}}$, that is, $\bar{a}a_{r+1} \mapsto \bar{b}b_{r+1} \in I$. Moreover, since τ is relational, the empty partial isomorphism is in I. Hence, $I : \mathcal{A} \cong_{\text{part}} \mathcal{B}$. By 3.2.8 we get:

– Any two models of T_{rand} are $L_{\infty\omega}$-equivalent and thus, for each $L_{\infty\omega}$-sentence φ, $T_{\text{rand}} \models \varphi$ or $T_{\text{rand}} \models \neg\varphi$.

Finally we show

– T_{rand} has a countable model and hence by 3.2.10, an (up to isomorphism) unique countable model $\mathcal{R}$, the so-called *infinite random structure.*

In fact, let $(\alpha_n)_{n\ge 0}$ be an enumeration of all pairs $(\overline{m}, \chi)$, where $\overline{m}$ is a tuple of distinct natural numbers and χ is an $(r+1)$-extension axiom, where $r :=$ length$(\overline{m})$; moreover suppose that for $\alpha_n = (\overline{m}, \chi)$ all entries of $\overline{m}$ are not greater than n. By induction on n we define structures $\mathcal{A}_n$ with

$$A_n = \{0, \ldots, n\} \quad \text{and} \quad \mathcal{A}_0 \subseteq \mathcal{A}_1 \subseteq \mathcal{A}_2 \subseteq \ldots$$

such that $\mathcal{A} := \bigcup_{n\ge 0} \mathcal{A}_n$ is a model of T_{rand}: Let $\mathcal{A}_0 = (A_0, (\emptyset)_{R\in\tau})$ (each relation symbol is interpreted by the empty set). Suppose $\mathcal{A}_n$ has been defined and $\alpha_n = (m_1, \ldots, m_r, \chi)$ with $\chi = \chi_\Phi$. Define $\mathcal{A}_{n+1}$ with universe A_{n+1} such that $\mathcal{A}_n \subseteq \mathcal{A}_{n+1}$ and such that for $\varphi \in \Delta_{r+1}$

$$\mathcal{A}_{n+1} \models \varphi[m_1, \ldots, m_r, n+1] \qquad \text{iff} \qquad \varphi \in \Phi$$

(note that v_{r+1} occurs in every formula of Δ_{r+1}). This ensures that $\mathcal{A} := \bigcup_{n\ge 0} \mathcal{A}_n$ is a model of χ. □

---

[2] In case $r = 0$ this sentence reduces to $\exists v_1 (\bigwedge_{\varphi\in\Phi} \varphi \wedge \bigwedge_{\varphi\in\Phi^c} \neg\varphi)$.

[3] The name will become clear in Chapter 4.

We close with a further application of 3.2.8. For a sentence φ let the *spectrum* $\mathrm{Spec}(\varphi)$ of φ be the set

$$\mathrm{Spec}(\varphi) := \{m \geq 1 \mid \text{ there is } \mathcal{A} \models \varphi \text{ with } \|A\| = m\}.$$

Proposition 3.2.12 *For any first-order sentence φ at least one of* $\mathrm{Spec}(\varphi)$ *or* $\mathrm{Spec}(\neg\varphi)$ *is cofinite (i.e., for some n_0, $\{n \mid n_0 \leq n\} \subseteq \mathrm{Spec}(\varphi)$ or $\{n \mid n_0 \leq n\} \subseteq \mathrm{Spec}(\neg\varphi)$).*

Proof. Let Φ be the set of the sentences

(i) $\varphi_{\geq m}$ for $m \geq 1$
(ii) $\forall \overline{x} R \overline{x}$ for $R \in \tau$
(iii) $c = d$ for $c, d \in \tau$.

Clearly, Φ is satisfiable and any two models $\mathcal{A}$ and $\mathcal{B}$ of Φ are partially isomorphic via $I := \{p \in \mathrm{Part}(\mathcal{A}, \mathcal{B}) \mid \mathrm{do}(p) \text{ finite}\}$, and hence by 3.2.8, elementarily equivalent. Therefore, given a first-order sentence φ, we have $\Phi \models \varphi$ or $\Phi \models \neg\varphi$, say, $\Phi \models \varphi$. By the compactness theorem there is a finite $\Phi_0 \subseteq \Phi$ such that $\Phi_0 \models \varphi$. Let n_0 be larger than any m such that $\varphi_{\geq m}$ is in Φ_0. Then Φ_0 and hence, φ has a model of cardinality n for each $n \geq n_0$. □

Exercise 3.2.13 Prove the previous result for "vocabularies" containing function symbols. Hint: Compare the passage from such a vocabulary to a function-free one given in 1.D. For any new relation symbol F corresponding to some $f \in \tau$, in (ii) above take the sentence $\forall \overline{x}(\exists^{=1} x F \overline{x} x \wedge F \overline{x} x_1)$ instead of $\forall \overline{x} x F \overline{x} x$. □

Exercise 3.2.14 For any $\mathcal{A}$ and $\mathcal{B}$ we have

$$W_0(\mathcal{A}, \mathcal{B}) \supseteq \ldots \supseteq W_m(\mathcal{A}, \mathcal{B}) \supseteq \ldots \supseteq W_\infty(\mathcal{A}, \mathcal{B}).$$

Show that in case A or B is finite there is an $m_0 \leq 1 + \min\{\|A\|, \|B\|\}$ such that

$$W_0(\mathcal{A}, \mathcal{B}) \supset \ldots \supset W_{m_0}(\mathcal{A}, \mathcal{B}) = W_\infty(\mathcal{A}, \mathcal{B}).$$ □

3.3 The Logics FO^s and $\mathrm{L}^s_{\infty\omega}$

In first-order logic FO, every finite structure $\mathcal{A}$ can be characterized up to isomorphism by a first-order sentence $\varphi_\mathcal{A}$ which, in general, needs $\|A\| + 1$ variables (compare, for example, the proof of 2.1.1). Hence, an arbitrary class K of finite structures can be axiomatized in $\mathrm{L}_{\infty\omega}$ by the sentence

$$\bigvee\{\varphi_\mathcal{A} \mid \mathcal{A} \in K\}$$

which, in general, contains infinitely many variables. Since every class of finite structures is axiomatizable in it, $\mathrm{L}_{\infty\omega}$ is too powerful in the finite to yield new general principles. The observation above motivates the restriction to formulas of $\mathrm{L}_{\infty\omega}$ containing only finitely many variables. In the present section we shall see that this approach really leads to logics which are interesting from the perspective of finite model theory.

Fix $s \geq 1$. We denote by $\mathrm{L}^s_{\infty\omega}$ and FO^s the fragments of $\mathrm{L}_{\infty\omega}$ and FO, respectively, containing only formulas, whose *free and bound* variables are among $v_1, \ldots, v_s$. Moreover, we set

$$\mathrm{L}^\omega_{\infty\omega} := \bigcup_{s\geq 1} \mathrm{L}^s_{\infty\omega}.$$

Whereas $\mathrm{FO} = \bigcup_{s\geq 1} \mathrm{FO}^s$, the formula $\bigvee\{\varphi_{=n} \mid n \geq 1\}$ belongs to $\mathrm{L}_{\infty\omega}$ but not to $\mathrm{L}^\omega_{\infty\omega}$.

Let $x = v_1, y = v_2$, and $z = v_3$ in examples.

Examples 3.3.1 (a) Let $\tau = \{<\}$. There are FO^2-formulas $\psi_n(x)$ and χ_n such that for orderings $\mathcal{A}$, $a \in A$, and $n \geq 0$,

$$\mathcal{A} \models \psi_n[a] \quad \text{iff} \quad a \text{ is the } n\text{-th element of } <^A,$$

and in case $n \geq 1$,

$$\mathcal{A} \models \chi_n \quad \text{iff} \quad \|A\| = n.$$

In fact, define inductively

$$\psi_0(x) := \forall y \neg y < x, \quad \psi_{n+1}(x) := \forall y(y < x \leftrightarrow \bigvee_{i\leq n} \exists x(x = y \wedge \psi_i(x))),$$

and set $\chi_n := \exists x \psi_{n-1}(x) \wedge \neg\exists x \psi_n(x)$.

(b) For each $n \geq 1$ there is an FO^3-formula $\varphi_n(x,y)$ that in digraphs expresses that there is a path of length at most n from x to y. In fact, let

$$\begin{aligned} \varphi_1(x,y) &:= Exy, \\ \varphi_{n+1}(x,y) &:= \varphi_n(x,y) \vee \exists z(Ezy \wedge \exists y(y = z \wedge \varphi_n(x,y))). \end{aligned}$$

Concerning the quantifier rank we can do better than in ψ_n and φ_n: Let φ be an $\mathrm{L}^s_{\infty\omega}$-formula and π a permutation of $1, \ldots, s$. By simultaneously replacing both the free and the bound occurrences of $v_1, \ldots, v_s$ by $v_{\pi(1)}, \ldots, v_{\pi(s)}$ one obtains a formula

$$\varphi\begin{pmatrix} v_{\pi(1)} & \ldots & v_{\pi(s)} \\ v_1 & \ldots & v_s \end{pmatrix}.$$

Clearly,

$$\mathcal{A} \models \varphi\begin{pmatrix} v_{\pi(1)} & \ldots & v_{\pi(s)} \\ v_1 & \ldots & v_s \end{pmatrix}[\bar{a}] \quad \text{iff} \quad \mathcal{A} \models \varphi[a_{\pi(1)}, \ldots, a_{\pi(s)}].$$

Now in the preceding examples we can replace ψ_n and φ_n by the formulas ψ'_n and φ'_n of quantifier rank $\leq n+1$, where

$$\psi'_0 := \psi_0, \quad \psi'_{n+1} := \forall y(y < x \leftrightarrow \bigvee_{i \leq n} \psi'_i \left(\begin{matrix} yx \\ xy \end{matrix}\right)),$$
$$\varphi'_1 := \varphi_1, \quad \varphi'_{n+1} := \varphi'_n \vee \exists z(Ezy \wedge \varphi'_n \left(\begin{matrix} xzy \\ xyz \end{matrix}\right)).$$

□

We write $\mathcal{A} \equiv^s \mathcal{B}$, $\mathcal{A} \equiv^{\mathrm{L}^s_{\infty\omega}} \mathcal{B}$, and $\mathcal{A} \equiv^s_m \mathcal{B}$ to express that $\mathcal{A}$ and $\mathcal{B}$ satisfy the same sentences in FO^s, in $\mathrm{L}^s_{\infty\omega}$, and in FO^s of quantifier rank $\leq m$, respectively.

In view of 3.2.2(b) together with the last remark of 3.2.2 we have

Proposition 3.3.2 *Assume $\mathcal{A}$ and $\mathcal{B}$ are finite. For every $\mathrm{L}^s_{\infty\omega}$-formula φ there is an FO^s-formula ψ with $\mathrm{free}(\psi) \subseteq \mathrm{free}(\varphi)$ such that*

$$\mathcal{A} \models \forall x_1 \dots \forall x_s(\varphi \leftrightarrow \psi) \quad \textit{and} \quad \mathcal{B} \models \forall x_1 \dots \forall x_s(\varphi \leftrightarrow \psi).$$

Corollary 3.3.3 *If $\mathcal{A}$ and $\mathcal{B}$ are finite then*

$$\mathcal{A} \equiv^s \mathcal{B} \quad \textit{implies} \quad \mathcal{A} \equiv^{\mathrm{L}^s_{\infty\omega}} \mathcal{B}.$$

Proof. Given an $\mathrm{L}^s_{\infty\omega}$-sentence φ, choose ψ according to the preceding proposition. Then

$$\begin{aligned} \mathcal{A} \models \varphi \quad &\text{iff} \quad \mathcal{A} \models \psi \\ &\text{iff} \quad \mathcal{B} \models \psi \\ &\text{iff} \quad \mathcal{B} \models \varphi. \end{aligned}$$

□

Clearly, if $\varphi \in \mathrm{FO}^s$ (or, $\varphi \in \mathrm{L}^s_{\infty\omega}$) then every subformula of φ contains at most s free variables (namely, at most $v_1, \dots, v_s$). This property characterizes the formulas of FO^s up to logical equivalence:

Proposition 3.3.4 *Assume $s \geq 1$. If every subformula of $\varphi(v_1, \dots, v_s) \in \mathrm{FO}$ has at most s free variables, then φ is logically equivalent to a formula of FO^s. The statement remains true, if we replace FO and FO^s by $\mathrm{L}_{\infty\omega}$ and $\mathrm{L}^s_{\infty\omega}$, respectively.*

Proof. By induction (on the quantifier rank) we associate with every formula $\varphi(v_1, \dots, v_s)$ all of whose subformulas have at most s free variables, a formula φ^* with

$$\models \varphi \leftrightarrow \varphi^*, \quad \mathrm{free}(\varphi) = \mathrm{free}(\varphi^*), \quad \varphi^* \in \mathrm{FO}^s.$$

For atomic φ set $\varphi^* := \varphi$, for $\varphi = \neg\chi$ and $\varphi = (\chi_1 \vee \chi_2)$ set $\varphi^* := \neg\chi^*$ and $\varphi^* = (\chi_1^* \vee \chi_2^*)$, respectively. Now, let $\varphi = \exists y \chi$. Then,

$$(*) \qquad \mathrm{free}(\chi) \subseteq \{v_1, \dots, v_s, y\} \text{ and } \|\mathrm{free}(\chi)\| \leq s.$$

If $y \notin \mathrm{free}(\chi)$ then $\chi = \chi(v_1, \ldots, v_s)$ and χ^* is defined by induction hypothesis. We set $\varphi^* := \chi^*$. If $y \in \mathrm{free}(\chi)$ and $y \in \{v_1, \ldots, v_s\}$ then, again, $\chi = \chi(v_1, \ldots, v_s)$ and χ^* is defined, and we set $\varphi^* := \exists y \chi^*$. Finally suppose $y \in \mathrm{free}(\chi)$ and $y \notin \{v_1, \ldots, v_s\}$. Then, by (*), there is an i such that $v_i \notin \mathrm{free}(\chi)$. Set $\chi_0 := \chi\begin{pmatrix} v_i & y \\ y & v_i \end{pmatrix}$ (as in the preceding example, $\chi_0 := \chi\begin{pmatrix} v_i & y \\ y & v_i \end{pmatrix}$ is obtained from χ by simultaneously replacing *all* occurrences of y and v_i by v_i and y, respectively). Then, $\models \varphi \leftrightarrow \exists v_i \chi_0$, $\mathrm{free}(\chi_0) \subseteq \{v_1, \ldots, v_s\}$, and every subformula of χ_0 has at most s free variables. Thus, χ_0^* is defined and we can set $\varphi^* := \exists v_i \chi_0^*$.

3.3.1 Pebble Games

Our next aim is an Ehrenfeucht-Fraïssé type characterization of the equivalence of structures in FO^s and in $\mathrm{L}^s_{\infty\omega}$. To motivate the intended games we look at $\varphi = \exists x \exists y (x < y \wedge \exists x\, y < x)$ and the orderings $\mathcal{A} := (\{a, b\}, <)$ and $\mathcal{B} := (\{c, d, e\}, <)$ where $a < b$ and $c < d < e$. Since $\mathcal{A} \models \neg\varphi$ and $\mathcal{B} \models \varphi$, the spoiler has a winning strategy in $\mathrm{G}_3(\mathcal{A}, \mathcal{B})$. How is the fact that φ only contains two variables reflected in the course of a play? A play won by the spoiler is given in the following table, where his selections have been underlined.

	$\mathcal{A}$	$\mathcal{B}$
first move	a	$\underline{c}$
second move	b	$\underline{d}$
third move	?	$\underline{e}$

There is no third move of the duplicator leading to a partial isomorphism. Apparently, the strategy of the spoiler consists in choosing, for the first two quantifiers $\exists x \exists y$, the elements c for x and d for y in $\mathcal{B}$ in order to have

$$\mathcal{B} \models (x < y \wedge \exists x\, y < x)[c, d].$$

The only selections for the duplicator leading to a partial isomorphism are a for x and b for y. Now, for the second quantifier $\exists x$, the spoiler selects in $\mathcal{B}$ the element e, thereby getting a witness for $\mathcal{B} \models \exists x\, y < x[d]$. Obviously the old value c for x is no longer relevant. Therefore, the play above may be represented more informatively by

	first move		second move		third move	
	$\mathcal{A}$	$\mathcal{B}$	$\mathcal{A}$	$\mathcal{B}$	$\mathcal{A}$	$\mathcal{B}$
x-box	a	$\underline{c}$	a	$\underline{c}$	?	$\underline{e}$
y-box	$*$	$*$	b	$\underline{d}$	b	$\underline{d}$

where the x-boxes and the y-boxes always contain the actual value for x and y, respectively, and $*$ stands for an empty box.

This example motivates how to adapt the Ehrenfeucht-Fraïssé method to our situation: As always, fix a vocabulary τ. By convention, $*$ will not belong to the universe of any structure. For $\overline{a} \in (A \cup \{*\})^s, \overline{a} = a_1 \dots a_s$, let $\text{supp}(\overline{a}) := \{i \mid a_i \in A\}$ be the *support* of $\overline{a}$, and if $a \in A$, let $\overline{a}\frac{a}{i}$ denote $a_1 \dots a_{i-1} a a_{i+1} \dots a_s$. For $\overline{a} \in (A \cup \{*\})^s$ and $\overline{b} \in (B \cup \{*\})^s$ we say that $\overline{a} \mapsto \overline{b}$ is an *s-partial isomorphism* from $\mathcal{A}$ to $\mathcal{B}$, if $\text{supp}(\overline{a}) = \text{supp}(\overline{b})$ and $\overline{a}' \mapsto \overline{b}'$ is a partial isomorphism from $\mathcal{A}$ to $\mathcal{B}$, where $\overline{a}'$ and $\overline{b}'$ are the subsequences of $\overline{a}$ and $\overline{b}$ with indices in the support.

Let $\mathcal{A}$ and $\mathcal{B}$ be structures, $\overline{a} \in (A \cup \{*\})^s$, $\overline{b} \in (B \cup \{*\})^s$ with $\text{supp}(\overline{a}) = \text{supp}(\overline{b})$. In the example above we had available a box for each relevant variable. It has become customary to replace the boxes by *pebbles*. Thus, in the *pebble game* $\text{G}^s_m(\mathcal{A}, \overline{a}, \mathcal{B}, \overline{b})$ we have s pebbles $\alpha_1, \dots, \alpha_s$ for $\mathcal{A}$ and s pebbles $\beta_1, \dots, \beta_s$ for $\mathcal{B}$. Initially, α_i is placed on a_i if $a_i \in A$, and off the board if $a_i = *$, and similarly, β_i is placed on $b_i \in B$ or off the board. Each play consists of m moves. In his j-th move, the spoiler selects a structure, $\mathcal{A}$ or $\mathcal{B}$, and a pebble for this structure (being off the board or already placed on an element). If he selects $\mathcal{A}$ and α_i, he places α_i on some element of $\mathcal{A}$, and then the duplicator places β_i on some element of $\mathcal{B}$. If the spoiler selects $\mathcal{B}$ and β_i, he places β_i on an element of $\mathcal{B}$ and the duplicator places α_i on some element of $\mathcal{A}$. (Note that there may be several pebbles on the same element.)

The duplicator wins the game if for *each* $j \leq m$ we have that $\overline{e} \mapsto \overline{f}$ is an s-partial isomorphism, where $\overline{e} = e_1 \dots e_s$ are the elements marked by $\alpha_1, \dots, \alpha_s$ after the j-th move ($e_i = *$ in case α_i is off the board) and where $\overline{f} = f_1 \dots f_s$ are the corresponding values given by $\beta_1, \dots, \beta_s$. For $j = 0$ this means that $\overline{a} \mapsto \overline{b}$ is an s-partial isomorphism.

The pebble game $\text{G}^s_\infty(\mathcal{A}, \overline{a}, \mathcal{B}, \overline{b})$ with infinitely many moves is defined similarly. We use $\text{G}^s_m(\mathcal{A}, \mathcal{B})$ as abbreviation for $\text{G}^s_m(\mathcal{A}, * \dots *, \mathcal{B}, * \dots *)$ and $\text{G}^s_\infty(\mathcal{A}, \mathcal{B})$ for $\text{G}^s_\infty(\mathcal{A}, * \dots *, \mathcal{B}, * \dots *)$.

The following theorem shows that the logics and the games fit together. Here and later, when writing $\mathcal{A} \models \varphi[\overline{a}]$ for $\overline{a} \in (A \cup \{*\})^s$ we tacitly assume that the free variables of φ have indices in $\text{supp}(\overline{a})$ (that is, $i \in \text{supp}(\overline{a})$ whenever $v_i \in \text{free}(\varphi)$).

Theorem 3.3.5 *For structures $\mathcal{A}$ and $\mathcal{B}$, and for $\overline{a} \in (A \cup \{*\})^s$ and $\overline{b} \in (B \cup \{*\})^s$ with* $\text{supp}(\overline{a}) = \text{supp}(\overline{b})$ *the following hold:*

(a) *$\overline{a}$ satisfies in $\mathcal{A}$ the same FO^s-formulas of quantifier rank $\leq m$ as $\overline{b}$ in $\mathcal{B}$ iff the duplicator wins $G^s_m(\mathcal{A}, \overline{a}, \mathcal{B}, \overline{b})$.*

(b) *$\overline{a}$ satisfies in $\mathcal{A}$ the same $\text{L}^s_{\infty\omega}$-formulas as $\overline{b}$ in $\mathcal{B}$ iff the duplicator wins $G^s_\infty(\mathcal{A}, \overline{a}, \mathcal{B}, \overline{b})$.*

In particular,

(c) *$\mathcal{A} \equiv^s_m \mathcal{B}$ iff the duplicator wins $G^s_m(\mathcal{A}, \mathcal{B})$.*

(d) *$\mathcal{A} \equiv^{\text{L}^s_{\infty\omega}} \mathcal{B}$ iff the duplicator wins $G^s_\infty(\mathcal{A}, \mathcal{B})$.*

Before proving the theorem we give some examples, whose claims are easily verified.

Examples 3.3.6 (a) Let $\tau = \emptyset$ and let $\mathcal{A}$ and $\mathcal{B}$ be τ-structures (i.e., sets). If $\|A\|, \|B\| \geq s$ then the duplicator wins $\mathrm{G}^s_\infty(\mathcal{A}, \mathcal{B})$ (or, equivalently, $\mathcal{A} \equiv^{\mathrm{L}^s_{\infty\omega}} \mathcal{B}$). Moreover, for arbitrary $\mathcal{A}$ and $\mathcal{B}$, the duplicator wins $\mathrm{G}^s_\infty(\mathcal{A}, \mathcal{B})$ iff he wins $\mathrm{G}^s_s(\mathcal{A}, \mathcal{B})$.

(b) For $l \geq 3$, let $\mathcal{G}_l$ and $\mathcal{G}_l \dot\cup \mathcal{G}_l$ be the graphs consisting of one and two cycles, respectively, of length $l+1$ (cf. 2.3.8). Then the duplicator wins $\mathrm{G}^2_\infty(\mathcal{G}_l, \mathcal{G}_l \dot\cup \mathcal{G}_l)$ and hence by the theorem, $\mathcal{G}_l \equiv^{\mathrm{L}^2_{\infty\omega}} \mathcal{G}_l \dot\cup \mathcal{G}_l$. Note that $\mathcal{G}_l \not\equiv^{\mathrm{L}^3_{\infty\omega}} \mathcal{G}_l \dot\cup \mathcal{G}_l$, since – in the notations of Example 3.3.1(b) – the sentence $\forall x \forall y (x = y \vee \bigvee_{n>0} \varphi_n(x,y))$ of $\mathrm{L}^3_{\infty\omega}$ expresses connectivity. In fact, the spoiler wins $\mathrm{G}^3_\infty(\mathcal{G}_l, \mathcal{G}_l \dot\cup \mathcal{G}_l)$. □

Further examples are contained in the following exercises.

Exercise 3.3.7 (a) Assume that the relation symbols in $\tau = \{<, \ldots\}$ are at most binary. Show that for finite ordered structures $\mathcal{A}$ and $\mathcal{B}$,

$$\mathcal{A} \cong \mathcal{B} \quad \text{iff} \quad \text{the duplicator wins } \mathrm{G}^2_\infty(\mathcal{A}, \mathcal{B}).$$

(b) For $m \geq s$ show that the duplicator wins $\mathrm{G}^s_m(\mathcal{A}, \mathcal{B})$ iff he wins the game $\mathrm{G}^s_m(\mathcal{A}, \mathcal{B})$ with the additional condition that in the first s moves distinct pebbles have to be chosen (so that after s moves all pebbles are placed on elements).

(c) Suppose that τ is relational and all its relation symbols are of arity $\leq s$. Assume that the duplicator wins $\mathrm{G}^s_\infty(\mathcal{A}, \mathcal{B})$ and that $\|A\| = \|B\| \leq s+1$. Show that $\mathcal{A} \cong \mathcal{B}$. □

A proof of Theorem 3.3.5 can easily be obtained from the corresponding proofs for FO and $\mathrm{L}_{\infty\omega}$: We introduce for FO^s and $\mathrm{L}^s_{\infty\omega}$ the further notions related to the Ehrenfeucht-Fraïssé method (namely, the back and forth properties, the m-isomorphism types, and the set of winning positions), and give the corresponding statements 3.3.9 and 3.3.10 which imply Theorem 3.3.5.

Definition 3.3.8 Structures $\mathcal{A}$ and $\mathcal{B}$ are *s-m-isomorphic*, $\mathcal{A} \cong^s_m \mathcal{B}$, iff there is a sequence $(I_j)_{j \leq m}$ of nonempty sets of s-partial isomorphisms with the following properties:

(*s-forth property*) For $j < m$, $\bar a \mapsto \bar b \in I_{j+1}$, $1 \leq i \leq s$, and $a \in A$ there is $b \in B$ such that $\bar a \frac{a}{i} \mapsto \bar b \frac{b}{i} \in I_j$.

(*s-back property*) For $j < m$, $\bar a \mapsto \bar b \in I_{j+1}$, $1 \leq i \leq s$, and $b \in B$ there is $a \in A$ such that $\bar a \frac{a}{i} \mapsto \bar b \frac{b}{i} \in I_j$.

We then write $(I_j)_{j \leq m} : \mathcal{A} \cong^s_m \mathcal{B}$.

The notions *s-partially isomorphic*, $\mathcal{A} \cong^s_{\text{part}} \mathcal{B}$, and $I : \mathcal{A} \cong^s_{\text{part}} \mathcal{B}$ are defined similarly. □

For $m \in \mathbb{N}$, any structure $\mathcal{A}$, and $\overline{a} \in (A \cup \{*\})^s$, the *s-m-isomorphism type* $\psi^m_{\overline{a}}$ $(= {}^s\psi^m_{\mathcal{A},\overline{a}})$ of $\overline{a}$ in $\mathcal{A}$ is given by:

$$\psi^0_{\overline{a}}(\overline{v}) \quad := \quad \bigwedge\{\psi \mid \psi \text{ atomic or negated atomic, and } \mathcal{A} \models \psi[\overline{a}]\}$$

(recall that when writing $\mathcal{A} \models \psi[\overline{a}]$ we assume that the free variables of ψ have indices in $\mathrm{supp}(\overline{a})$),

$$\psi^{m+1}_{\overline{a}}(\overline{v}) \quad := \quad \psi^0_{\overline{a}} \wedge \bigwedge_{1 \le i \le s} (\bigwedge_{a \in A} \exists v_i \psi^m_{\overline{a}\frac{a}{i}} \wedge \forall v_i \bigvee_{a \in A} \psi^m_{\overline{a}\frac{a}{i}}).$$

In particular, $\psi^m_{\mathcal{A}} := \psi^m_{*\ldots*}$ is an FO^s-sentence of quantifier rank m.

Finally we introduce the sets $W^s_m(\mathcal{A},\mathcal{B})$ and $W^s_\infty(\mathcal{A},\mathcal{B})$ of s-partial isomorphisms corresponding to winning positions in the respective games,

$$\begin{aligned} W^s_m(\mathcal{A},\mathcal{B}) \quad &:= \quad \{\overline{a} \mapsto \overline{b} \mid \text{ the duplicator wins } \mathrm{G}^s_m(\mathcal{A},\overline{a},\mathcal{B},\overline{b})\}, \\ W^s_\infty(\mathcal{A},\mathcal{B}) \quad &:= \quad \{\overline{a} \mapsto \overline{b} \mid \text{ the duplicator wins } \mathrm{G}^s_\infty(\mathcal{A},\overline{a},\mathcal{B},\overline{b})\}. \end{aligned}$$

Now the following two theorems can be proven completely parallel to those for FO and $\mathrm{L}_{\infty\omega}$.

Theorem 3.3.9 *Let* $\overline{a} \in (A \cup \{*\})^s$ *and* $\overline{b} \in (B \cup \{*\})^s$ *with* $\mathrm{supp}(\overline{a}) = \mathrm{supp}(\overline{b})$.

(a) *The following are equivalent:*
- (i) *The duplicator wins* $\mathrm{G}^s_m(\mathcal{A},\overline{a},\mathcal{B},\overline{b})$.
- (ii) $\overline{a} \mapsto \overline{b} \in W^s_m(\mathcal{A},\mathcal{B})$ *and* $(W^s_j(\mathcal{A},\mathcal{B}))_{j \le m} : \mathcal{A} \cong^s_m \mathcal{B}$.
- (iii) *There is* $(I_j)_{j \le m}$ *with* $\overline{a} \mapsto \overline{b} \in I_m$ *such that* $(I_j)_{j \le m} : \mathcal{A} \cong^s_m \mathcal{B}$.
- (iv) $\mathcal{B} \models \psi^m_{\overline{a}}[\overline{b}]$.
- (v) $\overline{a}$ *satisfies in* $\mathcal{A}$ *the same formulas of* FO^s *of quantifier rank* $\le m$ *as* $\overline{b}$ *in* $\mathcal{B}$.

(b) *The following are equivalent:*
- (i) *The duplicator wins* $\mathrm{G}^s_\infty(\mathcal{A},\overline{a},\mathcal{B},\overline{b})$.
- (ii) $\overline{a} \mapsto \overline{b} \in W^s_\infty(\mathcal{A},\mathcal{B})$ *and* $W^s_\infty(\mathcal{A},\mathcal{B}) : \mathcal{A} \cong^s_{\mathrm{part}} \mathcal{B}$.
- (iii) *There is* I *with* $\overline{a} \mapsto \overline{b} \in I$ *such that* $I : \mathcal{A} \cong^s_{\mathrm{part}} \mathcal{B}$.
- (iv) $\overline{a}$ *satisfies in* $\mathcal{A}$ *the same formulas of* $\mathrm{L}^s_{\infty\omega}$ *as* $\overline{b}$ *in* $\mathcal{B}$. □

Corollary 3.3.10 (a) *The following are equivalent:*
- (i) *The duplicator wins* $\mathrm{G}^s_m(\mathcal{A},\mathcal{B})$.
- (ii) $(W^s_j(\mathcal{A},\mathcal{B}))_{j \le m} : \mathcal{A} \cong^s_m \mathcal{B}$.
- (iii) $\mathcal{A} \cong^s_m \mathcal{B}$.
- (iv) $\mathcal{B} \models \psi^m_{\mathcal{A}}$.
- (v) $\mathcal{A} \equiv^s_m \mathcal{B}$.

(b) *The following are equivalent:*
- (i) *The duplicator wins* $\mathrm{G}^s_\infty(\mathcal{A},\mathcal{B})$.
- (ii) $W^s_\infty(\mathcal{A},\mathcal{B}) : \mathcal{A} \cong^s_{\mathrm{part}} \mathcal{B}$.
- (iii) $\mathcal{A} \cong^s_{\mathrm{part}} \mathcal{B}$.
- (iv) $\mathcal{A} \equiv^{\mathrm{L}^s_{\infty\omega}} \mathcal{B}$. □

Example 3.3.11 For $s \geq 1$ let ϵ_s be the conjunction of the finitely many r-extension axioms with $r \leq s$ (cf. 3.2.11). Clearly, $\epsilon_s \in \mathrm{FO}^s$. We have:

- Every model of ϵ_s has at least s elements.
- Every two models $\mathcal{A}$ and $\mathcal{B}$ of ϵ_s are s-partially isomorphic.

(The last statement is proved as the corresponding one in 3.2.11.) Therefore, by the corollary, any two models of ϵ_s are $\mathrm{L}^s_{\infty\omega}$-equivalent, and we get:

- For every $\mathrm{L}^s_{\infty\omega}$-sentence φ, either $\epsilon_s \models \varphi$ or $\epsilon_s \models \neg\varphi$. □

The next result provides us with an algebraic tool to show that a class of finite structures is not axiomatizable in $\mathrm{L}^s_{\infty\omega}$ or in $\mathrm{L}^\omega_{\infty\omega}(= \bigcup_{s\geq 1} \mathrm{L}^s_{\infty\omega})$.

Theorem 3.3.12 *Let K be a class of finite structures.*

(a) *For $s \geq 1$ the following are equivalent:*
  (i) *K is not axiomatizable in $\mathrm{L}^s_{\infty\omega}$.*
  (ii) *There are finite structures $\mathcal{A}$ and $\mathcal{B}$ such that*

$$\mathcal{A} \in K,\ \mathcal{B} \notin K,\ \text{and}\ \mathcal{A} \cong^s_{\mathrm{part}} \mathcal{B}.$$

(b) *The following are equivalent:*
  (i) *K is not axiomatizable in $\mathrm{L}^\omega_{\infty\omega}$.*
  (ii) *For every $s \geq 1$ there are finite structures $\mathcal{A}$ and $\mathcal{B}$ such that*

$$\mathcal{A} \in K,\ \mathcal{B} \notin K,\ \text{and}\ \mathcal{A} \cong^s_{\mathrm{part}} \mathcal{B}.$$

Proof. Clearly, (b) follows from (a). To show (ii) $\Rightarrow$ (i) in (a), suppose by contradiction, that $K = \mathrm{Mod}(\varphi)$ for some $\varphi \in \mathrm{L}^s_{\infty\omega}$. Choose $\mathcal{A}$ and $\mathcal{B}$ as given by (ii). Then $\mathcal{A} \models \varphi$ (since $\mathcal{A} \in K$), $\mathcal{B} \not\models \varphi$ (since $\mathcal{B} \notin K$), and $\mathcal{A} \equiv^{\mathrm{L}^s_{\infty\omega}} \mathcal{B}$ (since $\mathcal{A} \cong^s_{\mathrm{part}} \mathcal{B}$), a contradiction.

Conversely, suppose that the condition in (ii) is not satisfied. Then for all finite $\mathcal{A}$ and $\mathcal{B}$,

$$(*) \qquad \mathcal{A} \in K \text{ and } \mathcal{A} \equiv^{\mathrm{L}^s_{\infty\omega}} \mathcal{B} \ \text{ imply } \ \mathcal{B} \in K.$$

We show that $K = \mathrm{Mod}(\varphi)$ for the $\mathrm{L}^s_{\infty\omega}$-sentence $\varphi := \bigvee_{\mathcal{A}\in K} \bigwedge_{m\geq 0} \psi^m_{\mathcal{A}}$. Clearly, $K \subseteq \mathrm{Mod}(\varphi)$, since $\mathcal{B} \models \bigwedge_{m\geq 0} \psi^m_{\mathcal{B}}$ holds for any $\mathcal{B}$. To obtain the inclusion $\mathrm{Mod}(\varphi) \subseteq K$, assume that $\mathcal{B}$ is a finite model of φ. Then, for some $\mathcal{A} \in K$ and all $m \geq 0$, we have $\mathcal{B} \models \psi^m_{\mathcal{A}}$ and hence, $\mathcal{A} \equiv^s_m \mathcal{B}$. Thus, $\mathcal{A} \equiv^s \mathcal{B}$. By 3.3.3, we get $\mathcal{A} \equiv^{\mathrm{L}^s_{\infty\omega}} \mathcal{B}$, and hence by $(*)$, we obtain $\mathcal{B} \in K$. □

The corresponding nonaxiomatizability result for FO^s is part (a) of Exercise 3.3.14 below.

Example 3.3.13 Let τ be the empty vocabulary. The class EVEN[τ] is not $\mathrm{L}^\omega_{\infty\omega}$-axiomatizable, since for $s \geq 1$ and structures $\mathcal{A}$ and $\mathcal{B}$ with $\|A\| = s$ and $\|B\| = s+1$, we have $\mathcal{A} \in \mathrm{EVEN}[\tau]$ iff $\mathcal{B} \notin \mathrm{EVEN}[\tau]$; but $\mathcal{A} \equiv^{\mathrm{L}^s_{\infty\omega}} \mathcal{B}$ by 3.3.6(a). Extend the result to arbitrary τ. □

Exercises 3.3.14 Let K be a class of finite structures. Show:

(a) The following statements are equivalent:

(i) K is not axiomatizable in FO^s.

(ii) For $m \geq 1$ there are finite structures $\mathcal{A}$ and $\mathcal{B}$ such that

$$\mathcal{A} \in K, \mathcal{B} \notin K, \text{ and } \mathcal{A} \cong^s_m \mathcal{B}.$$

(b) For any global n-ary relation Γ on K with $n \leq s$ the following are equivalent:

(i) Γ is $\mathrm{L}^s_{\infty\omega}$-definable, i.e. there is an $\mathrm{L}^s_{\infty\omega}$-formula φ such that for $\mathcal{A} \in K$ and $\overline{a} \in A^n$,

$$\mathcal{A} \models \varphi[\overline{a}] \quad \text{iff} \quad \overline{a} \in \Gamma(\mathcal{A}).$$

(ii) Γ is "closed under the game G^s_∞", that means, if $\mathcal{A}, \mathcal{B} \in K$, $\overline{a} \in \Gamma(\mathcal{A})$, $\overline{b} \in B^n$ and the duplicator wins $\mathrm{G}^s_\infty(\mathcal{A}, \overline{a} * \ldots *, \mathcal{B}, \overline{b} * \ldots *)$, then $\overline{b} \in \Gamma(\mathcal{B})$. □

Exercise 3.3.15 Assume that τ is relational and contains only unary relation symbols. Fix $s \geq 1$ and let $\mathcal{A}$ and $\mathcal{B}$ be τ structures. Show that $W^s_s(\mathcal{A}, \mathcal{B}) = W^s_\infty(\mathcal{A}, \mathcal{B})$. □

Exercise 3.3.16 For an FO^s-formula $\varphi(\overline{x})$ of quantifier rank $\leq m$ show that

$$\models \varphi \leftrightarrow \bigvee \{\psi^m_{\mathcal{A},\overline{a}} \mid \mathcal{A} \text{ a structure}, \overline{a} \in A, \text{ and } \mathcal{A} \models \varphi[\overline{a}]\}$$

(argue as in the proof of 2.2.11).

3.3.2 The s-Invariant of a Structure

In this section we associate with every $s \geq 1$ and every structure $\mathcal{A}$ a structure $\mathcal{A}/s$, whose isomorphism type captures the $\mathrm{L}^s_{\infty\omega}$-type of $\mathcal{A}$ in a one-to-one fashion (see 3.3.17 below for the precise statement). Later we shall see that there is a uniform way to define an ordering on $\mathcal{A}/s$. This will enable us to translate some problems from arbitrary structures to ordered ones. For simplicity, let τ be relational (we encourage the reader to also treat vocabularies with constants).

Let $\mathcal{A}$ be a τ-structure. The binary relation $\sim$ defined on A^s by

$$(*) \qquad \overline{a} \sim \overline{b} \quad \text{iff} \quad \overline{a} \text{ and } \overline{b} \text{ satisfy the same } \mathrm{L}^s_{\infty\omega}\text{-formulas in } \mathcal{A}$$

is an equivalence relation on A^s. By 3.3.9(b), we have $\overline{a} \sim \overline{b}$ iff the duplicator wins $\mathrm{G}^s_\infty(\mathcal{A}, \overline{a}, \mathcal{A}, \overline{b})$. Let $[\overline{a}]$ denote the equivalence class of $\overline{a}$ and

$$A/s \; := \; \{[\overline{a}] \mid \overline{a} \in A^s\}$$

the set of equivalence classes. We endow A/s with a τ/s-structure $\mathcal{A}/s$: for every $[\overline{a}] \in A/s$, the relations on A/s capture the properties of $\overline{a}$ in any game $\mathrm{G}^s_\infty(\mathcal{A}, \overline{a}, \ldots)$. The relation symbols in τ/s (together with their meaning in $\mathcal{A}/s$) are:

- for every k-ary $R \in \tau \cup \{=\}$ and any $i_1, \ldots, i_k$ with $1 \le i_1, \ldots, i_k \le s$ a unary relation symbol $R_{i_1 \ldots i_k}$;

$$R^{A/s}_{i_1 \ldots i_k} \quad := \quad \{ \, [\overline{a}] \;\mid\; \overline{a} \in A^s,\; R^{\mathcal{A}} a_{i_1} \ldots a_{i_k} \}$$

(the $R^{A/s}_{i_1 \ldots i_k}$ capture the isomorphism type of $\overline{a}$);
- for $i = 1, \ldots, s$ a binary relation symbol S_i;

$$S^{A/s}_i := \{ \, ([\overline{a}], [\overline{a}']) \;\mid\; \overline{a}, \overline{a}' \in A^s, \text{ there is } a \in A \text{ such that } [\overline{a}'] = [\overline{a}\tfrac{a}{i}]\}$$

($S^{A/s}_i$ encodes the possible moves of the i-th pebble).

$\mathcal{A}/s$ is called the *s-invariant* of $\mathcal{A}$. It captures the $\mathrm{L}^s_{\infty\omega}$-theory of $\mathcal{A}$:

Theorem 3.3.17 *For structures $\mathcal{A}$ and $\mathcal{B}$,*

$$\mathcal{A} \equiv^{\mathrm{L}^s_{\infty\omega}} \mathcal{B} \qquad \textit{iff} \qquad \mathcal{A}/s \cong \mathcal{B}/s.$$

Proof. Suppose first that $\mathcal{A}/s$ and $\mathcal{B}/s$ are isomorphic and $\pi : \mathcal{A}/s \cong \mathcal{B}/s$. Set

$$I \quad := \quad \{\overline{a} \mapsto \overline{b} \mid \overline{a} \in A^s, \overline{b} \in B^s, \pi([\overline{a}]) = [\overline{b}]\}.$$

We show that $I : \mathcal{A} \cong^s_{\text{part}} \mathcal{B}$ and hence, $\mathcal{A} \equiv^{\mathrm{L}^s_{\infty\omega}} \mathcal{B}$. I is a nonempty set of s-partial isomorphisms (use the relations $R_{i_1 \ldots i_k}$). To show, say, that I has the s-forth property, assume that $\overline{a} \mapsto \overline{b} \in I, 1 \le i \le s$, and $a \in A$. Then $S^{A/s}_i[\overline{a}][\overline{a}\frac{a}{i}]$, hence $S^{B/s}_i[\overline{b}]\pi([\overline{a}\frac{a}{i}])$. By the definition of $S^{B/s}_i$, there is $b \in B$ such that $[\overline{b}\frac{b}{i}] = \pi([\overline{a}\frac{a}{i}])$, hence $\overline{a}\frac{a}{i} \mapsto \overline{b}\frac{b}{i} \in I$.

Conversely suppose that $\mathcal{A} \equiv^{\mathrm{L}^s_{\infty\omega}} \mathcal{B}$. Then $W^s_\infty(\mathcal{A}, \mathcal{B}) : \mathcal{A} \cong^s_{\text{part}} \mathcal{B}$. For $\overline{a} \in A^s$ and $\overline{b} \in B^s$ set

$$\pi([\overline{a}]) := [\overline{b}] \quad \text{iff} \quad \overline{a} \mapsto \overline{b} \in W^s_\infty(\mathcal{A}, \mathcal{B}).$$

Hence, by 3.3.9(b),

$$\pi([\overline{a}]) = [\overline{b}] \quad \text{iff} \quad \overline{a} \text{ in } \mathcal{A} \text{ satisfies the same } \mathrm{L}^s_{\infty\omega}\text{-formulas as } \overline{b} \text{ in } \mathcal{B}.$$

By this equivalence and by the definition $(*)$ of the equivalence relation, π is well-defined and injective; moreover, $\mathrm{do}(\pi) = A/s$ by the s-forth property of $W^s_\infty(\mathcal{A}, \mathcal{B})$ and $\mathrm{rg}(\pi) = B/s$ by the s-back property. Obviously, π is compatible with the interpretations of the $R_{i_1 \ldots i_k}$, and also with the interpretations of the S_i (use once more the s-back and s-forth properties of $W^s_\infty(\mathcal{A}, \mathcal{B})$). Therefore, $\pi : \mathcal{A}/s \cong \mathcal{B}/s$. □

We close this section by showing that for finite structures we can replace $\mathrm{L}^s_{\infty\omega}$ by FO^s. In fact, in view of 3.3.2 we have

Proposition 3.3.18 *Let $\mathcal{A}$ and $\mathcal{B}$ be finite structures, $\overline{a} \in (A \cup \{*\})^s$ and $\overline{b} \in (B \cup \{*\})^s$ with $\mathrm{supp}(\overline{a}) = \mathrm{supp}(\overline{b})$. If*

$$\textit{for all } \varphi \in \mathrm{FO}^s \ (\mathcal{A} \models \varphi[\overline{a}] \ \textit{ iff } \mathcal{B} \models \varphi[\overline{b}])$$

then

$$\textit{for all } \varphi \in \mathrm{L}^s_{\infty\omega} \ (\mathcal{A} \models \varphi[\overline{a}] \ \textit{ iff } \mathcal{B} \models \varphi[\overline{b}]). \qquad \square$$

Corollary 3.3.19 (a) *Let $\mathcal{A}$ be a finite structure and $\sim$ be defined as in $(*)$ above. Then, for $\overline{a}, \overline{b} \in A^s$,*

$$\overline{a} \sim \overline{b} \quad \textit{iff} \quad \overline{a} \textit{ and } \overline{b} \textit{ satisfy the same } \mathrm{FO}^s\textit{-sentences.}$$

(b) *For finite structures $\mathcal{A}$ and $\mathcal{B}$,*

$$\mathcal{A} \equiv^s \mathcal{B} \quad \textit{iff} \quad \mathcal{A}/s \cong \mathcal{B}/s.$$

Proof. (a) is immediate from the preceding proposition and (b) follows from 3.3.17 by 3.3.3. $\square$

3.3.3 Scott Formulas

For a finite structure $\mathcal{A}$ we saw in 2.2.10 that a single FO-sentence, e.g. $\varphi_{\mathcal{A}}^{\|A\|+1}$, characterizes $\mathcal{A}$ up to (isomorphisms and therefore up to) $\mathrm{L}_{\infty\omega}$-equivalence. We are going to show a similar result for FO^s and $\mathrm{L}^s_{\infty\omega}$, in this way strengthening 3.3.3.

Proposition 3.3.20 *Let $\mathcal{A}$ and $\mathcal{B}$ be structures. Then*

(a) $W_0^s(\mathcal{A},\mathcal{B}) \supseteq W_1^s(\mathcal{A},\mathcal{B}) \supseteq \dots$

(b) *If A and B are finite then there is an $m \leq (\|A\|+1)^s \cdot (\|B\|+1)^s$ such that $W^s_m(\mathcal{A},\mathcal{B}) = W^s_{m+1}(\mathcal{A},\mathcal{B})$.*

(c) *For $m \geq 0$, if $W^s_m(\mathcal{A},\mathcal{B}) = W^s_{m+1}(\mathcal{A},\mathcal{B})$ and $W^s_m(\mathcal{A},\mathcal{B})$ is nonempty, then $W^s_m(\mathcal{A},\mathcal{B}) : \mathcal{A} \cong_{\mathrm{part}} \mathcal{B}$.*

Proof. (a) follows immediately from the definition of the $W^s_j(\mathcal{A},\mathcal{B})$. (b) follows from (a), since there are at most $(\|A\|+1)^s \cdot (\|B\|+1)^s$ s-partial isomorphisms from $\mathcal{A}$ to $\mathcal{B}$.
(c) Suppose that $W^s_m(\mathcal{A},\mathcal{B}) = W^s_{m+1}(\mathcal{A},\mathcal{B})$. Then $W^s_m(\mathcal{A},\mathcal{B})$ has the s-back and the s-forth property: To show, say, the s-forth property, let $\overline{a} \mapsto \overline{b} \in W^s_m(\mathcal{A},\mathcal{B}), 1 \leq i \leq s$, and $a \in A$. By assumption, $\overline{a} \mapsto \overline{b} \in W^s_{m+1}(\mathcal{A},\mathcal{B})$, and therefore there is $b \in B$ such that $\overline{a}\frac{a}{i} \to \overline{b}\frac{b}{i} \in W^s_m(\mathcal{A},\mathcal{B})$. If, in addition, $W^s_m(\mathcal{A},\mathcal{B}) \neq \emptyset$, we altogether have $W^s_m(\mathcal{A},\mathcal{B}) : \mathcal{A} \cong_{\mathrm{part}} \mathcal{B}$. $\square$

Fix a finite structure $\mathcal{A}$ and let $\overline{a}, \overline{b}$ range over $(A \cup \{*\})^s$. By the proposition we know that

$$W_0^s(\mathcal{A},\mathcal{A}) \supseteq W_1^s(\mathcal{A},\mathcal{A}) \supseteq \dots \supseteq W_m^s(\mathcal{A},\mathcal{A}) \supseteq \dots$$

and that there must be a j with $W^s_j(\mathcal{A},\mathcal{A}) = W^s_{j+1}(\mathcal{A},\mathcal{A})$. The minimal j having this property is called the *s-rank* $r(\mathcal{A})$ of $\mathcal{A}$, $r(\mathcal{A}) = r(s,\mathcal{A})$. For given $\overline{a}$, the formula

$$\sigma_{\overline{a}} \ := \ \psi^{r(\mathcal{A})}_{\overline{a}} \wedge \bigwedge_{\overline{b}\in(\mathcal{A}\cup\{*\})^s} \forall v_1 \ldots \forall v_s (\psi^{r(\mathcal{A})}_{\overline{b}} \to \psi^{r(\mathcal{A})+1}_{\overline{b}})$$

(more exactly, $\sigma_{\overline{a}} = {}^s\sigma_{\mathcal{A},\overline{a}}$) is called the *s-Scott formula* of $\overline{a}$ in $\mathcal{A}$. It is an FO^s-formula of quantifier rank $r(\mathcal{A}) + 1 + s$. In particular, $\sigma_{\mathcal{A}} := \sigma_{*\ldots*}$ is an FO^s-sentence. It captures the whole $\mathrm{L}^s_{\infty\omega}$-theory of $\mathcal{A}$:

Theorem 3.3.21 *Let $\mathcal{A}$ be finite.*

(a) *For any structure $\mathcal{B}$,*

$$\mathcal{B} \models \sigma_{\mathcal{A}} \qquad \textit{iff} \qquad \mathcal{A} \equiv^{\mathrm{L}^s_{\infty\omega}} \mathcal{B}.$$

(b) *For $\overline{a} \in (\mathcal{A} \cup \{*\})^s$, any structure $\mathcal{B}$ and $\overline{b} \in (\mathcal{B} \cup \{*\})^s$ with* $\mathrm{supp}(\overline{a}) = \mathrm{supp}(\overline{b})$,

$$\mathcal{B} \models \sigma_{\overline{a}}[\overline{b}] \quad \textit{iff } \overline{a} \textit{ satisfies in } \mathcal{A} \textit{ the same } \mathrm{L}^s_{\infty\omega}\textit{-formulas as } \overline{b} \textit{ in } \mathcal{B}.$$

Proof. We restrict ourselves to (a). Since $\mathcal{A} \models \sigma_{\mathcal{A}}$, we have that $\mathcal{A} \equiv^{\mathrm{L}^s_{\infty\omega}} \mathcal{B}$ implies $\mathcal{B} \models \sigma_{\mathcal{A}}$. Now suppose that $\mathcal{B} \models \sigma_{\mathcal{A}}$, that is,

$$\mathcal{B} \models \psi^{r(\mathcal{A})}_{\mathcal{A}} \wedge \bigwedge_{\overline{b}\in(\mathcal{A}\cup\{*\})^s} \forall v_1 \ldots \forall v_s (\psi^{r(\mathcal{A})}_{\overline{b}} \to \psi^{r(\mathcal{A})+1}_{\overline{b}}).$$

Since $\mathcal{B} \models \psi^{r(\mathcal{A})}_{\mathcal{A}}$, we get $*\ldots* \mapsto *\ldots* \in W^s_{r(\mathcal{A})}(\mathcal{A},\mathcal{B})$ (by 3.3.9(a)). By the validity of the second conjunct in $\mathcal{B}$, we have $W^s_{r(\mathcal{A})}(\mathcal{A},\mathcal{B}) \subseteq W^s_{r(\mathcal{A})+1}(\mathcal{A},\mathcal{B})$ and hence, $W^s_{r(\mathcal{A})}(\mathcal{A},\mathcal{B}) = W^s_{r(\mathcal{A})+1}(\mathcal{A},\mathcal{B})$. Therefore, by 3.3.20(c), $W^s_{r(\mathcal{A})}(\mathcal{A},\mathcal{B}) : \mathcal{A} \cong^s_{\mathrm{part}} \mathcal{B}$ and thus, $\mathcal{A} \equiv^{\mathrm{L}^s_{\infty\omega}} \mathcal{B}$. □

Corollary 3.3.22 *In the finite, each $\mathrm{L}^s_{\infty\omega}$-formula φ is equivalent to a countable disjunction of FO^s-formulas. In fact, φ is equivalent to the $\mathrm{L}^s_{\infty\omega}$-formula $\bigvee\{\sigma_{\overline{a}} \mid \mathcal{A}$ finite, $\overline{a} \in A$, $\mathcal{A} \models \varphi[\overline{a}]\}$. Moreover, if K is any class of finite structures, then φ and $\bigvee\{\sigma_{\overline{a}} \mid \mathcal{A} \in K$, $\overline{a} \in A$, $\mathcal{A} \models \varphi[\overline{a}]\}$ are equivalent in all structures of K.* □

Exercise 3.3.23 For finite $\mathcal{A}$ and $\mathcal{B}$ show that $W^s_j(\mathcal{A},\mathcal{A}) = W^s_{j+1}(\mathcal{A},\mathcal{A})$ and $\mathcal{A} \equiv^{\mathrm{L}^s_{\infty\omega}} \mathcal{B}$ imply $W^s_j(\mathcal{B},\mathcal{B}) = W^s_{j+1}(\mathcal{B},\mathcal{B})$. Conclude: If $\mathcal{A} \equiv^{\mathrm{L}^s_{\infty\omega}} \mathcal{B}$ then $r(\mathcal{A}) = r(\mathcal{B})$. □

As an application of the formulas $\sigma_{\overline{a}}$ we give a condition for $\mathrm{L}^s_{\infty\omega}$ and FO^s to coincide in expressive power.

Let K be a class of finite structures. We say that K is *s-bounded* if the set $\{r(\mathcal{A}) \mid \mathcal{A} \in K\}$ of s-ranks of structures in K is bounded. The class K is *bounded* if it is s-bounded for every $s \geq 1$.

Theorem 3.3.24 *Let K be a class of finite structures.*

(a) *For $s \geq 1$ the following are equivalent:*

(i) *K is s-bounded.*

(ii) *On K, every $\mathrm{L}^s_{\infty\omega}$-formula is equivalent to an FO^s-formula.*

(iii) *On K, every $\mathrm{L}^s_{\infty\omega}$-formula is equivalent to an FO-formula.*

(b) *K is bounded iff FO and $\mathrm{L}^\omega_{\infty\omega}$ have the same expressive power on K.*

Proof. As (b) is a consequence of (a), it suffices to prove (a). First suppose that K is s-bounded and set $m := \sup\{r(\mathcal{A}) \mid \mathcal{A} \in K\} < \infty$. Thus, for $\mathcal{A} \in K$ and $\bar{a}$ in $\mathcal{A}$, the quantifier rank of $\sigma_{\bar{a}}$ is $\leq m + s + 1$. Let φ be any $\mathrm{L}^s_{\infty\omega}$-formula. Then the disjunction in the preceding corollary is a disjunction of formulas of quantifier rank $\leq m + s + 1$ and hence, it is a finite one. This shows that (i) implies (ii). The implication from (ii) to (iii) is trivial. To show that (iii) implies (i) assume, by contradiction, that K is not s-bounded. Let $\mathcal{A}_0, \mathcal{A}_1, \ldots$ be structures in K of pairwise distinct s-rank. For $M \subseteq \mathbb{N}$ let

$$\varphi_M \quad := \quad \bigvee\{\sigma_{\mathcal{A}_i} \mid i \in M\}.$$

By 3.3.23, if $L, M \subseteq \mathbb{N}$ and $L \neq M$ then not $K \models \varphi_L \leftrightarrow \varphi_M$.[4] Hence on K, $\mathrm{L}^s_{\infty\omega}$ contains uncountably many pairwise nonequivalent sentences and is therefore more expressive than FO. □

Example 3.3.25 (a) Suppose $\tau = \emptyset$ and let $\mathcal{A}$ be a τ-structure. Then $W^s_0(\mathcal{A},\mathcal{A}) = W^s_1(\mathcal{A},\mathcal{A})$. Hence, $r(\mathcal{A}) = 0$ and the class K of finite τ-structures is bounded. Therefore, FO and $\mathrm{L}^\omega_{\infty\omega}$ have the same expressive power on K.

(b) By Exercise 3.3.15, the class K of all finite τ-structures is bounded, if τ contains only unary relation symbols. Therefore, FO and $\mathrm{L}^\omega_{\infty\omega}$ have the same expressive power on K. □

Exercise 3.3.26 $\mathrm{L}^\omega_{\infty\omega}$ is stronger than FO on the class of finite orderings and on the class of graphs. (Hint: Use 3.3.1.) □

3.4 Logics with Counting Quantifiers

To express in first-order logic that there are, say, seven elements with the property $\varphi(x)$ we need, in general, at least seven quantifiers:

$$\exists x_1 \ldots \exists x_7 (\varphi(x_1) \wedge \ldots \wedge \varphi(x_7) \wedge \bigwedge_{1 \leq i < j \leq 7} \neg x_i = x_j)$$

(by 3.3.6(a) we see that in case $\varphi(x) := x = x$ we really need seven quantifiers).

[4] Recall that $K \models \varphi$ means that every structure in K is a model of φ.

Let FO(C), so-called *first-order logic with counting quantifiers*, and $L_{\infty\omega}(C)$ (by short, $C_{\infty\omega}$), so-called $L_{\infty\omega}$ *with counting quantifiers*, be the logics obtained from FO and $L_{\infty\omega}$, respectively, by adding, for every $l \geq 1$, a new quantifier $\exists^{\geq l}$ with the intended interpretation "there are at least l". More precisely, extend the calculus of formulas for first-order or infinitary logic by the following rule:

– If φ is a formula and $l \geq 1$ then $\exists^{\geq l} x \varphi$ is a formula

(here, $\exists^{\geq l} x$ is considered as a new quantifier (binding x) and not as an abbreviation as in 1.B). Fix the interpretation of these quantifiers by adding for $\varphi = \varphi(\overline{x}, x)$ and $\overline{a} \in A$ the clause

$$\mathcal{A} \models \exists^{\geq l} x \varphi[\overline{a}] \qquad \text{iff} \qquad \|\{b \in A \mid \mathcal{A} \models \varphi[\overline{a}, b]\}\| \geq l.$$

Since the quantifiers $\exists^{\geq l}$ are first-order definable, the languages FO(C) and $C_{\infty\omega}$ have the same expressive power as FO and $L_{\infty\omega}$, respectively. The situation changes if we restrict ourselves to $\mathrm{FO(C)}^s$ and $C^s_{\infty\omega}$, which are the fragments consisting of the formulas with variables among $v_1, \ldots, v_s$. For example, $\exists^{\geq 7} x\, x = x$ is a sentence in $\mathrm{FO(C)}^1$ not equivalent to any sentence in FO^1. And $\bigvee_{l \geq 1} (\exists^{\geq 2l} x\, x = x \wedge \neg \exists^{\geq 2l+1} x\, x = x)$ is a $C^1_{\infty\omega}$-sentence axiomatizing the class EVEN[τ] of structures of even cardinality that (by 3.3.13) is not equivalent to any sentence of $L^{\omega}_{\infty\omega}$. We set $C^{\omega}_{\infty\omega} := \bigcup_{s \geq 1} C^s_{\infty\omega}$.

For $l \geq 1$, set $\exists^{=l} x \varphi := \exists^{\geq l} x \varphi \wedge \neg \exists^{\geq l+1} x \varphi$, and let $\exists^{=0} x \varphi := \forall x \neg \varphi$. Then $\exists^{\geq l} x \varphi$ is equivalent to $\neg \bigvee_{j<l} \exists^{=j} x \varphi$. Hence, we would obtain logics of the same expressive power when adding the quantifiers $\exists^{=l}$ instead of $\exists^{\geq l}$.

Examples 3.4.1 (a) Suppose $\mathcal{A}$ and $\mathcal{B}$ are finite structures. If $\mathcal{A} \equiv^{\mathrm{FO(C)}^1} \mathcal{B}$, that is, if $\mathcal{A}$ and $\mathcal{B}$ satisfy the same sentences of $\mathrm{FO(C)}^1$, then $\|A\| = \|B\|$ (note that $\exists^{=\|A\|} x\, x = x$ is a sentence in $\mathrm{FO(C)}^1$).

(b) Let $\tau = \{<\}$. The finite models of the sentence of FO[τ]

$$\forall x \neg x < x \,\wedge\, \forall x \forall y \forall z ((x < y \wedge y < z) \rightarrow x < z) \,\wedge$$
$$\forall x \forall y \forall z ((y < x \wedge z < x) \rightarrow (y < z \vee y = z \vee z < y))$$

($<$ is irreflexive and transitive, and the predecessors of any element are linearly ordered) are called finite *$<$-forests.* For a $<$-forest $\mathcal{A}$ and $a \in \mathcal{A}$ the height $h_{\mathcal{A}}(a)$ is defined by

$$h_{\mathcal{A}}(a) \;:=\; \|\{b \in A \mid b < a\}\|,$$

and the height $h(\mathcal{A})$ by

$$h(\mathcal{A}) \;:=\; \max\{h_{\mathcal{A}}(a) \mid a \in A\}.$$

The element a is a *root* if $h_{\mathcal{A}}(a) = 0$. Every finite $<$-forest can be characterized, up to isomorphism, in $\mathrm{FO(C)}^2$, i.e., for every finite $<$-forest $\mathcal{A}$ there is a sentence φ in $\mathrm{FO(C)}^2$ such that for all finite $<$-forests $\mathcal{B}$,

$$\mathcal{B} \models \varphi \quad \text{iff} \quad \mathcal{B} \cong \mathcal{A}.$$

To prove this one shows by induction on the height, that for $<$-forests $\mathcal{A}$ with exactly one root there is a formula $\psi_{\mathcal{A}}(x)$ in $\mathrm{FO(C)}^2$ such that for any $<$-forest $\mathcal{B}$ and $b \in B$ one has

$$\mathcal{B} \models \psi_{\mathcal{A}}[b] \quad \text{iff} \quad \mathcal{B}_b \cong \mathcal{A},$$

where $\mathcal{B}_b$ is the substructure of $\mathcal{B}$ with universe $\{b' \in B \mid b = b' \vee b < b'\}$. In the induction step, for $\mathcal{A}$ with root a, $\psi_{\mathcal{A}}(x)$ gives the number of elements of A and, for any isomorphism type of some $\mathcal{A}_b$ with $b \in A \setminus \{a\}$, says how many trees $\mathcal{A}_c$ with $c \in A \setminus \{a\}$ are of this type.

(c) For $s \geq 1$ there are $<$-forests $\mathcal{A}$ and $\mathcal{B}$ that satisfy the same sentences in FO^s but are not isomorphic, e.g., $<$-forests consisting only of roots, the first one having s roots, the second one $s+1$ roots. □

We conclude this section by adapting the Ehrenfeucht-Fraïssé method to the case of counting quantifiers and by demonstrating with an example its value for investigating the expressive power of the logics $\mathrm{FO(C)}^s$ and $\mathrm{C}^s_{\infty\omega}$.

In the corresponding pebble games $\text{C-G}^s_m(\mathcal{A}, \overline{a}, \mathcal{B}, \overline{b})$ with m moves and $\text{C-G}^s_\infty(\mathcal{A}, \overline{a}, \mathcal{B}, \overline{b})$ with infinitely many moves, each move now consists of two steps:

1. The spoiler chooses one of the two structures, say $\mathcal{A}$, and a corresponding pebble, say α_i. He then chooses a subset X of A. The duplicator must answer with a subset Y of B of the same cardinality as X.
2. The spoiler places β_i on some element $b \in Y$. The duplicator answers by placing α_i on some $a \in X$ (X and Y can now be forgotten).

The definition for winning is given as in the previous pebble games; it only takes into consideration the chosen elements, not the subsets.

To understand what is going on in the two steps of a move suppose that the spoiler attempts to show that

$$\mathcal{A} \models \exists^{\geq l} x \varphi(x), \text{ but not } \mathcal{B} \models \exists^{\geq l} x \varphi(x).$$

He chooses a subset X consisting of l elements witnessing that $\mathcal{A} \models \exists^{\geq l} x \varphi(x)$. The duplicator claims that the elements of the subset Y witness that $\mathcal{B} \models \exists^{\geq l} x \varphi(x)$. According to the spoilers conviction there is a $b \in Y$ with not $\mathcal{B} \models \varphi[b]$. The duplicator means that some element a in X behaves as b.

Exercise 3.4.2 Let $\mathcal{A}$ and $\mathcal{B}$ be finite $<$-forests and assume that the duplicator has a winning strategy in $\text{C-G}^2_\infty(\mathcal{A}, \mathcal{B})$. Show that $\mathcal{A}$ and $\mathcal{B}$ are isomorphic. □

The reader will encounter no difficulties when trying to prove the following theorem like corresponding preceding results.

Theorem 3.4.3 *Let $\mathcal{A}$ and $\mathcal{B}$ be structures, $\overline{a} \in (A \cup \{*\})^s$ and $\overline{b} \in (B \cup \{*\})^s$ with* $\mathrm{supp}(\overline{a}) = \mathrm{supp}(\overline{b})$.

(a) *The following are equivalent:*
  (i) *For all $\varphi(\overline{x}) \in \mathrm{C}^s_{\infty\omega}$: $\mathcal{A} \models \varphi[\overline{a}]$ iff $\mathcal{B} \models \varphi[\overline{b}]$.*
  (ii) *The duplicator wins $\text{C-G}^s_\infty(\mathcal{A}, \overline{a}, \mathcal{B}, \overline{b})$.*

(b) $\mathcal{A} \equiv^{\mathrm{C}^s_{\infty\omega}} \mathcal{B}$ *iff the duplicator wins* $\text{C-G}^s_\infty(\mathcal{A}, \mathcal{B})$.[5] □

We now use this theorem to show that for coloured graphs the expressive power of $\mathrm{C}^2_{\infty\omega}$ corresponds to a natural graph-theoretical property.

Let $C_1, C_2, \ldots$ be unary relation symbols, the "colour relations". A *coloured graph* is, for some r, an $\{E, C_1, \ldots, C_r\}$-structure $\mathcal{G}$, where for $\mathcal{G} = (G, E^G, C_1^G, \ldots, C_r^G)$ the following holds:

- (G, E^G) is a graph;
- $C_1^G \dot{\cup} \ldots \dot{\cup} C_r^G = G$, that is, each vertex satisfies exactly one colour relation.

For $a \in G$ the *colour type* $\mathrm{ct}(a)$ is defined as

$$\mathrm{ct}(a) := (i, n_1, \ldots, n_r),$$

where

$$a \in C_i^G \quad \text{and} \quad n_j := \|\{b \in C_j^G \mid E^G ab\}\|.$$

$\mathcal{G}$ is *stable* if for $a, b \in G$,

$$\mathrm{ct}(a) = \mathrm{ct}(b) \quad \text{iff} \quad a, b \in C_i \text{ for some } i.$$

The proof of the following statement is straightforward.

Proposition 3.4.4 *Let $\mathcal{G} = (G, E^G, C_1^G, \ldots, C_r^G)$ be a stable coloured graph and $a, b \in G$. Then the following are equivalent:*

(i) *For $j = 1, \ldots, r$: $a \in C_j$ iff $b \in C_j$.*
(ii) *The duplicator has a winning strategy in the game $\text{C-G}^2_\infty(\mathcal{G}, a*, \mathcal{G}, b*)$.* □

We introduce a process of colour refinement leading from a coloured graph $\mathcal{G} = (G, E^G, C_1^G, \ldots, C_r^G)$ to a stable coloured graph: Let $m := \|\{\mathrm{ct}(a) \mid a \in G\}\|$ and order the set $\{\mathrm{ct}(a) \mid a \in G\}$ lexicographically. Set $\mathcal{G}' := (G, E^G, C'_1, \ldots, C'_m)$, where C'_k is the set of elements $a \in G$ such that $\mathrm{ct}(a)$ is the k-th element in this ordering. Clearly,

$$(*) \qquad \text{each } C_i^G \text{ is the union of some } C'_k.$$

Moreover, if C'_k is the set of elements of colour type $(i, n_1, \ldots, n_r)$, then $C'_k = \{a \in G \mid \mathcal{G} \models (C_i x \wedge \bigwedge_{j=1,\ldots,r} \exists^{=n_j} y (Exy \wedge C_j y))[a]\}$, that is

[5] Of course, the corresponding results for $\mathrm{FO(C)}^s$ hold, too.

$\binom{*}{*}$ each colour class of $\mathcal{G}'$ is definable in $\mathcal{G}$ by a formula of $\mathrm{C}^2_{\infty\omega}$ of quantifier rank ≤ 1,

where we extend the definition of quantifier rank for first-order logic by the clause $\mathrm{qr}(\exists^{\geq l} x\varphi) := 1 + \mathrm{qr}(\varphi)$.

Obviously, $\mathcal{G}$ is stable if $m = r$, i.e., if there is no proper colour refinement. If $\mathcal{G}'$ is not stable, define $\mathcal{G}^{(2)} := (\mathcal{G}')'$. By (*), we get that after finitely many steps, say after n steps, we must reach a stable coloured graph $\mathcal{G}^{(n)}$, the *stable coloured refinement* of $\mathcal{G}$. A simple induction, using $\binom{*}{*}$, shows that each colour class of $\mathcal{G}^{(n)}$ is definable by a $\mathrm{C}^2_{\infty\omega}$-formula of quantifier rank $\leq n$. This fact yields (ii) $\Rightarrow$ (i) of the following theorem.

Theorem 3.4.5 *For elements a and b of a coloured graph $\mathcal{G}$ the following are equivalent:*

(i) *a, b are in the same colour class of the stable coloured refinement of $\mathcal{G}$.*
(ii) *For all $\varphi(x) \in \mathrm{C}^2_{\infty\omega}$,*

$$\mathcal{G} \models \varphi[a] \quad \textit{iff} \quad \mathcal{G} \models \varphi[b].$$

To prove (i) $\Rightarrow$ (ii), use the preceding proposition (note that by (*), a winning strategy for the coloured refinement of $\mathcal{G}$ is a winning strategy for $\mathcal{G}$). $\square$

For a graph $\mathcal{G} = (G, E^G)$ let the stable coloured refinement be that of the coloured graph (G, E^G, G). Then

Corollary 3.4.6 *For elements a and b of a graph $\mathcal{G}$ the following are equivalent:*

(i) *a, b are in the same colour class of the stable coloured refinement of $\mathcal{G}$.*
(ii) *For all $\varphi(x) \in \mathrm{C}^2_{\infty\omega}$,*

$$\mathcal{G} \models \varphi[a] \quad \textit{iff} \quad \mathcal{G} \models \varphi[b]. \qquad \square$$

3.5 Failure of Classical Theorems in the Finite

We have seen that in the finite the compactness theorem for first-order logic fails and we shall see in Chapter 7 that there is no sound and complete proof calculus.

These facts raise the question whether and why first-order logic can serve as a useful means for the investigation of finite structures. The question demands for an answer even more as other central methods of classical model theory (such as the use of ultraproducts or saturated structures) become useless and, as we shall see in this section, further important results fail when restricted to finite structures. The examples we give include Beth's definability theorem, Craig's interpolation theorem, and some preservation theorems. Nevertheless, we hope to convince the reader that new methods and results intrinsic to the finite compensate for this loss. As a new aspect we mention the stronger impact of combinatorics which, in particular, will become apparent in connection with probabilities (see Chapter 4). Moreover, the restriction to the finite motivates the use of other languages, for example languages that are able to grasp notions of recursion or induction, in this way building a bridge to computational aspects (see Chapter 7).

Recall that $\models$ denotes the consequence relation with respect to arbitrary structures and $\models_{\text{fin}}$ that for finite structures; so $\varphi \models_{\text{fin}} \psi$ means that every finite model of φ is a model of ψ.

The Beth Property and the Interpolation Property

Let $\mathcal{L}$ be any logic considered so far, e.g. FO, $\mathrm{L}_{\infty\omega}$,... Let R be an n-ary relation symbol not contained in the vocabulary τ. An $\mathcal{L}[\tau \cup \{R\}]$-sentence φ *defines R implicitly* (*in the finite*), if every (finite) τ-structure $\mathcal{A}$ has at most one expansion $(\mathcal{A}, R^A)$ to a $\tau \cup \{R\}$-structure satisfying φ. We say that R is *explicitly definable* (*in the finite*) *relative to* φ, if there is an $\mathcal{L}[\tau]$-formula $\psi(\overline{x})$ such that

$$\varphi \models_{(\text{fin})} \forall \overline{x}(R\overline{x} \leftrightarrow \psi(\overline{x})).$$

Obviously, if R is explicitly definable relative to φ then φ defines R implicitly. We say that $\mathcal{L}$ has the *Beth property* (*in the finite*), if the converse holds, i.e., whenever an $\mathcal{L}$-sentence φ defines a relation symbol implicitly (in the finite), then there is an explicit definition of it (in the finite) relative to φ.

Beth's theorem states that first-order logic has the Beth property. It exhibits a certain balance between syntax and semantics: every implicitly definable, i.e., "semantically" definable, relation has an explicit, a "syntactic" definition.

Proposition 3.5.1 *First-order logic does not have the Beth property in the finite.*

Proof. We consider orderings in the vocabulary $\tau := \{<, S, \min, \max\}$. Let R be a unary relation symbol and let φ be the conjunction of the ordering axioms and of the following sentence fixing R as the set of even points,

$$\neg R \min \wedge \forall x \forall y (Sxy \rightarrow (Rx \leftrightarrow \neg Ry)).$$

Clearly, φ defines R implicitly in the finite. Suppose

$$\varphi \models_{\text{fin}} \forall x(Rx \leftrightarrow \psi(x))$$

for some FO[τ]-formula $\psi(x)$. Then $\psi(\max)$ together with the ordering axioms would define the class of finite orderings of even cardinality, contradicting 2.3.6. □

Closely related to the Beth property is the interpolation property (or, Craig property) (more precisely, the Beth property is a consequence of the interpolation property): The logic $\mathcal{L}$ has the *interpolation property* (*in the finite*) iff for all vocabularies σ and τ and any $\mathcal{L}$-sentences φ and ψ in the vocabularies σ and τ, respectively, such that $\varphi \models_{(\text{fin})} \psi$, there is an *interpolant*, that is, an $\mathcal{L}[\sigma \cap \tau]$-sentence χ such that

$$\varphi \models_{(\text{fin})} \chi \quad \text{and} \quad \chi \models_{(\text{fin})} \psi.$$

Craig's theorem states that first-order logic has the interpolation property. We shall see that in the finite the interpolation property fails for first-order logic. Our counterexample deals with a special case of the interpolation property that will be important later. Given a logic $\mathcal{L}$, some classes K of finite τ-structures happen to be axiomatizable in $\mathcal{L}$, if we equip the structures in K with an arbitrary ordering, that is, there is a sentence φ in the vocabulary $\tau\dot{\cup}\{<\}$ such that for

$$K_< \quad := \quad \{(\mathcal{A}, <) \mid \mathcal{A} \in K, \; < \text{ an ordering on } A\},$$

we have

$$K_< = \text{Mod}(\varphi).$$

The logic $\mathcal{L}$ is said to be *closed under order-invariant sentences in the finite,* whenever in this situation there is an $\mathcal{L}[\tau]$-sentence ψ such that $\text{Mod}(\psi) = K$.

If $\varphi = \varphi(<)$ axiomatizes $K_<$ and $<'$ is a new binary relation symbol, then

$$\varphi(<) \models_{\text{fin}} (\text{“} <' \text{ is an ordering”} \rightarrow \varphi(<')).$$

Clearly, if ψ is any interpolant then $\text{Mod}(\psi) = K$. Hence, a logic with the interpolation property is closed under order-invariant sentences in the finite. In particular, in the following proposition part (b) follows from part (a).

Proposition 3.5.2 (a) *First-order logic is not closed under order-invariant sentences in the finite.*
(b) *First-order logic does not have the interpolation property in the finite.*

We sketch a *proof* of (a). Let K be the class of finite Boolean algebras with an even number of atoms. Using the Ehrenfeucht-Fraïssé method, one can show that K is not axiomatizable in first-order logic. However, $K_<$ is axiomatizable in first-order logic. In fact, let φ be the conjunction of the axioms for Boolean algebras and the axioms for orderings and of a sentence expressing that there is an element containing exactly the atoms at an even position (in the ordering induced on the atoms) and containing the last atom. □

Exercise 3.5.3 In the finite, $\text{L}_{\omega_1\omega}$ has the Beth property, the interpolation property, and is closed under order-invariant sentences. Hint: Use the fact that every class of finite structures is axiomatizable in $\text{L}_{\omega_1\omega}$ (cf. 3.2.1(b)). E.g., for the interpolation property argue as follows: If $\varphi \models_{\text{fin}} \psi$ for the $\mathrm{L}_{\omega_1\omega}[\sigma]$-sentence φ and the $\mathrm{L}_{\omega_1\omega}[\tau]$-sentence ψ then every $\mathrm{L}_{\omega_1\omega}[\sigma \cap \tau]$-sentence χ axiomatizing the class

$$\{\mathcal{A}|(\sigma \cap \tau) \mid \mathcal{A} \text{ a finite } \sigma\text{-structure with } \mathcal{A} \models \varphi\}$$

is an interpolant. □

Exercise 3.5.4 (a) For $\tau = \{<\}$ let φ be a $\{<\}$-sentence expressing that $<$ is an ordering and, for $n \geq 1$, let χ_n be the sentence of FO^2 introduced in part (a) of 3.3.1, which says that an ordering has exactly n elements. Then (for both $\models$ and $\models_{\text{fin}}$),

$$\models_{(\text{fin})} (\varphi(<) \wedge \bigvee_{n\geq 1} \chi_{2\cdot n}(<)) \to (\varphi(<') \to \bigvee_{n\geq 1} \chi_{2\cdot n}(<')).$$

Show that there is no interpolant in $\mathrm{L}^{\omega}_{\infty\omega}$. Hence, $\mathrm{L}^{\omega}_{\infty\omega}$ does not have the interpolation property (in the finite).

(b) For $s \geq 2$ and unary P, let $\varphi(P)$ and $\psi(P)$ be the FO^s sentences

$$\begin{aligned} \varphi(P) &:= \text{"there is exactly one element satisfying } P\text{"} \\ \psi(P) &:= \text{"there are at least } s \text{ elements not satisfying } P\text{"}. \end{aligned}$$

Then for unary Q,

$$\models_{(\text{fin})} (\varphi(P) \wedge \psi(P)) \to (\varphi(Q) \to \psi(Q)),$$

but there is no interpolant in FO^s. Hence, FO^s does not have the interpolation property (in the finite).

(c) Show that FO^1 and $\mathrm{L}^1_{\infty\omega}$ have the interpolation property (in the finite) in case τ is relational. □

Preservation Theorems

In the model theory for arbitrary structures certain closure properties of the class of models of a sentence φ are reflected by syntactic properties of φ. Most of these so-called preservation theorems fail when literally translated to the finite. We give some examples.

Call a first-order formula *universal* (*existential*) if it is built up from atomic and negated atomic formulas using only the connectives $\wedge, \vee$ and the universal (existential) quantifier. If φ is a universal sentence, a simple inductive proof shows

$$\mathcal{B} \subseteq \mathcal{A} \text{ and } \mathcal{A} \models \varphi \text{ imply } \mathcal{B} \models \varphi,$$

that is, φ is *preserved under substructures*. Similarly, if φ is existential then

$$\mathcal{B} \subseteq \mathcal{A} \text{ and } \mathcal{B} \models \varphi \text{ imply } \mathcal{A} \models \varphi,$$

that is, φ is *preserved under extensions.* Moreover, in classical model theory one proves that every FO-sentence preserved under substructures is logically equivalent to a universal FO-sentence. We give an example of an FO-sentence that, in the finite, is preserved under substructures but is not equivalent to a universal first-order sentence.

Let the universal sentence φ_0 be the conjunction of the ordering axioms in $\{<, \min, \max\}$ and of the following sentence expressing that R is a "partial successor relation",

$$\forall x \forall y(Rxy \to x < y) \wedge \forall x \forall y \forall z((Rxy \wedge x < z) \to (y = z \vee y < z)).$$

Let φ_1 be the sentence $\forall x(\neg x = \max \to \exists y Rxy)$ expressing that R is the "total" successor relation. For finite structures $\mathcal{A}$ and $\mathcal{B}$,

$$(*) \qquad \mathcal{A} \models \varphi_0,\ \mathcal{B} \models (\varphi_0 \wedge \varphi_1),\ \text{and } \mathcal{B} \subseteq \mathcal{A} \ \text{ imply } \mathcal{A} = \mathcal{B}.$$

Using a new unary relation symbol Q, we set

$$\varphi \quad := \quad \varphi_0 \wedge (\varphi_1 \to \exists y Q y).$$

In finite models, φ is preserved under substructures: Suppose $(\mathcal{A}, Q^{\mathcal{A}}) \models \varphi$ and $(\mathcal{B}, Q^{\mathcal{B}}) \subseteq (\mathcal{A}, Q^{\mathcal{A}})$. Since φ_0 is universal, $\mathcal{B} \models \varphi_0$. If $\mathcal{B} \not\models \varphi_1$ then $(\mathcal{B}, Q^{\mathcal{B}}) \models \varphi$. If $\mathcal{B} \models \varphi_1$ then by $(*)$, $\mathcal{B} = \mathcal{A}$, therefore $(\mathcal{B}, Q^{\mathcal{B}}) = (\mathcal{A}, Q^{\mathcal{A}})$ and hence, $(\mathcal{B}, Q^{\mathcal{B}}) \models \varphi$.

Assume by contradiction, that ψ is a universal first-order sentence with $\models_{\text{fin}} \varphi \leftrightarrow \psi$, say, $\psi = \forall x_1 \ldots \forall x_n \chi$ with quantifier-free χ (every universal FO-formula is logically equivalent to one of this form!). Look at a $\{<, \min, \max, R\}$-structure $\mathcal{A}$ with $n+3$ elements, where $(A, <^A, \min^A, \max^A)$ is an ordering and where R^A is the successor relation. Set $Q^{\mathcal{A}} = \emptyset$. Then $(\mathcal{A}, Q^{\mathcal{A}}) \not\models \varphi$, and hence, $(\mathcal{A}, Q^{\mathcal{A}}) \models \exists x_1 ... \exists x_n \neg\chi$, say,

$$(\mathcal{A}, Q^{\mathcal{A}}) \models \neg\chi[a_1, \ldots, a_n].$$

Choose $a \in A \setminus \{a_1, ..., a_n, \min^A, \max^A\}$ and set $Q' = \{a\}$. Since χ is quantifier-free, $(\mathcal{A}, Q') \models \neg\chi[a_1, ..., a_n]$ and therefore, $(\mathcal{A}, Q') \not\models \forall x_1 ... \forall x_n \chi$. On the other hand, $(\mathcal{A}, Q') \models \varphi$. Hence φ and ψ are not equivalent in the finite.

This example gives part (a) of

Proposition 3.5.5 (a) *There is a first-order sentence which, in the finite, is preserved under substructures but not equivalent to a universal first-order sentence.*

(b) *There is a first-order sentence which, in the finite, is preserved under extensions but not equivalent to an existential first-order sentence.*

Proof. (b) Let φ be according to (a). Then $\neg\varphi$ is preserved under extensions and not equivalent to an existential sentence. □

We close with some remarks concerning "monotone" formulas. Fix a relation symbol R of τ of arity r. A sentence φ is *monotone in R (in the finite)* if

$$(\mathcal{A}, R_1) \models \varphi, \text{ (A finite) and } R_1 \subseteq R_2 \subseteq A^r \text{ imply } (\mathcal{A}, R_2) \models \varphi.$$

A first-order formula φ is *positive* in R if φ is built up from atomic formulas using $\neg, \wedge, \vee, \forall, \exists$ and any occurrence of the relation symbol R in φ is within the scope of an even number of negation symbols. An easy inductive argument shows that a sentence positive in R is monotone. While any first-order sentence monotone in R is logically equivalent to a formula positive in R, this is no longer true in the finite. We state this result; for a proof we refer to the example in Exercise 3.5.8.

Proposition 3.5.6 *There is a first-order sentence which, in the finite, is monotone in R, but not equivalent to a first-order sentence positive in R.*

As we have seen, preservation theorems of first-order logic fail when literally translated to the case of finite structures. However, it is still open whether – say, in the case of sentences preserved under substructures – there is a syntactically defined or, at least, recursive set of sentences preserved under substructures in the finite such that, in the finite, every first-order sentence preserved under substructures is equivalent to a sentence in this set.

Exercise 3.5.7 Define the notions of a universal $\mathrm{L}_{\omega_1\omega}$-sentence and of an $\mathrm{L}_{\omega_1\omega}$-sentence positive in R and show:

(a) In the finite, every $\mathrm{L}_{\omega_1\omega}$-sentence preserved under substructures is equivalent to a universal $\mathrm{L}_{\omega_1\omega}$-sentence.

(b) In the finite, every $\mathrm{L}_{\omega_1\omega}$-sentence monotone in R is equivalent to an $\mathrm{L}_{\omega_1\omega}$-sentence positive in R.

Hint for (a): Consider a variant of the Ehrenfeucht game by prescribing that in each move the spoiler has to choose an element of the second structure and argue with the corresponding isomorphism types defined by

$$\chi^0_{\overline{a}} := \varphi^0_{\overline{a}} \quad \text{and} \quad \chi^{j+1}_{\overline{a}} := \forall v \bigvee_{a \in A} \chi^j_{\overline{a}a}.$$

For (b) omit subformulas $\neg Rt_1 \ldots t_m$ in $\varphi^0_{\overline{a}}$, defining $\varphi^m_{\overline{a}}$ for $m > 0$ as usual. □

Exercise 3.5.8 (a) Assume that $\tau = \sigma \cup \{R\}$, where $R \notin \sigma$ is an r-ary relation symbol, and let $\mathcal{A}$ and $\mathcal{B}$ be τ-structures. Let $\mathrm{Part}^+(\mathcal{A}, \mathcal{B})$ be the set

$$\{p \in \mathrm{Part}(\mathcal{A}|\sigma, \mathcal{B}|\sigma) \mid \text{for all } \overline{a} \in \mathrm{do}(p), \text{ if } R^{\mathcal{A}}\overline{a} \text{ then } R^{\mathcal{B}}p(\overline{a})\}.$$

Write

$$\mathcal{A} \equiv^+_m \mathcal{B} \quad \text{iff} \quad \text{for all FO}[\tau]\text{-sentences } \varphi \text{ with } \mathrm{qr}(\varphi) \leq m \text{ that are positive in } R\text{: if } \mathcal{A} \models \varphi \text{ then } \mathcal{B} \models \varphi,$$

and

$$\mathcal{A} \cong_m^+ \mathcal{B} \quad \text{iff} \quad \text{there is a sequence } (I_j)_{j \leq m} \text{ of nonempty subsets of } \mathrm{Part}^+(\mathcal{A}, \mathcal{B}) \text{ with the back and forth properties.}$$

Show that $\mathcal{A} \equiv_m^+ \mathcal{B}$ iff $\mathcal{A} \cong_m^+ \mathcal{B}$.

(b) Let $\sigma = \{S\}$ and $\tau = \sigma \cup \{\prec\}$ with binary $S, \prec$. For $n, k \in \mathbb{N}$ let $\mathcal{G}_{n,k}$ be the structure with

$$\begin{aligned}
G_{n,k} &:= \{0, \ldots, n\} \times \{0, \ldots, k\}, \\
S^{G_{n,k}} &:= \{((i,j),(i,j+1)) \mid i \leq n, j < k\} \\
&\quad \cup \{((i,j),(i+1,j+1)) \mid i < n, j < k\} \\
&\quad \cup \{((n,j),(0,j+1)) \mid j < k\}, \\
\prec^{G_{n,k}} &:= \text{transitive closure of } S^{G_{n,k}}.
\end{aligned}$$

$\mathcal{G}_{n,k}$ can be pictured as a circular net of circumference $(n+1)$ and height k, where S-arrows are going from a point (that is not in the upper edge) to its upper and to its upper right neighbour. We say that a point (i,j) has (geographical) length i and (geographical) latitude j.

If $a \prec^{G_{n,k}} b$, let $\mathcal{G}_{n,k}^{(a,b)}$ be defined as $\mathcal{G}_{n,k}$, setting $\prec^{G_{n,k}^{(a,b)}} := \prec^{G_{n,k}} \setminus \{(a,b)\}$.

Show for $n = 4 \cdot (2^m+1)$: If $a \prec^{G_{n,n}} b$, a has latitude 2^m+1, b has latitude $3 \cdot (2^m+1)$, and if a and b have length difference $k := 2 \cdot (2^m+1)$, then

$$\mathcal{G}_{n,k} \cong_m^+ \mathcal{G}_{n,n}^{(a,b)}.$$

Hint: Let c and d in $G_{n,k}$ have latitude 2^m+1 and the lengths of a and b, respectively, and show that $(\mathcal{G}_{n,k}, c, d) \cong_m^+ (\mathcal{G}_{n,n}^{(a,b)}, a, b)$. For $l \leq m$, take as I_l the set of those $p \in \mathrm{Part}^+((\mathcal{G}_{n,k}, c, d), (\mathcal{G}_{n,n}^{(a,b)}, a, b))$ that

- preserve lengths;
- preserve the upper and the lower edge;
- preserve differences of latitudes up to 2^l for points whose lengths differ by at most 2^l and for points on the edges;
- do not decrease differences of latitudes.

(c) Set $\varphi = \forall x \forall y (Sxy \rightarrow (\neg x = y \wedge x \prec y)) \wedge (\varphi_1 \vee \varphi_2 \vee \varphi_3)$ where

$$\begin{aligned}
\varphi_1 &:= \text{“} \prec \text{ is transitive”} \\
\varphi_2 &:= \exists x \exists y (x \prec y \wedge y \prec x) \\
\varphi_3 &:= \exists x \exists y (x \prec y \wedge \forall z (Szy \rightarrow (\neg x \preceq z \wedge \kappa(z))))
\end{aligned}$$

with

$$\begin{aligned}
\kappa(z) := \; & \forall w \forall y \forall u \forall v ((y \preceq w \wedge Swz \wedge v \preceq u \wedge Suy) \rightarrow \\
& (v \prec y \wedge v \prec w \wedge v \prec z \wedge y \prec z)).
\end{aligned}$$

$\kappa(z)$ guarantees that $\prec$ is transitive on the set consisting of z and its iterated predecessors; more precisely, if $\mathcal{A} \models \kappa[a_0]$ and $S^{\mathcal{A}} a_{i+1} a_i$ for $i \geq 0$ then $a_i \prec a_j$ for $0 \leq j < i$.

Show that $\mathcal{G}_{n,k} \models \varphi$ and $\mathcal{G}_{n,n}^{(a,b)} \models \neg\varphi$, where n, k, a, b are as in (b).

(d) Show that, in the finite, φ is monotone in $\prec$ but not equivalent to a sentence positive in $\prec$. Hint for the monotonicity: Let $\mathcal{A}, \mathcal{B}$ be given such that $B = A$, $S^{\mathcal{B}} = S^{\mathcal{A}}$, and $\prec^{\mathcal{B}} \supseteq \prec^{\mathcal{A}}$ and such that $\mathcal{A} \models \varphi$. As φ_2 is positive in $\prec$, assume $\mathcal{A} \models \neg\varphi_2$. Consider the cases $\mathcal{A} \models \varphi_3$ and $\mathcal{A} \models \varphi_1 \wedge \neg\varphi_3$. In the first case, if (a, b) witnesses φ_3 in $\mathcal{A}$ but not in $\mathcal{B}$, find a chain $(a, b) = (a_0, b_0), (a_1, b_1), \ldots$ with $a_i \prec^{\mathcal{B}} b_i$, $(b_{i+1}, b_i) \in \mathrm{TC}(S^{\mathcal{A}})$, and $(a_i, b_i) \notin \mathrm{TC}(S^{\mathcal{A}})$ that ends with a pair witnessing φ_3 in $\mathcal{B}$.

(e) Show that φ is not monotone in $\prec$ for arbitrary structures. Hint: Start with a model of φ that consists of the negative integers with the natural successor relation and ordering together with an additional point. □

Notes 3.5.9 A reference for second-order logic is [133]; for the model theory of $\mathrm{L}_{\infty\omega}$ and $\mathrm{L}_{\omega_1\omega}$ we refer the reader to [8, 100]. The first study of fragments with finitely many variables is due to Henkin [84]. The languages $\mathrm{L}^s_{\infty\omega}$ and the corresponding pebble games were first introduced by Barwise [10] and then reinvented and successfully used by Immerman [90] and Poizat [130]. Further references are [96] and [40]. Compare [30] for most results in subsection 3.3.3. An exposition of the model theory of $\mathrm{L}^{\omega}_{\infty\omega}$ is contained in [86]. Surveys on recent results for finite variable logics are [58, 66]. (Infinite) counting quantifiers and the corresponding Ehrenfeucht-Fraïssé method have been considered in various contexts, e.g. in [145]. A thorough treatment in the context of finite model theory is contained in [125]. The results on graphs presented in section 3.4 are from [97]. The failure of many classical interpolation and preservation theorems was first presented explicitly in [70]. The example given in 3.5.8 is from [135], the first such example was presented in [6].

4. 0–1 Laws

So far we mainly looked at techniques and results which were developed for *arbitrary* structures, and we analyzed to what extent they remain valid if we restrict ourselves to *finite* ones. In the present chapter we study a concept that is genuine to the finite (even though there are extensions to arbitrary structures): the probability for a finite structure to be a model of a given sentence. A starting point was the observation that for a relational vocabulary τ and a first-order sentence φ, either nearly all finite structures are models of φ or nearly all finite structures are models of $\neg\varphi$ (for example, nearly all $\{E\}$-structures are not digraphs, that is, nearly all $\{E\}$-structures are models of $\neg\forall x \neg Exx$). One describes this property by saying that first-order logic satisfies the 0–1 law.

There are different ways of how to count structures and hence, there are different versions of 0–1 laws which we study for FO and for $\mathrm{L}^{\omega}_{\infty\omega}$ in the first and the third section. The second section treats relativized versions, for example, 0–1 laws inside a class of structures, say, inside the class of graphs. After some applications in section 4.4, we study the case of monadic second-order logic and disprove the 0–1 law for it in section 4.5.

In this chapter n is always a natural number greater than 0.

4.1 0–1 Laws for FO and $\mathrm{L}^{\omega}_{\infty\omega}$

As always we refer to a fixed vocabulary τ. For a class K of structures let $L_n(K)$ be the number of structures in K with universe $\{1, 2, ..., n\}$,

$$L_n(K) \;:=\; \|\{\mathcal{A} \in K \mid A = \{1, \ldots, n\}\}\|.$$

Sometimes, structures $\mathcal{A}$ with $A = \{1, \ldots, n\}$ are called *labeled* structures, since every element in such a structure is labeled with a natural number. Hence, $L_n(K)$ is the number of labeled structures in K of cardinality n. If K is the class of models of a sentence φ or the class of all τ-structures, we denote $L_n(K)$ also by $L_n(\varphi)$ and $L_n(\tau)$, respectively. Let $l_n(K)$ be the fraction of structures with universe $\{1, \ldots, n\}$ which are in K,

$$l_n(K) \;:=\; \frac{L_n(K)}{L_n(\tau)}.$$

In case it exists, $l(K) := \lim_{n\to\infty} l_n(K)$ is called the *labeled asymptotic probability* of K. Similarly as above, $l_n(\varphi)$ stands for $l_n(\mathrm{Mod}(\varphi))$ and $l(\varphi)$ for $l(\mathrm{Mod}(\varphi))$. If $l(\varphi) = 1$ we say that φ holds in *almost all* finite structures or that φ *almost surely* holds.

A class Ψ of sentences of a logic is said to satisfy the *labeled* 0–1 *law* if

$$l(\varphi) = 1 \qquad \text{or} \qquad l(\varphi) = 0$$

holds for every $\varphi \in \Psi$ (or, equivalently, for $\varphi \in \Psi$ either φ or $\neg\varphi$ holds in almost all finite structures).

Examples 4.1.1 (a) Suppose $\tau = \{P, c\}$, where P is unary. Since for any τ-structure (A, P^A, c^A),

$$(A, P^A, c^A) \models Pc \qquad \text{iff} \qquad (A, A \setminus P^A, c^A) \not\models Pc,$$

we see that $l_n(Pc) = \frac{1}{2}$ and thus, $l(Pc) = \frac{1}{2}$.

(b) For the "vocabulary" $\tau := \{f\}$ with a unary function symbol f consider the first-order sentence $\forall x f(x) \neq x$ expressing that f has no fixed-point. Then

$$l_n(\forall x f(x) \neq x) = \left(\frac{n-1}{n}\right)^n = \left(1 - \frac{1}{n}\right)^n$$

(since on the universe $\{1, \ldots, n\}$ one can fix the values of f independently and since for each argument i the $n-1$ possible values $\neq i$ do not lead to a fixed-point). Hence, $l(\forall x f(x) \neq x) = e^{-1}$.

(c) If K is the class EVEN$[\tau]$ of all τ-structures of even finite cardinality, then

$$l_n(K) = \begin{cases} 1 & \text{if } n \text{ is even} \\ 0 & \text{if } n \text{ is odd} \end{cases}$$

and hence, $l(K)$ does not exist. Therefore, $l(\varphi)$ does not exist for the second-order sentence φ expressing "there is a binary relation which is an equivalence relation having only equivalence classes with exactly two elements" or for the $\mathrm{L}_{\omega_1\omega}$-sentence $\bigvee_{k\geq 1} \varphi_{=2k}$.

(d) Let K be a class of structures for a relational τ. Construct a "random" structure of vocabulary τ on $\{1, \ldots, n\}$ by the following experiment: For every m-ary relation symbol R in τ and for every $i_1, \ldots, i_m \in \{1, \ldots, n\}$, toss a fair coin to decide whether $Ri_1 \ldots i_m$ is true. Then $l_n(K)$ is the probability for the outcome $\mathcal{A}$ of the experiment to belong to K. □

Examples (a) and (b) show that we can expect a labeled 0–1 law for first-order logic only for relational vocabularies. In fact, it then holds. The central point of the proof we give is the fact that the extension axioms hold almost surely and that for any φ, either φ or $\neg\varphi$ is a consequence of the extension axioms.

In the following suppose that τ is relational. Recall (cf. 3.2.11) that an $(r+1)$-extension axiom is a sentence $\chi_\Phi =$

$$\forall v_1 \ldots \forall v_r (\bigwedge_{1 \le i < j \le r} v_i \neq v_j \to \exists v_{r+1} (\bigwedge_{1 \le i \le r} v_i \neq v_{r+1} \land \bigwedge_{\varphi \in \Phi} \varphi \land \bigwedge_{\varphi \in \Phi^c} \neg \varphi))$$

where Φ is a subset of

$$\Delta_{r+1} = \{\varphi(v_1, \ldots, v_r, v_{r+1}) \mid \varphi \text{ has the form } R\overline{x}, \text{where } R \in \tau \text{ and where } v_{r+1} \text{ occurs in } \overline{x}\}.$$

Lemma 4.1.2 *Any extension axiom holds in almost all finite structures.*

Proof. Given Φ, we have to show that the asymptotic labeled probability $l(\chi_\Phi)$ equals 1. For any tuple $a_1, \ldots, a_r$ of distinct elements in a structure $\mathcal{A}$ and any further object a let δ be the probability that $a_1, \ldots, a_r, a$ satisfies $\Phi \cup \{\neg\varphi \mid \varphi \in \Phi^c\}$, when adding a to $\mathcal{A}$ as a new element and randomly fixing the truth values of $R\overline{b}$ for any R in τ and any sequence $\overline{b}$ in $A \cup \{a\}$ containing a. Clearly, if c is the number of subsets of Δ_{r+1}, then $\delta = \frac{1}{c}$; in particular, $\delta > 0$. Thus

$$\begin{aligned} l_n(\neg\chi_\Phi) &= l_n(\exists v_1 \ldots \exists v_r (\textstyle\bigwedge_{1 \le i < j \le r} v_i \neq v_j \land \\ &\quad \forall v_{r+1} (\textstyle\bigvee_{1 \le i \le r} v_i = v_{r+1} \lor \bigvee_{\varphi \in \Phi} \neg\varphi \lor \bigvee_{\varphi \in \Phi^c} \varphi))) \\ &\le n^r (\tfrac{c-1}{c})^{n-r} = n^r (1-\delta)^{n-r}. \end{aligned}$$

Therefore, $l(\neg\chi_\Phi) = \lim_{n \to \infty} l_n(\neg\chi_\Phi) = 0$. □

Recall that by T_{rand} $(= T_{\text{rand}}(\tau))$ we denote the set of extension axioms.

Corollary 4.1.3 *Let φ be a first-order sentence.*

(a) *If $T_{\text{rand}} \models \varphi$ then $l(\varphi) = 1$.* (b) *If $T_{\text{rand}} \models \neg\varphi$ then $l(\varphi) = 0$.*

Proof. (a) If $T_{\text{rand}} \models \varphi$ then by compactness, $T_0 \models \varphi$ for some finite subset T_0 of T_{rand}. Since T_0 is a set of extension axioms, $l(\bigwedge T_0) = 1$ by the preceding lemma. Hence, $l(\varphi) = 1$.

(b) If $T_{\text{rand}} \models \neg\varphi$ then by (a), $l(\neg\varphi) = 1$ and therefore, $l(\varphi) = 0$. □

For $s \ge 1$, let ϵ_s be the conjunction of the finitely many r-extension axioms with $r \le s$. Since $l(\epsilon_s) = 1$ we obtain

Corollary 4.1.4 *Let φ be an $\mathrm{L}^{\omega}_{\infty\omega}$-sentence.*

(a) *If $\epsilon_s \models \varphi$ then $l(\varphi) = 1$.* (b) *If $\epsilon_s \models \neg\varphi$ then $l(\varphi) = 0$.* □

Theorem 4.1.5 *Let τ be relational. Then both* FO$[\tau]$ *and* $\mathrm{L}^{\omega}_{\infty\omega}[\tau]$ *satisfy the labeled 0–1 law.*

Proof. The assertions follow from the above corollaries, since we know that for φ in FO$[\tau]$ (cf. 3.2.11)

$$T_{\text{rand}} \models \varphi \quad \text{or} \quad T_{\text{rand}} \models \neg\varphi$$

and that for $\varphi \in \mathrm{L}^{s}_{\infty\omega}[\tau]$ (cf. 3.3.11),

$$\epsilon_s \models \varphi \quad \text{or} \quad \epsilon_s \models \neg\varphi.$$

□

Exercise 4.1.6 Let τ be relational and $m \in \mathbb{N}$. For $n, k \geq 1$ let $f(n,k)$ be the probability that $\mathcal{A} \cong_m \mathcal{B}$ holds for random structures $\mathcal{A}$ and $\mathcal{B}$ with universe $\{1,\ldots,n\}$ and $\{1,\ldots,k\}$, respectively. Show that $\lim_{n,k\to\infty} f(n,k) = 1$. □

Exercise 4.1.7 Let $0 < p < 1$. Alter the construction of 4.1.1(d) of a random structure with universe $\{1,\ldots,n\}$ using a biased coin, so that $Ri_1 \ldots i_m$ now holds with probability p. For any extension axiom φ show that with probability one this random structure is a model of φ. The situation changes drastically if p is allowed to depend on the number n of elements (see [132]). □

Exercise 4.1.8 Let τ be relational. Extending the construction of 4.1.1(d), define an infinite random structure over $\{1,2,\ldots\}$ in the following way: For every P in τ, say of arity m, and every m-tuple $i_1,\ldots,i_m$ of positive integers toss a coin to decide whether $Pi_1 \ldots i_m$ is true. Show that with probability 1 (with respect to the canonical product measure) this infinite random structure is a model of T_{rand} and hence, it is the (up to isomorphism) uniquely determined countable model $\mathcal{R}$ of T_{rand} (cf. 3.2.11). Therefore, $\mathcal{R}$ is called the *infinite random structure.* □

Exercise 4.1.9 Show that for the infinite "vocabulary" $\tau = \{P_1, P_2, \ldots\}$ with unary P_i there is an $\mathrm{L}^{\omega}_{\omega_1\omega}[\tau]$-sentence without labeled asymptotic probability. □

4.2 Parametric Classes

Does every first-order sentence or its negation hold in almost all graphs? To treat such questions we introduce conditional probabilities. Slight modifications of the arguments given in the preceding section show that the 0–1 law holds for the conditional probabilities with respect to classes of structures which are axiomatizable by means of "parametric" axioms. The classes of graphs and digraphs are prominent examples.

We start with the definition of conditional probabilities: Suppose that K and H are classes of τ-structures. Define the *labeled probability* $l_n(K|H)$ by

$$l_n(K|H) \quad := \quad \frac{L_n(K \cap H)}{L_n(H)}.$$ [1]

If it exists, $l(K|H) := \lim_{n\to\infty} l_n(K|H)$ is called the *labeled asymptotic probability* of K *with respect to* H. Notations such as $l_n(\varphi|H)$ or $l_n(K|\tau)$ should be self-explaining.

Obviously, $l_n(K|\tau) = l_n(K)$, and $l_n(\mathrm{CONN}|\mathrm{GRAPH})$ is the number of connected graphs on $\{1,\ldots,n\}$ divided by the total number of graphs on $\{1,\ldots,n\}$.

[1] This fraction is defined only in case H contains a structure of cardinality n.

Let $\forall \text{distinct}\, x_1 \ldots x_s\, \psi$ abbreviate $\forall x_1 \ldots \forall x_s((\neg x_1 = x_2 \wedge \neg x_1 = x_3 \wedge \ldots \wedge \neg x_{s-1} = x_s) \rightarrow \psi)$ (which is $\forall x_1 \psi$ in case $s = 1$).

Definition 4.2.1 Let τ be relational. A first-order sentence φ is called *parametric*, if it is a conjunction of sentences $\forall \text{distinct}\, x_1 \ldots x_s\, \psi$, where $s \geq 1$ and ψ is a boolean combination of formulas of the form $Ry_1 \ldots y_n$ with $R \in \tau$ and $\{y_1, \ldots, y_n\} = \{x_1, \ldots, x_s\}$. (Note that s cannot exceed the maximum of the arities of relation symbols in τ.) A class K of structures is said to be *parametric*, if $K = \text{Mod}(\varphi)$ where φ is a parametric sentence. □

Examples 4.2.2 (a) $\forall x \neg Rxx \wedge \forall \text{distinct}\, xy(Rxy \rightarrow Ryx)$ and $\forall x \neg Rxx$ are parametric sentences axiomatizing the classes of graphs and digraphs, respectively.

(b) The class of *tournaments* is axiomatized by the parametric sentence

$$\forall x \neg Rxx \quad \wedge \forall \text{distinct}\, xy(Rxy \leftrightarrow \neg Ryx).$$

(c) For relational τ the class K of all τ-structures is parametric, since K is the class of models of the empty conjunction.

(d) Note that

$$\forall \text{distinct}\, xyz((Rxy \wedge Ryz) \rightarrow Rxz)$$

is not a parametric sentence (since, e.g., $\{x, y\} \neq \{x, y, z\}$). In fact, from our analysis of parametric classes it will become clear that, for example, the classes of transitive relations, equivalence relations, partial orderings, and orderings are not parametric, transitivity being the only obstacle.

(e) If R is k-ary, $\forall \text{distinct}\, x_1 \ldots x_k(Rx_1 \ldots x_k \wedge \neg Rx_1 \ldots x_k)$ is a parametric sentence that is true in all structures of cardinality $< k$, but has no model of cardinality $\geq k$. □

The following considerations will demonstrate that the sentence in (e) is a "sharpest" example of a parametric sentence having only models of bounded cardinality.

Let k be the maximum of the arities of the relation symbols in τ. A parametric sentence (and its model class) is called *nontrivial*, if it has a model of cardinality $\geq k$. We show that nontrivial parametric sentences have arbitrarily large models: Suppose that φ_0 is a nontrivial parametric sentence. If B is any nonempty set, then the following procedure, where we stepwise fix the relations on B, leads to a model of φ_0 with universe B: For any $s \leq k$ and any distinct $b_1, \ldots, b_s \in B$, we choose an arbitrary model $\mathcal{A}$ of φ_0 of cardinality $\geq s$ and distinct $a_1, \ldots, a_s \in A$ and define, for all R in τ, the "$\{b_1, \ldots, b_s\}$-part of R^B" as a copy of the "$\{a_1, \ldots, a_s\}$-part of R^A"; more precisely, we define this part of R^B such that for $\varphi(v_1, \ldots, v_s) = Ry_1 \ldots y_n$ with $\{y_1, \ldots, y_n\} = \{v_1, \ldots, v_s\}$

$$\mathcal{B} \models \varphi[b_1, \ldots, b_s] \quad \text{iff} \quad \mathcal{A} \models \varphi[a_1, \ldots, a_s].$$

Since in $\mathcal{B}$ every s-tuple of distinct elements behaves as some s-tuple in some model of φ_0 and as φ_0 is parametric, $\mathcal{B}$ itself is a model of φ_0. Thus we have proved:

(1) φ_0 has models in all cardinalities.

Call an $(r+1)$-extension axiom

$$\forall \text{distinct } v_1 \ldots v_r \, \exists v_{r+1} (\bigwedge_{1 \le i \le r} \neg v_i = v_{r+1} \wedge \bigwedge_{\varphi \in \Phi} \varphi \wedge \bigwedge_{\varphi \in \Phi^c} \neg\varphi)$$

compatible with φ_0 if

$$\{\varphi_0\} \cup \{\exists v_1 \ldots \exists v_r \exists v_{r+1} (\bigwedge_{1 \le i < j \le r+1} \neg v_i = v_j \wedge \bigwedge_{\varphi \in \Phi} \varphi \wedge \bigwedge_{\varphi \in \Phi^c} \neg\varphi)\}$$

is satisfiable. Let $T_{\text{rand}}(\varphi_0)$ be the set of sentences consisting of φ_0 and of all extension axioms compatible with φ_0. The proof in Example 3.2.11 showing that any two models of T_{rand} are partially isomorphic, also works for $T_{\text{rand}}(\varphi_0)$ and yields:

(2) Any two models of $T_{\text{rand}}(\varphi_0)$ are partially isomorphic and hence, $\mathrm{L}_{\infty\omega}$-equivalent. Therefore,

$$T_{\text{rand}}(\varphi_0) \models \psi \qquad \text{or} \qquad T_{\text{rand}}(\varphi_0) \models \neg\psi$$

holds for any $\mathrm{L}_{\infty\omega}$-sentence ψ.

Also the proof of 3.2.11 leading to a countable model of T_{rand} can be transferred to $T_{\text{rand}}(\varphi_0)$, since by the construction process described above the corresponding $\mathcal{A}_i$'s can be chosen as models of φ_0. Hence,

(3) $T_{\text{rand}}(\varphi_0)$ has an (up to isomorphism) unique countable model $\mathcal{R}(\varphi_0)$.

For $s \ge 1$, we denote by φ_0^s the conjunction of φ_0 with the finitely many r-extension axioms with $r \le s$ that are compatible with φ_0. By a similar argumentation, one obtains:

(4) Any two models of φ_0^s are s-partially isomorphic and hence, $\mathrm{L}^s_{\infty\omega}$-equivalent. Therefore, for any $\mathrm{L}^s_{\infty\omega}$-sentence ψ,

$$\varphi_0^s \models \psi \qquad \text{or} \qquad \varphi_0^s \models \neg\psi.$$

Finally, we have:

(5) If ψ is an extension axiom compatible with φ_0 then $l(\psi|\varphi_0) = 1$.

To give a proof, one can argue as for 4.1.2, but now restricting oneself to models of φ_0 and taking as c the number of subsets Φ of Δ_{r+1} that correspond to $(r+1)$-extension axioms compatible with φ_0. (Given Φ, distinct $a_1, \ldots, a_r$, and a new a, we can satisfy Φ by $a_1 \ldots a_r a$, applying the construction procedure described above to any $a_{i_1} \ldots a_{i_m} a$ with $1 \leq i_1 < i_2 < \ldots < i_m \leq r$.)

Let H be a class of structures and Ψ a class of sentences. We say that H *satisfies the labeled 0–1 law for* Ψ if

$$l(\psi|H) = 1 \qquad \text{or} \qquad l(\psi|H) = 0$$

holds for any $\psi \in \Psi$. From (2)–(5) we obtain:

Theorem 4.2.3 *Let H be a nontrivial parametric class. Then H satisfies the labeled 0–1 law for* $\mathrm{L}^{\omega}_{\infty\omega}$ *and hence, for* FO. □

4.3 Unlabeled 0–1 Laws

We saw in 4.1.1(d) that from a probabilistic point of view the definition of $l_n(K)$ is quite natural. But note that for $\tau := \{P\}$ with unary P and $i = 1, \ldots, n$ the structures $\mathcal{A}_i := (\{1, \ldots, n\}, P^{\mathcal{A}_i})$ with $P^{\mathcal{A}_i} := \{i\}$ are counted as n different structures in the definition of $l_n(K)$, even though they are isomorphic. In this section we study the so-called *unlabeled probability* $u_n(K)$, which is the proportion of *isomorphism types* of structures of cardinality n in K. Similarly, we define unlabeled conditional probabilities. It turns out that the labeled and unlabeled asymptotic conditional probabilities coincide in case almost all structures in the underlying class are rigid.

Fix a vocabulary τ. For a class K let $U_n(K)$ be the number of isomorphism types of structures of cardinality n in K or, equivalently,

$$U_n(K) \;:=\; \text{number of isomorphism types of structures in } K \text{ with universe } \{1, \ldots, n\}.$$

$U_n(\tau)$ and $U_n(\varphi)$ stand for $U_n(K)$, where K is the class of all τ-structures and the class of models of φ, respectively.

For arbitrary K we set

$$u_n(K) \;:=\; \frac{U_n(K)}{U_n(\tau)}$$

and denote (in case it exists) by $u(K) := \lim_{n\to\infty} u_n(K)$ the *unlabeled asymptotic probability*.

If K and H are classes of structures we call

$$u_n(K|H) \;:=\; \frac{U_n(K \cap H)}{U_n(H)}$$

the *unlabeled probability* and

$$u(K|H) \quad := \quad \lim_{n\to\infty} u_n(K|H)$$

the *unlabeled asymptotic probability* of K *with respect to* H.

Thus, $u_n(\text{CONN}|\text{GRAPH})$ is the number of isomorphism types of connected n vertex graphs divided by the total number of isomorphism types of n vertex graphs.

When studying the relationship between labeled and unlabeled probabilities, a prominent role is played by the class RIG (= RIG[τ]) of rigid structures. A structure $\mathcal{A}$ is called *rigid* if the identity on A is the only automorphism of $\mathcal{A}$.

Lemma 4.3.1 (a) $L_n(K) \le U_n(K)\cdot n!$
(b) *If* $K \subseteq \text{RIG}$ *then* $L_n(K) = U_n(K)\cdot n!$
(c) *If* $K \subseteq \text{RIG}^c$ [2] *then* $L_n(K) \le U_n(K)\cdot \frac{n!}{2}$.

Proof. Since $L_n(K) = L_n(K\cap \text{RIG}) + L_n(K \cap \text{RIG}^c)$ and similarly for U_n, it suffices to show (b) and (c).

Let $\mathcal{A}$ be a structure with universe $\{1,\dots,n\}$. There are $n!$ permutations of $\{1,\dots,n\}$. Every permutation π gives a structure $\mathcal{A}_\pi$ on $\{1,\dots,n\}$ such that $\pi : \mathcal{A} \cong \mathcal{A}_\pi$. Clearly, for permutations π and ρ of $\{1,\dots,n\}$ we have

$$(*) \qquad \mathcal{A}_\pi = \mathcal{A}_\rho \quad \text{iff} \quad \pi^{-1}\circ\rho : \mathcal{A} \cong \mathcal{A}.$$

Hence, if $\mathcal{A}$ is rigid, we have

$$\mathcal{A}_\pi = \mathcal{A}_\rho \quad \text{iff} \quad \pi = \rho.$$

Thus each rigid structure leads to $n!$ distinct structures on $\{1,\dots,n\}$. This shows (b).

If $\mathcal{A}$ is not rigid and ρ is a nontrivial automorphism of $\mathcal{A}$, then by $(*)$,

$$\mathcal{A}_\pi = \mathcal{A}_{\pi\circ\rho}$$

for any permutation π. Hence, any nonrigid structure leads to at most $\frac{n!}{2}$ distinct structures on $\{1,\dots,n\}$. This proves (c). □

Lemma 4.3.2 *For any class* H, $u_n(\text{RIG}|H) \le l_n(\text{RIG}|H)$. *In particular,*

$$u(\text{RIG}|H) = 1 \;\; \textit{implies} \;\; l(\text{RIG}|H) = 1.$$

Proof.

$$\begin{aligned} u_n(\text{RIG}|H) &= \frac{U_n(\text{RIG}\cap H)\cdot n!}{U_n(\text{RIG}\cap H)\cdot n! + U_n(\text{RIG}^c\cap H)\cdot n!} \\ &\le \frac{L_n(\text{RIG}\cap H)}{L_n(\text{RIG}\cap H) + L_n(\text{RIG}^c\cap H)} \quad \text{(by 4.3.1)} \\ &= l_n(\text{RIG}|H). \end{aligned}$$

□

[2] Recall that RIG^c denotes the complement of RIG, the class of τ-structures not in RIG.

As suggested by (b) and (c) of Lemma 4.3.1, almost all structures in a class H are rigid iff $L_n(H) \approx U_n(H) \cdot n!$. In fact:

Proposition 4.3.3 *Let H be a class of structures. Then*

$$u(\mathrm{RIG}|H) = 1 \qquad \textit{iff} \qquad \lim_{n\to\infty} \frac{L_n(H)}{U_n(H)\cdot n!} = 1.$$

Proof. As

$$\begin{aligned} \frac{L_n(H)}{U_n(H)\cdot n!} &= \frac{L_n(\mathrm{RIG}\cap H)}{U_n(H)\cdot n!} + \frac{L_n(\mathrm{RIG}^c\cap H)}{U_n(H)\cdot n!} \\ &= u_n(\mathrm{RIG}|H) + \frac{L_n(\mathrm{RIG}^c\cap H)}{U_n(H)\cdot n!} \qquad \text{(by 4.3.1(b))}, \end{aligned}$$

we get, using 4.3.1(c),

$$u_n(\mathrm{RIG}|H) \leq \frac{L_n(H)}{U_n(H)\cdot n!} \leq u_n(\mathrm{RIG}|H) + \frac{1}{2}u_n(\mathrm{RIG}^c|H) = 1 - \frac{1}{2}u_n(\mathrm{RIG}^c|H).$$

So, $u(\mathrm{RIG}|H) = 1$ implies $\lim_{n\to\infty} \frac{L_n(H)}{U_n(H)\cdot n!} = 1$, and $\lim_{n\to\infty} \frac{L_n(H)}{U_n(H)\cdot n!} = 1$ implies $u(\mathrm{RIG}^c|H) = 0$, that is, $u(\mathrm{RIG}|H) = 1$. □

The following theorem will be used to extend for parametric classes the 0–1 laws from the labeled to the unlabeled case.

Theorem 4.3.4 *Let H be a class of structures. If almost all structures in H are rigid, i.e., if $u(\mathrm{RIG}|H) = 1$, then for any class K the labeled and the unlabeled asymptotic probabilities with respect to H coincide, that is,*

$$l(K|H) = u(K|H) \quad (= u(K|\mathrm{RIG}\cap H)).^3$$

Proof. By assumption, $u(\mathrm{RIG}|H) = 1$ and therefore, $l(\mathrm{RIG}|H) = 1$ by 4.3.2. Immediately from the definitions we get

$$l(K|H) = l(K|\mathrm{RIG}\cap H) \qquad \text{and} \qquad u(K|H) = u(K|\mathrm{RIG}\cap H).$$

Therefore, the claim follows from

$$\begin{aligned} l_n(K|\mathrm{RIG}\cap H) &= \frac{L_n(K\cap \mathrm{RIG}\cap H)}{L_n(\mathrm{RIG}\cap H)} = \qquad \text{(by 4.3.1(b))} \\ & \frac{U_n(K\cap \mathrm{RIG}\cap H)\cdot n!}{U_n(\mathrm{RIG}\cap H)\cdot n!} = u_n(K|\mathrm{RIG}\cap H). \end{aligned}$$

□

[3] Clearly, $\lim_{n\to\infty} l_n(K|H) = \lim_{n\to\infty} u_n(K|H)$ means that both sides converge to the same number or else that both of them diverge.

Let H_0 be the parametric class consisting of all τ-structures $\mathcal{A}$ such that $R^A = \emptyset$ for all $R \in \tau$. Then $L_n(H_0) = U_n(H_0) = 1$ and $u_n(\mathrm{RIG}|H_0) = 0$ (for $n \geq 2$). On the other hand, almost all structures in a "free" parametric class are rigid. Intuitively speaking, a parametric class H is free if for some $r \geq 2$ there is a real choice when fixing the parts of the relations corresponding to r-tuples of distinct elements. More precisely: If $H = \mathrm{Mod}(\varphi_0)$ with a parametric φ_0, then H is *free*, if for some $m \geq 2$ there is a relation symbol R, say of arity r, and a surjection $i : \{1, \ldots, r\} \to \{1, \ldots, m\}$ such that

$$\varphi_0 \wedge \exists x_1 \ldots \exists x_m (Rx_{i(1)} \ldots x_{i(r)} \wedge \bigwedge_{1 \leq k < l \leq m} \neg x_k = x_l)$$

and

$$\varphi_0 \wedge \exists x_1 \ldots \exists x_m (\neg Rx_{i(1)} \ldots x_{i(r)} \wedge \bigwedge_{1 \leq k < l \leq m} \neg x_k = x_l)$$

are satisfiable.

The class H_0 introduced above is not free. The class of graphs is free, and for any relational τ containing at least one relation symbol of arity ≥ 2, the class of all τ-structures is a free parametric class.

Proposition 4.3.5 *Let H be a nontrivial free parametric class. Then almost all structures in H are rigid, that is, $u(\mathrm{RIG}|H) = 1$.*

For H the class of all structures we prove the proposition in the appendix of this section.

Together with Theorem 4.3.4 we now obtain the following

Corollary 4.3.6 *Let H be a nontrivial free parametric class. Then the labeled and the unlabeled asymptotic probabilities with respect to H coincide.* □

We extend this result to:

Proposition 4.3.7 *Let H be any nontrivial parametric class. Then the labeled and the unlabeled asymptotic probabilities with respect to H coincide.*

Before giving a proof we state a consequence. The class H satisfies the *unlabeled* 0–1 *law* for the set Φ of sentences if for all $\varphi \in \Phi$,

$$u(\varphi|H) = 1 \qquad \text{or} \qquad u(\varphi|H) = 0.$$

Then we have by 4.2.3:

Theorem 4.3.8 *Let H be a nontrivial parametric class. Then H satisfies the unlabeled 0–1 law for $\mathrm{L}^{\omega}_{\infty\omega}$ and hence, for* FO. □

Proof (of 4.3.7). By 4.3.6 we only have to consider the nonfree case. First we prove the claim for vocabularies containing only unary relation symbols. So suppose $\tau = \{R_1, \ldots, R_m\}$ with unary $R_1, \ldots, R_m$, and let φ_0 be a nontrivial parametric sentence. For $\alpha : \{1, \ldots, m\} \to \{0,1\}$ set $R^\alpha x := \varphi_1 \wedge \ldots \wedge \varphi_m$, where $\varphi_i = R_i x$ if $\alpha(i) = 1$, and $\varphi_i = \neg R_i x$ if $\alpha(i) = 0$. Then there are distinct $\alpha_1, \ldots, \alpha_t$ such that

$$\varphi_0 \text{ and } \forall x(R^{\alpha_1}x \vee \ldots \vee R^{\alpha_t}x)$$

are logically equivalent (note that any boolean combination of formulas $R_i x$ can be written as a disjunction of formulas $R^\alpha x$). We assume $t \geq 2$, as the case $t = 1$ is trivial. Since for $i = 1, \ldots, t$ and $k \geq 0$ the sentence $\exists^{>k} x\, R^{\alpha_i} x$ is a consequence of the set $T_{\text{rand}}(\varphi_0)$ of extension axioms compatible with φ_0, we have

(1) $l(\exists^{=k} x\, R^{\alpha_i} x \mid \varphi_0) = 0$.

The isomorphism type of a model $\mathcal{A}$ of φ_0 is determined by $(n_1, \ldots, n_t)$, where $n_i := \|\{a \in A \mid \mathcal{A} \models R^{\alpha_i} x[a]\}\|$. Using induction on n and – in the induction step – on t, it is easy to show that

(2) the number of t-tuples $(n_1, \ldots, n_t)$ such that $n_1 + \ldots + n_t = n$ equals the binomial coefficient $\binom{n+t-1}{t-1}$, a polynomial in n of degree $t-1$.

Hence, $U_n(\varphi_0) = \binom{n+t-1}{t-1}$.

By 2.3.12 and 3.3.25(b), any sentence of $L^\omega_{\infty\omega}[\tau]$ is equivalent to a finite boolean combination of sentences of the form $\exists^{=k} x\, R^\alpha x$ (where $k \geq 0$ and $\alpha : \{1, \ldots, m\} \to \{0,1\}$). Thus it suffices to show that

$$u(\exists^{=k} x\, R^\alpha x \mid \varphi_0) = l(\exists^{=k} x\, R^\alpha x \mid \varphi_0) \in \{0,1\}.$$

If $\alpha \neq \alpha_1, \ldots, \alpha \neq \alpha_t$, then $u(\exists^{=k} x\, R^\alpha x | \varphi_0) = l(\exists^{=k} x\, R^\alpha x | \varphi_0) = 0$ or 1 depending on whether $k > 0$ or $k = 0$. Let $\alpha = \alpha_i$. Then by (2), $U_n(\exists^{=k} x\, R^\alpha x \wedge \varphi_0) = \binom{n-k+t-2}{t-2}$, a polynomial in n of degree $t-2$, and $U_n(\varphi_0) = \binom{n+t-1}{t-1}$, a polynomial in n of degree $t-1$. Hence,

$$u(\exists^{=k} x\, R^\alpha x | \varphi_0) = \lim_{n\to\infty} \frac{U_n(\exists^{=k} x\, R^\alpha x \wedge \varphi_0)}{U_n(\varphi_0)} = 0,$$

and thus by (1), $u(\exists^{=k} x\, R^\alpha x \mid \varphi_0) = l(\exists^{=k} x\, R^\alpha x | \varphi_0) = 0$.

Finally, we turn to arbitrary vocabularies. Let H be a nontrivial parametric class which is not free. By the definition of freeness, if $\mathcal{A} \in H$ then any bijection of the universe that preserves the induced unary relations $\{a \mid R^A a \ldots a\}$ for $R \in \tau$, is an automorphism of $\mathcal{A}$. Hence, the counting arguments in the preceding special case remain valid. □

Exercise 4.3.9 Let $\sigma = \{P\}$ and $\tau = \{P, R\}$ where P, R are binary. Show for the FO$[\sigma]$-sentences $\varphi := \exists x \exists y\, Pxy$ and $\psi :=$ "P is an ordering" $\vee$ $\neg \exists x \exists y\, Pxy$ that $l^\sigma(\varphi|\psi) = 1$, $u^\sigma(\varphi|\psi) = \frac{1}{2}$ and $l^\tau(\varphi|\psi) = u^\tau(\varphi|\psi) = 1$. □

4.3.1 Appendix

In the following we prove (part of) Proposition 4.3.5. Fix a vocabulary τ and $n \geq 1$, and let $\mathcal{A}, \mathcal{B}, \ldots$ range over τ-structures with universe $\{1, \ldots, n\}$ and $\pi, \rho, \ldots$ over permutations of $\{1, \ldots, n\}$. We set

$$\begin{aligned} \mathrm{Aut}(\mathcal{A}) &:= \{\pi \mid \pi : \mathcal{A} \cong \mathcal{A}\} \\ \mathrm{Str}(\pi) &:= \{\mathcal{A} \mid \pi : \mathcal{A} \cong \mathcal{A}\}. \end{aligned}$$

Lemma 4.3.10 $U_n(\tau) \cdot n! = \sum_{\pi} \|\mathrm{Str}(\pi)\|$.

Proof. $U_n(\tau)$ is the number of equivalence classes of the relation $\sim$ where

$$\mathcal{A} \sim \mathcal{B} \quad \text{iff} \quad \mathcal{A} \cong \mathcal{B}.$$

Clearly,

$$\mathcal{A} \cong \mathcal{B} \quad \text{implies} \quad \|\mathrm{Aut}(\mathcal{A})\| = \|\mathrm{Aut}(\mathcal{B})\|.$$

Given $\mathcal{A}$ and π, once more we let $\mathcal{A}_\pi$ be the structure $\mathcal{A}$ with $\pi : \mathcal{A} \cong \mathcal{A}_\pi$. We have already remarked that

$$\mathcal{A}_\pi = \mathcal{A}_\rho \quad \text{iff} \quad \pi^{-1} \circ \rho \in \mathrm{Aut}(\mathcal{A}).$$

Thus, $\|\{\mathcal{B} \mid \mathcal{A} \cong \mathcal{B}\}\|$ is the index of the subgroup $\mathrm{Aut}(\mathcal{A})$ in the group of all permutations of $\{1, \ldots, n\}$; hence,

$$\|\mathrm{Aut}(\mathcal{A})\| = \frac{n!}{\|\{\mathcal{B} \mid \mathcal{A} \cong \mathcal{B}\}\|}.$$

Therefore, for fixed $\mathcal{A}$,

$$\sum_{\mathcal{B}, \mathcal{B} \cong \mathcal{A}} \|\mathrm{Aut}(\mathcal{B})\| = \sum_{\mathcal{B}, \mathcal{B} \cong \mathcal{A}} \|\mathrm{Aut}(\mathcal{A})\| = n!.$$

Taking into account that there are $U_n(\tau)$ many equivalence classes with respect to $\cong$, we have

$$\text{(1)} \qquad \sum_{\mathcal{B}} \|\mathrm{Aut}(\mathcal{B})\| = U_n(\tau) \cdot n!.$$

On the other hand,

$$\text{(2)} \qquad \sum_{\mathcal{B}} \|\mathrm{Aut}(\mathcal{B})\| = \|\{(\pi, \mathcal{B}) \mid \pi \in \mathrm{Aut}(\mathcal{B})\}\| = \sum_{\pi} \|\mathrm{Str}(\pi)\|.$$

(1) and (2) yield the desired equation. □

Let M be finite and $f : M \to M$ be a bijection. Denote by $c(f)$ the number of f-cycles,[4] by $\mathrm{spt}(f)$ the *support* of f,

$$\mathrm{spt}(f) := \{a \in M \mid f(a) \neq a\},$$

and set $s(f) := \|\mathrm{spt}(f)\|$. Since $a \in M \setminus \mathrm{spt}(f)$ gives rise to the f-cycle $\{a\}$ and since the f-cycle of any $a \in \mathrm{spt}(f)$ has at least two elements, we have

$$c(f) \leq \|M\| - s(f) + \tfrac{s(f)}{2} \leq \|M\| - \tfrac{s(f)}{2}. \tag{3}$$

Proposition 4.3.11 *Let τ be a relational vocabulary that contains at least one relation symbol of arity ≥ 2. Then $u(\mathrm{RIG}) = 1$.*

Proof. Obviously, we can assume $\|\tau\| = 1$. For simplicity, we restrict ourselves to $\tau = \{E\}$ with binary E. By 4.3.3 it suffices to show that $\lim_{n\to\infty} \frac{L_n(\tau)}{U_n(\tau)\cdot n!} = 1$, or equivalently, that

$$\lim_{n\to\infty} \frac{U_n(\tau)\cdot n!}{L_n(\tau)} = 1.$$

Clearly,

$$L_n(\tau) = 2^{n^2}. \tag{4}$$

Fix n and remind our convention that $\pi, \rho, \ldots$ denote permutations of $\{1, \ldots, n\}$ and $\mathcal{A}, \mathcal{B}, \ldots$ denote τ-structures with universe $\{1, \ldots, n\}$.

Each π induces a permutation $\tilde{\pi}$ of $\{1, \ldots, n\} \times \{1, \ldots, n\}$,

$$\tilde{\pi}((i,j)) \quad := \quad (\pi(i), \pi(j)).$$

If π is an automorphism of $\mathcal{A}$ and $\{\overline{a}, \tilde{\pi}(\overline{a}), \tilde{\pi}(\tilde{\pi}(\overline{a})), \ldots\}$ a $\tilde{\pi}$-cycle, then

$$E^{\mathcal{A}}\overline{a} \text{ iff } E^{\mathcal{A}}\tilde{\pi}(\overline{a}) \text{ iff } E^{\mathcal{A}}\tilde{\pi}(\tilde{\pi}(\overline{a}))\ldots \tag{5}$$

that is, we have $E^{\mathcal{A}}\overline{b}$ for all elements $\overline{b}$ of the $\tilde{\pi}$-cycle of $\overline{a}$, or for none. Therefore, π is an automorphism for exactly $2^{c(\tilde{\pi})}$ many τ-structures,

$$\|\mathrm{Str}(\pi)\| = 2^{c(\tilde{\pi})}. \tag{6}$$

By the preceding lemma, (4), and (6), we obtain

$$\frac{U_n(\tau)\cdot n!}{L_n(\tau)} = \frac{\sum_\pi \|\mathrm{Str}(\pi)\|}{2^{n^2}} = \sum_\pi 2^{c(\tilde{\pi})-n^2}.$$

We must show that $\sum_\pi 2^{c(\tilde{\pi})-n^2} \to 1$. Since for π the identity on $\{1, \ldots, n\}$ we have $c(\tilde{\pi}) = n^2$, this is equivalent to

[4] For $a \in M$ the set $\{a, f(a), f(f(a)), \ldots\}$ is the f-cycle of a.

$$\sum_{\pi \neq id} 2^{c(\tilde{\pi})-n^2} \to 0.$$

For any π,

$$\operatorname{spt}(\pi) \times \{1, \ldots, n\} \subseteq \operatorname{spt}(\tilde{\pi});$$

hence, by (3), $c(\tilde{\pi}) \leq n^2 - \frac{s(\pi)\cdot n}{2}$ and thus,

$$\sum_{\pi \neq id} 2^{c(\tilde{\pi})-n^2} \leq \sum_{\pi \neq id} 2^{\frac{-s(\pi)\cdot n}{2}}.$$

For $k = 2, \ldots, n$ the number of permutations π with $s(\pi) = k$ is $\leq \binom{n}{k} \cdot k! \leq n^k$. Therefore, for $n > 2 \cdot \log n$, we have

$$\begin{aligned} \textstyle\sum_{\pi \neq id} 2^{\frac{-s(\pi)\cdot n}{2}} \leq \sum_{k=2}^{n} n^k \cdot 2^{-\frac{k\cdot n}{2}} &= \textstyle\sum_{k=2}^{n} 2^{-\frac{1}{2}k\cdot(n - 2\cdot \log n)} \\ &\leq (n-1) \cdot 2^{-(n-2\cdot \log n)} \end{aligned}$$

(for the last inequality note that $k = 2$ gives the largest summand of the third sum). Since $(n-1)\cdot 2^{-(n-2\cdot\log n)} \to 0$ we obtain the desired result. □

4.4 Examples and Consequences

In the present section we give several examples concerning graphs and draw some consequences for fragments of second-order logic. We start with some general remarks.

Fix a nontrivial parametric sentence φ_0. By the results of section 4.2, $T_{\text{rand}}(\varphi_0)$ has a uniquely determined countable model $\mathcal{R}(\varphi_0)$, the random model of φ_0. Some results of the preceding sections are summarized in:

Proposition 4.4.1 *For an* $\mathrm{L}^{\omega}_{\infty\omega}$*-sentence* φ *the following are equivalent:*

(i) $T_{\text{rand}}(\varphi_0) \models \varphi$ (ii) $\mathcal{R}(\varphi_0) \models \varphi$

(iii) $l(\varphi|\varphi_0) = 1$ (iv) $u(\varphi|\varphi_0) = 1$. □

Using the extension axioms in $T_{\text{rand}}(\varphi_0)$ we show:

Proposition 4.4.2 (a) *Let* $\mathcal{B}$ *be a finite model of* φ_0*. Then almost all finite models of* φ_0 *contain a substructure isomorphic to* $\mathcal{B}$*.*

(b) *Let* $\mathcal{B}$ *be a finite model of* φ_0 *and* π_0 *an isomorphism of a substructure* $\mathcal{A}$ *of* $\mathcal{B}$ *into* $\mathcal{R}(\varphi_0)$*. Then* π_0 *can be extended to an isomorphism of* $\mathcal{B}$ *into* $\mathcal{R}(\varphi_0)$*.*

Proof. (a) Let $\mathcal{B}$ consist of s elements and let φ_0^s be the conjunction of φ_0 and the finitely many r-extension axioms of $T_{\text{rand}}(\varphi_0)$ with $r \leq s$. Clearly, any model of φ_0^s contains a substructure isomorphic to $\mathcal{B}$. As $l(\varphi_0^s|\varphi_0) = 1$

the claim follows.

(b) Let $A = \{\overline{a}\}$ and $B = \{\overline{a}, \overline{b}\}$. Then (note that $\varphi^0_{\mathcal{A},\overline{a}} = \varphi^0_{\mathcal{B},\overline{a}}$)

$$\forall \overline{v}(\varphi^0_{\mathcal{A},\overline{a}}(\overline{v}) \to \exists \overline{w} \varphi^0_{\mathcal{B},\overline{a}\overline{b}}(\overline{v}, \overline{w}))$$

is a consequence of the extension axioms in $T_{\mathrm{rand}}(\varphi_0)$ and hence, $\mathcal{R}(\varphi_0)$ is a model of it. Since $\mathcal{A} \models \varphi^0_{\mathcal{A},\overline{a}}[\overline{a}]$ and $\varphi^0_{\mathcal{A},\overline{a}}$ is quantifier-free, we have $\mathcal{R}(\varphi_0) \models \varphi^0_{\mathcal{A},\overline{a}}[\pi_0(\overline{a})]$, and therefore there are $\overline{d}$ in $\mathcal{R}(\varphi_0)$ such that $\mathcal{R}(\varphi_0) \models \varphi^0_{\mathcal{B},\overline{a}\overline{b}}[\pi_0(\overline{a}), \overline{d}]$. So $\overline{a}\overline{b} \mapsto \pi_0(\overline{a})\overline{d}$ is the required isomorphism. □

By a back and forth argument we get:

Proposition 4.4.3 *If $\mathcal{B}$ and $\mathcal{C}$ are isomorphic finite substructures of $\mathcal{R}(\varphi_0)$, then there is an automorphism of $\mathcal{R}(\varphi_0)$ mapping $\mathcal{B}$ onto $\mathcal{C}$.*

Proof. Assume that $\mathcal{R}(\varphi_0)$ has universe $\{a_0, a_1, a_2, \ldots\}$ and that $\pi : \mathcal{B} \cong \mathcal{C}$. Take the finite substructure $\mathcal{B}'_0$ of $\mathcal{R}(\varphi_0)$ with $B'_0 = B \cup \{a_0\}$. By part (b) of the preceding proposition, there is an isomorphism $\pi'_0 : \mathcal{B}'_0 \cong \mathcal{C}'_0$ for a suitable $\mathcal{C}'_0 \subseteq \mathcal{R}(\varphi_0)$ that extends π. Similarly, we get $\pi_0 : \mathcal{B}_0 \cong \mathcal{C}_0$ with $\pi'_0 \subseteq \pi_0$, $\mathcal{B}_0 \subseteq \mathcal{R}(\varphi_0)$, $\mathcal{C}_0 \subseteq \mathcal{R}(\varphi_0)$, and $C_0 = C'_0 \cup \{a_0\}$. Continuing this way we obtain a sequence $\pi_0 \subseteq \pi_1 \subseteq \pi_2 \subseteq \ldots$ with $a_i \in \mathrm{do}(\pi_i) \cap \mathrm{rg}(\pi_i)$. Thus, $\pi := \bigcup_{i \geq 0} \pi_i$ is the desired automorphism of $\mathcal{R}(\varphi_0)$. □

Graphs

Let φ_G be a parametric sentence axiomatizing the class of graphs. $T_{\mathrm{rand}}(\varphi_G)$ is equivalent to $T_{\mathrm{rand},G} := \{\varphi_G, \varphi_{\geq 2}\} \cup$

$$\{\forall \text{ distinct } x_1 \ldots x_n y_1 \ldots y_l \exists z (\bigwedge_{i=1}^{n} E x_i z \wedge \bigwedge_{i=1}^{l} (\neg E y_i z \wedge \neg y_i = z)) \mid n + l \geq 1\}.$$

In fact, the sentences in $T_{\mathrm{rand},G}$ are implied by the extension axioms in $T_{\mathrm{rand}}(\varphi_G)$; hence, any model of $T_{\mathrm{rand}}(\varphi_G)$ is a model of $T_{\mathrm{rand},G}$. On the other hand, a back and forth argument using the axioms in $T_{\mathrm{rand},G}$ shows that any two of its models and hence, any model of $T_{\mathrm{rand},G}$ and any model of $T_{\mathrm{rand}}(\varphi_G)$ are partially isomorphic; therefore, any model of $T_{\mathrm{rand},G}$ is a model of $T_{\mathrm{rand}}(\varphi_G)$.

It is well known that a graph cannot be planar if it contains the subgraph $\mathcal{K}_5$, a clique with 5 elements. Thus, by part (a) of 4.4.2, we get:

Proposition 4.4.4 *Almost all finite graphs are not planar.* □

Moreover, we have:

Proposition 4.4.5 *$\mathcal{R}(\varphi_G)$, the random graph, and almost all finite graphs $\mathcal{G}$ are connected, the diameter $D(\mathcal{G}) := \max\{d(a,b) \mid a, b \in G\}$ being 2.*

Proof. Note that, by the extension axioms, $\mathcal{R}(\varphi_G)$ and almost all finite graphs are models of

$$(*) \qquad \exists x \exists y \neg Exy \wedge \forall x \forall y \exists z (Exz \wedge Eyz).$$

But any graph $\mathcal{G}$ satisfying this sentence is connected with $D(\mathcal{G}) = 2$. □

Even though "connectedness" is not expressible in first-order logic, the first-order sentence $(*)$ of the preceding proof is a property of almost all graphs that implies "connectedness". The situation for "rigidity" is different:

Proposition 4.4.6 (a) *Almost all finite graphs are rigid.*
(b) $\mathcal{R}(\varphi_G)$ *is not rigid.*
(c) *No* $\mathrm{L}^{\omega}_{\infty\omega}$*-definable property of almost all graphs implies rigidity.*

Proof. For part (a) cf. 4.3.5 and 4.3.2. Part (b) is an immediate consequence of 4.4.3 (for $\mathcal{B}$ and $\mathcal{C}$ take two substructures of cardinality one). Part (c) follows from (b) because any $\mathrm{L}^{\omega}_{\infty\omega}$-definable property of almost all graphs holds in $\mathcal{R}(\varphi_G)$ (cf. 4.4.1). □

Fragments of Second-Order Logic

Connectedness of graphs can be expressed by a Π^1_1-sentence, for example by

$$\varphi_{\mathrm{CONN}} := \forall X \big(\forall x X x \vee \forall x \neg X x \vee \exists x \exists y\, (Xx \wedge \neg Xy \wedge Exy)\big).$$

Nonrigidity is expressible by

$$\exists X \forall x \forall y \forall u \forall v \exists z_1 \exists z_2 \exists w\, (X z_1 x \wedge X x z_2 \wedge \neg X w w \wedge \\ ((Xxy \wedge Xuv) \to ((x = u \leftrightarrow y = v) \wedge (Exu \leftrightarrow Eyv)))),$$

a so-called $\Sigma^1_1(\forall^*\exists^*)$-sentence , that is, a sentence of the form

$$\exists X_1 \ldots \exists X_s \forall y_1 \ldots \forall y_m \exists z_1 \ldots \exists z_l \chi$$

with $s, m, l \in \mathbb{N}$ and χ quantifier-free. Similarly, $\Sigma^1_1(\exists^*\forall^*)$-sentences have the form

$$\exists X_1 \ldots \exists X_s \exists y_1 \ldots \exists y_m \forall z_1 \ldots \forall z_l \chi$$

where χ is quantifier-free.

Part (a) of the following proposition generalizes the fact mentioned above that connectedness is implied by a first-order property of almost all graphs. Part (b) shows that nonrigidity cannot be expressed by a $\Sigma^1_1(\exists^*\forall^*)$-sentence (otherwise almost all finite graphs would be nonrigid).

Proposition 4.4.7 *Suppose that* φ_0 *is nontrivial parametric.*

(a) *Let* φ *be a* Π^1_1*-sentence. If* $\mathcal{R}(\varphi_0) \models \varphi$ *then there is a first-order sentence* ψ *such that*

$$u(\psi | \varphi_0) = 1 \qquad \textit{and} \qquad \models \psi \to \varphi.$$

(b) *Let φ be a $\Sigma_1^1(\exists^*\forall^*)$-sentence. If $\mathcal{R}(\varphi_0) \models \varphi$ then there is a first-order sentence ψ such that*

$$u(\psi|\varphi_0) = 1 \qquad \textit{and} \qquad \models_{\text{fin}} \psi \to \varphi.$$

Proof. (a) Assume that $\mathcal{R}(\varphi_0) \models \varphi$ holds for the Π_1^1-sentence

$$\varphi = \forall X_1 \dots \forall X_s \chi,$$

where χ contains no second-order quantifiers. Then the set $T_{\text{rand}}(\varphi_0) \cup \{\neg\chi\}$ of $\tau \cup \{X_1, \dots, X_s\}$-sentences has no model (otherwise, by the Löwenheim-Skolem theorem, it would have a countable model whose τ-reduct would be (isomorphic to) the unique countable model $\mathcal{R}(\varphi_0)$ of $T_{\text{rand}}(\varphi_0)$; but then $\mathcal{R}(\varphi_0) \models \exists X_1 \dots \exists X_s \neg\chi$, contrary to $\mathcal{R}(\varphi_0) \models \forall X_1 \dots \forall X_s \chi$). By compactness, there is a finite subset T_0 of $T_{\text{rand}}(\varphi_0)$ such that $T_0 \cup \{\neg\chi\}$ is not satisfiable. Let ψ be the conjunction of the sentences in T_0. Then $u(\psi|\varphi_0) = 1$ and $\models \psi \to \chi$, hence $\models \psi \to \forall X_1 \dots \forall X_s \chi$.

(b) Suppose that for the $\Sigma_1^1(\exists^*\forall^*)$-sentence

$$\varphi = \exists X_1 \dots \exists X_s \exists \overline{x} \forall \overline{y} \chi$$

with quantifier-free χ we have $\mathcal{R}(\varphi_0) \models \varphi$, say, $(\mathcal{R}(\varphi_0), X_1, \dots, X_s) \models \exists \overline{x} \forall \overline{y} \chi$. Choose $\overline{a}$ in $\mathcal{R}(\varphi_0)$ such that

$$(*) \qquad (\mathcal{R}(\varphi_0), X_1, \dots, X_s) \models \forall \overline{y} \chi[\overline{a}]$$

and denote by $\mathcal{A}_0$ the submodel of $\mathcal{R}(\varphi_0)$ with universe $\{\overline{a}\}$. Since $\exists \overline{x} \varphi^0_{\overline{a}}(\overline{x})$ holds in $\mathcal{R}(\varphi_0)$, there is a ψ that is the conjunction of φ_0 and of finitely many extension axioms compatible with φ_0 such that $\models \psi \to \exists \overline{x} \varphi^0_{\overline{a}}(\overline{x})$. Obviously, $u(\psi|\varphi_0) = 1$. We show that $\models_{\text{fin}} \psi \to \varphi$.

So let $\mathcal{B}$ be a finite model of ψ. Choose $\overline{d}$ in B such that $\mathcal{B} \models \varphi^0_{\overline{a}}[\overline{d}]$. Then $\overline{d} \mapsto \overline{a}$ is an isomorphism from the substructure of $\mathcal{B}$ with universe $\{\overline{d}\}$ into $\mathcal{R}(\varphi_0)$. By 4.4.2(b), there is an extension π of $\overline{d} \mapsto \overline{a}$ that is an isomorphism of $\mathcal{B}$ onto a substructure $\mathcal{B}'$ of $\mathcal{R}(\varphi_0)$. It suffices to show that $\mathcal{B}'$ is a model of φ. By $(*)$, as $\forall \overline{y} \chi$ is universal,

$$(\mathcal{B}', X_1 \cap B', \dots, X_s \cap B') \models \forall \overline{y} \chi[\overline{a}],$$

hence $(\mathcal{B}', X_1 \cap B', \dots, X_s \cap B') \models \exists \overline{x} \forall \overline{y} \chi$, thus $\mathcal{B}' \models \exists X_1 \dots \exists X_s \exists \overline{x} \forall \overline{y} \chi$. □

As a consequence we get:

Theorem 4.4.8 *Let φ_0 be nontrivial parametric and let φ be a $\Sigma_1^1(\exists^*\forall^*)$-sentence.*

(a) *If $\mathcal{R}(\varphi_0) \models \varphi$ then $u(\varphi|\varphi_0) = 1$.*
(b) *If $\mathcal{R}(\varphi_0) \not\models \varphi$ then $u(\varphi|\varphi_0) = 0$.*

Proof. (a) If $\mathcal{R}(\varphi_0) \models \varphi$ then, by part (b) of the proposition, there is a first-order sentence ψ such that $u(\psi|\varphi_0) = 1$ and $\models_{\text{fin}} \psi \to \varphi$. In particular, $u(\varphi|\varphi_0) = 1$.

For (b), assume that $\mathcal{R}(\varphi_0) \models \neg\varphi$. Since $\neg\varphi$ is (logically equivalent to) a Π_1^1-sentence there is, by part (a) of the proposition, a first-order sentence ψ such that $u(\psi|\varphi_0) = 1$ and $\models \psi \to \neg\varphi$. Therefore, $u(\neg\varphi|\varphi_0) = 1$ and hence, $u(\varphi|\varphi_0) = 0$. □

By 4.3.7, the preceding results remain valid for the labeled probability as well. Hence, we have:

Corollary 4.4.9 $\Sigma_1^1(\exists^*\forall^*)$ *satisfies the labeled and the unlabeled* 0–1 *law with respect to nontrivial parametric classes.* □

The satisfiability problem for $\exists^*\forall^*$-sentences is decidable, and we have just seen that $\Sigma_1^1(\exists^*\forall^*)$ has the 0–1 law. This is a special case of a general phenomenon: It has turned out that the satisfiability problem for a prefix class Φ of first-order logic is decidable just in case the 0–1 law holds for $\Sigma_1^1\Phi$ $(:= \{\exists\overline{R}\varphi \mid \varphi \in \Phi\})$, cf. [126].

Orderings

In the preceding exposition we dealt with parametric classes; the following examples show, among other things, that the class of orderings is not parametric.

Examples 4.4.10 (a) Let $\tau = \{<, P\}$ with unary P and let $\mathcal{O} = \mathcal{O}[\tau]$ be the class of τ-structures ordered by $<$. Denote by φ a first-order sentence expressing that the first element of the ordering is in P, say, $\varphi = \exists x(Px \land \forall y\, \neg y < x)$. Then

$$u(\varphi|\mathcal{O}) = l(\varphi|\mathcal{O}) = \frac{1}{2}$$

(argue as in 4.1.1(a)). By 4.2.3, $\mathcal{O}$ is not parametric.

(b) For $\tau = \{<\}$ consider the class ORD of orderings. By 2.3.6, for any first-order sentence φ we have $u(\varphi|\text{ORD}) = l(\varphi|\text{ORD}) = 1$ iff $(\{0, \ldots, 2^k\}, <) \models \varphi$, where k is the quantifier rank of φ. Therefore, ORD satisfies the labeled and the unlabeled 0–1 law for FO. On the other hand, ORD does not satisfy the (labeled or unlabeled) 0–1 law for $\text{L}^2_{\infty\omega}$. In fact, the probabilities $l(\varphi|\text{ORD})$ and $u(\varphi|\text{ORD})$ do not exist for any $\text{L}^2_{\infty\omega}$-sentence φ expressing that the ordering has an even number of elements. (As φ one can take $\bigvee_{n\geq 1} \chi_{2n}$, where χ_{2n} are the FO^2-formulas introduced in Example 3.3.1(a).) □

4.5 Probabilities of Monadic Second Order Properties

Already in 4.1.1(c) we saw that, in general, the (labeled or unlabeled) asymptotic probabilities do not always exist for second-order sentences. In the present section we show that this is true even for monadic Σ_1^1-sentences, that is, for sentences of the form

$$\exists X_1 \dots \exists X_m \psi$$

where $X_1, \dots, X_m$ are unary and ψ is first-order. Moreover, we shall show the corresponding result for nontrivial free parametric classes.

Theorem 4.5.1 *Let φ_0 be a nontrivial free parametric sentence in a relational vocabulary τ_0. Then there is a monadic Σ_1^1-sentence φ of vocabulary τ_0 such that the labeled asymptotic probability $l(\varphi|\varphi_0)$ does not exist.*

Since the class of all finite τ-structures is a nontrivial parametric class in case τ contains an at least binary relation symbol, we obtain

Corollary 4.5.2 *Let τ be a vocabulary that contains an at least binary relation symbol. Then there is a monadic Σ_1^1-sentence of vocabulary τ without labeled asymptotic probability.* □

Proof (of 4.5.1). Let φ_0 be a nontrivial free parametric sentence and H its class of finite models. Then the following lemma 4.5.3 tells us that there is a first-order formula $\kappa(x, y, X, Y, Z, U)$ with unary relation variables X, Y, Z, U such that

$$\exists X \exists Y \exists Z \exists U (\text{“}\kappa(\_, \_, X, Y, Z, U) \text{ is an ordering of the universe”})$$

almost surely holds on H. Therefore, with a new unary relation variable V, the sentence

> $\exists X \exists Y \exists Z \exists U \exists V$(“$\kappa(\_, \_, X, Y, Z, U)$ is an ordering of the universe whose first element belongs to V and whose last element does not belong to V” $\wedge \forall x \forall y$ (“if y is the κ-successor of x then $(Vx \leftrightarrow \neg Vy)$”)

almost surely holds for the structures in H of even cardinality and almost surely gets wrong for the structures in H of odd cardinality. So it has no labeled asymptotic probability (recall that φ_0 is nontrivial and thus, has models in all cardinalities). □

Lemma 4.5.3 *There is a first-order formula $\kappa(x, y, X, Y, Z, U)$ with unary relation variables X, Y, Z, U such that with*

$$\varphi \;:=\; \exists X \exists Y \exists Z \exists U (\text{“}\kappa(\_, \_, X, Y, Z, U) \textit{ is an ordering of the universe”})$$

the labeled asymptotic probability $l(\varphi|\varphi_0)$ equals 1.

Proof. As φ_0 is free, there is a relation symbol R in τ_0 of arity $r \geq 2$, an $m \geq 2$, and a surjection i: $\{1, \dots, r\} \to \{1, \dots, m\}$ such that both

$$\varphi_0 \wedge \exists x_1 \dots \exists x_m (Rx_{i(1)} \dots x_{i(r)} \wedge \bigwedge_{1 \leq k < l \leq m} x_k \neq x_l)$$

and

$$\varphi_0 \wedge \exists x_1 \dots \exists x_m (\neg R x_{i(1)} \dots x_{i(r)} \wedge \bigwedge_{1 \le k < l \le m} x_k \neq x_l)$$

are satisfiable. We set

$$\psi(x_1, \dots, x_m) \quad := \quad R x_{i(1)} \dots x_{i(r)} \wedge \bigwedge_{1 \le k < l \le m} x_k \neq x_l$$

and

$$\psi^{\neg}(x_1, \dots, x_m) \quad := \quad \neg R x_{i(1)} \dots x_{i(r)} \wedge \bigwedge_{1 \le k < l \le m} x_k \neq x_l.$$

Then, among the 0-isomorphism types of m distinct elements satisfiable in some model of φ_0, there are types implying $\psi(x_1, \dots, x_m)$ and types implying $\psi^{\neg}(x_1, \dots, x_m)$. Hence, the fraction p of types implying $\psi(x_1, \dots, x_m)$ satisfies $0 < p < 1$. Interchanging ψ and $\psi^{\neg}$, if necessary, we may assume that

$$0 < p \le \frac{1}{2}.$$

Intuitively, if we take distinct elements $a_1, \dots, a_m$ in a model $\mathcal{A}$ of φ_0, the probability for $\psi[a_1, \dots, a_m]$ to hold in $\mathcal{A}$ is p.

We first consider the case $m = 2$ and write $\psi(x, y)$ for $\psi(x_1, x_2)$. The idea behind our argumentation is as follows: We show that with high probability models of φ_0 contain a subset of logarithmic size ordered by $\psi(x, y)$. The ordering can suitably be extended to an ordering of the whole universe; this will be done in two steps.

We come to the details. Given n, we set

$$\begin{aligned} c &:= -\log_p n, \\ r &:= \lfloor \sqrt{\tfrac{1}{2} c} \rfloor, \text{ the integer part of } \sqrt{\tfrac{1}{2} c}, \\ s &:= \lfloor 3c \rfloor, \\ t &:= \lfloor \frac{n}{2\sqrt{c}} \rfloor - 1, \end{aligned}$$

$$X := \{1, \dots, r\} \quad \text{and} \quad Z := \{n - s + 1, \dots, n\}.$$

Note that for sufficiently large n, the sets X and Z are disjoint.

In the following we equip the set of models of φ_0 over $\{1, \dots, n\}$ with the uniform probability distribution.

Claim 1. Almost surely in models of φ_0 over $\{1, \dots, n\}$ there is a set Y of power r disjoint from $X \cup Z$ such that

$$\psi^Y_<(x, y) \quad := \quad Yx \wedge Yy \wedge \exists z (Xz \wedge \psi(z, x) \wedge \neg \psi(z, y))$$

defines an ordering on Y.

Claim 2. Almost surely in models of φ_0 over $\{1,\dots n\}$, for any subset Y of $\{1,\dots,n\}$ of cardinality r which is disjoint from $X \cup Z$, we have

(1) $\forall x \forall y (Zx \wedge Zy \wedge x \neq y \rightarrow \exists z (Yz \wedge \neg(\psi(z,x) \leftrightarrow \psi(z,y))))$;
(2) with $U := (Y \cup Z)^c$:

$$\forall x \forall y ((Ux \wedge Uy \wedge x \neq y) \rightarrow \exists z (Zz \wedge \neg(\psi(z,x) \leftrightarrow \psi(z,y)))).$$

Then we are done: If $\psi^Y_<(x,y)$ defines an ordering on Y and (1), (2) of Claim 2 are valid, then

$$\psi^Z_<(x,y) := x \neq y \wedge Zx \wedge Zy \wedge \text{ “for the } \psi^Y_<\text{-smallest element } z \text{ of } Y \text{ with } \neg(\psi(z,x) \leftrightarrow \psi(z,y)) \text{ we have } \psi(z,x)\text{”}$$

defines an ordering on Z and

$$\psi^U_<(x,y) := x \neq y \wedge Ux \wedge Uy \wedge \text{ “for the } \psi^Z_<\text{-smallest element } z \text{ of } Z \text{ with } \neg(\psi(z,x) \leftrightarrow \psi(z,y)) \text{ we have } \psi(z,x)\text{”}$$

defines an ordering on U.

Therefore, with

$$\begin{aligned}\kappa(x,y,X,Y,Z,U) := {} & \text{“}X,Y,Z \text{ are disjoint and } U = (Y \cup Z)^c\text{”} \wedge \\ & (\psi^Y_<(x,y) \vee \psi^Z_<(x,y) \vee \psi^U_<(x,y) \\ & \vee (Yx \wedge \neg Yy) \vee (Zx \wedge Uy)),\end{aligned}$$

claims 1 and 2 yield that

$$l(\exists X \exists Y \exists Z \exists U \text{“}\kappa(\_,\_,X,Y,Z,U) \text{ defines a linear ordering”} \mid \varphi_0) = 1.$$

So, we have φ as claimed in the statement of the lemma.

We give the proofs of Claim 1 and Claim 2. To prove Claim 1, for $i \in \{1,\dots,t\}$ we set

$$Y(i) := \{r \cdot i + 1, \dots, r \cdot i + r\}.$$

Note that the $Y(i)$ are pairwise disjoint and disjoint from both X and Z (as $r \cdot t + r < \frac{n}{2} < n - s + 1$). We show that almost surely in models of φ_0 over $\{1,\dots,n\}$ there is an $i \in \{1,\dots,t\}$ such that

(*) for all $a, b \in X$, $\psi(a, r \cdot i + b)$ holds iff $a \leq b$.

Then almost surely there is a Y, namely $Y = Y(i)$ for an i satisfying (*), such that for all $d_1, d_2 \in Y$

> there is an $a \in X$ such that $\psi(a, d_1)$ holds and $\neg\psi(a, d_2)$ holds iff $d_2 < d_1$,

that is, $\psi^Y_<(x, y)$ defines an ordering on Y.

For $i \in \{1, \ldots, n\}$ let q be the probability that i satisfies $(*)$ (where we refer to the uniform probability distribution on the set of models of φ_0 over $\{1, \ldots, n\}$). Then

$$q \geq p^{r^2} \geq p^{\frac{1}{2}c} = \frac{1}{\sqrt{n}}.$$

Hence, the probability that no $i \in \{1, \ldots, t\}$ satisfies $(*)$ is

$$\begin{aligned} &\leq \ (1 - \frac{1}{\sqrt{n}})^t \ \leq \ (1 - \frac{1}{\sqrt{n}})^{\frac{n}{\sqrt{2c}} - 2} \\ &\leq \ ((1 - \frac{1}{\sqrt{n}})^{\sqrt{n}})^{\frac{\sqrt{n}}{2\sqrt{c}} - \frac{2}{\sqrt{n}}} \longrightarrow_{n \to \infty} 0. \end{aligned}$$

We now come to a proof of Claim 2. Given Y, the probability of the failure of (1) is

$$\leq \binom{s}{2} \cdot (p^2 + (1-p)^2)^r \leq 5 \cdot c^2 \cdot (p^2 + (1-p)^2)^{\sqrt{\frac{1}{2}c} - 1} \longrightarrow_{n \to \infty} 0,$$

and the probability for the failure of (2) is

$$\begin{aligned} &\leq \ \binom{n}{2} \cdot (p^2 + (1-p)^2)^s \\ &\leq \ n^2 \cdot (p^2 + (1-p)^2)^{\log_p n^{-3} - 1} \\ &\leq \ n^2 \cdot (p^2 + (1-p)^2)^{\log_{p^2 + (1-p)^2} n^{-3} - 1} \quad \text{(note that } p^2 + (1-p)^2 \geq p\text{)} \\ &\leq \ \frac{1}{n \cdot (p^2 + (1-p)^2)} \longrightarrow_{n \to \infty} 0. \end{aligned}$$

Altogether the case $m = 2$ is settled.

We finally sketch a proof for the case $m > 2$. The model construction method for parametric classes given in section 4.2 shows that for distinct elements $a_1, \ldots, a_m$ the probability for $\psi(a_1, \ldots, a_m)$ to hold in a model over $\{1, \ldots, n\}$ remains the same if we fix $a_3, \ldots, a_m$. Therefore, the preceding proof for the case $m = 2$ remains valid with the following changes:

(i) We consider structures over $\{1, \ldots, n, n+1, \ldots, n+m-2\}$.
(ii) We fix x_3 by $n+1, \ldots, x_m$ by $n+m-2$.
(iii) We replace $U = (Y \cup Z)^c$ by the complement of $Y \cup Z$ within $\{1, \ldots, n\}$.
(iv) We put $n+1, \ldots, n+m-2$ at the end of the ordering defined on $\{1, \ldots, n\}$ for the case $m = 2$.

This leads to the following changes in the definition of φ:

(i) We add "$\exists x_3 \ldots \exists x_m$" to the prefix.
(ii) We add the conjunct "$\bigwedge_{3 \leq k < l \leq m} x_k \neq x_l$" to the kernel.
(iii) We define $\kappa(x, y, x_3, \ldots, x_m, X, Y, Z, U)$ as in case $m = 2$ with $x_3, \ldots, x_m$ as parameters stemming from $\psi(x, y, x_3, \ldots, x_m)$ and then replace it by the disjunction of the following formulas

$$\bigwedge_{3 \leq k \leq m} x \neq x_k \wedge \bigvee_{3 \leq k \leq m} y = x_k,$$

$$\bigvee_{3 \leq k < m} \bigvee_{k < j \leq m} (x = x_k \wedge y = y_j),$$

$$\bigwedge_{3 \leq k \leq m} (x \neq x_k \wedge y \neq x_k) \wedge \kappa(x, y, x_3, \ldots, x_m, X', Y', Z', U')$$

where, e.g., X' is $X \setminus \{x_3, \ldots, x_k\}$.

□

Exercise 4.5.4 Let τ contain the relation symbol $<$ and at least one further relation symbol of arity ≥ 2. Let $\mathcal{O}[\tau]$ be the class of ordered τ-structures. Then there is a first-order φ such that $l(\varphi|\mathcal{O}[\tau])$ does not exist. (Hint: Argue as in the preceding proof, but now let intervals take over the role of the subsets X, Y, Z and U.) □

Notes 4.5.5 The 0–1 law for first-order logic was independently proven by Glebskij, Logan, Liogonkij, Talanov [47] and Fagin [37], its extension to parametric classes is due to Oberschelp [123], and its extension to $L^{\omega}_{\infty\omega}$ to Kolaitis and Vardi [109]. The papers [107, 108] contain, among other things, the results on fragments of second-order logic presented in section 4.4. The last section is based on [98] (cf. also [99, 141]) and the Appendix 4.3.1 on [38]. Survey articles on 0–1 laws are [21, 72, 111, 146].

5. Satisfiability in the Finite

A classical question having its origin in the decision problem for first-order logic, asks for specific classes Φ of sentences whether they have the finite model property, i.e., whether every satisfiable sentence of Φ has a finite model. We are going to consider two examples where we can demonstrate the methodological usefulness of techniques and results we have developed so far.

A first-order sentence φ expressing that $<$ is a partial ordering without maximal elements is satisfiable but has no finite model. As φ we can take (1) or (2):

(1) $\forall x \neg x < x \wedge \forall x \forall y \forall z((x < y \wedge y < z) \rightarrow x < z) \wedge \forall x \exists y\, x < y$

(2) $\forall x \forall y \forall z \exists u(\neg x < x \wedge ((x < y \wedge y < z) \rightarrow x < z) \wedge x < u)$.

The sentence in (1) uses only three variables, the sentence in (2) is a $\forall^3\exists$-sentence. These sentences are "best" possible ones: We show in section 5.1 that every satisfiable sentence with at most two variables has a finite model and in section 5.2 that the same holds for every satisfiable $\forall^2\exists^*$-sentence without the equality sign.[1]

5.1 Finite Model Property of FO^2

We prove that every satisfiable sentence with at most two variables in a relational vocabulary has a finite model. As a consequence, every sentence of modal logic has a finite model (cf. 5.1.8). We remark that one can remove the restriction on constants (cf. 5.1.7), but the result is not valid for "vocabularies" with function symbols: Consider the sentence with two variables

$$\forall x \forall y (f(x) = f(y) \rightarrow x = y) \wedge \exists y \forall x \neg f(x) = y$$

expressing that f is injective but not surjective.

Fix a relational vocabulary τ and let $x := v_1$ and $y := v_2$. For the purpose of this section we say that a first-order formula (possibly containing second-order variables) is *normal*, if it has the form

[1] Furthermore, any satisfiable universal sentence has a finite model.

$$\forall x \forall y \psi \wedge \bigwedge_{i=1}^{r} \forall x \exists y \psi_i$$

where $\psi, \psi_i \in \mathrm{FO}^2$ are quantifier-free. We shall use the next lemma to restrict ourselves to normal sentences when proving the finite satisfiability of satisfiable FO^2-sentences.

Lemma 5.1.1 *Every sentence $\exists \overline{X}(\varphi \wedge \forall x \forall y \psi)$, where φ is normal and $\psi \in \mathrm{FO}^2$, is equivalent to a sentence of the form $\exists \overline{Y} \chi$, where χ is normal.*

Proof. We proceed by induction on the number of quantifiers in ψ. If ψ contains no quantifiers, the result is immediate. In the induction step we show how to eliminate a quantifier in ψ. So let, say, $\forall x \psi_0$ be a subformula of ψ with quantifier-free ψ_0. Then, ψ is logically equivalent to

$$\exists X(\forall y(Xy \leftrightarrow \forall x \psi_0) \wedge \forall x \forall y\, \psi')$$

where ψ' results from ψ by replacing $\forall x \psi_0$ by Xy, and hence, it is logically equivalent to

$$\exists X(\forall x \forall y(Xy \rightarrow \psi_0) \wedge \forall y \exists x(\neg\psi_0 \vee Xy) \wedge \forall x \forall y \psi')$$

and thus to

$$\exists X(\forall x \forall y(Xy \rightarrow \psi_0) \wedge \forall x \exists y(\neg\psi_0 \binom{yx}{xy} \vee Xx) \wedge \forall x \forall y \psi')$$

($\psi_0 \binom{yx}{xy}$ is obtained from ψ_0 by simultaneously replacing *all* occurrences of x and y by y and x, respectively).

Altogether, $\exists \overline{X}(\varphi \wedge \forall x \forall y \psi)$ is equivalent to

$$\exists \overline{X} \exists X(\varphi \wedge \forall x \forall y(Xy \rightarrow \psi_0) \wedge \forall x \exists y(\neg\psi_0 \binom{yx}{xy} \vee Xx) \wedge \forall x \forall y \psi')$$

where the first conjunct is normal and ψ' has less quantifiers than ψ. By induction hypothesis, we obtain our claim. □

Corollary 5.1.2 *Every sentence of* FO^2 *is equivalent to a sentence of the form $\exists \overline{Y} \chi$, where χ is normal.*

Proof. Given an FO^2-sentence ψ, apply the preceding lemma to $\exists \overline{X}(\varphi \wedge \forall x \forall y \psi)$, where $\overline{X}$ is the empty sequence and $\varphi := \forall x \forall y x = x$. □

We now formulate the main result of this section.

Theorem 5.1.3 *Every satisfiable first-order sentence with at most two variables in a relational vocabulary has a finite model.*

Proof. Let φ be such a sentence. We apply the corollary above. Since $\exists \overline{Y}\chi$ and χ have models over the same universes, we may assume that φ has the form

$$\varphi = \forall x \forall y \psi \wedge \bigwedge_{i=1}^{r} \forall x \exists y \psi_i$$

where $\psi, \psi_i \in \mathrm{FO}^2$ are quantifier-free. Moreover, we may suppose that

$$\text{for } i = 1, \ldots, r, \quad \psi_i \models x \neq y$$

since over structures with at least two elements $\forall x \exists y \psi_i(x, y)$ is equivalent to

$$\forall x \exists y (x \neq y \wedge (\psi_i(x, y) \vee \psi_i(x, x))).$$

Let $\mathcal{A}$ be a model of φ. An element $a \in A$ is a *king* (in $\mathcal{A}$) if there is no other element b of A with the same 0-isomorphism type, i.e., $\varphi^0_{\mathcal{A},a} = \varphi^0_{\mathcal{A},b}$. If $\mathcal{A} \models \psi_i[a, b]$, we call b a *child* of a (in $\mathcal{A}$), more exactly, an *i-child*. For $a \in A$ and $i \in \{1, \ldots, r\}$ we let a^i be a fixed i-child of a. Then, $a \neq a^i$. We set

$$C := \bigcup_{a \in A,\, a \text{ king}} \{a, a^1, \ldots, a^r\}.$$

Clearly, C is finite. We show that there is a $\mathcal{B}$ such that

(i) $B = C \cup (\{\varphi^0_{\mathcal{A},a} \mid a \in A \text{ no king }\} \times \{1, \ldots, r\} \times \{0, 1, 2\})$.
(ii) Each 0-isomorphism type of a pair of elements of B is realized in $\mathcal{A}$.
(iii) For $i = 1, \ldots, r$, all elements of B have an i-child in $\mathcal{B}$.

Then $\mathcal{B}$ is a model of $\forall x \forall y\, \psi$ by (ii), and of $\bigwedge_{i=1}^{r} \forall x \exists y \psi_i$ by (iii); thus by (i), $\mathcal{B}$ is a finite model of φ.

To define $\mathcal{B}$, we fix the 0-isomorphism type of all pairs of elements of B in a suitable way to ensure (ii) and (iii). The rest – in case τ contains relation symbols of arity ≥ 3 – can be fixed in an arbitrary way.

Step 1. For $a, b \in C, a \neq b$, we set

$$\varphi^0_{\mathcal{B},a,b} \quad := \quad \varphi^0_{\mathcal{A},a,b}.$$

Step 2. Let $b \in B$. We aim at providing children for b in $\mathcal{B}$. So let $i \in \{i, \ldots, r\}$.

- If $b \in C$ and b is a king or b has an i-child in $\mathcal{A}$ that lies in C, b has an i-child in $\mathcal{B}$ because of Step 1.
- If $b = a^j$ for a king a, but b has no i-child in C, we let $b' := (\varphi^0_{\mathcal{A},(a^j)^i}, i, 0)$ be an i-child of b in $\mathcal{B}$ by setting

$$\varphi^0_{\mathcal{B},b,b'} \quad := \quad \varphi^0_{\mathcal{A},a^j,(a^j)^i}.$$

(In case there are several possibilities for a and j, we fix one choice; and we also do so in similar situations.)

– If $b = (\varphi^0_{\mathcal{A},a}, j, k)$ (and, hence, $a \in A$ is not a king in $\mathcal{A}$) and a^i is a king in $\mathcal{A}$, we let a^i be an i-child of b in $\mathcal{B}$ by setting

$$\varphi^0_{\mathcal{B},b,a^i} \quad := \quad \varphi^0_{\mathcal{A},a,a^i}.$$

– If $b = (\varphi^0_{\mathcal{A},a}, j, k)$ and if a^i is not a king in $\mathcal{A}$, we let $b' := (\varphi^0_{\mathcal{A},a^i}, i, (k+1) \pmod 3))$ be an i-child of b in $\mathcal{B}$ by setting

$$\varphi^0_{\mathcal{B},b,b'} \quad := \quad \varphi^0_{\mathcal{A},a,a^i}.$$

In all of these cases, by fixing a type $\varphi^0_{\mathcal{B},a,b}$, we of course at the same time also fix $\varphi^0_{\mathcal{B},b,a}$.

Step 3. If, e.g., for $d \in C$, $b := (\varphi^0_{\mathcal{A},a}, j, k)$, and $b' := (\varphi^0_{\mathcal{A},a'}, j', k')$ the 0-isomorphism type of (d, b) or of (b, b') has not been fixed in the first two steps, we set

$$\varphi^0_{\mathcal{B},d,b} := \varphi^0_{\mathcal{A},d,a} \quad \text{or} \quad \varphi^0_{\mathcal{B},b,b'} := \varphi^0_{\mathcal{A},a,a'},$$

respectively. Note that the definitions we have made do not contradict each other, as for $c \in C$ we have $\varphi^0_{\mathcal{B},c} = \varphi^0_{\mathcal{A},c}$ and for $b = (\varphi^0_{\mathcal{A},a}, j, k)$ we have $\varphi^0_{\mathcal{B},b} = \varphi^0_{\mathcal{A},a}$. By construction, (ii) and (iii) are satisfied. □

Corollary 5.1.4 *For any relational vocabulary τ, the set Φ of logically valid first-order sentences with at most two variables is decidable.*

Proof. By the completeness theorem for first-order logic the set Φ is enumerable. For its "complement"

$$\Phi^{\text{nv}} \quad := \quad \{\varphi \ \mathrm{FO}^2[\tau]\text{-sentence} \mid \varphi \text{ is not logically valid}\}$$

we have by the preceding theorem that

$$\Phi^{\text{nv}} \quad = \quad \{\varphi \ \mathrm{FO}^2[\tau]\text{-sentence} \mid \neg\varphi \text{ has a finite model}\}.$$

Therefore, Φ^{nv} is enumerable too, and hence, Φ is decidable. □

We remarked after Corollary 4.4.9 that the classical prefix classes with a decidable satisfaction problem correspond to fragments of Σ^1_1 with a 0–1 law. Even though we just have shown that FO^2 has a decidable satisfaction problem, the 0–1 law does not hold for the class $\Sigma^1_1\text{-}\mathrm{FO}^2 := \{\exists \overline{X}\varphi \mid \varphi \in \mathrm{FO}^2\}$, cf. [114].

An analysis of the proof of the main theorem leads to the statement of the following exercise.

Exercise 5.1.5 Show that to any first-order sentence φ in at most two variables one can effectively assign a natural number m such that:

- φ has a model if and only if φ has a model of power $\leq m$.
- If φ has a model of power $\geq m$ then φ has a model in each cardinality $\geq m$. □

Exercise 5.1.6 Let $\tau = \{<\}$ with $<$ binary. Show that for all $n \geq 1$ there is a satisfiable $\mathrm{FO}^2[\tau]$-sentence having only models of cardinality $\geq n$. (Hint: Use the formulas introduced in part (a) of 3.3.1, but note that the class of orderings is not axiomatizable in FO^2.) □

Exercise 5.1.7 Show that any satisfiable sentence with at most two variables in a vocabulary possibly containing constants has a finite model. (Hint: Let φ be such a sentence and denote by C the set of constants. For any n-ary relation symbol P and any map $\pi : \{1,\ldots,n\} \to C \cup \{x,y\}$ introduce a new 2-ary relation symbol P^π. Apply the theorem to the sentence obtained from φ by replacing any subformula $P\pi(1)\ldots\pi(n)$ by $P^\pi xy$ and treating equations and quantifiers appropriately.) □

Exercise 5.1.8 Show that any satisfiable sentence of modal logic has a finite model. The sentences of modal logic are built up from propositional variables $p_1, p_2, \ldots$ by means of the propositional connectives, say $\neg$ and $\vee$, and the "*necessity operator*" $\Box$. A *structure* (or *frame*) $\mathcal{M}$ is a triple $(A, R, \mathcal{F})$ where A is a nonempty set, the set of "*states*", R is a binary relation on A, and $\mathcal{F} : \{p_1, p_2, \ldots\} \to \mathrm{Pow}(A)$, $\mathrm{Pow}(A)$ being the power set of A. For $s \in A$, the satisfaction relation $\mathcal{M} \models_s \alpha$ is defined by induction on the complexity of α:

$$
\begin{array}{lll}
\mathcal{M} \models_s p_i & \text{iff} & s \in \mathcal{F}(p_i) \\
\mathcal{M} \models_s \neg\alpha & \text{iff} & \text{not } \mathcal{M} \models_s \alpha \\
\mathcal{M} \models_s (\alpha \vee \beta) & \text{iff} & \mathcal{M} \models_s \alpha \ \text{ or } \ \mathcal{M} \models_s \beta \\
\mathcal{M} \models_s \Box\alpha & \text{iff} & \mathcal{M} \models_t \alpha \ \text{ for all } t \text{ such that } Rst.
\end{array}
$$

If $\mathcal{M} \models_s \alpha$ for all $s \in A$, then $\mathcal{M}$ is said to be a *model* of α. To solve the exercise, let the propositional variables occurring in α be among $p_1, \ldots, p_n$. Associate with α a formula $\varphi_\alpha(x) \in \mathrm{FO}^2[\{R, P_1, \ldots, P_n\}]$, where $P_1, \ldots, P_n$ are unary, such that for all $(A, R, \mathcal{F})$ and $s \in A$,

$$(A, R, \mathcal{F}) \models_s \alpha \quad \text{iff} \quad (A, R, \mathcal{F}(p_1), \ldots, \mathcal{F}(p_n)) \models \varphi_\alpha[s]. \qquad \Box$$

5.2 Finite Model Property of $\forall^2\exists^*$-Sentences

We fix a relational vocabulary τ. In the preceding section we have proved that every FO^2-sentence has models in the same cardinalities as a sentence of the form

$$\forall x \forall y\, \psi \wedge \bigwedge_{i=1}^{r} \forall x \exists y\, \psi_i$$

with quantifier-free ψ, ψ_i. This sentence is equivalent to

$$(*) \qquad \forall x \forall y \exists y_1 \ldots \exists y_r (\psi(x,y) \wedge \bigwedge_{i=1}^{r} \psi_i(x, y_i)).$$

We then have proved the finite model property for these sentences.

By a $\forall^2\exists^*$-sentence we mean a first-order sentence of the form

$$\forall x_1 \forall x_2 \exists y_1 \ldots \exists y_k \psi'$$

where $k \geq 0$ and ψ' is quantifier-free. In particular, the sentence $(*)$ is a $\forall^2\exists^*$-sentence. In this section we extend the result about the finite model property for sentences of the form $(*)$ to arbitrary $\forall^2\exists^*$-sentences, however under the proviso that they have models without kings. (Recall that $a \in A$ is a king in the structure $\mathcal{A}$ if for no $b \in A, b \neq a$, we have $\varphi_a^0 = \varphi_b^0$.)

Theorem 5.2.1 *Suppose that τ is a relational vocabulary. If ψ is a $\forall^2\exists^*$-sentence which has a model without kings, then it has a finite model.*

We first draw a consequence. For a structure $\mathcal{A}$ in a relational vocabulary τ and $l \geq 2$, denote by $\mathcal{A} \times l$ the structure which for every element of A contains l duplicates; more precisely: $\mathcal{A} \times l$ is the τ-structure with universe $A \times \{0, \ldots, l-1\}$ such that for any n-ary R in τ,

$$R^{A \times l} \quad := \quad \{((a_1, i_1), \ldots, (a_n, i_n)) \mid R^A a_1 \ldots a_n,\ 0 \leq i_1, \ldots, i_n \leq l-1\}.$$

Clearly, $\mathcal{A} \times l$ is a structure without kings, and a routine proof by induction shows that

$$\mathcal{A} \models \psi \qquad \text{iff} \qquad \mathcal{A} \times l \models \psi$$

holds for all sentences ψ without (the) equality (sign). As a corollary of the above theorem we therefore obtain:

Corollary 5.2.2 *Suppose that τ is a relational vocabulary and ψ is a $\forall^2\exists^*$-sentence without equality. If ψ is satisfiable then it has a finite model.*

As in 5.1.4 for FO^2 we now get:

Corollary 5.2.3 *The set of logically valid $\forall^2\exists^*$-sentences without equality in a relational vocabulary is decidable.* □

Proof (of 5.2.1). In models with at least two elements a $\forall^2\exists^*$-sentence

$$\forall v_1 \forall v_2 \exists v_3 \ldots \exists v_k \psi'(v_1, \ldots, v_k)$$

is equivalent to the sentence

$$\forall v_1 \forall v_2 \exists x_3 \ldots \exists x_k \exists z_3 \ldots \exists z_k (\neg v_1 = v_2 \rightarrow (\psi'(v_1, v_1, x_3, \ldots, x_k) \\ \wedge \psi'(v_1, v_2, z_3, \ldots, z_k))),$$

and, for example, a sentence of the form

$$\forall v_1 \forall v_2 \exists x \exists y \psi'(v_1, v_2, x, y)$$

is equivalent to

$$\forall v_1 \forall v_2 \exists x \exists y ((\psi'(v_1, v_2, x, x) \vee \psi'(v_1, v_2, x, y)) \wedge \neg x = y).$$

Therefore, we can assume that our $\forall^2\exists^*$-sentence ψ has the form

$$\forall v_1 \forall v_2 \exists v_3 \ldots \exists v_k (\neg v_1 = v_2 \rightarrow (\psi'(v_1, \ldots, v_k) \wedge \bigwedge_{3 \leq i < j \leq k} \neg v_i = v_j))$$

where ψ' is quantifier-free.

Choose a model $\mathcal{A}$ of ψ without kings and let

$$S := \{\varphi_a^0 \mid a \in A\} \qquad \text{and} \qquad T := \{\varphi_{ab}^0 \mid a, b \in A, a \neq b\}$$

be the 0-isomorphism types of elements and of pairs of elements of $\mathcal{A}$, respectively. If $\rho(v_1, \ldots, v_l)$ is a 0-isomorphism type of any l-tuple and $1 \leq m, n \leq l$ with $m \neq n$, let $\rho_m(v_1)$ and $\rho_{m,n}(v_1, v_2)$ be the induced 0-isomorphism types of v_m and of v_m, v_n, respectively; in particular, for any $\mathcal{B}$ and $b_1, \ldots, b_l \in B$,

$$\mathcal{B} \models \rho[b_1, \ldots, b_l] \text{ implies } \varphi_{b_m}^0 = \rho_m \text{ and } \varphi_{b_m b_n}^0 = \rho_{m,n}.$$

As $\mathcal{A}$ has no kings, we get:

(1) For all $\varphi, \varphi' \in S$ there is a $\chi \in T$ such that $\varphi = \chi_1$ and $\varphi' = \chi_2$.

Moreover, since $\mathcal{A}$ is a model of ψ, we have:

(2) For every $\chi \in T$ there is a 0-isomorphism type $\rho(v_1, \ldots, v_k)$ with

(a) $\rho_i \in S$ for $i = 1, \ldots, k$
(b) $\rho_{m,n} \in T$ for $1 \leq m < n \leq k$ and $\rho_{1,2} = \chi$
(c) $\models \rho \rightarrow \psi'(v_1, \ldots, v_k)$.

To get the statement of the theorem it suffices to show:

(+) Suppose that S and T are nonempty sets of 0-isomorphism types of elements and of pairs of elements, respectively, satisfying (1) and (2). Then ψ has a finite model.

Let $s := \|S\|$ and $t := \|T\|$ and fix an ordering on S. To prove (+), we give a method to construct, for every $n \geq k$, structures $\mathcal{B}$ with universe $\{1, 2, \ldots, n \cdot s\}$ (afterwards we show that with nonvanishing probability these structures are models of ψ). The 0-isomorphism types of elements are fixed by a deterministic algorithm while the 0-isomorphism types of tuples of more than one element are chosen randomly. The exact construction procedure of $\mathcal{B}$ reads as follows:

(i) If $a \in \{1, 2, \ldots, n \cdot s\}$ and $a = i \cdot s + j$ for some i, j such that $0 \leq i < n$ and $1 \leq j \leq s$, ensure that φ^0_a is equal to the j-th element in S.
(ii) If $a, b \in \{1, 2, \ldots, n \cdot s\}$, $1 \leq a < b \leq n \cdot s$, choose at random a χ in $\{\chi \in T \mid \chi_1 = \varphi^0_a, \chi_2 = \varphi^0_b\}$ (this set is nonempty by (1)) and ensure that $\varphi^0_{ab} = \chi$.
(iii) If R is an m-ary relation symbol in τ, define the truth value of $Ra_1 \ldots a_m$ at random for any $a_1, \ldots, a_m \in \{1, 2, \ldots, n \cdot s\}$ containing at least three and at most k distinct members.
(iv) If R is an m-ary relation symbol in τ, define the truth value of $Ra_1 \ldots a_m$ to be "false" if $a_1 \ldots a_m$ contains more than k distinct members.

Let $\mathrm{Str}(n)$ be the collection of possible values of $\mathcal{B}$ with $\{1, 2, \ldots, n \cdot s\}$ as universe. Equip $\mathrm{Str}(n)$ with the uniform probability distribution μ. Let $\bar{a} = a_1 \ldots a_k$ denote pairwise distinct elements of $\{1, 2, \ldots, n \cdot s\}$ and let d be the number of formulas $Rv_{i_1} \ldots v_{i_m}$ where $R \in \tau$ and $\{v_{i_1}, \ldots, v_{i_m}\}$ contains at least three and at most k distinct variables.

Claim 1. Suppose that $\chi \in T$ and that the 0-isomorphism type $\rho(v_1, \ldots, v_k)$ satisfies (2) with respect to χ. Then the conditional probability

$$\mu(\varphi^0_{\bar{a}} = \rho \mid \varphi^0_{a_1 a_2} = \chi, \varphi^0_{a_i} = \rho_i \text{ for } i = 3, \ldots, k)$$

is $\geq \delta$, where $\delta = \left(\frac{1}{t}\right)^{\binom{k}{2}-1} \cdot \left(\frac{1}{2}\right)^d$, that is,

$$\mu(\{\mathcal{B} \mid \mathcal{B} \models \rho[\bar{a}]\} \mid \{\mathcal{B} \mid \mathcal{B} \models \chi[a_1, a_2], \mathcal{B} \models \rho_i[a_i] \text{ for } i = 3, \ldots, k\}) \geq \delta.$$

Proof. The proof is immediate by the definition of d and the fact that, once $\varphi^0_{a_i}$ for $i = 1, \ldots, k$ and $\varphi^0_{a_1 a_2}$ are fixed, 0-isomorphism types according to (ii) have to be chosen for $\binom{k}{2} - 1$ pairs of distinct elements.

Claim 2. Fix a_1, a_2. Then

$$\mu(\{\mathcal{B} \mid \mathcal{B} \not\models \exists v_3 \ldots \exists v_k (\psi'(a_1, a_2, v_3, \ldots, v_k) \wedge \bigwedge_{1 \leq i < j \leq k} v_i \neq v_j)\}) \leq (1-\delta)^f,$$

where f is the integer part of $\frac{n-2}{k-2}$.

Proof. Let $\chi \in T$ and choose (for χ) a corresponding $\rho(v_1, \ldots, v_k)$ according to (2). It suffices to prove that the conditional probability

$$(3) \quad \mu(\{\mathcal{B} \mid \mathcal{B} \not\models \exists v_3 \ldots \exists v_k \rho(a_1, a_2, v_3, \ldots, v_k)\} \mid \varphi^0_{a_1 a_2} = \chi) \leq (1-\delta)^f.$$

By (i), in any $\mathcal{B}$ every 0-isomorphism type in S is realized by n ($\geq 2 + f \cdot (k-2)$) distinct elements. Therefore, for $i = 3, \ldots, k$ and $j = 1, \ldots, f$ there are pairwise distinct elements $a_i^j \in \{1, 2, \ldots, n \cdot s\} \setminus \{a_1, a_2\}$ with $\varphi^0_{a_i^j} = \rho_i$. Under the given conditions the events $\mathcal{B} \models \rho(a_1, a_2, a_3^j, \ldots, a_k^j)$ for $1 \leq j \leq f$ are independent (compare the construction procedure). Now,

$$\begin{aligned}&\{\mathcal{B} \mid \mathcal{B} \not\models \exists v_3 \dots \exists v_k \rho(a_1, a_2, v_3, \dots, v_k)\}\\ \subseteq\ &\{\mathcal{B} \mid \text{ for } j = 1, \dots, f,\ \mathcal{B} \not\models \rho(a_1, a_2, a_3^j, \dots, a_k^j)\}.\end{aligned}$$

Therefore, by Claim 1 we obtain (3).

Since $\{\mathcal{B} \mid \mathcal{B} \not\models \psi\} =$

$$\bigcup_{\substack{a_1,a_2\\ a_1\neq a_2}} \{\mathcal{B} \mid \mathcal{B} \not\models \exists v_3 \dots \exists v_k (\psi'(a_1, a_2, v_3, \dots, v_k) \ \wedge \bigwedge_{1\leq i<j\leq k} v_i \neq v_j)\},$$

we obtain by Claim 2 that

$$\mu(\{\mathcal{B} \mid \mathcal{B} \not\models \psi\}) \leq n \cdot s \cdot (n \cdot s - 1) \cdot (1 - \delta)^f.$$

As f is the integer part of $\frac{n-2}{k-2}$, for n big enough we have $n \cdot s \cdot (n \cdot s - 1) \cdot (1-\delta)^f < 1$. Then the probability that $\mathcal{B}$ satisfies ψ is positive, and therefore some member of $\mathrm{Str}(n)$ satisfies ψ. □

Exercise 5.2.4 Let τ be a vocabulary (which perhaps contains constants). Show that any satisfiable $\forall^2\exists^*$-sentence without equality has a finite model (argue as in 5.1.7). Conclude that any satisfiable $\exists^*\forall^2\exists^*$-sentence without equality has a finite model. □

Notes 5.2.5 Theorem 5.1.3 is due to Mortimer [121], Corollary 5.1.4 for sentences without equality was already known before (cf. [131]). The proof of 5.1.3 given here incorporates simplifications from [55]. Together with [44] it shows that the satisfiability problem for FO^2 is NEXPTIME-complete. The papers [59, 60, 127, 144] discuss the decidability of extensions and fragments of FO^2. Corollary 5.2.2 goes back to Gödel [48]; the proof given for 5.2.1 is taken from [75]. For formulas with equality the corresponding satisfiability problem is undecidable ([49]; see also [50]). For thorough information on the decidability and the finite model property of fragments of first-order logic we refer the reader to [13].

6. Finite Automata and Logic: A Microcosm of Finite Model Theory

One of the major aims of finite model theory consists in characterizing the queries in a given complexity class by means of a logic in which they can be described. In this way one obtains a new measure of complexity: the complexity of formal descriptions. Moreover, the characterizations allow to translate problems on complexity classes into purely modeltheoretic problems, thus opening them to modeltheoretic methods. In general, this relationship between complexity classes and logics turns out to be useful for both, complexity theory and logic.

Our considerations will show that first-order logic which has played a central role so far is not adequate for descriptive complexity. What is missing? Which features of computations cannot be expressed in first-order logic and which natural extensions of first-order logic capture these features? Once these questions are clear they show the way to stronger systems such as logics with fixed-point operators that we shall treat in the next chapters. These extensions have natural modeltheoretic properties and thus may turn out to be useful in other areas of model theory, too.

The present chapter deals with the – historically first – logical characterization of a complexity class: the characterization of the class of languages recognized by automata by means of monadic second-order logic. It turns out that this simple case already reflects some of the crucial aspects encountered later, at the same time witnessing that the interplay between model theory and complexity theory is fruitful for both sides.

Even though we include all definitions and proofs, a certain familiarity with automata will be helpful.

6.1 Languages Accepted by Automata

We recall some definitions and notations. Let $\mathbb{A}$ be a finite alphabet and $\mathbb{A}^*$ and $\mathbb{A}^+$ the set of strings (or words) and the set of nonempty strings over $\mathbb{A}$, respectively. Thus, $\mathbb{A}^* = \mathbb{A}^+ \cup \{\lambda\}$, where λ is the empty word. In automata theory subsets of $\mathbb{A}^+$ are called *languages*.[1] A *nondeterministic automaton* M, in short, an NDA (over the alphabet $\mathbb{A}$) is given by a tuple

$$M = (S, q_0, \delta, F),$$

where S is a finite set, the set of *states*, $q_0 \in S$ is the *initial state*, $F \subseteq S$ is the set of (*accepting* or) *final states* and $\delta \subseteq S \times \mathbb{A} \times S$ is the *transition relation* (intuitively, $(q, a, p) \in \delta$ means: if M is in state q and reads a, then M can pass into p). This relation induces a function $\tilde{\delta} : S \times \mathbb{A}^* \to \mathrm{Pow}(S)$, where $\mathrm{Pow}(S)$ denotes the power set of S; $\tilde{\delta}$ is given by

$$\begin{aligned} \tilde{\delta}(q, \lambda) &:= \{q\} \\ \tilde{\delta}(q, wa) &:= \{p \mid (r, a, p) \in \delta \text{ for some } r \in \tilde{\delta}(q, w)\}. \end{aligned}$$

In particular, $\tilde{\delta}(q, a) = \{p \mid (q, a, p) \in \delta\}$ for $a \in \mathbb{A}$. If $\tilde{\delta}(q, a)$ is a singleton for every $a \in \mathbb{A}$, then M is said to be a *deterministic automaton* or, by short, an *automaton*. Clearly, in this case $\tilde{\delta}(q, w)$ is a singleton for any $w \in \mathbb{A}^*$. If $\tilde{\delta}(q, w) = \{p\}$, we simply write $\tilde{\delta}(q, w) = p$ (and similarly, $\delta(q, a) = p$ stands for $\tilde{\delta}(q, a) = \{p\}$).

The language *recognized* (or *accepted*) by the NDA M is defined by

$$L(M) \quad := \quad \{w \in \mathbb{A}^+ \mid \tilde{\delta}(q_0, w) \cap F \neq \emptyset\}.$$

Hence, in case M is deterministic,

$$L(M) \quad = \quad \{w \in \mathbb{A}^+ \mid \tilde{\delta}(q_0, w) \in F\}.$$

Let us state the main result of this chapter, even though we still have not introduced all concepts appearing in it. With respect to the logical characterizations we have in mind, we are mainly interested in the equivalence of (ii) and (vi). The other equivalences are not only useful for the proof, but also interesting in their own.

For a language $L \subseteq \mathbb{A}^+$ the following are equivalent:

(i) *L is the union of equivalence classes of an invariant equivalence relation on $\mathbb{A}^+$ of finite index.*
(ii) *L is recognized by an automaton.*
(iii) *L is recognized by an* NDA.
(iv) *L is regular.*
(v) *L is definable in monadic second-order logic by a Σ_1^1-sentence.*
(vi) *L is definable in monadic second-order logic.*

---

[1] In the literature languages, in particular languages recognized by automata, may contain the empty word. Since to any word we will assign a structure whose cardinality coincides with the length of the word, it is more convenient for our purposes to exclude the empty string.

Note that (ii) $\Rightarrow$ (iii) and (v) $\Rightarrow$ (vi) are trivial. We prove the other implications by a series of propositions starting with those which mainly deal with automata; monadic second-order logic will be treated in the next section.

An equivalence relation $\sim$ on $\mathbb{A}^+$ is called *invariant* if

$$u, v, w \in \mathbb{A}^+ \text{ and } u \sim v \quad \text{imply} \quad uw \sim vw.$$

Denote by $[u]$ the equivalence class of u and by $\mathbb{A}^+/_\sim$ the set of equivalence classes. The *index* of $\sim$ is the cardinality of $\mathbb{A}^+/_\sim$.

Proposition 6.1.1 *Let $\sim$ be an invariant equivalence relation on $\mathbb{A}^+$ of finite index. Suppose that the language $L \subseteq \mathbb{A}^+$ is the union of equivalence classes,*

$$L = [u_1] \cup \ldots \cup [u_r]$$

for some $u_1, \ldots, u_r \in \mathbb{A}^+$. Then L is recognized by an automaton.

Proof. Add $[\lambda]$, "the equivalence class of λ", as a new object to $\mathbb{A}^+/_\sim$ and define the automaton $M = (S, q_0, \delta, F)$ by

$$S := \mathbb{A}^+/_\sim \cup \{[\lambda]\}, \quad q_0 := [\lambda], \quad \delta([u], a) := [ua], \quad F := \{[u_1], \ldots, [u_r]\}.$$

By invariance of $\sim$, the transition function δ is well-defined. For $u, v \in \mathbb{A}^*$, a simple induction on the length of v shows that $\tilde{\delta}([u], v) = [uv]$, in particular, $\tilde{\delta}([\lambda], v) = [v]$. Therefore,

$$L(M) = \{v \in \mathbb{A}^+ \mid \tilde{\delta}(q_0, v) \in F\} = \{v \in \mathbb{A}^+ \mid [v] \in F\} = [u_1] \cup \ldots \cup [u_r] = L.$$

□

For later applications we state the following result:

Lemma 6.1.2 (Pumping Lemma) *Let $\sim$ be an invariant equivalence relation on $\mathbb{A}^+$ of finite index. Then there is an $n \geq 0$ such that for any word $u \in \mathbb{A}^+$ with $|u| \geq n$* [2] *there exist $v, w \in \mathbb{A}^+$ and $x \in \mathbb{A}^*$ with*

$$u = vwx, \ |vw| \leq n, \quad \text{and} \quad vw^k \sim vw \text{ for all } k \geq 0.$$

Hence, by invariance, $vw^k y \sim vwy$ for all $k \geq 0$ and $y \in \mathbb{A}^*$.

Proof. Let l be the index of $\sim$ and set $n := l + 1$. Suppose that $u \in \mathbb{A}^+, u = a_1 \ldots a_s$, where $a_1, \ldots, a_s \in \mathbb{A}$ and $s \geq n$. Then, for some i and j with $1 \leq i < j \leq n$, we have $a_1 \ldots a_i \sim a_1 \ldots a_j$. Let $v = a_1 \ldots a_i$ and $w = a_{i+1} \ldots a_j$. Thus $v \sim vw$, and by invariance of $\sim$, $vw \sim vw^2 \sim vw^3 \sim \ldots$. □

We now turn to a description of the languages recognized by automata in terms of simple, so-called regular expressions. First some notations: The *concatenation* of languages L_1 and L_2, denoted by $L_1 L_2$, is the set $\{uv \mid u$ is

[2] $|u|$ denotes the length of u.

in L_1 and v is in $L_2\}$. Define $L^1 := L$ and $L^n := L^{n-1}L$ for $n > 1$. The *plus* (or *positive*) *closure* L^+ of L is the set $L^+ := \bigcup_{n\geq 1} L^n$. The *regular expressions* (over $\mathbb{A}$) are strings over the alphabet $\{\emptyset\} \cup \{\, \mathbf{a} \mid a \in \mathbb{A}\} \cup \{\cup, {}^+,), (\}$. Together with the languages they denote, the regular expressions are defined recursively as follows:

(a) $\emptyset$ is a regular expression and denotes the empty set.
(b) $\mathbf{a}$ is a regular expression and denotes the set $\{a\}$.
(c) If r and s are regular expressions denoting the languages R and S, respectively, then $(r \cup s)$, (rs), and r^+ are regular expressions that denote the sets $R \cup S, RS$, and R^+, respectively.

A language is *regular* if it is denoted by some regular expression. For convenience we often omit parentheses in regular expressions (such as in $r_1 \cup \ldots \cup r_k$) that have no influence on the language they denote. Moreover, we assume that $^+$ has higher precedence than concatenation or $\cup$, and that concatenation has higher precedence than $\cup$.

Proposition 6.1.3 *If L is recognized by an* NDA *then L is regular.*

Proof. Let L be recognized by the NDA $M = (S, q_0, \delta, F)$ with $S = \{q_0, \ldots, q_n\}$. Let L_k^{ij} be the set of all nonempty strings that M can read starting in q_i and ending in q_j without going through any state numbered $\geq k$,

$$\begin{aligned} L_k^{ij} \;:=\; & \{b_1 \ldots b_s \mid s \geq 1,\, b_1, \ldots, b_s \in \mathbb{A}, \text{ there are } i_0, \ldots, i_s \text{ such that} \\ & i_1, \ldots, i_{s-1} < k,\, i_0 = i,\, i_s = j, \text{ and } (q_{i_m}, b_{m+1}, q_{i_{m+1}}) \in \delta \text{ for } m < s\}. \end{aligned}$$

Since $L(M) = \bigcup_{q_j \in F} L_{n+1}^{0j}$, it suffices to show that all L_k^{ij} are regular. We proceed by induction on k. As $L_0^{ij} = \{a \in \mathbb{A} \mid (q_i, a, q_j) \in \delta\}$ is a subset of $\mathbb{A}$, say $L_0^{ij} = \{a_1, \ldots, a_r\}$, L_0^{ij} is denoted by $(\mathbf{a}_1 \cup \ldots \cup \mathbf{a}_r)$, or by $\emptyset$ in case $r = 0$.

For the induction step note that a nonempty string is in L_{k+1}^{ij}, if it can be read without visiting any state numbered $\geq k+1$, thereby starting in q_i, ending in q_j, and passing through q_k zero times or one or more than one time. Hence,

$$L_{k+1}^{ij} = L_k^{ij} \cup L_k^{ik} L_k^{kj} \cup L_k^{ik} (L_k^{kk})^+ L_k^{kj}.$$

By induction hypothesis, for all i', j' there is a regular expression $r_k^{i'j'}$ denoting $L_k^{i'j'}$. Therefore, L_{k+1}^{ij} is denoted by

$$r_k^{ij} \cup r_k^{ik} r_k^{kj} \cup r_k^{ik} (r_k^{kk})^+ r_k^{kj}. \qquad \square$$

6.2 Word Models

We turn to languages definable in monadic second-order logic. In order to bridge the gap between automata and logic, we introduce a correspondence between words and structures. Once more we fix an alphabet $\mathbb{A}$ and let $\tau(\mathbb{A})$ be the vocabulary $\{<\} \cup \{P_a \mid a \in \mathbb{A}\}$, where $<$ is binary and the P_a are unary. For a given $u \in \mathbb{A}^+$, say $u = a_1 \dots a_n$, we consider structures of the form

$$(B, <, (P_a)_{a \in \mathbb{A}})$$

where the cardinality of B equals the length of u, $<$ is an ordering of B, and P_a corresponds to the positions in u carrying an a,

$$P_a \quad := \quad \{b \in B \mid \text{ for some } j,\ b \text{ is the j-th element of } < \text{ and } a_j = a\}.$$

We call them *word models* for u and denote the class of word models for u by K_u. For example, if $\mathbb{A} = \{a, b\}$ and $u = abbab$, the structure

$$(\{1, \dots, 5\}, <, P_a, P_b),$$

where $<$ is the natural ordering on $\{1, \dots, 5\}, P_a = \{1, 4\}$, and $P_b = \{2, 3, 5\}$ is a word model for u.

Any two word models for u are isomorphic. For simplicity we therefore often speak of *the* word model for u, denoting it by $\mathcal{B}_u$.

Note that for $u, v \in \mathbb{A}^+$ we obtain a word model for uv by forming the ordered sum (cf. 1.A3) $\mathcal{B}_u \lhd \mathcal{B}_v$.

A language $L \subseteq \mathbb{A}^+$ is *definable in monadic second-order logic*, if there is a sentence φ in MSO$[\tau(\mathbb{A})]$ such that $\text{Mod}(\varphi) = \bigcup_{u \in L} K_u$, or, more succinctly (but not fully correct), $\text{Mod}(\varphi) = \{\mathcal{B}_u \mid u \in L\}$. The notion of definability in FO is introduced similarly.

Let φ_W be the first-order sentence

$$\varphi_W \quad := \quad \text{“} < \text{ is a total ordering”} \wedge \forall x \bigvee_{a \in \mathbb{A}} P_a x \wedge \bigwedge_{\substack{a, b \in \mathbb{A} \\ a \neq b}} \forall x \neg (P_a x \wedge P_b x).$$

Then, $\text{Mod}(\varphi_W)$ is the class of all word models, $\text{Mod}(\varphi_W) = \{\mathcal{B}_u \mid u \in \mathbb{A}^+\}$, and therefore the language $\mathbb{A}^+$ is definable in first-order logic.

As the next step in the proof of the main result stated at the beginning of this chapter we shall show that any regular language is definable in monadic second-order logic. First we introduce some notations: Let $\psi_{\min}(x)$ and $\psi_{\max}(x)$ be first-order formulas defining the first and the last element of the ordering, respectively:

$$\psi_{\min}(x) := \forall y \neg y < x, \qquad \psi_{\max}(x) := \forall y \neg x < y.$$

For any formula φ of MSO and variables x and y let $\varphi^{[x,y]}$ be a formula expressing that the closed interval $[x, y]$ satisfies φ, and similarly, $\varphi^{]x,y]}$ a formula expressing that the half-open interval $]x, y]$ satisfies φ. Such formulas can be obtained from φ by relativizing the first-order quantifiers to the interval, the main clause of an inductive definition being (for $z \neq x, z \neq y$)

$$(\exists z\varphi)^{[x,y]} := \exists z(x \leq z \wedge z \leq y \wedge \varphi^{[x,y]})$$
$$(\exists z\varphi)^{]x,y]} := \exists z(x < z \wedge z \leq y \wedge \varphi^{]x,y]}).$$

Proposition 6.2.1 *Any regular language is definable in monadic second-order logic by a Σ_1^1-sentence.*

Proof. First we prove by induction on the length of the regular expression r that there is a sentence φ_r of MSO defining the language denoted by r. Afterwards we show that we can replace φ_r by a Σ_1^1-sentence.

$$\varphi_\emptyset := \exists x \neg x = x, \quad \varphi_{\mathrm{a}} := \varphi_W \wedge \exists x \forall y(y = x \wedge P_a x)$$
$$\varphi_{(r \cup s)} := \varphi_W \wedge (\varphi_r \vee \varphi_s)$$
$$\varphi_{(rs)} := \varphi_W \wedge \text{"the universe is partitioned into two intervals satisfying } \varphi_r \text{ and } \varphi_s\text{, respectively"}$$
$$= \varphi_W \wedge \exists x \exists y \exists z(\psi_{\min}(x) \wedge y < z \wedge \psi_{\max}(z) \wedge \varphi_r^{[x,y]} \wedge \varphi_s^{]y,z]})$$
$$\varphi_{r^+} := \varphi_W \wedge \text{"there is a set of right endpoints of intervals, which partition the universe, all parts satisfying } \varphi_r\text{"}$$
$$= \varphi_W \wedge \exists X(\exists y(Xy \wedge \psi_{\max}(y)) \wedge \exists x \exists y(\psi_{\min}(x) \wedge Xy \wedge \forall z(z < y \rightarrow \neg Xz) \wedge \varphi_r^{[x,y]}) \wedge \forall x \forall y ((x < y \wedge Xx \wedge Xy \wedge \forall z(x < z < y \rightarrow \neg Xz)) \rightarrow \varphi_r^{]x,y]})).$$

Now, one obtains a Σ_1^1-sentence by inductively bringing all existential second-order quantifiers to the front. In general, a monadic second-order formula $\forall \overline{x} \exists Y \chi$ with first-order χ is not equivalent to a monadic Σ_1^1-formula (cf. 2.4.8). However, in the case of the formula in the last two lines of φ_{r^+} we can argue as follows: Suppose that φ_r is equivalent to $\exists Y_1 \ldots \exists Y_m \chi$. In models of φ_W (and only these are relevant) the formula in the last two lines is equivalent to

$$\exists Y_1 \ldots \exists Y_m \forall x \forall y((x < y \wedge Xx \wedge Xy \wedge \forall z(x < z < y \rightarrow \neg Xz)) \rightarrow \chi^{]x,y]})$$

(for the nontrivial implication of the equivalence piece $Y_1, \ldots, Y_m$ together from corresponding subsets chosen in the (disjoint) intervals). □

With the next proposition we close the last gap in the proof of the main result.

Proposition 6.2.2 *Let $L \subseteq \mathbb{A}^+$ be definable in monadic second-order logic. Then there is an invariant equivalence relation on $\mathbb{A}^+$ of finite index such that L is a union of equivalence classes.*

Proof. Assume that $\mathrm{Mod}(\varphi) = \{\mathcal{B}_u \mid u \in L\}$ where φ is a sentence of MSO. Let m be the quantifier rank of φ and define $\sim$ on $\mathbb{A}^+$ by

$$u \sim v \quad \text{iff} \quad \mathcal{B}_u \equiv_m^{\mathrm{MSO}} \mathcal{B}_v$$

(recall that $\mathcal{A} \equiv_m^{\text{MSO}} \mathcal{B}$ means that $\mathcal{A}$ and $\mathcal{B}$ satisfy the same sentences of MSO of quantifier rank $\leq m$). Clearly, $\sim$ is an equivalence relation. Since, up to logical equivalence, there are only finitely many sentences of quantifier rank $\leq m$ (cf. 3.1.3), the relation $\sim$ is of finite index. By definition of m,

$$\mathcal{B}_u \models \varphi \text{ and } u \sim v \text{ imply } \mathcal{B}_v \models \varphi.$$

Thus, $L = \bigcup\{[u] \mid u \in \mathbb{A}^+, \mathcal{B}_u \models \varphi\}$. Finally, $\sim$ is invariant: Assume $u \sim v$ and $w \in \mathbb{A}^+$. Then $\mathcal{B}_u \equiv_m^{\text{MSO}} \mathcal{B}_v$. Since $\equiv_m^{\text{MSO}}$ is preserved by ordered sums (cf. 3.1.4), we obtain

$$\mathcal{B}_{uw} \cong \mathcal{B}_u \lhd \mathcal{B}_w \equiv_m^{\text{MSO}} \mathcal{B}_v \lhd \mathcal{B}_w \cong \mathcal{B}_{vw}$$

that is, $uw \sim vw$. □

Summing up we have shown the result stated at the beginning of section 6.1:

Theorem 6.2.3 *For a language $L \subseteq \mathbb{A}^+$ the following are equivalent:*

(i) *L is the union of equivalence classes of an invariant equivalence relation on $\mathbb{A}^+$ of finite index.*
(ii) *L is recognized by an automaton.*
(iii) *L is recognized by an* NDA.
(iv) *L is regular.*
(v) *L is definable in monadic second-order logic by a Σ_1^1-sentence.*
(vi) *L is definable in monadic second-order logic.* □

In particular, this theorem shows that a language is accepted by an automaton just in case it is definable in monadic second-order logic. Another description of languages accepted by automata is given by means of the regular expressions. Is that a logical description too? What makes a logic? We shall address this problem later.

6.3 Examples and Applications

By the following examples and applications we try to exhibit the methodological usefulness of Theorem 6.2.3.

Proposition 6.3.1 (a) *The class of languages over $\mathbb{A}$ accepted by automata is closed under the boolean operations (complementation and union).*

(b) (Pumping Lemma). *Let L be accepted by an automaton. Then there is $n \geq 0$ such that for any $u \in \mathbb{A}^+$ with $|u| \geq n$ there exist $v, w \in \mathbb{A}^+$ and $x \in \mathbb{A}^*$ with*

$$u = vwx, |vw| \leq n, \text{ and for } k \geq 0 \text{ and } y \in \mathbb{A}^* : vw^k y \in L \text{ iff } vwy \in L.$$

Proof. Part (a) holds, since monadic second-order logic is closed under the boolean connectives $\neg$ and $\vee$, and part (b) is a reformulation of the Pumping Lemma 6.1.2. Both results also have simple direct proofs. □

Example 6.3.2 Let $\mathbb{A} = \{a\}$. Identify $\underbrace{a \dots a}_{\text{n-times}}$ with the natural number n and thus, $\mathbb{A}^+$ with the set $\mathbb{N}_+$ of positive natural numbers. A subset L of $\mathbb{N}_+$ is accepted by an automaton iff L is *ultimately periodic*, that is, if there are $p, r \in \mathbb{N}_+$ such that for all $m \geq p$,

$$m + r \in L \quad \text{iff} \quad m \in L.$$

In fact, assume first that L is accepted by an automaton. By the Pumping Lemma there are $n, j, r \in \mathbb{N}_+$ and $l \geq 0$ with $n = j + r + l$ such that for all $k \geq 0$ and $s \in \mathbb{N}$,

$$j + kr + s \in L \quad \text{iff} \quad j + r + s \in L.$$

In particular, if $m \geq p := j + r$, say $m = j + r + s$, then (take $k = 2$)

$$m + r \in L \quad \text{iff} \quad m \in L.$$

Now let L be ultimately periodic. Choose corresponding $p, r \in \mathbb{N}_+$. Set $L_1 := \{m \in L \mid m < p\}$ and $L_2 := \{m \in L \mid p \leq m < p + r\}$. Then, by periodicity, $L = L_1 \cup L_2 \cup \{m + kr \mid m \in L_2, k \geq 1\}$. So L is the union of the finite (and hence regular) sets L_1 and L_2 and of the languages denoted by the regular expressions $\mathbf{a}^m(\mathbf{a}^r)^+$ with $m \in L_2$. Thus L is regular. □

As a consequence, the classes of finite ordered structures of vocabulary $\{<\}$ that are axiomatizable in MSO coincide with the ultimately periodic ones.

Example 6.3.3 For $\mathbb{A} = \{a, b\}$ the set

$$L \;:=\; \{u \in \mathbb{A}^+ \mid \text{the number of } a\text{'s in } u \text{ equals the number of } b\text{'s in } u\}$$

is not accepted by an automaton. Otherwise, choose n and a representation vwx of $a^n b^n$ according to the Pumping Lemma. Since $|vw| \leq n$, we have $w \in \{a\}^+$. Hence the string vw^2x contains more a's than b's and therefore, $vw^2x \notin L$ (while $vwx = a^n b^n \in L$), a contradiction. □

We use this example to prove nonaxiomatizability results for monadic second-order logic: A graph (G, E^G) is *bipartite*, if there is an $X \subseteq G$ such that $E^G \subseteq (X \times (G \setminus X)) \cup ((G \setminus X) \times X)$, and it is *balanced*, if the set X can be chosen such that, moreover, $\|X\| = \|G \setminus X\|$. Denote by BAL the class of finite balanced graphs and by $\text{BAL}_<$ the class of finite balanced graphs carrying an arbitrary ordering on their universe,

$$\text{BAL}_< \;:=\; \{(\mathcal{G}, <) \mid \mathcal{G} \in \text{BAL}, \ < \text{ an ordering of } G\}.$$

Proposition 6.3.4 *The class* $\mathrm{BAL}_<$ *– and hence the class* BAL *– is not axiomatizable in monadic second-order logic.*

Proof. Suppose that $\mathrm{BAL}_< = \mathrm{Mod}(\varphi)$ for a sentence φ of MSO. Let $\mathbb{A} = \{a, b\}$ and let L be as in the preceding example. For $u \in \mathbb{A}^+$ recall that $\mathcal{B}_u = (B_u, <, P_a, P_b)$ denotes a word model associated with u, say, with $B_u = \{1, \ldots, |u|\}$ and $<$ the natural ordering. Let $\mathcal{G}_u = (B_u, R_u)$ be the bipartite graph given by

$$R_u \;:=\; \{(i, j) \in B_u \times B_u \mid P_a i \text{ iff } P_b j\}.$$

Then,

$$(\mathcal{G}_u, <) \in \mathrm{BAL}_< \quad \text{iff} \quad u \in L$$

and

$$(\mathcal{G}_u, <) \models \varphi \quad \text{iff} \quad \mathcal{B}_u \models \varphi \frac{(P_a \ldots \leftrightarrow P_b \,\_)}{E \ldots \_}$$

(where $\varphi \frac{(P_a \ldots \leftrightarrow P_b \,\_)}{E \ldots \_}$ is obtained by replacing any subformula in φ of the form Exy by $(P_a x \leftrightarrow P_b y)$). Therefore, $\mathrm{Mod}(\varphi \frac{(P_a \ldots \leftrightarrow P_b \,\_)}{E \ldots \_}) = \{\mathcal{B}_u | u \in L\}$. Theorem 6.2.3 now implies that L is accepted by an automaton, which contradicts the preceding example. □

Corollary 6.3.5 (a) *Let* HAM *be the class of finite graphs with a Hamiltonian circuit. Then,* HAM *and* $\mathrm{HAM}_<$ *are not axiomatizable in* MSO.
(b) *Let* CHS *be the set of finite graphs which contain a clique of at least half their size. Then,* CHS *and* $\mathrm{CHS}_<$ *are not axiomatizable in* MSO.

Proof. (a) A graph of the form $(X \dot{\cup} Y, E)$ with $E = \{(a, b) \mid (a \in X, b \in Y)$ or $(a \in Y, b \in X)\}$ has a Hamiltonian circuit iff it is balanced. If $\mathrm{HAM}_< = \mathrm{Mod}(\varphi)$ for an MSO-sentence φ, then the sentence

$$\exists X (\forall x \forall y (Exy \rightarrow (Xx \leftrightarrow \neg Xy)) \wedge \varphi \frac{(X \ldots \leftrightarrow \neg X \_)}{E \ldots \_})$$

would axiomatize the class $\mathrm{BAL}_<$.

(b) Suppose that $\mathrm{CHS}_< = \mathrm{Mod}(\varphi)$ for some φ of MSO. Then an axiomatization of $\mathrm{BAL}_<$ in MSO would be given by

$$\exists X (\forall x \forall y (Exy \rightarrow (Xx \leftrightarrow \neg Xy))$$
$$\wedge \;\varphi \frac{(X \ldots \wedge X \_ \wedge \neg \ldots = \_)}{E \ldots \_} \;\;\wedge\;\; \varphi \frac{(\neg X \ldots \wedge \neg X \_ \wedge \neg \ldots = \_)}{E \ldots \_})$$

(note that the conjunction in the last line implies that both X and its complement have size at least half of the universe). □

Exercise 6.3.6 Show that a graph $\mathcal{G}$ is bipartite iff it contains no cycle of odd length (i.e., for no $n \geq 0$ there are $a_1, \ldots, a_{2n+1}$ such that $a_1 E^G a_2, a_2 E^G a_3, \ldots, a_{2n} E^G a_{2n+1}$, and $a_{2n+1} E^G a_1$). Conclude that the class of bipartite graphs is not first-order axiomatizable (of course, it is axiomatizable in MSO). □

6.4 First-Order Definability

We turn to the problem of characterizing the languages that are accepted by automata and are first-order definable. The proof of 6.2.1 shows that second-order quantifiers are only needed for the positive closure, that is, in the transition from a regular expression r to r^+. Therefore, if r does not contain the symbol $^+$, the language L denoted by r is first-order definable. But a simple induction on the length of such an r shows that then L must be finite.

Example 6.4.1 Let $\mathbb{A}$ be an alphabet. For $a \in \mathbb{A}$ the language $\mathbb{A}^+ \setminus \{a\}$ is infinite but first-order definable by

$$\varphi_W \wedge (\exists x \neg \psi_{\text{min}}(x) \vee \exists x(\psi_{\text{min}}(x) \wedge \neg P_a x))$$

(for the definition of φ_W and ψ_{min} compare section 6.2). □

We thus see that the class of languages denoted by regular expressions without $^+$ is not closed under complementation whereas the class of first-order definable languages is. Therefore, we add closure under complementation in the definition of *plus free* regular expressions:[3]

- $\emptyset$, **a** (for $a \in A$) are plus free regular expressions.
- If r and s are plus free regular expressions then so are $\sim r$, $(r \cup s)$, (rs).

If r denotes the language L, then $\sim r$ denotes $\mathbb{A}^+ \setminus L$. A language is said to be *plus free regular* if it is denoted by a plus free regular expression.

Theorem 6.4.2 *A language is plus free regular iff it is definable in first-order logic.*

Proof. If the language L is definable by the first-order sentence φ, then $\mathbb{A}^+ \setminus L$ is definable by $(\varphi_W \wedge \neg\varphi)$. This observation together with the corresponding parts of the proof of 6.2.1 shows that any plus free regular language is first-order definable.

We turn to the other direction. Recall that $\tau(\mathbb{A}) = \{<\} \cup \{P_a \mid a \in \mathbb{A}\}$. For convenience, we add a constant min to this vocabulary, which henceforth will always denote the first element; more precisely, we shall only look at models of $\varphi_W \wedge \psi_{\text{min}}(\text{min})$. By induction on the quantifier rank of the FO$[\tau(\mathbb{A}) \cup \{\text{min}\}]$-sentence φ we show for a language L that if

$$\text{Mod}(\varphi_W \wedge \psi_{\text{min}}(\text{min}) \wedge \varphi) = \{(\mathcal{B}_u, \text{min}^{\mathcal{B}_u}) \mid u \in L\}$$

then L is plus free regular. (This gives the claim.)

First assume that φ is atomic. Then φ is min $=$ min or P_a min for some $a \in \mathbb{A}$. In the first case, L is $\mathbb{A}^+$, which can be denoted by the plus free

[3] In the literature languages and, in particular, languages accepted by automata, may contain the empty word. Then the operation $^+$ is replaced by the operation *, where $L^* := L^+ \cup \{\lambda\}$ and the role of 'plus free' is taken over by 'star free'.

regular expression $\sim \emptyset$. Let φ be $P_a \min$. Then $L = \{a\} \cup \{a\}\mathbb{A}^+$. Therefore, L is denoted by $\mathbf{a} \cup \mathbf{a}(\sim \emptyset)$.

If the languages defined by the sentences φ and ψ are denoted by the plus free expressions r and s, then $\sim r$ and $r \cup s$ correspond to the sentences $\neg\varphi$ and $(\varphi \vee \psi)$, respectively.

Let $\varphi = \exists x \psi(x)$. Then $\mathrm{Mod}(\varphi_W \wedge \psi_{\min}(\min) \wedge \exists x \psi(x)) =$

$$(*) \qquad \begin{aligned} & \mathrm{Mod}(\varphi_W \wedge \psi_{\min}(\min) \wedge \psi(\min)) \\ \cup \; & \mathrm{Mod}(\varphi_W \wedge \psi_{\min}(\min) \wedge \exists x(\neg x = \min \wedge \psi(x))). \end{aligned}$$

By induction hypothesis, the first class of structures in $(*)$ corresponds to a plus free regular language. Concerning the second class, let c be a new constant. Then the finite models of $\varphi_W \wedge \psi_{\min}(\min) \wedge \exists x(\neg x = \min \wedge \psi(x))$ are the $\tau(\mathbb{A}) \cup \{\min\}$-reducts of the finite structures $(\mathcal{A}, \min^A, c^A)$ such that

$$(\mathcal{A}, \min^A, c^A) \models \varphi_W \wedge \psi_{\min}(\min) \wedge \neg c = \min \wedge \psi(c).$$

Note that any such structure can be written in the form

$$(\mathcal{A}, \min^A, c^A) = (\mathcal{A}_1 \lhd \mathcal{A}_2, \min^A, c^A),$$

where $\lhd$ denotes the ordered sum and where $(\mathcal{A}_1, \min^A) \models (\varphi_W \wedge \psi_{\min}(\min))$ and $(\mathcal{A}_2, c^A) \models (\varphi_W \wedge \psi_{\min}(c))$. Let m be the quantifier rank of ψ. Choose the – up to logical equivalence – finite set $\{(\psi_i(\min), \chi_i(c)) \mid i \in I\}$ of pairs of FO-sentences of quantifier rank $\leq m$ such that

$$\begin{aligned} & (\mathcal{A}_1, \min^{A_1}) \models (\varphi_W \wedge \psi_{\min}(\min) \wedge \psi_i(\min)) \\ \text{and} \quad & (\mathcal{A}_2, c^{A_2}) \models (\varphi_W \wedge \psi_{\min}(c) \wedge \chi_i(c)) \\ \text{imply} \quad & (\mathcal{A}_1, \min^{A_1}) \lhd (\mathcal{A}_2, c^{A_2}) \models \psi(c). \end{aligned}$$

By induction hypothesis there are plus free regular expressions r_i and s_i denoting the languages defined by $\varphi_W \wedge \psi_{\min}(\min) \wedge \psi_i(\min)$ and $\varphi_W \wedge \psi_{\min}(\min) \wedge \chi_i(\min)$, respectively. Then the plus free regular expression $\bigcup_{i \in I}(r_i s_i)$ denotes the language defined by $(\varphi_W \wedge \psi_{\min}(\min) \wedge \exists x(\neg x = \min \wedge \psi(x)))$. Note that, if $(\mathcal{A}_1 \lhd \mathcal{A}_2, \min^{A_1}, c^{A_2}) \models \psi(c)$ then (by 2.3.11) the pair $(\varphi^m_{(\mathcal{A}_1, \min^{A_1})}, \varphi^m_{(\mathcal{A}_2, c^{A_2})})$ of m-isomorphism types belongs (up to logical equivalence) to $\{(\psi_i(\min), \chi_i(c)) \mid i \in I\}$. □

Again, for $\mathbb{A} = \{a\}$ identify $\mathbb{A}^+$ with the set $\mathbb{N}_+$ of positive natural numbers. Examples 6.3.2 and 6.3.3 show that automata do not have the ability to count; for instance, they cannot recognize if a given string has prime length; more explicitly, the set $\{p \mid p \text{ a prime}\}$ is not accepted by an automaton. On the other hand, we saw in Example 6.3.2 that automata are capable to count modulo a natural number, e.g. the set $\{5n \mid n \geq 1\}$ is accepted by an automaton. But first-order logic even lacks this restricted counting ability;

in fact, it is an immediate consequence of 2.3.6 that a subset L of $\mathbb{N}_+$ is first-order definable iff for some $n \geq 1$

$$\{m \mid m \geq n\} \cap L = \emptyset \qquad \text{or} \qquad \{m \mid m \geq n\} \subseteq L.$$

Before we state the corresponding theorem for word models over *arbitrary* alphabets, we remark that we encounter this noncounting feature of first-order logic at various points in the book.

Theorem 6.4.3 *For a language $L \subseteq \mathbb{A}^+$ accepted by an automaton the following are equivalent:*

(i) *L is definable in first-order logic.*

(ii) *L is noncounting in the sense that there is an integer $k \geq 1$ such that for every $y \in \mathbb{A}^+$ and $x, z \in \mathbb{A}^*$,*

$$xy^k z \in L \qquad \textit{iff} \qquad xy^{k+1} z \in L.$$

Proof. We only prove the implication from (i) to (ii) and refer the reader to the literature for the involved proof of the other half (cf. 6.4.4). Suppose $\{\mathcal{B}_u \mid u \in L\} = \mathrm{Mod}(\varphi)$ for $\varphi \in \mathrm{FO}[\tau(\mathbb{A})]$. Let $k := 2^m + 1$, where m is the quantifier rank of φ. Then by 2.3.13, for any $y \in \mathbb{A}^+$ we have

$$\mathcal{B}_{y^k} \cong \lhd^k \mathcal{B}_y \equiv_m \lhd^{k+1} \mathcal{B}_y \cong \mathcal{B}_{y^{k+1}}.$$

Using 2.3.10, we obtain

$$\mathcal{B}_{xy^kz} \cong \mathcal{B}_x \lhd \mathcal{B}_{y^k} \lhd \mathcal{B}_z \equiv_m \mathcal{B}_x \lhd \mathcal{B}_{y^{k+1}} \lhd \mathcal{B}_z \cong \mathcal{B}_{xy^{k+1}z}.$$

In particular,

$$\mathcal{B}_{xy^kz} \models \varphi \qquad \text{iff} \qquad \mathcal{B}_{xy^{k+1}z} \models \varphi$$

and hence,

$$xy^k z \in L \qquad \text{iff} \qquad xy^{k+1} z \in L.$$

□

As the results of this section show, the plus operation cannot be captured in first-order logic. An instance of this operation can be viewed as the fixed-point of a monotone operation. In fact, let $L \subseteq \mathbb{A}^+$ be a language. Define $C_L : \mathrm{Pow}(\mathbb{A}^*) \to \mathrm{Pow}(\mathbb{A}^*)$ by

$$C_L(M) \quad := \quad L \cup ML.$$

Then

(a) C_L is monotone, that is, $M_1 \subseteq M_2$ implies $C_L(M_1) \subseteq C_L(M_2)$.

(b) For $n \geq 1$, $\underbrace{C_L(\ldots(C_L(\emptyset))\ldots)}_{\text{n-times}} = L \cup L^2 \cup \ldots \cup L^n$.

M is a *fixed-point* of C_L if $C_L(M) = M$. It can easily be proved that the least – with respect to set-theoretical inclusion – fixed-point of C_L is given by

$$C_L(\emptyset) \cup C_L(C_L(\emptyset)) \cup C_L(C_L(C_L(\emptyset))) \cup \ldots$$

Hence by (b),

$$L^+ \text{ is the least fixed-point of } C_L.$$

Fixed-points of monotone operations play a prominent role in the general theory. We shall study them in Chapters 7 and 8. In particular, we shall see that polynomial time queries on structures with an underlying ordering can be captured in a logic extending first-order logic by adding the ability to express fixed-points of definable monotone operations.

Notes 6.4.4 The connection between finite automata and monadic second-order logic as expressed in Theorem 6.2.3 is due to [15, 140]. For a proof of Theorem 6.4.3 and references on finite automata compare [136] or [138]. Some examples are taken from [17].

7. Descriptive Complexity Theory

In Chapter 1 we gave the example of a database D that contains the names of the main cities in the world and the pairs (a, b) of cities such that a given airline offers service from a to b without stopover. D may be interpreted as a first-order structure, more precisely, as a digraph $\mathcal{G} = (G, E^G)$, where G is the set of cities and $E^G ab$ means that there is a flight from a to b without stopover. Now, first-order logic may be viewed as a query language. For example, let

$$\varphi(x, y) \quad := \quad Exy \lor \exists z(Exz \land Ezy).$$

If φ is thought of as a query to D, the response will consist in the set of pairs (a, b) of cities such that a can be reached from b with at most one stop. First-order logic provides a rich class of database queries. However, there are plausible queries which are not first-order expressible. The query Q "Is it possible to fly from x to y using only the airline in question?" to databases of type D is an example. Also from a computational point of view answers to Q are more complex than answers to φ. Descriptive complexity theory analyzes the complexity of all queries definable in a given logic, the central question being the following: Given a complexity class $\mathcal{C}$, is there a logic $\mathcal{L}$ such that the queries definable in $\mathcal{L}$ are precisely the queries in $\mathcal{C}$? For short: Given a complexity class $\mathcal{C}$, is there a logic that captures $\mathcal{C}$? In the positive case, there mostly will be an effective procedure translating every formula of such a logic into a program for the corresponding query. In this sense the logic can be viewed as a higher programming language for $\mathcal{C}$.

In the first section we briefly introduce some extensions of first-order logic which have been considered with this aim in mind. Their model theory will be studied in Chapter 8. The second section is devoted to complexity classes of structures. In particular, we describe how structures may be viewed as inputs to Turing machines. In section 7.3 we show how certain logics can be used to describe computations of a given complexity class. In section 7.4 we look at the other direction and study the complexity of the satisfaction relation of such logics. By bringing together both sides we are lead to the descriptive characterizations of complexity classes we aim at, thus constituting a bridge between complexity theory and model theory. First consequences are given in the last section of this chapter.

Throughout this chapter all structures will be finite.

7.1 Some Extensions of First-Order Logic

We introduce inflationary fixed-point logic, partial fixed-point logic, deterministic transitive closure logic, and transitive closure logic. These logics are obtained from first-order logic by adding operations well-suited to describe iterative and recursive procedures, for example, the behaviour of computing machines. As already remarked in the introduction to this chapter, their model theory will be studied in Chapter 8.

Let M be a finite nonempty set. Denote by $\mathrm{Pow}(M)$ the power set of M. A function $F : \mathrm{Pow}(M) \to \mathrm{Pow}(M)$ induces a sequence

$$\emptyset,\ F(\emptyset),\ F(F(\emptyset)),\ldots$$

of subsets of M. For its members we write $F_0, F_1, \ldots$. So $F_0 = \emptyset$ and $F_{n+1} = F(F_n)$. Suppose there is an $n_0 \geq 0$ such that $F_{n_0+1} = F_{n_0}$, that is, $F(F_{n_0}) = F_{n_0}$. Then $F_m = F_{n_0}$ for all $m \geq n_0$. We denote F_{n_0} by F_∞ and say that the *fixed-point* F_∞ of F *exists*. In case the fixed-point F_∞ does not exist, we agree to set $F_\infty := \emptyset$.

F is said to be *inflationary* if for all $X \subseteq M$,

$$X \subseteq F(X).$$

Lemma 7.1.1 (a) *The sequence $(F_n)_{n\geq 0}$ is periodic, more precisely: There are $m < 2^{\|M\|}$ and $l \geq 1$ such that $F_k = F_{k+l}$ for all $k \geq m$.*
(b) *If F_∞ exists then $F_\infty = F_{2^{\|M\|}-1}$.*
(c) *If F is inflationary then F_∞ exists and $F_\infty = F_{\|M\|}$.*

Proof. (a) As $\mathrm{Pow}(M)$ has $2^{\|M\|}$ elements, there are $m < 2^{\|M\|}$ and $l \geq 1$ such that $F_m = F_{m+l}$. Therefore, $F_{m+1} = F(F_m) = F(F_{m+l}) = F_{m+1+l}$, $F_{m+2} = F_{m+2+l}, \ldots$.

(b) Choose $m < 2^{\|M\|}$ and $l \geq 1$ according to (a). If $F_m = F_{m+1}$ then $F_m = F_{2^{\|M\|}-1} = F_\infty$. If $F_m \neq F_{m+1}$ then, by (a), for $s \geq m$, we get $(F_m =)\, F_{m+s\cdot l} \neq F_{m+1+s\cdot l}\, (= F_{m+1})$ and hence, F_∞ does not exist.

(c) By assumption, $F_0 \subseteq F_1 \subseteq \ldots \subseteq M$. Since M has $\|M\|$ elements, this sequence must get constant not later than with $F_{\|M\|}$. □

Let $\varphi(x_1, \ldots, x_k, \overline{u}, X, \overline{Y})$ be a formula in the vocabulary τ, where the relation variable X has arity k; moreover, let $\mathcal{A}$ be a τ-structure, $\overline{b}$ an interpretation of $\overline{u}$ in A, and $\overline{S}$ an interpretation of $\overline{Y}$ over A. Then $\varphi, \mathcal{A}, \overline{b}$, and $\overline{S}$ give rise to an operation $F^\varphi : \mathrm{Pow}(A^k) \to \mathrm{Pow}(A^k)$ defined by

$$F^\varphi(R) \quad := \quad \{(a_1, \ldots, a_k) \mid \mathcal{A} \models \varphi[a_1, \ldots, a_k, \overline{b}, R, \overline{S}]\}.$$

(The notation F^φ does not make explicit all relevant data.)

To give an example, let $\mathcal{G} = (G, E^G)$ be a graph and

$$\varphi_0(x, y, X) \quad := \quad (Exy \vee \exists z(Xxz \wedge Ezy))$$

with xy corresponding to $\overline{x}$ above. Then, $F_0^{\varphi_0} = \emptyset$, $F_1^{\varphi_0} = F^{\varphi_0}(\emptyset) = E^G$, $F_2^{\varphi_0} = F^{\varphi_0}(E^G) = E^G \cup \{(a,b) \mid (E^G ac \text{ and } E^G cb) \text{ for some } c \in G\}$. By induction on n one shows that

$$F_n^{\varphi_0} = \{(a,b) \mid \text{there is a path of length} \leq n \text{ from } a \text{ to } b\},$$

and hence,

$$F_\infty^{\varphi_0} = \{(a,b) \mid \text{there is a path from a to b}\}.$$

Note that for φ as above the function $F^{(X\overline{x} \vee \varphi)}$ (where $\overline{x} = x_1 \dots x_k$) is inflationary.

One obtains *Inflationary Fixed-Point Logic* FO(IFP) and *Partial Fixed-Point Logic* FO(PFP) by closing first-order logic FO under inflationary and arbitrary fixed-points of definable operations, respectively. We state the precise definitions.

For a vocabulary τ the class FO(IFP)$[\tau]$ of formulas of FO(IFP) of vocabulary τ is given by the calculus (we use the succinct notation

$$\frac{\varphi_1, \dots, \varphi_n}{\varphi}$$

for the clause "If $\varphi_1, \dots, \varphi_n$ are formulas then φ is a formula")

- $\dfrac{\quad}{\varphi}$ where φ is an atomic second-order formula over τ
- $\dfrac{\varphi}{\neg\varphi}$; $\dfrac{\varphi, \psi}{(\varphi \vee \psi)}$; $\dfrac{\varphi}{\exists x \varphi}$
- $\dfrac{\varphi}{[\text{IFP}_{\overline{x},X}\varphi]\,\overline{t}}$ where the lengths of $\overline{x}$ and $\overline{t}$ are the same and coincide with the arity of X.

For FO(PFP) we replace the last rule by

- $\dfrac{\varphi}{[\text{PFP}_{\overline{x},X}\varphi]\,\overline{t}}$ where the lengths of $\overline{x}$ and $\overline{t}$ are the same and coincide with the arity of X.

Sentences are formulas without free first-order and second-order variables, where the free occurrence of variables is defined in the standard way, adding, for example, for FO(IFP) the clause

$$\text{free}([\text{IFP}_{\overline{x},X}\varphi]\,\overline{t}) \quad := \quad \text{free}(\overline{t}) \cup (\text{free}(\varphi) \setminus \{\overline{x}, X\}).$$

The semantics is defined inductively w.r.t. the calculus above,

$$[\text{IFP}_{\overline{x},X}\varphi]\,\overline{t} \quad \text{meaning that} \quad \overline{t} \in F_\infty^{(X\overline{x} \vee \varphi)}$$

and

$$[\text{PFP}_{\overline{x},X}\varphi]\,\overline{t} \quad \text{meaning that} \quad \overline{t} \in F_\infty^\varphi.$$

More precisely: If X is k-ary and if the variables free in $[\text{IFP}_{\overline{x},X}\varphi]\,\overline{t}$ are among $\overline{u}$ and $\overline{Y}$, and $\overline{b}$ and $\overline{S}$ are interpretations in $\mathcal{A}$ of $\overline{u}$ and $\overline{Y}$, respectively, then

$$\mathcal{A} \models [\text{IFP}_{\overline{x},X}\varphi]\,\overline{t}\,[\overline{b},\overline{S}] \qquad \text{iff} \qquad (t_1[\overline{b}],\ldots,t_k[\overline{b}]) \in F_\infty^{(X\overline{x}\vee\varphi)},$$

$$\mathcal{A} \models [\text{PFP}_{\overline{x},X}\varphi]\,\overline{t}\,[\overline{b},\overline{S}] \qquad \text{iff} \qquad (t_1[\overline{b}],\ldots,t_k[\overline{b}]) \in F_\infty^{\varphi}.$$

Examples 7.1.2 (a) In the language of graphs the formula

$$\psi_0(x,y) \quad := \quad [\text{IFP}_{xy,X}\,(Exy \,\vee\, \exists z(Xxz \wedge Ezy))]\,xy$$

of FO(IFP) expresses that x, y are connected by a path. Hence, the class CONN of connected graphs is axiomatizable in FO(IFP) by $\forall x\forall y(\neg x = y \rightarrow \psi_0(x,y))$ (and the graph axioms). It is not axiomatizable in FO (cf. 2.3.8).

(b) For $\tau = \{<, S, \min, \max\}$ the sentence

$$\neg[\text{IFP}_{x,X}\,(x = \min \vee \exists y\exists z(Xy \wedge Syz \wedge Szx))]\max$$

of FO(IFP) together with the ordering axioms axiomatizes the class of orderings of even cardinality. The same holds for the class of orderings of odd cardinality and the FO(PFP)-sentence $\exists x\,[\text{PFP}_{x,X}\,\psi(x,X)]\,x$ where

$$\psi(x,X) = (\forall y\,\neg Xy \wedge x = \min) \vee (X\max \wedge x = \max) \vee \exists y(Xy \wedge \exists u(Syu \wedge Sux)).$$

□

To compare the expressive power of logics we introduce the following relations.

Definition 7.1.3 Let $\mathcal{L}_1$ and $\mathcal{L}_2$ be logics.

(a) $\mathcal{L}_1 \leq \mathcal{L}_2$ (read: *$\mathcal{L}_1$ is at most as expressive as $\mathcal{L}_2$*) if for every τ and every sentence $\varphi \in \mathcal{L}_1[\tau]$ there is a sentence $\psi \in \mathcal{L}_2[\tau]$ such that $\text{Mod}(\varphi) = \text{Mod}(\psi)$.[1]

(b) $\mathcal{L}_1 \equiv \mathcal{L}_2$ (read: *$\mathcal{L}_1$ and $\mathcal{L}_2$ have the same expressive power*) if $\mathcal{L}_1 \leq \mathcal{L}_2$ and $\mathcal{L}_2 \leq \mathcal{L}_1$.

(c) $\mathcal{L}_1 < \mathcal{L}_2$ if $\mathcal{L}_1 \leq \mathcal{L}_2$ and not $\mathcal{L}_2 \leq \mathcal{L}_1$. □

In most cases, a proof of $\mathcal{L}_1 \leq \mathcal{L}_2$ even yields that every *formula* of $\mathcal{L}_1$ is equivalent to a formula of $\mathcal{L}_2$. In particular, $\mathcal{L}_1 \leq \mathcal{L}_2$ implies that $\mathcal{L}_1 \leq \mathcal{L}_2$ holds for all formulas of $\mathcal{L}_1$ containing only free *individual* variables (one simply replaces these variables by new constants). Thus, $\mathcal{L}_1 \leq \mathcal{L}_2$ will imply that every global relation definable in $\mathcal{L}_1$ is also definable in $\mathcal{L}_2$.

By Example 7.1.2(a), we have FO < FO(IFP).

---

[1] $\mathcal{L}[\tau]$ denotes the class of formulas of $\mathcal{L}$ of vocabulary τ.

Proposition 7.1.4 FO(IFP) $\leq$ FO(PFP).

Proof. Note that $[\mathrm{IFP}_{\overline{x},X}\varphi]\,\overline{t}$ is equivalent to $[\mathrm{PFP}_{\overline{x},X}(X\overline{x} \vee \varphi)]\,\overline{t}$. $\square$

Let R be a binary relation on a set M, $R \subseteq M^2$. The *transitive closure* TC(R) of R is defined by

$$\begin{aligned}\mathrm{TC}(R) \ := \ & \{(a,b) \in M^2 \mid \text{there exist } n > 0 \text{ and } e_0,\ldots,e_n \in M \text{ such} \\ & \text{that } a = e_0, b = e_n, \text{ and for all } i < n,\ (e_i, e_{i+1}) \in R\}.\end{aligned}$$

And the *deterministic transitive closure* DTC(R) is defined by

$$\begin{aligned}\mathrm{DTC}(R) \ := \ & \{(a,b) \in M^2 \mid \text{there exist } n > 0 \text{ and } e_0,\ldots,e_n \in M \text{ such} \\ & \text{that } a = e_0, b = e_n, \text{ and for all } i < n,\ e_{i+1} \text{ is the} \\ & \text{unique } e \text{ for which } (e_i, e) \in R\}.\end{aligned}$$

Transitive Closure Logic FO(TC) and *Deterministic Transitive Closure Logic* FO(DTC) are obtained by closing FO under the transitive closure and the deterministic transitive closure of definable relations, respectively. More precisely: For a vocabulary τ the class of formulas of FO(TC) of vocabulary τ is given by the calculus

- $\dfrac{}{\varphi}$ where φ is an atomic first-order formula over τ
- $\dfrac{\varphi}{\neg\varphi}$; $\dfrac{\varphi, \psi}{(\varphi \vee \psi)}$; $\dfrac{\varphi}{\exists x \varphi}$
- $\dfrac{\varphi}{[\mathrm{TC}_{\overline{x},\overline{y}}\,\varphi]\,\overline{s}\overline{t}}$ where the variables in $\overline{x}\,\overline{y}$ are pairwise distinct and where the tuples $\overline{x}, \overline{y}, \overline{s}$, and $\overline{t}$ are all of the same length, $\overline{s}$ and $\overline{t}$ being tuples of terms.

For FO(DTC) the last rule is replaced by $\dfrac{\varphi}{[\mathrm{DTC}_{\overline{x},\overline{y}}\,\varphi]\,\overline{s}\overline{t}}$ with the same side conditions. We define

$$\mathrm{free}([\mathrm{TC}_{\overline{x},\overline{y}}\,\varphi]\,\overline{s}\overline{t}) \ := \ \mathrm{free}(\overline{s}) \cup \mathrm{free}(\overline{t}) \cup (\mathrm{free}(\varphi) \setminus \{\overline{x}, \overline{y}\}),$$

and similarly for FO(DTC).

The meaning of $[\mathrm{TC}_{\overline{x},\overline{y}}\,\varphi(\overline{x},\overline{y},\overline{u})]\,\overline{s}\overline{t}$ is

$$(\overline{s},\overline{t}) \in \mathrm{TC}(\{(\overline{x},\overline{y}) \mid \varphi(\overline{x},\overline{y},\overline{u})\}),$$

and the meaning of $[\mathrm{DTC}_{\overline{x},\overline{y}}\,\varphi(\overline{x},\overline{y},\overline{u})]\,\overline{s}\overline{t}$ is

$$(\overline{s},\overline{t}) \in \mathrm{DTC}(\{(\overline{x},\overline{y}) \mid \varphi(\overline{x},\overline{y},\overline{u})\})$$

(here $\{(\overline{x},\overline{y}) \mid \varphi(\overline{x},\overline{y},\overline{u})\}$ is considered as a binary relation on the set of length($\overline{x}$)-tuples of the universe).

Thus a graph is connected if it is a model of $\forall x \forall y(\neg x = y \rightarrow [\mathrm{TC}_{x,y} Exy]\,xy)$.

Proposition 7.1.5 (a) FO(DTC) $\leq$ FO(TC).
(b) FO(TC) $\leq$ FO(IFP).

Proof. For (a) note that

$$\models_{\text{fin}} [\text{DTC}_{\overline{x},\overline{y}}\varphi(\overline{x},\overline{y},\overline{u})]\,\overline{s}\overline{t} \leftrightarrow [\text{TC}_{\overline{x},\overline{y}}(\varphi(\overline{x},\overline{y},\overline{u}) \wedge \forall\overline{z}(\varphi(\overline{x},\overline{z},\overline{u}) \rightarrow \overline{z}=\overline{y}))]\,\overline{s}\overline{t}$$

and for (b) that

$$\models_{\text{fin}} [\text{TC}_{\overline{x},\overline{y}}\varphi(\overline{x},\overline{y},\overline{u})]\,\overline{s}\overline{t} \leftrightarrow [\text{IFP}_{\overline{x}\,\overline{y},X}(\varphi(\overline{x},\overline{y},\overline{u}) \vee\ \exists\overline{v}(X\overline{x}\,\overline{v} \wedge \varphi(\overline{v},\overline{y},\overline{u})))]\,\overline{s}\overline{t}.$$

□

In the following, axiomatizability in a logic (in the sense of the following definition) will be a major issue.

Definition 7.1.6 Let K be a class of τ-structures and $\mathcal{L}$ a logic. K is *axiomatizable* in $\mathcal{L}$, if there is a sentence of $\mathcal{L}$ of vocabulary τ such that $K = \text{Mod}(\varphi)$. □

Sometimes, when relating logics and complexity classes it is convenient to restrict oneself to sufficiently large structures. This does not affect problems of axiomatizability. In fact, for a class K of structures and $m \geq 1$ denote by K_m the subclass of K of structures of cardinality $\geq m$,

$$K_m \quad := \quad \{\mathcal{A} \mid \mathcal{A} \in K, \|A\| \geq m\}.$$

For every finite structure $\mathcal{A}$ there is a sentence $\varphi_{\mathcal{A}}$ of FO characterizing $\mathcal{A}$ up to isomorphism (cf. 2.1.1), that is,

$$\mathcal{B} \models \varphi_{\mathcal{A}} \ \text{ iff } \ \mathcal{B} \cong \mathcal{A}$$

holds for all $\mathcal{B}$. Hence, for any logic $\mathcal{L}$ with FO $\leq \mathcal{L}$,

$$K \text{ is axiomatizable in } \mathcal{L} \ \text{ iff } \ K_m \text{ is axiomatizable in } \mathcal{L}.$$

In fact, setting $\varphi_m := \bigvee\{\varphi_{\mathcal{A}} \mid \mathcal{A} \in K, \|A\| < m\}$ we have that $K = \text{Mod}(\varphi)$ implies $K_m = \text{Mod}(\varphi \wedge \neg\varphi_m)$, and $K_m = \text{Mod}(\psi)$ implies $K = \text{Mod}(\psi \vee \varphi_m)$.

7.2 Turing Machines and Complexity Classes

The aim of the present section is to recall basic definitions and results from computation theory, to fix our computation model for structures as inputs, and to introduce the corresponding complexity classes. It is not intended as a substitute for a course in computation theory, and some familiarity with the basic notions will be helpful. For references compare the notes 7.5.26.

We adopt a computation model which belongs to the most popular ones in theoretical computer science, the Turing machine model. Our choice is motivated by the fact that Turing machine computations allow for simple descriptions and for natural definitions of complexity classes. However, as Turing machines deal only with strings, any other data must be coded by strings. The tedious work of encoding is one price we have to pay for the advantages. Moreover, the model does not well reflect parallel, distributed, or real time computations, as only local changes are made in a step. However, these shortages are not really important for our aims.

For the following, we fix a finite alphabet $\mathbb{A}$. A *Turing machine* M is a finite device that performs operations on a tape which is bounded to the left and unbounded to the right and divided into *squares* (or *cells*). The machine operates stepwise, each step leading from one situation to a new one. In any situation every square of the tape either contains a single symbol from $\mathbb{A}$ or is blank. In the latter case we say that it contains the symbol "blank". There is one exception: the leftmost or "virtual" cell always contains an endmark, the "virtual" letter α (which is not in $\mathbb{A}$). M has a head which, in any situation, scans a single square of the tape and, in any step of a computation, erases or replaces the scanned symbol by another one and moves one cell to the left or to the right or remains at its place (we speak of a *read-and-write head*).

In every situation, M is in one of the states of a finite set State(M), the *set of states* of M. State(M) contains a special state s_0, the *initial* state, and special states s_+, the *accepting* state, and s_-, the *rejecting* state. We assume that s_0, s_+, and s_- are pairwise distinct. The action or behaviour of M in a situation depends on the current state of M and on the symbol currently being scanned by the head. It is given by Instr(M), the *set of instructions* of M. Each instruction has the form

$$(*) \qquad sa \to s'bh$$

where

- $s, s' \in \text{State}(M), \; s \neq s_+, \; s \neq s_-$
- $a, b \in \mathbb{A} \cup \{\alpha, \text{blank}\}$ and $(a = \alpha \text{ iff } b = \alpha)$
- $h \in \{-1, 0, 1\}$, and if $a = \alpha$ then $h \neq -1$.

The instruction $(*)$ means: If you are in state s and your head scans a cell with symbol a, replace a by b, move your head one cell to the left ($h = -1$), or to the right ($h = 1$), or don't move ($h = 0$); finally, change to state s'.

A machine M is *deterministic*, if for all $s \in S$ and $a \in \mathbb{A} \cup \{\alpha, \text{blank}\}$ there is at most one instruction of the form $(*)$ in Instr(M). In order to emphasize that a machine is not required to be deterministic we sometimes call it *nondeterministic*.

As usual, denote by $\mathbb{A}^*$ the set of words over $\mathbb{A}$ and by $\mathbb{A}^+$ the set of nonempty words over $\mathbb{A}$. Let $u \in \mathbb{A}^*$, $u = a_1 \ldots a_r$ with $a_i \in \mathbb{A}$. M is *started* with u if M begins a computation (or *run*) in state s_0 in the situation

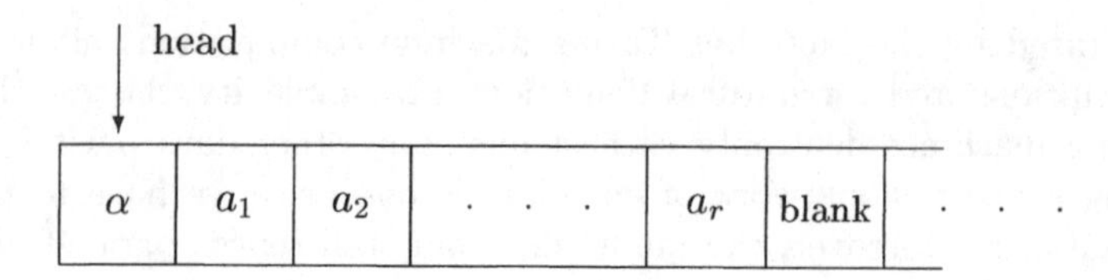

The computation proceeds stepwise, each step corresponding to the execution of one instruction of M. The machine stops when it is in a state s scanning a symbol $a \in \mathbb{A} \cup \{\alpha, \text{blank}\}$ such that there is no instruction of the form $(*)$ in $\text{Instr}(M)$. If then $s = s_+$ we speak of an *accepting run*, and if $s = s_-$ of a *rejecting run. M accepts u* if there is at least one accepting run of M started with u, and *M rejects u* if all runs started with u are finite and rejecting.

Subsets of $\mathbb{A}^+$ are called *languages.* A language $L \subseteq \mathbb{A}^+$ is *accepted* by M if for all $u \in \mathbb{A}^+$,

$$M \text{ accepts } u \quad \text{iff} \quad u \in L.$$

L is *decided* by M if, in addition,

$$M \text{ rejects } u \quad \text{iff} \quad u \notin L.$$

Clearly, if M decides L then M accepts L. L is said to be *decidable* if it is decided by some deterministic Turing machine, and *acceptable* or *enumerable* if it is accepted by some nondeterministic Turing machine.

For a function $f : \mathbb{N} \to \mathbb{N}$ we say that M is *f time-bounded*, if for all $u \in \mathbb{A}^+$ accepted by M there is an accepting run of M started with u which has length at most $f(|u|)$ (recall that $|u|$ denotes the length of the word u). And M is *f space-bounded*, if for all $u \in \mathbb{A}^+$ accepted by M there is an accepting run which uses at most $f(|u|)$ cells before stopping.

Denote by $\mathbb{N}[x]$ the set of polynomials with coefficients from $\mathbb{N}$. A language $L \subseteq \mathbb{A}^+$ is in PTIME ("polynomial time") or in PSPACE ("polynomial space"), if it is accepted by a deterministic machine that is p time-bounded or p space-bounded, respectively, for some polynomial $p \in \mathbb{N}[x]$. The classes NPTIME ("nondeterministic polynomial time") and NPSPACE ("nondeterministic polynomial space") are defined similarly, now allowing nondeterministic machines.

Immediately from the definitions one gets

$$\text{PTIME} \subseteq \text{NPTIME} \quad \text{and} \quad \text{PTIME} \subseteq \text{PSPACE} \subseteq \text{NPSPACE},$$

and one can show that

$$\text{NPTIME} \subseteq \text{PSPACE} \quad \text{and} \quad \text{PSPACE} = \text{NPSPACE}.$$

Hence,

$$\text{PTIME} \subseteq \text{NPTIME} \subseteq \text{PSPACE} \ (= \text{NPSPACE}).$$

Let $\mathbb{A} = \{a, b\}$ and $L := \{u \in \mathbb{A}^+ \mid u \text{ contains an even number of } a\text{'s}\}$. One can easily design a machine that accepts L and is time-bounded by the polynomial $x + 2$: the head just runs over the string, the state being "even" or "odd" depending on whether the number of a's already scanned is even or odd, respectively. Essentially, this machine does not need any "working space" but only the space for the input. In order to measure only the working space and to introduce complexity classes like LOGSPACE and NLOGSPACE, where the working space needed is smaller than the input space, it is convenient to separate the input and to introduce machines with an input tape and a work tape. For other purposes it might be useful to introduce several input tapes and several work tapes, maybe also (if calculating a function, for example) one or more output tapes, with a head for each tape, where the heads can move independently of each other. It turns out that the definition of the usual complexity classes does not depend on the number of tapes or on other peculiarities such as the form of the tape, i.e., whether it is unbounded to both sides or not. We shall often use this robustness, choosing, for example, the number of input and work tapes according to needs and to convenience.

7.2.1 Digression: Trahtenbrot's Theorem

We have already remarked that first-order logic, when restricted to the finite, does not admit a complete proof calculus. In the following we give a proof of Trahtenbrot's Theorem which immediately implies this fact.

Fix an alphabet $\mathbb{A}$. It is well-known that it is not decidable whether a deterministic Turing machine M accepts the empty word, by short, whether *M halts* ("Undecidability of the Halting Problem").[2] We use this result to show for a certain vocabulary $\sigma(\mathbb{A})$:

Theorem 7.2.1 (Trahtenbrot's Theorem) *Finite Satisfiability is not decidable, that is, the set*

$$\mathrm{Sat}[\sigma(\mathbb{A})] := \{\varphi \mid \varphi \textit{ is a sentence of } \mathrm{FO}[\sigma(\mathbb{A})] \textit{ satisfiable in the finite}\}$$

is not decidable.

Proof. We assign, in an effective way, to every deterministic machine M over $\mathbb{A}$ a sentence φ_M of $\mathrm{FO}[\sigma(\mathbb{A})]$ such that

$$(*) \qquad \varphi_M \text{ is satisfiable in the finite iff } M \text{ halts.}$$

This gives the claim by the result just mentioned. Without loss of generality we may restrict ourselves to deterministic Turing machines M whose set of states is an initial segment $\{0, \ldots, s_M\}$ of the natural numbers, $s_0 := 0$ being the initial state and 1 the accepting state, and which stop only in the accepting state. We number the cells of the tape as indicated by

[2] More exactly: One encodes Turing machines M in some natural way as words w_M. Then the language $\{w_M \mid M \text{ halts}\}$ is not decidable.

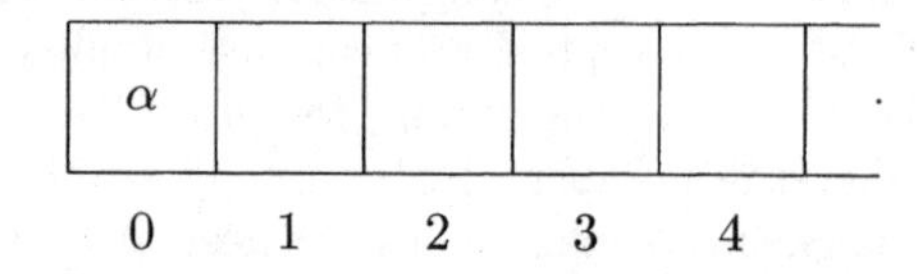

In particular, the number 0 is given to the virtual cell. If the machine M makes at least n steps, we let C_n denote the n-th *configuration*. C_n contains the following data: the state, the number of the cell scanned by the head, and the tape inscription after n steps.

Let $\tau_0 = \{<, S, \min, \max\}$ be the vocabulary for orderings. In the following we sometimes write 0 instead of min. In addition to the symbols of τ_0 the vocabulary $\sigma(\mathbb{A})$ contains relation symbols State (binary), Head (binary), and for every $a \in \mathbb{A} \cup \{\alpha, \text{blank}\}$ a binary relation symbol Letter_a.

For every natural number $n \geq s_M$ (recall that $\{0, \ldots, s_M\}$ is the set of states of M) we define a structure $\mathcal{A}_n$ with universe $\{0, \ldots, n\}$, which reflects the initial segment $C_0, \ldots, C_n$ of the computation of M started with the empty word (however, only $C_0, \ldots, C_k$ with $k < n$ if M stops after k steps). For $s, t \leq n$,

$$\begin{array}{lll} \text{State}^{\mathcal{A}_n}\, s\, t & \text{iff} & \text{according to } C_t \text{ the state is } s \\ \text{Head}^{\mathcal{A}_n}\, i\, t & \text{iff} & \text{according to } C_t \text{ the head is in cell } i \\ \text{Letter}_a^{\mathcal{A}_n}\, i\, t & \text{iff} & \text{according to } C_t \text{ the letter } a \text{ is in cell } i. \end{array}$$

The sentence φ_M will have the properties (a) and (b) which immediately give the equivalence $(*)$ stated above:

(a) If M, started with the empty word, stops after k steps (in the accepting state) and $n \geq s_M, k$, then $\mathcal{A}_n \models \varphi_M$.
(b) If $\mathcal{A}$ is a finite model of φ_M and M, started with the empty word, runs at least k steps, then $\|A\| \geq k$.

As φ_M we take the conjunction of the $\{<, S, \min, \max\}$-ordering axioms together with the conjunction of the sentences in (1)-(4) (where we write 0 for min, 1 for the successor of min, etc.):

(1) "The universe has at least $s_M + 1$ elements."
(2) $\text{State}\, 0\, 0 \wedge \text{Head}\, 0\, 0 \wedge \text{Letter}_\alpha\, 0\, 0 \wedge \forall x(\neg x = 0 \to \text{Letter}_{\text{blank}}\, x\, 0)$
(at time 0 the state is 0, the head scans the virtual cell, the virtual cell contains α, and all other cells are empty).
(3) For each instruction $sa \to s'bh$ a conjunct $\varphi_{sa \to s'bh}$, which describes the changes due to this instruction. For example, if $h = 0$ then $\varphi_{sa \to s'bh}$ is the sentence

$$\begin{aligned} &\forall y \forall t((\text{State}\, s\, t \wedge \text{Head}\, y\, t \wedge \text{Letter}_a\, y\, t) \\ &\qquad \to \exists t'(S t\, t' \wedge \text{State}\, s'\, t' \wedge \text{Head}\, y\, t' \wedge \text{Letter}_b\, y\, t' \\ &\qquad \wedge \forall v(\neg v = y \to \bigwedge_{a \in \mathbb{A} \cup \{\alpha, \text{blank}\}} (\text{Letter}_a\, v\, t \to \text{Letter}_a\, v\, t')))). \end{aligned}$$

We leave it to the reader to write down the sentence $\varphi_{sa\to s'bh}$ for $h=-1$ and $h=1$.

(4) $\exists t$ State $1\,t$
(the accepting state is reached). □

Coding $\sigma(\mathbb{A})$-structures by graphs (cf. 11.2.5), one obtains from the undecidability of Sat$[\sigma(\mathbb{A})]$, that for a binary relation symbol E the set

$$\text{Graph-Sat} \;:=\; \{\varphi \mid \varphi \text{ is an FO}[\{E\}]\text{-sentence satisfiable in a finite graph}\}$$

is not decidable. Hence, for any vocabulary τ containing an at least binary relation symbol, the set

$$\text{Sat}[\tau] \;:=\; \{\varphi \mid \varphi \text{ is an FO}[\tau]\text{-sentence satisfiable in the finite}\}$$

is not decidable.

Clearly, there is a decision procedure that, for any finite structure $\mathcal{A}$ whose universe is a set of natural numbers and for any first-order sentence φ, checks whether $\mathcal{A} \models \varphi$. Since φ is satisfiable in the finite iff there is a model $\mathcal{A}$ of φ with $A = \{0,\ldots,n\}$ for some n, we can use this decision procedure to enumerate Sat$[\tau]$. Together with Trahtenbrot's Theorem this shows:

Theorem 7.2.2 *If τ contains an at least binary relation symbol then the set*

$$\text{Val}[\tau] \;:=\; \{\varphi \in \text{FO}[\tau] \mid \varphi \textit{ is a sentence valid in all finite structures}\}$$

of sentences valid in all finite structures is not enumerable.

Proof. For any sentence φ we have

$$(+) \qquad \varphi \notin \text{Sat}[\tau] \quad \text{iff} \quad \neg\varphi \in \text{Val}[\tau].$$

If Val$[\tau]$ would be enumerable, then by (+), FO$[\tau]\setminus$ Sat$[\tau]$ would be enumerable, too. This would lead to a decision procedure for Sat$[\tau]$ (contradicting Trahtenbrot's Theorem): Given a sentence φ, start enumeration procedures for Sat$[\tau]$ and FO$[\tau] \setminus$ Sat$[\tau]$ until one of them yields φ. □

Now we have the negative result concerning the proof calculus: If there would be a complete proof calculus for first-order logic in the finite, we could effectively enumerate all possible formal proofs and hence, also the sentences valid in the finite, contradicting Theorem 7.2.2.

7.2.2 Structures as Inputs

We fix the conventions which allow to regard finite structures as inputs to Turing machines. Note that, in general, structures are abstract objects and the same holds for their elements, so that there is no canonical way of representing structures by strings (or by sequences of strings). If the same structure has different representations, the machine should produce the same answer on all the representations. We postpone this problem and first make life easier by restricting our attention to ordered structures. Recall the definition (cf. 1.A2):

Definition 7.2.3 Let $\{<\} \subseteq \tau_0 \subseteq \{<, S, \min, \max\}$ and $\tau_0 \subseteq \tau$. A τ-structure $\mathcal{A}$ is *ordered* if the reduct $\mathcal{A}|\tau_0$ is an ordering (that is, $<^A$ is an ordering and S, min, and max, if present, are interpreted by the successor relation, the least and the last element of the ordering, respectively). $\mathcal{O}[\tau]$ is the class of ordered τ-structures. If ψ is a sentence in the vocabulary τ, ordMod(ψ) denotes the class of ordered models of ψ, or equivalently, ordMod(ψ) = Mod($\psi \wedge \psi_0$), where ψ_0 is the conjunction of the ordering axioms for the vocabulary τ_0. □

Let $\mathcal{A} \in \mathcal{O}[\tau]$ be an ordered structure with $\|A\| = n$. By passing to an isomorphic copy we can assume – and always will tacitly do so – that $A = \{0, \ldots, n-1\}$ and that $<^A$ is the natural ordering on this set; that is, we identify or "label" the least element of $<^A$ in $\mathcal{A}$ with 0, its successor with 1, etc.[3]

Suppose $\tau = \tau_0 \dot\cup \tau_1$, say $\tau_1 = \{R_1, \ldots, R_k, c_1, \ldots, c_l\}$ (when writing τ_1 in this way we tacitly assume that the symbols in τ_1 are given in the order $R_1, \ldots, R_k, c_1, \ldots, c_l$), and τ_0 as in the preceding definition.

A *Turing machine for τ-structures* will have $1 + k + l$ *input tapes* and m *work tapes* for some $m \geq 1$. All tapes are bounded to the left and unbounded to the right. Their cells are numbered as indicated by[4]

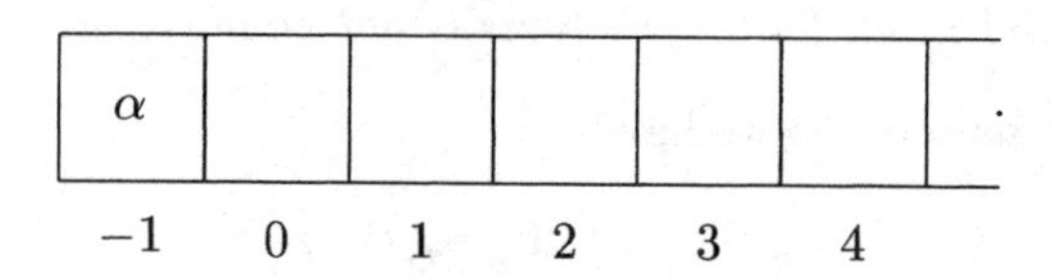

The "virtual" cell is numbered by -1 and always contains the virtual letter α. All input tapes will contain an input word followed by the virtual letter ω indicating the end of the input word. Each tape has its own head. The heads can move independently of each other. Those on input tapes are *read-only* heads (i.e., they do not change the scanned letter), while those on the work tapes are *read-and-write* heads. The alphabet only contains the symbol "1". Moreover, we identify "0" with "blank"; thus during a computation almost all cells contain 0.

[3] By this convention, for isomorphic $\mathcal{A}, \mathcal{B} \in \mathcal{O}[\tau]$ a Turing machine M started with $\mathcal{A}$ will have the same input inscriptions as M started with $\mathcal{B}$.

[4] For the purposes of the next sections this numbering is more appropriate than that of the section on Trahtenbrot's Theorem, where we numbered the virtual cell by 0.

With an ordered τ-structure $\mathcal{A}$ we associate the following input inscriptions on the $1+k+l$ input tapes (numbered from 0 to $k+l$): The 0-th tape, the "universe tape", contains a sequence of $1's$ of length $n := \|A\|$,

α	1	1	. . .	1	ω
-1	0	1		$n-1$	n

For $1 \leq i \leq k$, the i-th input tape contains the information about $R := R_i$ coded as follows: Suppose R is r-ary, that is, $R^A \subseteq \{0,\ldots,n-1\}^r$. Of course, $\|\{0,\ldots,n-1\}^r\| = n^r$. Then, for $j < n^r$, the j-th cell will contain "1" just in case the j-th r-tuple in the lexicographic ordering of $\{0,\ldots,n-1\}^r$ is in R. More formally: For $j < n^r$, let $|j|_r$ be the j-th r-tuple in the lexicographic ordering of $\{0,\ldots,n-1\}^r$, i.e., look at the unique n-adic representation of j,

$$j = j_1 \cdot n^{r-1} + j_2 \cdot n^{r-2} + \ldots + j_{r-1} \cdot n + j_r \quad \text{with} \quad 0 \leq j_i < n,$$

and set $|j|_r := (j_1,\ldots,j_r)$. (If n is not clear from the context, we write $|j|_r^n$ instead of $|j|_r$.) Then the i-th input tape has the inscription

α	a_0	a_1	a_2	a_3	. . .	a_{n^r-1}	ω
-1	0	1	2	3		n^r-1	n^r

where

$$a_j = 1 \quad \text{iff} \quad R^A|j|_r$$

(and hence, $a_j = 0$ iff not $R^A|j|_r$).

For $1 \leq i \leq l$, the $(k+i)$-th input tape contains the binary representation of $j := c_i^A$ without leading zeros.

We say that a Turing machine M is *started with* $\mathcal{A}$, if the input tapes contain the information on $\mathcal{A}$ in the way just described, the work tapes are empty, and each head scans the cell numbered 0 of its tape. As in the case of one-tape machines, M has finite sets State(M) of states and Instr(M) of instructions. State(M) contains an *initial* (or *starting*) state s_0, an *accepting* state s_+, and a *rejecting* state s_-. *Instructions* now have the form

$$(*) \qquad s\, b_0 \ldots b_{k+l}\, c_1 \ldots c_m \rightarrow s' c_1' \ldots c_m'\, h_0 \ldots h_{k+l+m}$$

with the meaning

> If you are in state s, your heads scan $b_0, \ldots, b_{k+l}$ on the input tapes and $c_1, \ldots, c_m$ on the work tapes, replace $c_1, \ldots, c_m$ by $c'_1, \ldots, c'_m$, move the i-th head according to h_i and, finally, change to state s'.

Here, $s, s' \in \text{State}(M)$, $b_0, \ldots, b_{k+l} \in \{0, 1, \alpha, \omega\}$, $c_1, \ldots, c_m$, $c'_1, \ldots, c'_m \in \{0, 1, \alpha\}$, and $h_0, \ldots, h_{k+l+m} \in \{-1, 0, 1\}$. Moreover,

- if $b_j = \alpha$ then $h_j \neq -1$
  (if the head scans the leftmost square it cannot move left)
- if $b_j = \omega$ then $h_j \neq 1$
- if $c_j = \alpha$ then $h_{k+l+j} \neq -1$ and $c'_j = \alpha$
- if $c_j \in \{0, 1\}$ then $c'_j \in \{0, 1\}$
- $s \neq s_+$ and $s \neq s_-$.

The *base* of the instruction in (*) is given by

$$s\, b_0 \ldots b_{k+l}\, c_1 \ldots c_m.$$

M is said to be *deterministic*, if no two distinct instructions in Instr(M) have the same base. Sometimes, to emphasize that we do not require a machine to be deterministic, we speak of a *nondeterministic* machine.

The notions of accepting run, rejecting run, and of "M accepts $\mathcal{A}$" are adapted from section 7.2 in the obvious way. For a function $f : \mathbb{N} \to \mathbb{N}$ we say that M is f *time-bounded*, if for any $\mathcal{A}$ accepted by M there is an accepting run of M, started with $\mathcal{A}$, of length at most $f(\|A\|)$. And M is f *space-bounded*, if for all $\mathcal{A}$ accepted by M there is an accepting run which uses at most $f(\|A\|)$ squares on each work tape before stopping.

Let K be a class of ordered τ-structures. M *accepts* K if M accepts exactly those ordered τ-structures that lie in K. For classes of structures the definitions of PTIME ("polynomial time"), NPTIME ("nondeterministic polynomial time"), PSPACE ("polynomial space") are introduced in the obvious way. For example,

- K is in PTIME iff there is a deterministic machine M and a polynomial $p \in \mathbb{N}[x]$ such that M accepts K and M is p time-bounded.

And we define

- K is in NLOGSPACE, "nondeterministic logarithmic space" (LOGSPACE, "deterministic logarithmic space") iff there is a (deterministic) machine M and $d \geq 1$ such that M accepts K and is $d \cdot \log$ space-bounded ($\log n$ stands for the least natural number $\geq \log_2 n$).

Exercise 7.2.4 Denote by $l(\mathcal{A})$ the sum of the lengths of the input words on the input tapes of a machine started with $\mathcal{A}$. Show that in the definitions of the complexity classes mentioned above we can replace $\|A\|$ by $l(\mathcal{A})$. □

At the end of section 7.1 we have observed that for a class K of ordered structures and $m \geq 1$ the class

$$K_m \quad := \quad \{\mathcal{A} \in K \mid \|A\| \geq m\}$$

is axiomatizable in a logic $\mathcal{L}$ iff K is axiomatizable in $\mathcal{L}$. Analogously, for any of the complexity classes $\mathcal{C}$ introduced so far, we have

$$K \in \mathcal{C} \quad \text{iff} \quad K_m \in \mathcal{C}$$

(since we can change a machine, without essentially effecting its time and space bounds, in such a way that it runs on a given finite set of inputs in a prescribed form; cf. Exercise 7.2.5 below). We give an application of this fact. Suppose that K is in PSPACE. Then K_2 is in PSPACE, too, say, it is accepted by a machine M that is q space-bounded, where $q(x) = a_s x^s + a_{s-1}x^{s-1} + \ldots + a_1 x + a_0$. For suitable d, $q(n) \leq n^d$ for all $n \geq 2$. Thus M is x^d space-bounded. These observations will allow us to restrict ourselves to monic polynomials $p(x) = x^d$ when considering PSPACE (or PTIME, NPTIME).

Exercise 7.2.5 For fixed τ and $m \geq 1$ there is $c_m \in \mathbb{N}$ such that for any $K \subseteq \mathcal{O}[\tau]$ and any $f : \mathbb{N} \to \mathbb{N}$:

(a) If K is accepted by an f time-bounded machine, then K_m is accepted by an $f + 2\cdot(m+1)$ time-bounded machine, where $f + 2\cdot(m+1)$ denotes the function whose value for n is $f(n) + 2\cdot(m+1)$.
(b) If K_m is accepted by an f time-bounded machine, then K is accepted by an $f + c_m$ time-bounded machine.
(c) K is accepted by an f space-bounded machine iff K_m is accepted by an f space-bounded machine. □

7.3 Logical Descriptions of Computations

Let K be a class of ordered τ-structures, $K \subseteq \mathcal{O}[\tau]$. We write $K \in$ IFP if K is axiomatizable in FO(IFP), and use similar notations for the other logics. Our main goal is to show

$K \in$ LOGSPACE	iff	$K \in$ DTC
$K \in$ NLOGSPACE	iff	$K \in$ TC
$K \in$ PTIME	iff	$K \in$ IFP
$K \in$ NPTIME	iff	$K \in \Sigma_1^1$
$K \in$ PSPACE	iff	$K \in$ PFP.

(Σ_1^1 denotes the fragment of second-order logic consisting of the sentences of the form $\exists X_1 \ldots \exists X_m \psi$ where ψ is first-order, cf. section 3.1).

These results provide the bridge between logic and complexity theory we have aimed at, by relating purely machine-oriented characterizations and characterizations by means of logical definability. Section 7.5 exemplifies the methodological possibilities they offer.

In this section we prove the implications from left to right; the other implications will be settled in the next section.

Let $\mathcal{C}$ be one of the complexity classes above and $\mathcal{L}$ the logic associated to $\mathcal{C}$ by the corresponding equivalence. Assume that $K \in \mathcal{C}$, and let M be a Turing machine witnessing that $K \in \mathcal{C}$. We are going to describe the behaviour of M by a formula φ_M of $\mathcal{L}$ in such a way that for any ordered structure $\mathcal{A}$,

$$\mathcal{A} \models \varphi_M \quad \text{iff} \quad M \text{ accepts } \mathcal{A}$$

and hence,

$$K = \text{ordMod}(\varphi_M).$$

The reader familiar with classical recursion theory will observe the analogy of the proofs given below with usual proofs showing the μ-recursiveness of Turing computable functions. There, the transition from one configuration to the next is coded by "simple" (e.g. primitive recursive) functions, and the μ-operator is used to detect the end of a computation. Similarly here, using the ordering relation, we can describe the transition from one configuration to the next by means of a "simple" logic, mostly first-order logic. The additional expressive power, for example the operator IFP, is used to find out whether the computation stops and to get its outcome. In recursion theory the proof of the equivalence shows that one application of the μ-operator suffices. Similarly we shall obtain, say, that every formula of FO(IFP) is equivalent to a formula with only one occurrence of IFP.

For the following we fix a vocabulary $\tau = \tau_0 \dot{\cup} \tau_1$ where, for simplicity, we assume that $\tau_0 = \{<, S, \min, \max\}$ and τ_1 is relational, $\tau_1 = \{R_1, \ldots, R_k\}$ with r_i-ary R_i. For convenience we set $r_0 = 1$.

A Turing machine M for τ-structures has $1 + k$ input tapes and a certain number m of work tapes. In order to describe computations that M performs when started with a structure, we introduce *configurations* which contain the relevant data of possible situations of a computation. These data are:

- the current state
- the current inscriptions of the work tapes
- the current position of the heads on both the input and the work tapes.

An *accepting configuration* is a configuration with state s_+. A configuration CONF$'$ is said to be a *successor* of the configuration CONF, if an instruction of M allows M to pass from CONF to CONF$'$ in one step. For convenience, we extend this definition by saying that any *accepting* configuration is a successor of itself. If M is deterministic then every configuration has at most one successor.

Let M be a (nondeterministic) Turing machine for τ-structures which is x^d space-bounded, that is, if M accepts an ordered structure $\mathcal{A}$, then there is an accepting run that scans at most n^d squares on each work tape, where $n := \|A\|$. We may assume that $r_i \leq d$ for $i = 1, \ldots, k$ (r_i being the arity of R_i). Fix a structure $\mathcal{A}$. By the remarks at the end of the last two sections when proving that a class of structures is axiomatizable in a logic or acceptable by a Turing machine of a certain complexity bound, we can restrict ourselves to sufficiently large finite structures. Here we look at structures $\mathcal{A}$ such that for $n := \|A\|$ we have $n > k+m$ and $n > \|\mathrm{State}(M)\|$. We assume that State($M$), the set of states of M, is an initial segment of the natural numbers and that $s_0 = 0$ is the starting state.

Let CONF be a configuration, where at most the n^d first cells of each work tape are not empty and where the heads scan one of these cells. A first attempt to code the contents of these cells could consist in dividing the relevant part of each work tape into $\frac{n^d}{\log n} =: r$ blocks of length log n and reading each block as a natural number $< n$ in binary representation. This would require variables $x_1, \ldots, x_r$ for each tape. Then a formula bearing the information on successive configurations would contain at least the variables $x_1, \ldots, x_r$ and would, therefore, depend on the cardinality n of the universe. We overcome this difficulty for PTIME, NPTIME, and PSPACE by using relation variables instead of individual variables.

To code the data of CONF we first introduce the "state relation" $\mathrm{ST}^{\mathrm{CONF}}$, the "end-of-tape relations" $\mathrm{E}_j^{\mathrm{CONF}}$, the "head relations" $\mathrm{H}_j^{\mathrm{CONF}}$, and the "inscription relations" $\mathrm{I}_j^{\mathrm{CONF}}$. The last ones are only introduced for the work tapes, since the inscriptions of input tapes are given by the input structure and kept fixed during the whole computation. We introduce: The unary relation

$$\mathrm{ST}^{\mathrm{CONF}} := \{s\} \text{ where } s \text{ is the state of CONF;}$$

for $0 \leq j \leq k+m$, the unary

$$\mathrm{E}_j^{\mathrm{CONF}} := \begin{cases} \{0\}, & \text{if the } j\text{-th head faces } \alpha \\ \{n-1\}, & \text{if the } j\text{-th head faces } \omega \\ \emptyset, & \text{otherwise;} \end{cases}$$

for $0 \leq j \leq k$, the r_j-ary[5]

$$\mathrm{H}_j^{\mathrm{CONF}} := \{|e|_{r_j} \mid 0 \leq e, \text{ the } j\text{-th head scans the } e\text{-th square and this does not contain } \omega\};$$

for $k+1 \leq j \leq k+m$, the d-ary

$$\mathrm{H}_j^{\mathrm{CONF}} := \{|e|_d \mid 0 \leq e, \text{ the } j\text{-th head scans the } e\text{-th square}\};$$

for $k+1 \leq j \leq k+m$, the d-ary

[5] Recall that $|e|_{r_j}$ is the r_j-tuple given by the n-adic representation of e and that the square containing α is numbered -1.

$$\mathrm{I}_j^{\mathrm{CONF}} := \{|e|_d \mid 0 \le e < n^d \text{ and the } e\text{-th square of the } j\text{-th work tape contains the symbol } 1\}.$$

Obviously, CONF is uniquely determined by these relations. The starting configuration CONF_0, for example, is given by

$$\mathrm{ST}^{\mathrm{CONF}_0} = \{0\},\ \ \mathrm{E}_j^{\mathrm{CONF}_0} = \emptyset,\ \ \mathrm{H}_j^{\mathrm{CONF}_0} = \{(0,\dots,0)\},\ \text{and } \mathrm{I}_j^{\mathrm{CONF}_0} = \emptyset.$$

For technical convenience we encode CONF in a single $(d+2)$-ary relation $C^{\mathrm{CONF}} \subseteq \{0,\dots,n-1\}^{d+2}$ by joining the preceding relations as follows (we add two first coordinates to distinguish these relations and fill up with zeroes in the middle to get arity $(d+2)$):[6]

$$\begin{aligned} C^{\mathrm{CONF}} \;:=\; & \{(0,0)\} \times \{\tilde{0}\} \times \mathrm{ST}^{\mathrm{CONF}} \\ \cup\; & \bigcup_{0\le j\le k+m} \{(1,j)\} \times \{\tilde{0}\} \times \mathrm{E}_j^{\mathrm{CONF}} \\ \cup\; & \bigcup_{0\le j\le k+m} \{(2,j)\} \times \{\tilde{0}\} \times \mathrm{H}_j^{\mathrm{CONF}} \\ \cup\; & \bigcup_{k+1\le j\le k+m} \{(3,j)\} \times \mathrm{I}_j^{\mathrm{CONF}}. \end{aligned}$$

Clearly, given $C \subseteq \{0,\dots,n-1\}^{d+2}$, we can easily decide whether there is a configuration CONF of M, where only the first n^d cells of each work tape are relevant, such that $C = C^{\mathrm{CONF}}$, or as we shortly say, whether C is an n^d-bounded configuration.

The following lemma now states how the behaviour of M can be described by suitable formulas.

Lemma 7.3.1 *Let M be a Turing machine which is x^d space-bounded. There is a first-order formula $\varphi_{\mathrm{start}}(\overline{x})$ and there are first-order formulas $\varphi_{\mathrm{succ}}(\overline{x}, X)$ and $\psi_{\mathrm{succ}}(X,Y)$ (more precisely, second-order formulas without second-order quantifiers) such that for all sufficiently large $\mathcal{A} \in \mathcal{O}[\tau]$ and $\overline{a} \in A^{d+2}$ we have:*

(a) *"$\varphi_{\mathrm{start}}(\overline{x})$ describes the starting configuration": If C_0 denotes the starting configuration of M started with $\mathcal{A}$ then*

$$\mathcal{A} \models \varphi_{\mathrm{start}}[\overline{a}] \quad \text{iff} \quad \overline{a} \in C_0.$$

(b) *"$\varphi_{\mathrm{succ}}(\overline{x}, X)$ describes the successor of X": If M is deterministic and C is an n^d-bounded configuration of M (where $n := \|A\|$) then*

$$\mathcal{A} \models \varphi_{\mathrm{succ}}[\overline{a}, C] \ \text{ iff } \ C \text{ has an } n^d\text{-bounded successor } C' \text{ and } \overline{a} \in C'.$$

[6] By $\tilde{0}$ we denote constant sequences $0\dots0$ of appropriate length; in general, $\tilde{c}$ will denote a sequence of the form $c\dots c$.

(c) *"$\psi_{\text{succ}}(X,Y)$ expresses that Y is a successor of X": If C_1 is an n^d-bounded configuration of M and C_2 a further $(d+2)$-ary relation on A then*

$$\mathcal{A} \models \psi_{\text{succ}}[C_1, C_2] \text{ iff } C_2 \text{ is an } n^d\text{-bounded configuration of } M \text{ which is a successor of } C_1.$$

We postpone the proof of this lemma and first draw some consequences.

Theorem 7.3.2 *Let $K \subseteq \mathcal{O}[\tau]$ be a class of ordered structures. If K is in* PSPACE *then K is axiomatizable in* FO(PFP).

Proof. Let M be a deterministic machine witnessing $K \in$ PSPACE. By previous remarks we can assume that M is x^d space-bounded for suitable d. We set

$$\varphi(\overline{x}, X) \quad := \quad (\neg\exists\overline{y}X\overline{y} \wedge \varphi_{\text{start}}(\overline{x})) \vee (\exists\overline{y}X\overline{y} \wedge \varphi_{\text{succ}}(\overline{x}, X)),$$

where φ_{start} and φ_{succ} are the formulas assigned to M in the preceding lemma. Let $\mathcal{A}$ be an ordered structure and $n := \|A\|$. By this lemma, $F_0^\varphi, F_1^\varphi, F_2^\varphi, \ldots$ is the sequence $\emptyset$, C_0, $C_1, \ldots$ where

- C_0 is the starting configuration
- if C_i is an n^d-bounded configuration of M with an n^d-bounded successor configuration C then $C_{i+1} = C$. In particular, if C_i is accepting then $C_i = C_{i+1} = C_{i+2} \cdots$
- if C_i is an n^d-bounded configuration without a successor configuration or with a successor configuration which is not n^d-bounded, then $C_{i+1} = \emptyset$, $C_{i+2} = C_0$, $C_{i+3} = C_1, \ldots$, that is, the sequence has no fixed-point.

Summarizing, we have

$$\begin{aligned} M \text{ accepts } \mathcal{A} \quad &\text{iff} \quad F_\infty^\varphi \text{ is an accepting configuration} \\ &\text{iff} \quad F_\infty^\varphi \text{ is a configuration with state } s_+. \end{aligned}$$

"F_∞^φ is a configuration with state s_+" is expressed by the formula (recall the ST^{CONF}-part of C^{CONF})

$$\exists y(\text{"}y \text{ is the } s_+\text{-th element of } <\text{"} \wedge [\text{PFP}_{\overline{x},X}\varphi] \min \min \widetilde{\min}\, y).$$ [7]

We abbreviate it by $[\text{PFP}_{\overline{x},X}\varphi] \min \min \widetilde{\min}\, s_+$. Then,

$$\begin{aligned} \mathcal{A} \in K \quad &\text{iff} \quad M \text{ accepts } \mathcal{A} \\ &\text{iff} \quad \mathcal{A} \models [\text{PFP}_{\overline{x},X}\varphi] \min \min \widetilde{\min}\, s_+, \end{aligned}$$

that is, $K = \text{ordMod}([\text{PFP}_{\overline{x},X}\varphi] \min \min \widetilde{\min}\, s_+)$. □

[7] Compare footnote 6.

Corollary 7.3.3 *Let $K \subseteq \mathcal{O}[\tau]$ be in* PSPACE. *Then K is axiomatizable by a sentence of* FO(PFP) *with only one occurrence of* PFP. □

We now turn to PTIME and will realize that we can do better. Whereas accepting runs of polynomially space-bounded machines may have exponential length, we here only have to consider runs of polynomial length. This enables us to code the whole run into a single relation which can be obtained as the fixed-point of an inflationary process. On the logical side we therefore can replace PFP by IFP.

Consider a (finite or infinite) run $C_0, C_1, \dots$ of an x^d time-bounded (and hence, x^d space-bounded) deterministic machine started with a structure of cardinality n. If the run accepts the structure, C_{n^d-1} must be an accepting configuration. The inflationary process indicated above is given by a formula $\varphi(\overline{v}, \overline{x}, Z)$ with

$$F_i^{(Z\overline{v}\,\overline{x} \vee \varphi)} = \bigcup_{\substack{m<i \\ C_m \text{ defined}}} \{|m|_d\} \times C_m, \qquad F_\infty^{(Z\overline{v}\,\overline{x} \vee \varphi)} = \bigcup_{\substack{m<n^d \\ C_m \text{ defined}}} \{|m|_d\} \times C_m,$$

that is, we use the first d coordinates as time stamps when coding the run in one relation (as above, $|m|_d$ denotes the m-th tuple of $\{0, \dots, n-1\}^d$ in the lexicographic ordering).

Theorem 7.3.4 *Let $K \subseteq \mathcal{O}[\tau]$ be a class of ordered structures. If K is in* PTIME *then K is axiomatizable in* FO(IFP).

Proof. Let M be a deterministic machine witnessing $K \in$ PTIME. We can assume that, for suitable d, M is x^d time-bounded. For $\overline{v} = v_0 \dots v_{d-1}$ we set

$$\varphi(\overline{v}, \overline{x}, Z) \;\; := \;\; (\overline{v} = \widetilde{\min} \wedge \varphi_{\text{start}}(\overline{x})) \vee \exists \overline{u}(S^d \overline{u}\,\overline{v} \wedge \varphi_{\text{succ}}(\overline{x}, Z\overline{u}\_)).$$

Here, $\overline{v} = \widetilde{\min}$ abbreviates $v_0 = \min \wedge \dots \wedge v_{d-1} = \min$; $S^d\overline{u}\,\overline{v}$ stands for "$\overline{v}$ is the successor of $\overline{u}$ in the lexicographic ordering", and $\varphi_{\text{succ}}(\overline{x}, Z\overline{u}\_)$ is obtained from $\varphi_{\text{succ}}(\overline{x}, X)$ by replacing subformulas $X\overline{t}$ by $Z\overline{u}\,\overline{t}$. Then we have for $\mathcal{A} \in \mathcal{O}[\tau]$ with $n := \|A\|$:

$$\begin{aligned} \mathcal{A} \in K \quad & \text{iff} \quad M \text{ accepts } \mathcal{A} \\ & \text{iff} \quad \text{the } (n^d-1)\text{-th configuration of } M, \text{ started} \\ & \qquad\;\; \text{with } \mathcal{A}, \text{ is defined and has state } s_+ \\ & \text{iff} \quad \mathcal{A} \models [\text{IFP}_{\overline{v}\,\overline{x}, Z}\varphi]\widetilde{\max} \min \min \widetilde{\min}\, s_+, \end{aligned}$$

that is, K is the class of ordered models of a sentence of FO(IFP). □

Corollary 7.3.5 *Let $K \subseteq \mathcal{O}[\tau]$ be in* PTIME. *Then K is axiomatizable in* FO(IFP) *by a sentence with only one occurrence of* IFP. □

It is now easy to extend the previous result to NPTIME.

Theorem 7.3.6 *Let $K \subseteq \mathcal{O}[\tau]$ be a class of ordered structures. If K is in* NPTIME *then K is axiomatizable in* SO *by a Σ_1^1-sentence.*

Proof. Choose M witnessing $K \in$ NPTIME and assume that M is x^d time-bounded. Then, for $\mathcal{A} \in \mathcal{O}[\tau]$ with $n := \|A\|$,

$\mathcal{A} \in K$ iff there is a run of M, started with $\mathcal{A}$, of length $\leq n^d$ that accepts $\mathcal{A}$
iff there is a sequence $C_0, \ldots, C_{n^d-1}$ of n^d-bounded configurations of M, started with $\mathcal{A}$, such that C_0 is the starting configuration, C_{i+1} is a successor configuration of C_i, and s_+ is the state of C_{n^d-1}
iff $\mathcal{A} \models \varphi$,

where φ is the sentence (the intended meaning of the second-order variable Z being $\bigcup_{m<n^d} \{|m|_d\} \times C_m$):

$$\varphi := \exists Z(\forall \overline{x}(Z\widetilde{\min}\overline{x} \leftrightarrow \varphi_{\text{start}}(\overline{x})) \land \forall \overline{u} \forall \overline{v}(S^d \overline{u}\,\overline{v} \to \psi_{\text{succ}}(Z\overline{u}_-, Z\overline{v}_-)) \land Z\widetilde{\max} \min \min \widetilde{\min}\, s_+).$$

□

We now come back to a proof of the key lemma 7.3.1.

Proof (of 7.3.1). Let M be an x^d space-bounded machine for τ-structures. Recall the specific coding of an n^d-bounded configuration CONF in a single relation C^{CONF} comprising relations ST^{CONF}, E_j^{CONF}, H_j^{CONF}, and I_j^{CONF} that contain the information on the state, the endmarks, the head positions, and the inscription of the work tapes. Let $\overline{x}$ be the sequence of variables $xyx_1 \ldots x_d$.

For (a) we can set

$$\begin{aligned} \varphi_{\text{start}}(\overline{x}) \;:=\;& \overline{x} = \tilde{0} && \text{"the state is } s_0\text{"} \\ & \lor (x = 2 \land 0 \leq y \leq k+m \land x_1 \ldots x_d = \tilde{0}) && \text{"the heads scan the 0-th cell"} \end{aligned}$$

where, for example, $0 \leq y \leq k+m$ means that y is equal to or less than the $(k+m)$-th element in the ordering $<$ and 0 stands for min. Also in the following we shall use such self-explanatory abbreviations.

For parts (b) and (c) we first introduce, for every instruction instr $\in$ Instr(M),

$$\text{instr} = sb_0 \ldots b_{k+l}c_1 \ldots c_m \to s'c'_1 \ldots c'_m h_0 \ldots h_{k+l+m},$$

a formula $\varphi_{\text{instr}}(\overline{x}, X)$ which for n^d-bounded configurations X expresses

"X has base $s\overline{b}\overline{c}$, and if the successor configuration according to instr is not n^d-bounded, then $\{\overline{x} \mid \varphi_{\text{instr}}(\overline{x}, X)\} = \emptyset$, otherwise $\{\overline{x} \mid \varphi_{\text{instr}}(\overline{x}, X)\}$ is this successor configuration".

Moreover, we need a formula $\varphi_{\text{acc}}(X)$ which for n^d-bounded configurations X expresses

"X is an accepting configuration".

Using $\varphi_{\text{instr}}(\overline{x}, X)$ and $\varphi_{\text{acc}}(X)$, we get the desired formulas φ_{succ} and ψ_{succ} of parts (b) and (c) as (recall that we agreed to set $C_{m+1} = C_m$ for accepting configurations C_m).

$$\varphi_{\text{succ}}(\overline{x}, X) \quad := \quad (\varphi_{\text{acc}}(X) \wedge X\overline{x}) \vee \bigvee_{\text{instr} \in \text{instr}(M)} \varphi_{\text{instr}}(\overline{x}, X),$$

and

$$\begin{aligned}\psi_{\text{succ}}(X, Y) \quad := \quad & (\varphi_{\text{acc}}(X) \wedge \forall \overline{x}(Y\overline{x} \leftrightarrow X\overline{x})) \vee \\ & \bigvee_{\text{instr} \in \text{instr}(M)} (\exists \overline{x}\, \varphi_{\text{instr}}(\overline{x}, X) \wedge \forall \overline{x}(Y\overline{x} \leftrightarrow \varphi_{\text{instr}}(\overline{x}, X))).\end{aligned}$$

It remains to give $\varphi_{\text{instr}}(\overline{x}, X)$ and $\varphi_{\text{acc}}(X)$. We set

$$\varphi_{\text{acc}}(X) \quad := \quad X00\tilde{0}s_+.$$

The formula $\varphi_{\text{instr}}(\overline{x}, X)$ has the form

$$\varphi_{\text{instr}}(\overline{x}, X) \quad := \quad \varphi_{s,\overline{b},\overline{c}}(X) \wedge \varphi_{s',\overline{c}',\overline{h}}(\overline{x}, X),$$

where, for an n^d-bounded configuration X, the formula $\varphi_{s,\overline{b},\overline{c}}(X)$ expresses

"X has base $s, \overline{b}, \overline{c}$",

and $\varphi_{s',\overline{c}',\overline{h}}(\overline{x}, X)$ expresses

"if the successor Y of X according to $s', \overline{c}', \overline{h}$ is not n^d-bounded then $\{\overline{x} \mid \varphi_{s',\overline{c}',\overline{h}}(\overline{x}, X)\} = \emptyset$, else $\{\overline{x} \mid \varphi_{s',\overline{c}',\overline{h}}(\overline{x}, X)\} = Y$".

For easier reading of the formulas below we introduce the following abbreviations:

ENDMARK $yz := X1y\tilde{0}z$	"the y-th head faces the endmark z" (which is 0 for α and $n-1$ for ω)
HEAD $y\overline{z} := X2y\tilde{0}\overline{z}$	"the y-th head is on position $\lvert\overline{z}\rvert$"
ONE $y\overline{z} := X3y\overline{z}$	"the y-th work tape contains 1 on position $\lvert\overline{z}\rvert$".

Now we take as $\varphi_{s,\overline{b},\overline{c}}(X)$ the conjunction of

$X00\tilde{0}s$

"s is the state"

$\bigwedge_{b_j=\alpha} \text{ENDMARK } j \min \wedge \bigwedge_{c_j=\alpha} \text{ENDMARK } (k+j) \min$

"heads at the left end of a tape"

$\bigwedge_{b_j=\omega} \text{ENDMARK } j \max$

"heads at the right end on input tapes"

$\bigwedge_{b_j=1} \exists x_1 \ldots \exists x_{r_j} (\text{HEAD } j\tilde{0}x_1 \ldots x_{r_j} \wedge R_j x_1 \ldots x_{r_j})$

"heads of input tapes facing a 1"

$\bigwedge_{b_j=0} \exists x_1 \ldots \exists x_{r_j} (\text{HEAD } j\tilde{0}x_1 \ldots x_{r_j} \wedge \neg R_j x_1 \ldots x_{r_j})$

"heads of input tapes facing a 0"

$\bigwedge_{c_j=1} \exists x_1 \ldots \exists x_d (\text{HEAD } (k+j)x_1 \ldots x_d \wedge \text{ONE } (k+j)x_1 \ldots x_d)$

"heads of work tapes facing a 1"

$\bigwedge_{c_j=0} \exists x_1 \ldots \exists x_d (\text{HEAD } (k+j)x_1 \ldots x_d \wedge \neg\text{ONE } (k+j)x_1 \ldots x_d)$

"heads of work tapes facing a 0".

Finally, we take as $\varphi_{s',\overline{c}',\overline{h}}(\overline{x}, X)$ the conjunction $(\varphi_1 \wedge \varphi_2)$, where φ_1 is

$$\bigwedge_{\substack{h_j=1 \\ k+1 \le j \le k+m}} \neg\text{HEAD } j\widetilde{\max}$$

"heads of work tapes moving to the right do not face the (n^d-1)-th square"

and where φ_2 is the disjunction of

$(x = y = 0 \wedge x_1 \ldots x_{d-1} = \tilde{0} \wedge x_d = s')$

"s' is the new state"

$\bigvee_{k+1 \le j \le k+m} (\neg\text{HEAD } jx_1 \ldots x_d \wedge \text{ONE } jx_1 \ldots x_d \wedge x = 3 \wedge y = j)$

"work tape content unchanged on squares not scanned"

$\bigvee_{c'_j=1} (\text{HEAD } jx_1 \ldots x_d \wedge x = 3 \wedge y = k + j)$

"new content 1 on scanned squares of a work tape"

$\bigvee_{h_j=1} (\text{ENDMARK } j0 \wedge x = 2 \wedge y = j \wedge x_1 \ldots x_d = \tilde{0})$

"heads scanning α and moving to the right come to position 0"

$\bigvee_{h_j=0} (\text{ENDMARK } j0 \wedge x = 1 \wedge y = j \wedge x_1 \ldots x_d = \tilde{0})$

"unchanged position of heads facing α"

$$\bigvee_{h_j=-1} (\text{ENDMARK } j \max \wedge x = 2 \wedge y = j \wedge x_1 \ldots x_{d-r_j} = \tilde{0}$$
$$\wedge x_{d-r_j+1} \ldots x_d = \widetilde{\max})$$

"heads scanning ω and moving to the left come to position $n^{r_j}-1$"

$$\bigvee_{h_j=0} (\text{ENDMARK } j \max \wedge x = 1 \wedge y = j \wedge x_1 \ldots x_{d-1} = \tilde{0} \wedge x_d = \max)$$

"unchanged position of heads facing ω"

$$\bigvee_{h_j=-1} (\text{HEAD } j\tilde{0} \wedge x = 1 \wedge y = j \wedge x_1 \ldots x_d = \tilde{0})$$

"heads scanning α from their new position"

$$\bigvee_{\substack{j \leq k \\ h_j=1}} (\text{HEAD } j\tilde{0}\underbrace{\max \ldots \max}_{r_j\text{-times}} \wedge x = 1 \wedge y = j \wedge x_1 \ldots x_{d-1} = \tilde{0}$$
$$\wedge x_d = \max)$$

"heads of input tapes scanning ω from their new position"

$$\bigvee_{j \leq k+m} \exists u_1 \ldots \exists u_d (\text{“}x_1 \ldots x_d = u_1 \ldots u_d + h_j\text{”} \wedge \text{HEAD } j u_1 \ldots u_d \wedge x = 2$$
$$\wedge y = j)$$

"new head position of heads on 'interior' squares." □

We come to the corresponding results for LOGSPACE and NLOGSPACE. Recall that $K \in$ NLOGSPACE means that there is a (nondeterministic) machine M and some $d \geq 1$ such that M accepts K and is $d \cdot \log$ space-bounded. Every natural number $i < n$ codes a word over $\{0, 1\}$ of length $\log n$, namely its binary representation $|i|^2_{\log n}$ of length $\log n$. Thus, using d variables, we can represent the relevant contents of a work tape. Moreover, by restricting ourselves to sufficiently large structures $\mathcal{A}$, we can assume that $d \cdot \log n < n$ (where $n := \|A\|$). Hence, each head position can be represented by a single number $< n$. Altogether, we can describe the data of a configuration by a sequence of natural numbers $< n$ of length independent of n, where we agree to use the first number to represent the state. The exact definitions will be given later when proving the following lemma.

Lemma 7.3.7 *Let M be $d \cdot \log$ space-bounded. Then there are formulas $\chi_{\text{start}}(\overline{x})$ of* FO *and $\chi_{\text{succ}}(\overline{x}, \overline{x}')$ of* FO(DTC) *such that for all sufficiently large $\mathcal{A} \in \mathcal{O}[\tau]$ and $\overline{a}$ in A,*

(a) *$\mathcal{A} \models \chi_{\text{start}}[\overline{a}]$ iff $\overline{a}$ is the (description of the) starting configuration.*
(b) *For any $(d \cdot \log \|A\|)$-bounded configuration $\overline{a}$ and any $\overline{b}$,*

$$\mathcal{A} \models \chi_{\text{succ}}[\overline{a}, \overline{b}] \quad \text{iff} \quad \overline{b} \text{ is a } (d \cdot \log \|A\|)\text{-bounded successor configuration of } \overline{a}.$$

Before we give a proof we derive some consequences.

Theorem 7.3.8 *Let $K \subseteq \mathcal{O}[\tau]$ be a class of ordered structures. If $K \in$ LOGSPACE then K is axiomatizable in* FO(DTC).

Proof. Let M be a deterministic machine witnessing $K \in$ LOGSPACE, say, M is d·log space-bounded. Let χ_{start} and χ_{succ} be the formulas corresponding to M according to the preceding lemma. Then, by (a) and (b) of this lemma, we have for $\mathcal{A} \in \mathcal{O}[\tau]$:

M accepts $\mathcal{A}$

iff there is a sequence $\overline{a}_0, \ldots, \overline{a}_k$ of $(d \cdot \log \|A\|)$-bounded configurations such that $\overline{a}_0$ is the starting configuration, $\overline{a}_{i+1}$ is the successor configuration of $\overline{a}_i$, and $\overline{a}_k$ is an accepting configuration

iff $\mathcal{A} \models \exists \overline{x}(\chi_{\text{start}}(\overline{x}) \wedge \exists \overline{x}'([\text{DTC}_{\overline{x},\overline{x}'} \chi_{\text{succ}}(\overline{x}, \overline{x}')]\, \overline{x}\, \overline{x}' \wedge x_1' = s_+))$.

Hence, K is the class of ordered models of a sentence of FO(DTC). □

Theorem 7.3.9 *Let* $K \subseteq \mathcal{O}[\tau]$ *be a class of ordered structures. If* $K \in$ NLOGSPACE *then* K *is axiomatizable in* FO(TC).

Proof. Let M be a machine witnessing $K \in$ NLOGSPACE. Since M is nondeterministic we just have to replace DTC by TC in the last proof. In fact, we have for $\mathcal{A} \in \mathcal{O}[\tau]$

M accepts $\mathcal{A}$

iff $\mathcal{A} \models \exists \overline{x}(\chi_{\text{start}}(\overline{x}) \wedge \exists \overline{x}'([\text{TC}_{\overline{x},\overline{x}'} \chi_{\text{succ}}(\overline{x}, \overline{x}')]\, \overline{x}\, \overline{x}' \wedge x_1' = s_+))$.

Since FO(DTC) $\leq$ FO(TC), we have obtained a sentence of FO(TC) axiomatizing K. □

Denote by FO(posTC) the class of formulas of FO(TC) which only contain positive occurrences of TC, that is, each such occurrence is in the scope of an even number of negation symbols. In 8.6.12 we will see that FO(DTC) $\leq$ FO(posTC). Thus the preceding proof yields

Corollary 7.3.10 *If a class of ordered structures is in* NLOGSPACE *then it is axiomatizable by a sentence of* FO(posTC). □

We turn to a proof of Lemma 7.3.7. As already remarked above, in the case of log space-bounded machines we code configurations not by relations but by numbers. This change requires to define some arithmetical functions and predicates in FO(DTC). The next two lemmas serve this purpose.

Lemma 7.3.11 *There are* FO(DTC)*-formulas*

$$\varphi_+(x, y, z),\ \varphi_\cdot(x, y, z),\ \varphi_2(x, y),\ \text{and}\ \varphi_{\log(\text{universe})}(x)$$

such that for any ordered structure $\mathcal{A}$ *with* $A = \{0, \ldots, \|A\| - 1\}$ *and any* $a, b, c \in A$,

$$\begin{array}{lll} \mathcal{A} \models \varphi_+[a, b, c] & \textit{iff} & a + b = c \\ \mathcal{A} \models \varphi_\cdot[a, b, c] & \textit{iff} & a \cdot b = c \\ \mathcal{A} \models \varphi_2[a, b] & \textit{iff} & 2^a = b \\ \mathcal{A} \models \varphi_{\log(\text{universe})}[a] & \textit{iff} & a = \log \|A\|. \end{array}$$

Proof. For better readability, instead of describing the natural numbers in terms of the ordering, we use constants 1, 2, ...

Given numbers x and y, the path

$$(0, x) \to (1, x+1) \to (2, x+2) \to \ldots \to (y, x+y)$$

from $(0, x)$ to $(y, x+y)$ shows that as $\varphi_+(x, y, z)$ we can take the formula

$$(y = \min \land z = x) \lor [\mathrm{DTC}_{uv,u'v'}\,(Suu' \land Svv')] \min xyz.$$

Similarly, the path

$$(0, 0) \to (1, x) \to (2, 2 \cdot x) \to (3, 3 \cdot x) \to \ldots \to (y, y \cdot x)$$

shows that we can set

$$\begin{aligned} \varphi_\cdot(x, y, z) \quad := \quad & (y = \min \land z = \min) \\ & \lor [\mathrm{DTC}_{uv,u'v'}\,(Suu' \land \varphi_+(v, x, v'))] \min\min yz \end{aligned}$$

and the path

$$(0, 1) \to (1, 2) \to (2, 4) \to \ldots \to (x, 2^x)$$

shows that we can set

$$\begin{aligned} \varphi_2(x, y) \quad := \quad & (x = \min \land y = 1) \\ & \lor [\mathrm{DTC}_{uv,u'v'}\,(Suu' \land \varphi_\cdot(v, 2, v'))] \min 1xy. \end{aligned}$$

Finally, let

$$\varphi_{\log(\text{universe})}(x) := \neg\exists y \varphi_2(x, y) \land \forall z(z < x \to \exists y \varphi_2(z, y)).$$ □

Let $l := \log n - 1$. Then $2^l < n$. Recall that for m with $m < 2^l$ and $m_0, \ldots, m_{l-1} \in \{0, 1\}$ we have

$$|m|_l^2 = m_0 \ldots m_{l-1} \quad \text{iff} \quad m = m_0 \cdot 2^{l-1} + m_1 \cdot 2^{l-2} + \ldots + m_{l-2} \cdot 2 + m_{l-1}.$$

We then say that m_k is the k-th digit of $|m|_l^2$. If n, and hence l, is clear from the context, we denote $|m|_l^2$ by $[m]$. We write $\overline{u}$ for $u_0 \ldots u_d$ and $\overline{u} < 2^l$ for $u_0 < 2^l \land \ldots \land u_d < 2^l$; similarly with $\overline{u}', \overline{x}$, and $\overline{x}'$.

Lemma 7.3.12 *There are formulas of* FO(DTC) *which, in ordered structures* $\mathcal{A}$, *define the following relations (where* $n = \|A\|$, $l = \log n - 1$) :

One mk	*iff*	$m < 2^l$, $k < l$, *and the* k*-th digit of* $[m]$ *is* 1
Zero mk	*iff*	$m < 2^l$, $k < l$, *and the* k*-th digit of* $[m]$ *is* 0
One$_d$ $\overline{u}k$	*iff*	$\overline{u} < 2^l$, $k < (d+1) \cdot l$, *and the* k*-th digit of the concatenation* $[u_0] \ldots [u_d]$ *is* 1
Zero$_d$ $\overline{u}k$	*iff*	$\overline{u} < 2^l$, $k < (d+1) \cdot l$, *and the* k*-th digit of the concatenation* $[u_0] \ldots [u_d]$ *is* 0
Equal$_d$ $\overline{u}k\overline{u}'$	*iff*	$\overline{u}, \overline{u}' < 2^l$, $k < (d+1) \cdot l$, *and the words* $[u_0] \ldots [u_d]$ *and* $[u'_0] \ldots [u'_d]$ *differ at most at the* k*-th position.*

We denote the corresponding formulas by $\varphi_{\text{one}}(x, z)$, $\varphi_{\text{zero}}(x, z)$, $\varphi_{d\text{-one}}(\overline{x}, z)$, $\varphi_{d\text{-zero}}(\overline{x}, z)$, and $\varphi_{d\text{-equal}}(\overline{x}, z, \overline{x}')$, respectively.

Proof. Note that for $m < 2^l$ and $k < l$ the k-th digit of $[m]$ is 1 iff

$$\exists y \in \mathbb{N}\, \exists z \in \mathbb{N}(m = y \cdot 2^{l-k} + z \wedge 2^{l-k-1} \leq z < 2^{l-k}).$$

Using this fact one easily shows, for example, the following equivalences, which immediately can be formalized with the help of the formulas of the preceding lemma.

$$\begin{array}{lll}
\text{One}\, mk & \text{iff} & m < 2^l \wedge k < l \wedge \\
& & \exists y \exists z(m = y \cdot 2^{l-k} + z \wedge 2^{l-k-1} \leq z < 2^{l-k}) \\
\text{One}_1\, u_0 u_1 k & \text{iff} & (u_0 < 2^l \wedge u_1 < 2^l \wedge k < l \wedge \text{One}\, u_0 k) \vee \\
& & (l \leq k < 2 \cdot l \wedge \text{One}\, u_1 (k - l)) \\
\text{Equal}_d\, \overline{u} k \overline{u}' & \text{iff} & \overline{u} < 2^l \wedge \overline{u}' < 2^l \wedge k < (d+1) \cdot l \wedge \\
& & \forall i((i < (d+1) \cdot l \wedge i \neq k) \rightarrow (\text{One}_d\, \overline{u} i \leftrightarrow \text{One}_d\, \overline{u}' i)).
\end{array}$$

□

Finally, we come to the

Proof (of 7.3.7). Let M be a logarithmically space-bounded machine for τ-structures, say, M is $d \cdot \log$ space-bounded. For simplicity we assume that $\tau = \{R\}$ with binary R and that M only has one work tape. We restrict ourselves to structures of cardinality n with $n > d \cdot \log n$, $(d+1) \cdot l \geq d \cdot \log n$, and $n > s_M + 1$ (recall that $l := \log n - 1$ and that $\{0, \ldots, s_M\}$ is the set of states of M). When M is started with a structure $\mathcal{A}$, where $A = \{0, \ldots, n-1\}$, we can code the data of a resulting configuration by a tuple

$$(z, u_\alpha, u_\omega, u, v_\alpha, v_\omega, v_0, v_1, w_\alpha, w, y_0, \ldots, y_d)$$

where

- z is the state
- u_α, u_ω, u code the position of the head on the 0-th input tape (the "universe tape") by

$$u_\alpha = \begin{cases} 0 & \text{if the head does not face } \alpha \\ n-1 & \text{if the head faces } \alpha \end{cases}$$

$$u_\omega = \begin{cases} 0 & \text{if the head does not face } \omega \\ n-1 & \text{if the head faces } \omega \end{cases}$$

  and u is the number of the cell faced by the head if it is an interior one, otherwise $u = 0$
- similarly, $v_\alpha, v_\omega, v_0, v_1$ code the position of the head on the first input tape, the tape for the binary relation R; the variables v_0, v_1 represent the head position $v_0 \cdot n + v_1$

- w_α, w code the position of the head on the work tape (note that $d \cdot \log n < n$)
- the concatenation $[y_0] \dots [y_d]$ is the inscription of the first $(d+1) \cdot l$ cells of the work tape.

Sometimes, for notational simplicity, we denote the sequence $z u_\alpha u_\omega \dots y_d$ simply by $\overline{x}$.

Concerning part (a) of the lemma, we can set

$$\chi_{\text{start}}(\overline{x}) \quad := \quad \overline{x} = \tilde{0}.$$

Concerning part (b), we define $\chi_{\text{succ}}(\overline{x}, \overline{x}')$ by

$$\chi_{\text{succ}}(\overline{x}, \overline{x}') \quad := \quad \chi_{\text{acc}}(\overline{x}, \overline{x}') \vee \bigvee_{\text{instr} \in \text{instr}(M)} \chi_{\text{instr}}(\overline{x}, \overline{x}')$$

where

$$\chi_{\text{acc}}(\overline{x}, \overline{x}') \quad := \quad (x_1 = s_+ \wedge \overline{x}' = \overline{x})$$

(which in case $\overline{x}$ is a configuration expresses that $\overline{x}$ is accepting and $\overline{x}' = \overline{x}$) and where for every instruction

$$\text{instr} = s b_0 b_1 c_1 \to s' c_1' h_0 h_1 h_2$$

$\chi_{\text{instr}}(\overline{x}, \overline{x}')$ is a formula which, in case $\overline{x}$ is a $(d \cdot \log)$-bounded configuration, expresses that $\overline{x}$ has base $s b_0 b_1 c_1$, the successor configuration of $\overline{x}$ according to instr is $(d \cdot \log)$-bounded and is $\overline{x}'$. As an example, we explicitly give $\chi_{\text{instr}}(\overline{x}, \overline{x}')$ for

$$\text{instr} \quad = \quad s 1 \alpha 1 \to s' 0 (-1) 11,$$

namely as the conjunction of

$z = s$

"s is the state"

$u_\alpha = \min \wedge u_\omega = \min$

"the head of the 0-th input tape faces an interior cell"

$v_\alpha = \max \wedge v_\omega = \min \wedge v_0 = \min \wedge v_1 = \min$

"the head of the first input tape faces α"

$w_\alpha = \min \wedge \ \text{One}_d\, y_0 \dots y_d w$

"the head of the work tape faces a 1"

$z' = s'$

"s' is the new state"

$u'_\omega = \min \wedge ((u > 0 \wedge Su'u \wedge u'_\alpha = \min) \vee (u = 0 \wedge u' = 0 \wedge u'_\alpha = \max))$

"new head position of the 0-th input tape"

$v'_\alpha = \min \wedge v'_0 = \min \wedge v'_1 = \min \wedge v'_\omega = \min$

"new head position of the first input tape is cell 0"

$w'_\alpha = \min \wedge Sww' \wedge w' < d \cdot \log n$,
i.e. $w'_\alpha = \min \wedge Sww' \wedge \exists x(\varphi_{\log(\text{universe})}(x) \wedge w' < d \cdot x)$

"new head position of the work tape is within the bounds"

$\text{Zero}_d\, y'_0 \ldots y'_d w$

"new content of cell scanned on the work tape"

$\text{Equal}_d\, y_0 \ldots y_d w y'_0 \ldots y'_d$

"work tape content unchanged on cells not scanned". □

7.4 The Complexity of the Satisfaction Relation

Suppose, for example, that the class K of ordered structures is axiomatizable by the FO(IFP)-sentence φ, $K = \{\mathcal{A} \in \mathcal{O}[\tau] \mid \mathcal{A} \models \varphi\}$. We aim at showing that $K \in$ PTIME, that is, for fixed φ we want to prove that the satisfaction relation $\mathcal{A} \models \varphi$ can be decided in time polynomially bounded in $\|A\|$. One also says that φ has a polynomial time *model-checker*. To simplify the corresponding algorithms we start with some general remarks. Note that manipulations in algorithms of the kind we now describe do not destroy polynomial time and logarithmic space bounds.

(1) Using an additional work tape W' it is possible at any time of a computation to move the head of a given work tape W to the rightmost square which the head of the tape has scanned so far. (In fact, change the given program so that the head on W' moves in the same way as the head on W, but always prints the symbol 1.)
(2) By (1) it is possible at any time of a computation to erase the content of a work tape (note that the additional work tape used in (1) can be cleared in a trivial way); in particular, one can change a program – without changing the accepted class – such that all work tapes are empty whenever the program stops.
(3) The content of a worktape W can be copied to an empty tape W_1 (using (1), bring the corresponding heads H and H_1 to the rightmost cell scanned by H and copy the content cell by cell).
(4) In our applications the 0-th input tape has the inscription $\underbrace{1\ldots1}_{n-\text{times}}$, where n is the cardinality of the structure we consider. One can write the binary representation of n (of length $\leq \log n$) on a work tape. We say "a counter is set to n". Similarly, a counter can be set to n^d for any fixed $d \geq 1$.

Let $\mathcal{L}$ be one of the logics considered in the preceding section and $\mathcal{C}$ the corresponding complexity class. We want to show that for any sentence of

$\mathcal{L}$ the class K of its ordered models is in $\mathcal{C}$. We even show that there is a machine M *strongly witnessing* $K \in \mathcal{C}$, that is,

- M accepts K;
- for any $\mathcal{A} \in \mathcal{O}[\tau]$ every run of M, started with $\mathcal{A}$, stops at s_+ or s_-; in particular, if M is deterministic then M decides K;
- for any $\mathcal{A} \in \mathcal{O}[\tau]$ every run of M satisfies the time or space bounds characteristic for $\mathcal{C}$.

The proof that the class of ordered models of a sentence φ of $\mathcal{L}$ is in $\mathcal{C}$ proceeds by induction on φ. So we have to deal with formulas, too. We therefore introduce the following notation: For a formula $\varphi(x_1, \ldots, x_l, Y_1, \ldots, Y_r)$ we let

$$\text{ordMod}(\varphi) \quad := \quad \{(\mathcal{A}, a_1, \ldots, a_l, P_1, \ldots, P_r) \mid \mathcal{A} \in \mathcal{O}[\tau],\ \mathcal{A} \models \varphi[\overline{a}, \overline{P}]\},$$

that is, we consider the ordered models of $\varphi(c_1, \ldots, c_l, P_1, \ldots, P_r)$, a sentence in an enlarged vocabulary.

Theorem 7.4.1 *Let $K \subseteq \mathcal{O}[\tau]$ be a class of ordered structures.*
(a) *If $K \in$ DTC then $K \in$ LOGSPACE.*
(b) *If $K \in$ posTC then $K \in$ NLOGSPACE.*

Proof. By induction on the corresponding formulas φ we show that the class of ordered models of φ is in LOGSPACE and NLOGSPACE, respectively, and that there exists a machine strongly witnessing this fact. We handle both cases simultaneously. By passing to an equivalent formula we can assume that in formulas of FO(posTC) the TC operation does not occur in the scope of any negation symbol (the new formula may also contain $\wedge, \forall$).

Suppose that φ is atomic, say for simplicity, $\varphi = Rxy$. We show that there is a machine M strongly witnessing that

$$\{(\mathcal{A}, i, j) \mid \mathcal{A} \in \mathcal{O}[\tau], R^A ij\} \in \text{LOGSPACE}.$$

Let $(\mathcal{A}, i, j) \in \mathcal{O}[\tau \cup \{c, d\}]$ with $A = \{0, 1, \ldots, n-1\}$ be given. Note that the information, whether $R^A ij$ holds, is to be found in the $(i \cdot n + j)$-th square of the input tape corresponding to R, and (the binary representations of) i and j are available on the input tapes corresponding to c and d. Now it should be clear how a machine strongly witnessing that $\text{ordMod}(Rxy) \in$ LOGSPACE can be designed.

$\varphi = \neg\psi$: By the remarks above, ψ does not contain TC and hence, is in FO(DTC). By induction hypothesis, there is a machine M strongly witnessing that $\text{ordMod}(\psi) \in$ LOGSPACE. For φ just interchange the roles of s_+ and s_- in M.

$\varphi(x_1, \ldots, x_l) = (\psi \vee \chi)$: By induction hypothesis, there are corresponding machines M_ψ for $\psi(x_1, \ldots, x_l)$ and M_χ for $\chi(x_1, \ldots, x_l)$. Let M be a machine that first carries out the computation of M_ψ and then that ofM_χ accepting the input in case at least one, M_ψ or M_χ, accepts, and rejecting otherwise. (After the computation of M_ψ the work tapes are erased as explained in remark (2) above.)

$\varphi = (\psi \wedge \chi)$: similarly.

$\varphi(x_1, \ldots, x_l) = \exists x\psi$: By induction hypothesis, there is a corresponding machine M_0 for $\psi(x_1, \ldots, x_l, x)$. A machine M for φ operates as follows: Suppose M is started with an ordered structure $(\mathcal{A}, a_1, \ldots, a_l)$ where $A = \{0, \ldots, n-1\}$. Then, for $i = 0, \ldots, n-1$, M writes the binary representation of i on a work tape and checks, using M_0, whether $\mathcal{A} \models \psi[a_1, \ldots, a_l, i]$. (Note that the binary representation of i on the work tape does not carry an endmark ω as required on the corresponding input tape of M_0. To remind the end of the representation of i, proceed as explained in remark (1) above.) If the answer is positive at least once, M stops in state s_+, otherwise in s_-.

$\varphi = \forall x\psi$: similarly.

$\varphi = [\mathrm{DTC}_{\overline{x},\overline{y}}\, \psi]\, \overline{s}\overline{t}$, where ψ is a formula of FO(DTC): For simplicity, we assume that the free variables of ψ are among $\overline{x}, \overline{y}$ and that $\overline{x} = x$, $\overline{y} = y$, $\overline{s} = s$, and $\overline{t} = t$. Choose a machine M_0 strongly witnessing that $\mathrm{ordMod}(\psi) \in$ LOGSPACE. Given $\mathcal{A}$ with $A = \{0, \ldots, n-1\}$, if there is a ψ-path from s to t, there is one of length $\leq n$. Therefore, the machine M, we aim at, can be organized as follows: It writes $i := s$ on a work tape and sets a counter to n, which is used to invoke a subroutine at most n times. M rejects in case the counter becomes negative. Using M_0, the subroutine checks for $j = 0, \ldots, n-1$ whether $\mathcal{A} \models \psi[i, j]$ holds for exactly one j; if not, M rejects. Otherwise, M checks whether j equals t; in the affirmative case M accepts. In the negative case, M sets $i := j$ and reduces the counter by one.

$\varphi = [\mathrm{TC}_{\overline{x},\overline{y}}\, \psi]\, \overline{s}\overline{t}$: Once more, for simplicity, we assume that the free variables of ψ are among $\overline{x}, \overline{y}$ and that $\overline{x} = x$, $\overline{y} = y$, $\overline{s} = s$, and $\overline{t} = t$. Choose a machine M_0 strongly witnessing $\mathrm{ordMod}(\psi) \in$ NLOGSPACE. We give the basic idea underlying the construction of a machine M for φ: A counter is set to n (again, we assume that $A = \{0, 1, \ldots, n-1\}$), which is used to carry out a subroutine at most n times. M will stop in the rejecting state in case the counter becomes negative. M writes $i := s$ on a work tape. The subroutine starts choosing an element $j \in \{0, \ldots, n-1\}$ nondeterministically (by using a counter to randomly write a $\{0,1\}$ word of length $\log n$ on a tape). Then it checks, using M_0, whether $\psi(i, j)$ holds. If not, M stops in state s_-. Otherwise, if $j = t$, M stops in s_+, and if $j \neq t$, it sets $i := j$. □

Theorem 7.4.2 *Let $K \subseteq \mathcal{O}[\tau]$ be a class of ordered structures.*

(a) *If $K \in$ IFP then $K \in$ PTIME.*

(b) *If $K \in$ PFP then $K \in$ PSPACE.*

Proof. Again, we give the proof by induction on the formula φ axiomatizing the class K. The cases where φ is atomic, $\neg\psi$, $(\psi \vee \chi)$, or $\exists x\psi$ are handled as in the preceding proof. The corresponding machines are polynomially time-bounded or space-bounded if the machines used in the induction hypotheses are. Now, for part (a), suppose that $\varphi = [\text{IFP}_{\overline{x},X}\,\psi(\overline{x},X)]\,\overline{t}$ where X is r-ary, and that M_0 is a machine strongly witnessing that

$$\{(\mathcal{A},\overline{a},R) \mid \mathcal{A} \in \mathcal{O}[\tau], \mathcal{A} \models \psi[\overline{a},R]\} \in \text{PTIME}$$

(for simplicity, we assume that the free variables of ψ are among $\overline{x}, X$). The machine M, we look for, contains a subroutine that uses work tapes W and W'. If started with a word of length n^r on W – the code of an r-ary relation R – and empty W', the subroutine writes, invoking the machine M_0, the code of

$$R' \quad := \quad \{\overline{a} \mid \mathcal{A} \models (X\overline{x} \vee \psi)\,[\overline{a},R]\}$$

on the tape W' without changing the content of W.

The machine M for φ operates as follows: It sets $R := \emptyset$ and uses the subroutine to calculate R'. If $R = R'$ it checks whether $R\overline{t}$ or not $R\overline{t}$ and accepts or rejects, respectively. Otherwise, it sets $R := R'$, erases the content of W', and again starts the subroutine. Note that $R = R'$ will be achieved after at most n^d calls to the subroutine (by 7.1.1(c)).

For part (b) assume that $\varphi = [\text{PFP}_{\overline{x},X}\,\psi(\overline{x},X)]\,\overline{t}$ with r-ary X and that M_0 is a machine strongly witnessing $\{(\mathcal{A},\overline{a},R) \mid \mathcal{A} \in \mathcal{O}[\tau], \mathcal{A} \models \psi[\overline{a},R]\} \in$ PSPACE. Given a structure $\mathcal{A}$, the operation F^{ψ} associated with ψ satisfies $F^{\psi}_{2^{n^r}-1} = F^{\psi}_{2^{n^r}}$ (and this set is the fixed-point F^{ψ}_{∞}) or $F^{\psi}_{\infty} = \emptyset$ (cf. 7.1.1(b)). The machine M for φ starts its computation on $\mathcal{A}$ by setting a counter to $2^{n^r} - 1$ (that is, it writes the word $1 \ldots 1$ of length n^r on a work tape). Then it proceeds as in the IFP-case, but now using the counter to ensure that the subroutine which here evaluates

$$R' \quad := \quad \{\overline{a} \mid \mathcal{A} \models \psi[\overline{a},R]\},$$

is invoked at most $2^{n^r} - 1$ times. When the counter gets negative, it checks whether $R = R'$ and whether Rt. If both questions are answered positively it accepts, otherwise it rejects. □

Theorem 7.4.3 *Let $K \subseteq \mathcal{O}[\tau]$ be a class of ordered structures.*

(a) *If $K \in \Sigma^1_1$ then $K \in$ NPTIME.*

(b) *If $K \in$ SO then $K \in$ PSPACE.*

Proof. (a) Let $K = \text{Mod}(\varphi)$ where $\varphi = \exists X_1 \ldots \exists X_l \psi$, ψ is first-order, and the arity of X_i is r_i. By 7.4.1(a), there is a machine M_0 strongly witnessing that $\text{Mod}(\psi(X_1, \ldots, X_l))$ is in PTIME (even in LOGSPACE). The machine M for φ, started with $\mathcal{A} \in \mathcal{O}[\tau]$, nondeterministically writes words over $\{0,1\}$ of length $n^{r_1}, \ldots, n^{r_l}$ on different work tapes, which are intended as codes of interpretations $P_1, \ldots, P_l$ of $X_1, \ldots, X_l$. Then, using M_0, it checks whether $\mathcal{A} \models \psi[P_1, \ldots, P_l]$ or not, and stops in an accepting or rejecting state, respectively.

(b) Let $K = \mathrm{Mod}(\varphi)$ for the formula φ of SO. To gain a machine M witnessing $K \in$ PSPACE we proceed by induction on φ. For φ atomic or of the form $\neg\psi, (\psi \vee \chi)$, or $\exists x\psi$ we argue as in the proof of 7.4.1. For $\varphi = \exists X\psi$ with r-ary X the machine M writes the word $1\ldots1$ of length n^r on a work tape W; then it systematically decreases this word, checking in each case with a polynomially space-bounded machine for ψ, whether ψ holds, if the interpretation of X is given by the tape W. □

7.5 The Main Theorem and Some Consequences

To summarize the results of the preceding sections we introduce the following notion:

Definition 7.5.1 A logic $\mathcal{L}$ *captures* a complexity class $\mathcal{C}$ if for all τ with $< \in \tau$ and $K \subseteq \mathcal{O}[\tau]$ we have

$$K \in \mathcal{C} \quad \text{iff} \quad K \text{ is axiomatizable in } \mathcal{L}. \qquad \square$$

Theorem 7.5.2 (Main Theorem) (a) FO(DTC) *captures* LOGSPACE.
(b) FO(TC) *captures* NLOGSPACE.
(c) FO(IFP) *captures* PTIME.
(d) Σ_1^1 *captures* NPTIME.
(e) FO(PFP) *captures* PSPACE.

Note that we have proved the theorem up to part (b). We only have shown that

$$\text{FO(posTC) captures NLOGSPACE.}$$

At the end of this section we will prove that

$$\text{FO(posTC)} \equiv \text{FO(TC) on ordered structures.}$$

Then the proof will be finished.

The main theorem gives descriptive characterizations of some important complexity classes by certain extensions of first-order logic. Is there a well-known complexity class that, in this sense, corresponds to first-order logic itself? We have mentioned already several times that first-order logic has rather limited posssibilities to speak about inductive or recursive procedures. In fact, it has been proved that first-order logic captures a very low complexity class, namely AC^0, a class which is defined in terms of circuits; cf. [95].

The study of the complexity of evaluating a formula φ of a logic $\mathcal{L}$ in a structure $\mathcal{A}$ arises in various contexts. For example, $\mathcal{A}$ may be a database instance and φ a corresponding query, or $\mathcal{A}$ may represent the state space of a program and φ a desired property. When considering such evaluations the following kinds of complexities have been treated (the first one being the subject of this chapter):

- *data complexity* of $\mathcal{L}$: For a fixed sentence, we measure the complexity as a function of the size of the structure;
- *expression complexity* of $\mathcal{L}$: For a fixed structure, we measure the complexity as a function of the length of the formula;
- *combined complexity* of $\mathcal{L}$: It is measured as a function of both the size of the structure and the length of the formula.

In this book we mainly concentrate on data complexity. The following exercise contains two results on combined complexity. Part (a) illustrates a general phenomenon, namely a considerable increase against data complexity (cf. [52]); part (b) shows that it may pay to look for axiomatizations with few variables. For a proof the reader should carefully analyze the proof of 7.4.1.

Exercise 7.5.3 (a) The combined complexity of FO is in PSPACE. (In fact, it is PSPACE-complete; see [142].)

(b) For $s \geq 1$, the combined complexity of FO^s is in PTIME. (In fact, it is PTIME-complete for $s \geq 2$; see [143].) □

The descriptive characterizations of complexity classes given by the main theorem are of importance in various respects:

- They may help to recognize that a concrete problem is in a given complexity class (by expressing it in the corresponding logic).
- They allow to view the logics involved as higher programming languages for problems of the corresponding complexity class. (Note that the proofs of the preceding section show how to convert a sentence φ into an algorithm accepting the class of models of φ and satisfying the required resource restrictions.)
- Characteristic features of the logic may be seen as characteristic features of the complexity class described by it and may add to a better understanding. (For instance, the result about FO(IFP) and PTIME shows us that inflationary inductions are an essential ingredient of PTIME algorithms.)
- The descriptive characterizations allow to convert problems, methods, and results of complexity theory into logic and vice versa, thus widening the methodological possibilities for both sides.

In the remainder of this section we give some illustrating examples for the last point. We first stay with ordered structures and subsequently discuss the role of order.

Sentences φ and ψ in a vocabulary τ with $< \in \tau$ are said to be *equivalent on ordered structures* if for all ordered τ-structures $\mathcal{A}$,

$$\mathcal{A} \models \varphi \quad \text{iff} \quad \mathcal{A} \models \psi.$$

Corollary 7.5.4 *On ordered structures, every* FO(IFP)*-sentence is equivalent to an* FO(IFP)*-sentence in which* IFP *occurs at most once. The same applies to* FO(PFP) *and* PFP.

Proof. Suppose $\varphi \in \mathrm{FO(IFP)}[\tau]$ with $< \in \tau$. Then $\mathrm{ordMod}(\varphi) \in$ PTIME. Now the claim follows from 7.3.5. For FO(PFP) cf. 7.3.3. □

The next corollary also has a simple direct proof.

Corollary 7.5.5 *Let $\mathcal{C}$ be one of the complexity classes mentioned in the Main Theorem. If K is a class of ordered structures in $\mathcal{C}$ then there is a Turing machine M strongly witnessing $K \in \mathcal{C}$, that is,*

- *M accepts K*
- *every run stops in the accepting or in the rejecting state*
- *every run fulfills the time or space bounds characteristic for $\mathcal{C}$.*

Proof. Let $\mathcal{L}$ be the logic capturing $\mathcal{C}$. Then there is a sentence of $\mathcal{L}$ axiomatizing K. By the results of section 7.4 we know that for every class K axiomatizable in $\mathcal{L}$ there is a machine strongly witnessing $K \in \mathcal{C}$. □

An immediate consequence of the Main Theorem is the equivalence of (i) and (ii):

(i) PTIME = PSPACE
(ii) FO(IFP) $\equiv$ FO(PFP) on ordered structures.

Note, however, that here PTIME and PSPACE are understood as classes of ordered structures and not as languages over alphabets. Does (i) mean the same as PTIME = PSPACE in complexity theory? We want to show this by making clear that here and in complexity theory we deal only with different presentations of a complexity class.

If $\mathcal{C}$ is a complexity class of complexity theory, we denote by $\mathcal{C}'$ the corresponding complexity class of structures. For example, for PTIME we have

> $\mathcal{C}$ consists of all languages L, $L \subseteq \mathbb{A}^+$ for some alphabet $\mathbb{A}$, such that there exists a deterministic Turing machine accepting L in polynomial time

and

> $\mathcal{C}'$ consists of all classes K, $K \subseteq \mathcal{O}[\tau]$ for some τ with $< \in \tau$, such that there is a deterministic Turing machine M accepting K in polynomial time.

In the following let $\mathcal{C}, \mathcal{C}_1, \mathcal{C}_2$ range over the complexity classes LOGSPACE, NLOGSPACE, PTIME, NPTIME, and PSPACE of complexity theory.

Our first consideration will show that $\mathcal{C} \subseteq \mathcal{C}'$ up to a natural transition from words to ordered structures, thereby using word models (already introduced in section 6.2).

Fix an alphabet $\mathbb{A}$ and let $\tau(\mathbb{A})$ be the vocabulary $\{<\} \cup \{P_a \mid a \in \mathbb{A}\}$ with unary P_a. If $u \in \mathbb{A}^+$, denote by K_u the class of structures of the form

$$(B, <, (P_a)_{a \in \mathbb{A}}),$$

where the cardinality of B equals the length of u, $<$ is an ordering of B, and P_a corresponds to the positions in u carrying an a. For $L \subseteq \mathbb{A}^+$ set $K(L) := \bigcup_{u \in L} K_u$. Clearly, $K(L) \subseteq \mathcal{O}[\tau(\mathbb{A})]$ and, since

$$K(\mathbb{A}^+) = \mathrm{ordMod}(\forall x(\bigvee_{a \in \mathbb{A}} (P_a x \wedge \bigwedge_{\substack{b \in \mathbb{A} \\ b \neq a}} \neg P_b x))),$$

we have

$$(1) \qquad K(\mathbb{A}^+) \in \mathrm{LOGSPACE}.$$

One can easily show that

$$(2) \qquad \text{for } L \subseteq \mathbb{A}^+, \;\; L \in \mathcal{C} \text{ iff } K(L) \in \mathcal{C}',$$

thus obtaining "$\mathcal{C} \subseteq \mathcal{C}'$ up to transitions".

Now we turn to a statement showing $\mathcal{C}' \subseteq \mathcal{C}$ up to a transition from ordered structures to words. Let τ with $< \in \tau$ be given. Set $\mathbb{A}_0 := \{0, 1, \alpha, \omega\}$. For $\mathcal{A} \in \mathcal{O}[\tau]$ let $u_\mathcal{A}$ be the word in $\mathbb{A}_0^+$ obtained by concatenating the inscriptions on all input tapes of a Turing machine started with input $\mathcal{A}$ (where we include the "virtual letters" α and ω). For a class $K \subseteq \mathcal{O}[\tau]$ set

$$L(K) \;\; := \;\; \{u_\mathcal{A} \mid \mathcal{A} \in K\}.$$

Clearly, given τ, there is a polynomial $p \in \mathbb{N}[x]$ such that for all $\mathcal{A} \in \mathcal{O}[\tau]$ we have $\|A\| \leq |u_\mathcal{A}| \leq p(\|A\|)$. In particular, for $p(x) := x^d$ we have that $\log \|A\| \leq \log |u_\mathcal{A}| \leq d \cdot \log \|A\|$. Invoking these relations one shows

$$(3) \qquad L(\mathcal{O}[\tau]) \in \mathrm{LOGSPACE}$$

$$(4) \qquad \text{for } K \subseteq \mathcal{O}[\tau], \;\; K \in \mathcal{C}' \text{ iff } L(K) \in \mathcal{C}$$

((3) is immediate, and (4) is left as an exercise to the reader). Thus, "$\mathcal{C}' \subseteq \mathcal{C}$ up to transitions".

From (2) and (4) we infer

Proposition 7.5.6 $\mathcal{C}_1 \subseteq \mathcal{C}_2$ *iff* $\mathcal{C}'_1 \subseteq \mathcal{C}'_2$.

Proof. First, suppose $\mathcal{C}_1 \subseteq \mathcal{C}_2$ and let $K \in \mathcal{C}'_1$ where $K \subseteq \mathcal{O}[\tau]$. Then by (4), $L(K) \in \mathcal{C}_1$ and by hypothesis, $L(K) \in \mathcal{C}_2$. Therefore, $K \in \mathcal{C}'_2$ (again by (4)). Now assume $\mathcal{C}'_1 \subseteq \mathcal{C}'_2$ and let $L \in \mathcal{C}_1$. Then, $K(L) \in \mathcal{C}'_1$ by (2), and hence, $K(L) \in \mathcal{C}'_2$. Therefore, $L \in \mathcal{C}_2$ (again by (2)). □

Together with the Main Theorem we have

Corollary 7.5.7 (a) FO(IFP) $\equiv$ FO(PFP) *on ordered structures iff* PTIME = PSPACE (*in complexity theory*).

(b) FO(IFP) $\equiv \Sigma_1^1$ *on ordered structures*
iff PTIME = NPTIME (*in complexity theory*). □

Moreover, the preceding argument shows that in order to get FO(IFP) $\equiv$ FO(PFP) on ordered structures (or, equivalently, PTIME = PSPACE) it suffices to prove that FO(IFP) $\equiv$ FO(PFP) holds for ordered τ-structures where, besides $<$, only unary relation symbols are in τ. Since one need only consider languages over $\{0, 1\}$, even a single unary relation symbol suffices.

Corollary 7.5.8 *The following are equivalent:*
(i) PTIME = NPTIME
(ii) FO(IFP) $\equiv$ SO *on ordered structures.*

Proof. If (ii) holds then $\Sigma_1^1 \leq$ FO(IFP) on ordered structures, thus NPTIME $\leq$ PTIME. Conversely, if NPTIME = PTIME then, on ordered structures, $\Sigma_1^1 \equiv$ FO(IFP). As Σ_1^1 is closed under existential quantifications and FO(IFP) under boolean operations, an easy induction yields SO $\equiv$ FO(IFP). □

Whereas the preceding corollaries contain the translation of *problems* from complexity theory to logics we now turn to the translation of a *result.* In complexity theory one shows

$$\text{LOGSPACE} \subseteq \text{NLOGSPACE} \subseteq \text{PTIME} \subseteq \text{NPTIME} \subseteq \text{PSPACE}$$

and

$$\text{LOGSPACE} \neq \text{PSPACE}.$$

Hence, by the Main Theorem,

Corollary 7.5.9 *On ordered structures,*
(a) FO(DTC) $\leq$ FO(TC) $\leq$ FO(IFP) $\leq \Sigma_1^1 \leq$ FO(PFP).
(b) FO(DTC) $\not\equiv$ FO(PFP). □

Note that most of the $\leq$-relations in (a) are immediate; the remaining ones and part (b) will be obtained in chapters 8 and 9 by purely modeltheoretic means.

For any of the complexity classes $\mathcal{C}$ introduced so far, in complexity theory one defines the class co-$\mathcal{C}$ to be the class of complements of languages in $\mathcal{C}$, that is, for any alphabet $\mathbb{A}$ and $L \subseteq \mathbb{A}^+$,

$$L \in \text{co-}\mathcal{C} \quad \text{iff} \quad (\mathbb{A}^+ \setminus L) \in \mathcal{C}.$$

Clearly, any deterministic class $\mathcal{C}$ is closed under complements, that is, $\mathcal{C} =$ co-$\mathcal{C}$. Similarly, we define the class co-$\mathcal{C}'$ as the class of complements of classes of structures in $\mathcal{C}'$; more precisely, for τ with $< \in \tau$ and $K \subseteq \mathcal{O}[\tau]$, we set

$$K \in \text{co-}\mathcal{C}' \quad \text{iff} \quad (\mathcal{O}[\tau] \setminus K) \in \mathcal{C}'.$$

Exercise 7.5.10 Show that Proposition 7.5.6 remains true if one allows $\mathcal{C}_1$ and $\mathcal{C}_2$ to range over the complexity classes mentioned in the Main Theorem and their complements. (Use (1)-(4) before 7.5.6.) □

We now are going to discuss the role of order and to get information whether and to what extent orderings can be avoided. Let $\mathcal{A}$ be a not necessarily ordered structure. We already mentioned that in order to consider $\mathcal{A}$ as an input for a Turing machine, we have to represent it as a string (or a sequence of strings), for example, by labeling the elements of $\mathcal{A}$ in some way. Taking, say, the lexicographic ordering of the labels, we get an ordering on A and hence, an ordered structure. Now, if $\mathcal{A}$ is a graph we can state questions such as "Is there a path from the 5-th to the 28-th element?" whose answer depends on this ordering and is senseless for $\mathcal{A}$ itself. The following framework enables us to concentrate on questions intrinsic to $\mathcal{A}$.

Definition 7.5.11 Let K be a class of (unordered) τ-structures. Set $\tau_< := \tau \dot{\cup} \{<\}$. The class $K_<$ of *ordered representations* of structures in K is given by

$$K_< := \{(\mathcal{A},<) \mid \mathcal{A} \in K, < \text{ an ordering of } A\}.$$ □

If $\mathcal{L}$ is a logic capturing the complexity class $\mathcal{C}$ we have

$$K_< \in \mathcal{C} \quad \text{iff} \quad \text{there is } \varphi \in \mathcal{L}[\tau_<] \text{ such that } K_< = \text{Mod}(\varphi).$$

The sentence φ on the right side of the equivalence is order-invariant in the finite, since for every $\mathcal{A}$ and any orderings $<_1$ and $<_2$ of A we have

$$(\mathcal{A},<_1) \in K_< \quad \text{iff} \quad (\mathcal{A},<_2) \in K_<$$

and thus,

$$(\mathcal{A},<_1) \models \varphi \quad \text{iff} \quad (\mathcal{A},<_2) \models \varphi.$$

Now, if $\mathcal{L}$ would be closed under order-invariant sentences in the finite (by 3.5.2, FO does not have this property), we would have

$$K_< \in \mathcal{C} \text{ iff there is } \psi \in \mathcal{L}[\tau] \text{ such that } K = \text{Mod}^\tau(\psi).$$ [8]

In general, this does not hold. To give a counterexample for FO(DTC), let $K = \text{EVEN}[\tau]$ with $\tau = \emptyset$ be the class of sets of even cardinality. Since $K_< \in$ LOGSPACE, there is a sentence φ of FO(DTC)$[\tau_<]$ such that

$$K_< = \text{Mod}(\varphi);$$

for example, as φ we can take the sentence $\neg[\text{DTC}_{x,y}\, y = x + 2]$ min max, where we use self-explanatory abbreviations. This sentence is order-invariant

[8] We use $\text{Mod}^\tau(\ldots)$ with upper index τ to rule out any ambiguity on the vocabulary considered.

in the finite: the evaluation of φ in a structure $(A, <^A)$ makes use of the ordering $<^A$, but the outcome of this evaluation does not depend on the specific ordering $<^A$ we have chosen. We shall see in 8.4.4 that for no sentence ψ of FO(DTC)[τ], even of FO(PFP)[τ],

$$K = \mathrm{Mod}(\psi).$$

These observations lead to a stronger notion of what it means to capture a complexity class.

Definition 7.5.12 Let $\mathcal{L}$ be a logic and $\mathcal{C}$ a complexity class. $\mathcal{L}$ *strongly captures* $\mathcal{C}$ if for all vocabularies τ and all classes K of τ-structures,

$$K_< \in \mathcal{C} \text{ iff } K \text{ is axiomatizable in } \mathcal{L}. \qquad \square$$

The following proposition holds for all complexity classes $\mathcal{C}$ considered so far; essentially one needs that $\mathcal{C}$ contains LOGSPACE. The proof is left to the reader as an exercise.

Proposition 7.5.13 *If $\mathcal{L}$ strongly captures $\mathcal{C}$ then $\mathcal{L}$ captures $\mathcal{C}$.* $\square$

The converse is false: The counterexample given before the definition shows that FO(PFP) does not capture PSPACE strongly. For the class EVEN[τ] we used as a counterexample, we have EVEN$[\tau]_< \in$ LOGSPACE, and EVEN[τ] is not axiomatizable in FO(PFP). Since FO(DTC) $\leq$ FO(TC) $\leq$ FO(IFP) $\leq$ FO(PFP) holds for arbitrary structures, we see that none of these logics strongly captures the complexity class corresponding to it by the Main Theorem. The result cannot be extended to Σ_1^1 and NPTIME, since we know $\Sigma_1^1 \leq$ FO(PFP) only on *ordered* structures. In fact, we have

Theorem 7.5.14 Σ_1^1 *strongly captures* NPTIME.

Proof. Let τ be arbitrary and K be a class of τ-structures. Assume $K = \mathrm{Mod}^\tau(\varphi)$ for some $\Sigma_1^1[\tau]$-sentence φ, say, $\varphi = \exists X_1 \ldots \exists X_m \psi$ with first-order ψ. Set $\chi := \exists X_1 \ldots \exists X_m(\psi \wedge$ "$<$ is an ordering"). Then, $\chi \in \Sigma_1^1[\tau_<]$, $\mathrm{Mod}(\chi) = K_<$, and hence, $K_< \in$ NPTIME by the Main Theorem. Conversely, if $K_< \in$ NPTIME then, again by the Main Theorem, there is a sentence $\varphi \in \Sigma_1^1[\tau_<]$ such that $K_< = \mathrm{Mod}^{\tau_<}(\varphi)$. Set $\psi := \exists < \varphi$. Then $\psi \in \Sigma_1^1[\tau]$, and for any τ-structure $\mathcal{A}$ we have

$$\begin{array}{lll} \mathcal{A} \models \psi & \text{iff} & \text{there is } <^A \text{ with } (\mathcal{A}, <^A) \models \varphi \\ & \text{iff} & \text{there is } <^A \text{ with } (\mathcal{A}, <^A) \in K_< \\ & \text{iff} & \mathcal{A} \in K, \end{array}$$

that is, $K = \mathrm{Mod}^\tau(\psi)$. $\square$

Theorem 7.5.15 Π_1^1 *strongly captures* co-NPTIME.

Proof. Let τ be arbitrary and K be a class of τ-structures. If $\mathrm{Str}[\tau]$ denotes the class of all τ-structures then

$$(+) \qquad (\mathrm{Str}[\tau] \setminus K)_< = \mathcal{O}[\tau_<] \setminus K_<.$$

Therefore,

$$\begin{array}{lll} K \in \Pi^1_1 & \text{iff} & \text{there is } \psi(X_1,\ldots,X_m) \in \mathrm{FO}[\tau] \text{ such that} \\ & & K = \mathrm{Mod}(\forall X_1 \ldots \forall X_m \psi) \\ & \text{iff} & \text{there is } \chi(X_1,\ldots,X_m) \in \mathrm{FO}[\tau] \text{ such that} \\ & & \mathrm{Str}[\tau] \setminus K = \mathrm{Mod}(\exists X_1 \ldots \exists X_m \chi) \\ & \text{iff} & (\mathrm{Str}[\tau] \setminus K)_< \in \mathrm{NPTIME} \qquad \text{(by 7.5.14)} \\ & \text{iff} & (\mathcal{O}[\tau_<] \setminus K_<) \in \mathrm{NPTIME} \qquad \text{(by (+))} \\ & \text{iff} & K_< \in \text{co-NPTIME}. \end{array}$$

□

Corollary 7.5.16 NPTIME = co-NPTIME *iff* $\Sigma^1_1 \equiv \Pi^1_1$.

Proof. As an example we show the implication from left to right. Let K be a class of structures. Then

$$\begin{array}{lll} K \in \Sigma^1_1 & \text{iff} & K_< \in \mathrm{NPTIME} \\ & \text{iff} & K_< \in \text{co-NPTIME} \quad \text{(by hypothesis)} \\ & \text{iff} & K \in \Pi^1_1. \end{array}$$

□

Corollary 7.5.17 NPTIME = co-NPTIME *iff* $\mathrm{SO} \equiv \Sigma^1_1$.

Proof. By the preceding corollary it suffices to prove

$$\Pi^1_1 \equiv \Sigma^1_1 \quad \text{iff} \quad \mathrm{SO} \equiv \Sigma^1_1.$$

For a logic $\mathcal{L}$ we write $\varphi \underset{\sim}{\in} \mathcal{L}$ to express that the sentence φ is equivalent to an $\mathcal{L}$-sentence. Clearly,

$$\varphi \in \Sigma^1_1 \quad \text{implies} \quad \neg\varphi \underset{\sim}{\in} \Pi^1_1,$$
$$\varphi \in \Pi^1_1 \quad \text{implies} \quad \neg\varphi \underset{\sim}{\in} \Sigma^1_1.$$

Now suppose that $\mathrm{SO} \equiv \Sigma^1_1$ and therefore, $\Pi^1_1 \leq \Sigma^1_1$. Let $\varphi \in \Sigma^1_1$. Then, $\neg\varphi \underset{\sim}{\in} \Pi^1_1$ and hence, $\neg\varphi \underset{\sim}{\in} \Sigma^1_1$. Therefore, $\varphi \underset{\sim}{\in} \Pi^1_1$. Conversely, assume that $\Pi^1_1 \equiv \Sigma^1_1$. One easily shows that the class Σ^1_1 is closed – up to equivalence – under $\vee$ and existential first-order and second-order quantifications. For closure under $\neg$ argue as follows: Suppose $\varphi = \neg\psi$ and $\psi \in \Sigma^1_1$. Then $\neg\psi \underset{\sim}{\in} \Pi^1_1 \equiv \Sigma^1_1$. □

Exercise 7.5.18 Show for $k \geq 1$ that $\Sigma_k^1 \equiv \Sigma_{k+1}^1$ implies $\Sigma_k^1 \equiv$ SO. The question, whether $\Sigma_k^1 \equiv \Sigma_{k+1}^1$ for some k, is open. However, it is conjectured that

$$\Sigma_1^1 < \Sigma_2^1 < \dots .$$

For $k \geq 1$, the class of ordered structures axiomatizable in Σ_k^1 forms the k-th stage PH_k of the *polynomial hierarchy* PH. Its initial stage PH_0 is defined to be PTIME. In 12.4.8 we shall give a definition of PH in terms of complexity. □

Exercise 7.5.19 Define the logic SO(PFP). Show that it captures PSPACE strongly. In 8.5.1 we shall implicitly show by a purely modeltheoretic proof that SO(PFP) $\equiv$ FO(PFP) on ordered structures. □

The question whether LOGSPACE, NLOGSPACE, and PTIME can be strongly captured by a logic is a major issue of descriptive complexity theory. Of course, one has to give a precise definition of what is meant by a "logic" in this context. Moreover, to rule out trivial positive answers, one has to require some conditions of effectiveness. Note that in the capturing results obtained in the preceding sections, we effectively assigned to every sentence a Turing machine accepting its models. And to every Turing machine accepting a class of models we effectively assigned a sentence axiomatizing this class. Up to now all attempts to find a logic that strongly captures PTIME in such an effective way have failed. In Chapter 11 we shall come back to such attempts.

We close this section by a result which completes the proof of the Main Theorem (compare 8.6.15 for a modeltheoretic proof).

Theorem 7.5.20 *On ordered structures,* FO(posTC) $\equiv$ FO(TC).

Proof. We make use of the fact (shown in 8.6.14) that, on ordered structures, every formula of FO(posTC) is equivalent to a formula of the form

$$[\mathrm{TC}_{\overline{x},\overline{y}}\, \psi]\, \widetilde{\min}\widetilde{\max}$$

with first-order ψ.

The proof of the theorem proceeds by induction on FO(TC)-formulas, the only nontrivial case being the negation step. By the induction hypothesis and the fact just mentioned we may assume that

$$\varphi = \neg\, [\mathrm{TC}_{\overline{x},\overline{y}}\, \psi]\, \overline{s}\overline{t}$$

with first-order ψ. For simplicity, we assume that $\psi = \psi(\overline{x}, \overline{y})$. By the Main Theorem (more precisely, by 7.4.1) there is a Turing machine M_0 strongly witnessing that $\mathrm{ordMod}(\psi) \in$ LOGSPACE.

Suppose $\overline{x} = x_1 \dots x_r$. Given a structure $\mathcal{A}$ and $\overline{a}, \overline{b} \in A^r$, let $d_\psi(\overline{a}, \overline{b})$ be the length of the shortest ψ-path connecting $\overline{a}$ and $\overline{b}$,

$$d_\psi(\overline{a},\overline{b}) := \min\{k > 0 \mid \text{there exist } \overline{a}_0 = \overline{a}, \overline{a}_1, \ldots, \overline{a}_k = \overline{b} \text{ such that } \mathcal{A} \models \psi[\overline{a}_i, \overline{a}_{i+1}] \text{ for } i < k\},$$

where $d_\psi(\overline{a},\overline{b}) := \infty$ in case the set on the right side is empty. Note that

– if $d_\psi(\overline{a},\overline{b}) < \infty$ then $0 < d_\psi(\overline{a},\overline{b}) \leq \|A\|^r$
– $d_\psi(\overline{a},\overline{c}) \leq d_\psi(\overline{a},\overline{b}) + d_\psi(\overline{b},\overline{c})$.

Moreover, $\neg\,[\mathrm{TC}_{\overline{x},\overline{y}}\,\psi(\overline{x},\overline{y})]\,\overline{s}\overline{t}$ is equivalent to

$$\text{“}\|\{\overline{v} \mid d_\psi(\overline{s},\overline{v}) < \infty\}\| = \|\{\overline{v} \mid d_{(\psi(\overline{x},\overline{y})\,\wedge\,\neg\overline{y}=\overline{t})}(\overline{s},\overline{v}) < \infty\}\|\text{”}.$$

We first show that there is a nondeterministic log space-bounded[9] machine M such that for any ordered structure $(\mathcal{A},\overline{a},\overline{l},\overline{w},\overline{w}')$,

if $\|\{\overline{e} \mid d_\psi(\overline{a},\overline{e}) \leq \overline{l}\}\| = \overline{w}$ then

$$M \text{ accepts } (\mathcal{A},\overline{a},\overline{l},\overline{w},\overline{w}') \;\text{ iff }\; \|\{\overline{e} \mid d_\psi(\overline{a},\overline{e}) \leq \overline{l}+1\}\| = \overline{w}',$$

where the corresponding natural numbers $\leq \|A\|^r$ are given by their $\|A\|$-adic representations $\overline{l} = l_0 \ldots l_r$, $\overline{w} = w_0 \ldots w_r$, $\overline{w}' = w_0' \ldots w_r'$.

We present the machine M. When during its computation M checks whether $\mathcal{A} \models \psi[\overline{c},\overline{d}]$ holds or not, this is done by invoking M_0. Started with $(\mathcal{A},\overline{a},\overline{l},\overline{w},\overline{w}')$, M first sets a counter to $\overline{w}'$. Then, for every $\overline{b} \in A^r$, it carries out either (1) or (2), the choice being done nondeterministically:

(1) M nondeterministically guesses a path witnessing $d_\psi(\overline{a},\overline{b}) \leq \overline{l}+1$ and decreases the counter by one; in case the counter is zero it rejects.
(2) Using an additional counter with the initial value $\overline{w}$, M nondeterministically guesses $\overline{w}$ many distinct tuples $\overline{c} \in A^r$ together with a proof that $d_\psi(\overline{a},\overline{c}) \leq \overline{l}$; for each such $\overline{c}$ it shows that $\overline{c} \neq \overline{b}$ and $\neg\psi[\overline{c},\overline{b}]$ (thus, in case $\|\{\overline{e} \mid d_\psi(\overline{a},\overline{e}) \leq \overline{l}\}\| = \overline{w}$, proving that $d_\psi(\overline{a},\overline{b}) > \overline{l}+1$). In case $\overline{l} = 0$ it shows $\neg\psi[\overline{a},\overline{b}]$.

Finally, if all $\overline{b} \in A^r$ have been dealt with and the counter is 0, M accepts.

Since M is log space-bounded there is, by the Main Theorem, a formula $\chi_\psi(\overline{u},\overline{y},\overline{w},\overline{w}') \in \mathrm{FO(posTC)}$ axiomatizing the class accepted by M. Then, noting that $1\underbrace{0\ldots0}_{r \text{ times}}$ is (the representation of) $\|A\|^r$, we get a formula $\rho_\psi(\overline{u},\overline{w}) \in \mathrm{FO(posTC)}$ with the meaning

$$\text{“}\|\{\overline{v} \mid d_\psi(\overline{u},\overline{v}) < \infty\}\| = \overline{w}\text{”}$$

or, equivalently, with the meaning

$$\text{“}\|\{\overline{v} \mid d_\psi(\overline{u},\overline{v}) \leq \|A\|^r\}\| = \overline{w}\text{”}$$

[9] log space-bounded means $c\cdot$ log space-bounded for some $c > 0$.

by setting

$$\rho_\psi(\overline{u},\overline{w}) \quad := \quad [\mathrm{TC}_{\overline{y}\,\overline{w},\overline{y}'\overline{w}'}(\chi_\psi(\overline{u},\overline{y},\overline{w},\overline{w}') \land \overline{y}' = \overline{y}+1)]\,\widetilde{\min}\widetilde{\min}1\underbrace{0\ldots0}_{r \text{ times}}\overline{w}.$$

Since we saw above that $\varphi = \neg\,[\mathrm{TC}_{\overline{x},\overline{y}}\,\psi(\overline{x},\overline{y})]\,\overline{s}\overline{t}$ is equivalent to

$$\text{“}\|\{\overline{v} \mid d_\psi(\overline{s},\overline{v}) < \infty\}\| = \|\{\overline{v} \mid d_{(\psi(\overline{x},\overline{y})\land\neg\overline{y}=\overline{t})}(\overline{s},\overline{v}) < \infty\}\|\text{”},$$

we obtain that φ is equivalent to the formula

$$\exists\overline{z}\,(\rho_\psi(\overline{s},\overline{z}) \land \rho_{(\psi\land\neg\overline{y}=\overline{t})}(\overline{s},\overline{z})),$$

a formula of FO(posTC). □

As a consequence of part (b) of the Main Theorem whose proof we just have completed we get, since FO(TC) is closed under negation:

Corollary 7.5.21 NLOGSPACE = co-NLOGSPACE. □

In the preceding considerations, when treating questions of effectiveness, we have restricted ourselves to ordered structures $\mathcal{A}$, as they offer a natural way of encoding by passing to the unique isomorphic copy whose universe is $\{0,\ldots,\|A\|-1\}$ and whose order is the natural order on $\{0,\ldots,\|A\|-1\}$. There is an alternative way, namely the direct restriction to structures whose universe forms an initial segment of the natural numbers. Of course, both frameworks are equivalent:

Exercise 7.5.22 Let τ be an arbitrary vocabulary. A *numerical* τ-structure is a τ-structure whose universe is an initial segment of the natural numbers, that is, the universe is the set $\{0,\ldots,n\}$ for some n. For a class K of τ-structures let

$$K_{\mathrm{num}} \quad := \quad \{\mathcal{A} \in K \mid \mathcal{A} \text{ is numerical}\}$$

be the set of numerical structures in K, and for a sentence φ let

$$\mathrm{Mod}_{\mathrm{num}}(\varphi) \quad := \quad \{\mathcal{A} \mid \mathcal{A} \models \varphi, \mathcal{A} \text{ numerical}\}$$

be the set of its numerical models.

Let $\mathcal{C}$ and $\mathcal{L}$ be any of the complexity classes and logics, respectively, considered so far. Show

(a) For any class K, we have $K_< \in \mathcal{C}$ iff $K_{\mathrm{num}} \in \mathcal{C}$ (that is, there is a Turing machine according to $\mathcal{C}$ that, started with any numerical $\mathcal{A}$ accepts $\mathcal{A}$ iff $\mathcal{A} \in K_{\mathrm{num}}$).

(b) $\mathcal{L}$ strongly captures $\mathcal{C}$ iff for any τ and any class K of τ-structures, $K_{\mathrm{num}} \in \mathcal{C}$ iff there is a sentence φ of $\mathcal{L}[\tau]$ such that $K_{\mathrm{num}} = \mathrm{Mod}_{\mathrm{num}}(\varphi)$.

(c) $\mathcal{L}$ captures $\mathcal{C}$ iff for any τ and any class K of τ-structures, $K_{\mathrm{num}} \in \mathcal{C}$ iff there is an $\mathcal{L}$-sentence φ of $\mathcal{L}[\tau_<]$ with $K_{\mathrm{num}} = \{\mathcal{A} \mid (\mathcal{A},<^A) \models \varphi$, $A = \{0,\ldots,n\}$ for some n, and $<^A$ is the natural ordering on $A\}$. □

7.5.1 Appendix[10]

In the preceding sections Turing machines have only been used as *acceptors*, that is, given a machine M and an input structure $\mathcal{A}$, the question was whether M would accept $\mathcal{A}$. However, Turing machines can also be used to calculate functions. For later purposes we are interested in machines that map structures to structures, so-called transducers.

Let τ and σ be vocabularies containing $<$. We consider (τ, σ)-*Turing machines*, that is, deterministic Turing machines M that have input tapes for encoding τ-structures, work tapes, and in addition so-called *output tapes*, namely as many as we need for our way of encoding σ-structures as inputs. The output tapes are write-only tapes: In the beginning of a calculation they are empty, and their heads scan the first cell. In a step of M the heads on the output tapes either are not active or write a letter on the scanned cell and then move to the right neighbour. (So the final positions of the heads on the output tapes take over the role of the endmark ω since they determine the portion of the tape visited by the heads during the computation.) Moreover, the behaviour of M does not depend on the actual inscriptions and head positions of the output tapes. Altogether it is made precise by the form of the instructions that look like (cf. $(*)$ in subsection 7.2.2):

$$(+) \qquad s b_0 \ldots b_{k+l} c_1 \ldots c_m \to s' c'_1 \ldots c'_m h_0 \ldots h_{k+l+m} d_0 \ldots d_p$$

(in case M has $p+1$ output tapes $\mathrm{OP}_0, \ldots, \mathrm{OP}_p$), where $d_i \in \{00, 10, 11\}$. The meaning of $(+)$ is fixed as for $(*)$ in subsection 7.2.2 adding that

$d_i = 00$	means	"Do nothing on OP_i",
$d_i = 10$	means	"Write 0 on the scanned cell of OP_i and move the head to the cell on the right",
$d_i = 11$	means	"Write 1 on the scanned cell of OP_i and move the head to the cell on the right".

A (τ, σ)-Turing machine M is a (τ, σ)-*transducer* if M, started with an ordered τ-structure $\mathcal{A}$, finally stops and the output tapes then carry the encoding of an ordered σ-structure $\mathcal{B}$ (where, up to the missing endmark ω, the same encoding as for input structures is used). To be definite, we let $\{0, 1, \ldots, \|B\|-1\}$ be the universe of $\mathcal{B}$.

When considering space-bounded transducers, the space used on the output tapes (similar to that used on the input tapes) does not count. Hence, a log space-bounded transducer may fill a number of cells on the output tapes that is not log space-bounded in the size of the input.

Without changing the space needed and, up to a constant factor, the time needed, we may assume that transducers are *normalized* in the sense that in the instructions $(+)$ either (i) or (ii), where

[10] The notions and results of this appendix are only needed in Chapters 11 and 12. They may be skipped in a first reading.

(i) all d_i are $= 00$
(ii) exactly one d_i is $\neq 00$, $c_1 \dots c_m = c'_1 \dots c'_m$, and $h_0 = \dots = h_{k+l+m} = 0$.

So in a step of a normalized transducer either no output tape is involved or only one output tape, but no other tape. In case (ii), we speak of an $(i, 0)$-instruction if $d_i = 10$, and of an $(i, 1)$-instruction if $d_i = 11$.

Fix τ, σ as above and let M be a normalized (τ, σ)-transducer. If we forget the output tapes and the d_i, we get a usual deterministic Turing machine M'. Thus, if

$$sb_0 \dots b_{k+l}\, c_1 \dots c_m \to s'c'_1 \dots c'_m h_0 \dots h_{k+l+m} d_0 \dots d_p \tag{1}$$

is an instruction of M then

$$s_0 b_0 \dots b_{k+l} c_1 \dots c_m \to s'c_1 \dots c'_m h_0 \dots h_{k+l+m} \tag{2}$$

is an instruction of M'; we call it an (i, j)-instruction of M' if (1) is an (i, j)-instruction of M. When carrying out an (i, j)-instruction, M' at most changes its state.

Now, start M' with an ordered τ-structure $\mathcal{A}$. Then, for every point of the run of M started with $\mathcal{A}$, the inscription of the output tapes of M (and hence, the output structure of M) can be reconstructed from the run of M' as follows:

> OP_i has 0 in cell l if the corresponding sequence of configurations of M' contains at least l many $(i, 0)$- or $(i, 1)$-configurations, the l-th one being an $(i, 0)$-configuration.[11] And similarly, OP_i has 1 in cell l if the l-th $(i, 0)$- or $(i, 1)$-configuration is an $(i, 1)$-configuration.

This observation enables us to describe the behaviour of M and, in particular, the output structure via the behaviour of M', using formulas of a suitable logic as we have done it for acceptors in section 7.3. As an example that we shall need later we consider the case that M is log space-bounded and the logic is FO(DTC), recalling that the behaviour of log space-bounded acceptors can be described by formulas of FO(DTC) (cf. 7.3.8).

Theorem 7.5.23 *Let M be a log space-bounded (τ, σ)-transducer. There are an $e \geq 1$, an* FO(DTC)*-formula $\varphi_{\mathrm{uni}}(x_1, \dots, x_e)$, and for every, say r-ary, $R \in \sigma$ an* FO(DTC)*-formula $\varphi_R(\overline{x}_1, \dots, \overline{x}_r)$ where* $\mathrm{length}(\overline{x}_1) = \dots = \mathrm{length}(\overline{x}_r) = e$ *such that for every ordered τ-structure $\mathcal{A}$ the machine M, started with $\mathcal{A}$, leads to an output structure isomorphic to*

$$(\chi^{\mathcal{A}}_{\mathrm{uni}}(-), (\varphi^{\mathcal{A}}_R(-, \dots, -))_{R \in \sigma}).$$

We leave the details of a proof to the reader. One uses an FO(DTC)-description of the run of M. The following exercise contains the key parts

[11] An (i, j)-configuration is a configuration calling an (i, j)-instruction.

needed to obtain the formulas mentioned in the theorem. Note that, when started with an ordered structure $\mathcal{A}$, both M and M' stop after the same number of steps. As the number of different, say, $d{\cdot}\log\|A\|$-bounded configurations of M' is polynomial in $\|A\|$, both M and M' are (n^e-1) time-bounded for a suitable e. We may assume that $e = d$.

For notational simplicity let $\sigma = \{E, <\}$ with binary E.

Exercise 7.5.24 Referring to M' and using the formulas $\chi_{\text{start}}(\overline{x})$ and $\chi_{\text{succ}}(\overline{x}, \overline{x}')$ from Lemma 7.3.7, show for sufficiently large τ-structures $\mathcal{A}$:

(a) For $i \leq p$ there is an FO(DTC)-formula $\chi_{i\text{-free}}(\overline{x}, \overline{x}')$ such that for $\overline{a}, \overline{a}'$ in A:

$$\mathcal{A} \models \chi_{i\text{-free}}[\overline{a}, \overline{a}'] \quad \text{iff} \quad \begin{array}{l}\overline{a} \text{ and } \overline{a}' \text{ are } (d \cdot \log\|A\|)\text{-bounded configurations} \\ \text{of } M', \text{ and } M', \text{ with } \mathcal{A} \text{ written on its input tapes,} \\ \text{runs from } \overline{a} \text{ to } \overline{a}' \text{ without passing an} \\ (i,0)\text{-configuration or an } (i,1)\text{-configuration.}\end{array}$$

(b) For $i \leq p$ there is an FO(DTC)-formula $\chi_{i\text{-count}}(\overline{x}, \overline{y})$ with length$(\overline{y}) = e$ such that for $\overline{a}, \overline{b}$ in A:

$$\mathcal{A} \models \chi_{i\text{-count}}[\overline{a}, \overline{b}] \quad \text{iff} \quad \begin{array}{l}\overline{a} \text{ is the } \overline{b}\text{-th } (i,0)\text{-configuration or} \\ (i,1)\text{-configuration of } M' \text{ started with } \mathcal{A} \\ \text{(we refer to the lexicographical ordering of } A^e).\end{array}$$

(c) Let $\mathcal{B} = (B, E^B, <^B)$ be the output structure that M gives on $\mathcal{A}$. There are FO(DTC)-formulas $\chi_{\text{uni}}(\overline{y}), \chi_E(\overline{y}, \overline{z})$, and $\chi_<(\overline{y}, \overline{z})$ with length$(\overline{y})$ = length$(\overline{z}) = e$ such that

$$\begin{array}{rcl} B & = & \{l \mid \mathcal{A} \models \chi_{\text{uni}}[\overline{b}],\ \overline{b} \text{ is the } l\text{-th element in the lexicographical} \\ & & \hfill \text{ordering of } A^e\}; \\ E^B & = & \{(l,m) \mid \mathcal{A} \models \chi_E[\overline{b}, \overline{c}],\ \overline{b} \text{ is the } l\text{-th and } \overline{c} \text{ the } m\text{-th element in} \\ & & \hfill \text{the lexicographical ordering of } A^e\}; \\ <^B & = & \{(l,m) \mid \mathcal{A} \models \chi_<[\overline{b}, \overline{c}],\ \overline{b} \text{ is the } l\text{-th, } \overline{c} \text{ the } m\text{-th element}\} \text{ is the} \\ & & \hfill \text{natural ordering on } B. \end{array}$$

□

Exercise 7.5.25 Formulate and prove the analogue of Theorem 7.5.23 for polynomially time-bounded (τ, σ)-transducers and FO(IFP). □

Notes 7.5.26 As a reference for computation and complexity theory we mention the books [87, 128, 46]. Theorem 6.2.1 goes back to [139]. The characterizations of complexity classes given by 7.5.2 are due to Immerman [93, 94] (LOGSPACE, NLOGSPACE), Immerman [91], Livchak [119], and Vardi [142] (PTIME), Fagin [36] (NPTIME), Abiteboul and Vianu [2] (PSPACE). Corollary 7.5.21 was independently proved in [94] and [137]. An important topic we have not treated here is circuit complexity. We refer the reader to [95] for information and further references.

8. Logics with Fixed-Point Operators

In Chapter 7 we have introduced logics with fixed-point operators in the context of computational complexity and shown that they capture important complexity classes. In the present chapter we study the (finite) model theory of fixed-point logics. Though we sometimes refer to Chapter 7, we repeat the relevant definitions and results to provide an independent approach.

In the rest of the book all structures will be finite, unless stated otherwise explicitly. Equivalence of formulas means equivalence with respect to all finite structures.

8.1 Inflationary and Least Fixed-Points

We prove some basic facts about fixed-points and, besides inflationary fixed-point logic, present an important sublogic, namely least fixed-point logic. We state two major results, the proofs being postponed to the next section: Inflationary and least fixed-point logic have the same expressive power, and every formula is equivalent to a formula containing at most one fixed-point application – a fact that extends Corollary 7.5.4 from ordered structures to arbitrary ones.

We observed in Chapter 2 that many important notions and global relations cannot be expressed by first-order formulas. An example is given by the reflexive and transitive closure of the edge relation in a graph (cf. 2.3.8). In the vocabulary $\tau := \{E\}$ for graphs consider the formula

$$\chi(x, y, X) \quad := \quad (x = y \vee \exists z(Xxz \wedge Ezy)).$$

It gives rise to a sequence of sets defined by

$$(*) \qquad X_0 := \emptyset, \qquad X_{n+1} := \{(x, y) \mid \chi(x, y, X_n)\}.$$

In other terms: Let $\mathcal{G} = (G, E^G)$ be a graph. Look at the function F^χ on the power set of $G \times G$, $F^\chi : \text{Pow}(G^2) \to \text{Pow}(G^2)$, given by

$$F^\chi(U) \quad := \quad \{(a, b) \in G^2 \mid \mathcal{G} \models \chi[a, b, U]\}$$

for $U \subseteq G \times G$. Then, $(*)$ corresponds to the sequence of sets

$$F_0^{\mathcal{X}} := \emptyset, \qquad F_{n+1}^{\mathcal{X}} := F^{\mathcal{X}}(F_n^{\mathcal{X}}).$$

Note that

$$F_n^{\mathcal{X}} = \{(a,b) \in G^2 \mid d(a,b) < n\},$$

where $d(a,b)$ denotes the distance between a and b in $\mathcal{G}$ (see 1.A1). For $F_\infty^{\mathcal{X}} := \bigcup_{n \geq 0} F_n^{\mathcal{X}}$ we have $F(F_\infty^{\mathcal{X}}) = F_\infty^{\mathcal{X}}$, that is, $F_\infty^{\mathcal{X}}$ is a fixed-point of $F^{\mathcal{X}}$ and

$$F_\infty^{\mathcal{X}} = \{(a,b) \in G^2 \mid d(a,b) < \infty\}.$$

In particular, for $a \neq b$,

$$a, b \text{ are connected by a path in } \mathcal{G} \qquad \text{iff} \qquad (a,b) \in F_\infty^{\mathcal{X}},$$

and

$$\mathcal{G} \text{ is connected} \qquad \text{iff} \qquad F_\infty^{\mathcal{X}} = G \times G.$$

The relations $F_n^{\mathcal{X}}$ that lead to the fixed-point $F_\infty^{\mathcal{X}}$ are first-order definable, but in general, $F_\infty^{\mathcal{X}}$ is not. In the fixed-point logics we are going to introduce the relation $F_\infty^{\mathcal{X}}$ will be definable, too; in particular, the connectivity of graphs will be expressible.

Let us first study some aspects of the process above on a more abstract level. Fix a finite set M. A function $F : \text{Pow}(M) \to \text{Pow}(M)$ gives rise to a sequence of sets

$$\emptyset,\ F(\emptyset),\ F(F(\emptyset)), \ldots.$$

Denote its members by $F_0, F_1, \ldots$, i.e., $F_0 = \emptyset$ and $F_{n+1} = F(F_n)$. F_n is called the n-th *stage* of F. Suppose that there is an $n_0 \in \mathbb{N}$ such that $F_{n_0+1} = F_{n_0}$, that is, $F(F_{n_0}) = F_{n_0}$. Then, $F_m = F_{n_0}$ for all $m \geq n_0$. We denote F_{n_0} by F_∞ and say that the *fixed-point* F_∞ of F *exists*. In case the fixed-point F_∞ does not exist, we agree to set $F_\infty := \emptyset$.

F is *inductive* if $F_0 \subseteq F_1 \subseteq \ldots$.

Lemma 8.1.1 (a) *If F_∞ exists then $F_\infty = F_{2^{\|M\|}-1}$.*
(b) *If $F : \text{Pow}(M) \to \text{Pow}(M)$ is inductive then F_∞ exists and $F_\infty = F_{\|M\|}$.*

Proof. Part (a) coincides with (b) of 7.1.1.
(b) By assumption, $F_0 \subseteq F_1 \subseteq \ldots \subseteq M$. Since M has $\|M\|$ elements, this sequence must get constant not later than with $F_{\|M\|}$. □

F is *inflationary* if for all $X \subseteq M$

$$X \subseteq F(X),$$

and *monotone* if for all $X, Y \subseteq M$

$$X \subseteq Y \text{ implies } F(X) \subseteq F(Y).$$

Lemma 8.1.2 (a) *If F is inflationary or monotone then F is inductive.*

(b) *If F is monotone then F_∞ is the* least fixed-point *of F, i.e., $F(F_\infty) = F_\infty$ and $F(X) = X$ implies $F_\infty \subseteq X$; even $F(X) \subseteq X$ implies $F_\infty \subseteq X$.*

(c) *If $F : \text{Pow}(M) \to \text{Pow}(M)$ is arbitrary and $F' : \text{Pow}(M) \to \text{Pow}(M)$ is given by $F'(X) := X \cup F(X)$, then F' is inflationary. In case F is inductive we have $F'_n = F_n$ for all $n \geq 0$ and hence, $F'_\infty = F_\infty$.*

Proof. (a) If F is inflationary then $F_n \subseteq F(F_n) = F_{n+1}$; thus F is inductive. Suppose that F is monotone. We have $F_0 = \emptyset \subseteq F(\emptyset) = F_1$ and therefore, by monotonicity, $F_1 = F(F_0) \subseteq F(F_1) = F_2$. Going on this way we obtain $F_n \subseteq F_{n+1}$ for all n.

(b) By part (a) and 8.1.1(b) the fixed-point F_∞ exists. Suppose $F(X) \subseteq X$. We have $F_0 = \emptyset \subseteq X$. Suppose $F_n \subseteq X$. Then, by monotonicity, $F_{n+1} = F(F_n) \subseteq F(X) \subseteq X$. Therefore, $F_\infty \subseteq X$.

(c) Clearly, F' is inflationary. Let F be inductive. We show by induction on n that $F'_n = F_n$. Obviously, $F'_0 = F_0$. Suppose $F'_n = F_n$. Then $F'_{n+1} = F'_n \cup F(F'_n) = F_n \cup F(F_n) = F_n \cup F_{n+1} = F_{n+1}$. □

Exercise 8.1.3 Let $M := \{0, 1, 2\}$ and $F : \text{Pow}(M) \to \text{Pow}(M)$. Show:

(a) If $F(X) = \emptyset$ for all $X \subseteq M$, then F is monotone but not inflationary.

(b) If

$$F(X) = \begin{cases} X & \text{for } X = M \\ X \cup \{\|X\|\} & \text{otherwise} \end{cases}$$

then F is inflationary but not monotone.

(c) If

$$F(X) = \begin{cases} X & \text{for } X = \emptyset \\ X \setminus \{\|X\|\} & \text{otherwise} \end{cases}$$

then F is inductive but neither inflationary nor monotone. □

Exercise 8.1.4 Let $F : \text{Pow}(M) \to \text{Pow}(M)$ be monotone. Show that F has a *greatest fixed-point*, i.e., there is $F_g \subseteq M$ such that

- $F(F_g) = F_g$
- $F(Y) = Y$ implies $Y \subseteq F_g$.

Define $G : \text{Pow}(M) \to \text{Pow}(M)$ by $G(X) := M \setminus F(M \setminus X)$. Prove that

- G is monotone
- $F_g = M \setminus G_\infty$, i.e., the greatest fixed-point of F is the complement of the least fixed-point of G. □

Exercise 8.1.5 (a) Assume $F : \text{Pow}(M) \to \text{Pow}(M)$ is *antitone*, that is, $X \subseteq Y$ implies $F(Y) \subseteq F(X)$. Show that

$$F_0 \subseteq F_2 \subseteq F_4 \subseteq \ldots \subseteq F_5 \subseteq F_3 \subseteq F_1.$$

Define $H : \text{Pow}(M) \to \text{Pow}(M)$ by $H(X) := F(F(X))$. Show that H is monotone, that $\bigcup_{n\geq 0} F_{2\cdot n}$ is the least fixed-point H_∞ of H and $\bigcap_{n\geq 0} F_{2\cdot n+1}$ is the greatest fixed-point H_g of H. Conclude:

- F_∞ exists iff $H_\infty = H_g$;
- if F_∞ exists then $F_\infty = H_\infty$ and F_∞ is the unique fixed-point of H (and hence of F).

(b) For a concrete example of an antitone operation consider a digraph $\mathcal{G} = (G, E^G)$. The *game associated with* $\mathcal{G}$ is played between two players I and II. The game starts in an arbitrary point of G. It is played in rounds. Each *round* consists of two moves, a move of I followed by a move of II. A player can *move* from point a to point b if $E^G ab$. A player *looses* if he cannot move. A point $a \in G$ is *won* for a player, if he has a winning strategy for games started at a. The point a is *drawn*, if for no player it is a won point. Define $F : \text{Pow}(G) \to \text{Pow}(G)$ by

$$F(X) \quad := \quad \{a \in G \mid \exists b \in G \setminus X : E^G ab\}.$$

Show that F is antitone and that for $n \geq 0$

- $a \in F_{2\cdot n}$ iff a is won for I in $\leq n$ rounds;
- $a \notin F_{2\cdot n+1}$ iff a is won for II in $\leq n+1$ rounds.

Conclude:

- $\bigcup_{n\geq 0} F_{2\cdot n}$ is the set of points won for I;
- $(\bigcap_{n\geq 0} F_{2\cdot n+1})^c$ is the set of points won for II;
- F_∞ exists iff there are no drawn points;
- If F_∞ exists then F_∞ is the set of points won for I. □

A formula $\varphi(x_1, \ldots, x_k, \overline{u}, X, \overline{Y})$ in the vocabulary τ, where the relation variable X has arity k, together with a τ-structure $\mathcal{A}$ and interpretations $\overline{b}$ and $\overline{S}$ of $\overline{u}$ and $\overline{Y}$, respectively, gives rise to an operation $F^\varphi : \text{Pow}(A^k) \to \text{Pow}(A^k)$ defined by

$$F^\varphi(R) \quad := \quad \{(a_1, \ldots, a_k) \mid \mathcal{A} \models \varphi[a_1, \ldots, a_k, \overline{b}, R, \overline{S}]\}.$$

(The notation F^φ does not make explicit all relevant data.) At the beginning of this section we have already given a concrete example using this notation. A further example: For the function F introduced in part (b) of the preceding exercise for the game associated with a digraph $\mathcal{G}$ we have $F = F^\varphi$ where

$$\varphi(x, X) \quad := \quad \exists y(\neg Xy \wedge Exy).$$

And for the function $(F^\varphi)'$ associated with F^φ as in part (c) of the preceding lemma we have

$$(+) \qquad (F^\varphi)' = F^{(X\overline{x} \vee \varphi)}.$$

We now introduce *inflationary fixed-point logic* FO(IFP). It contains first-order logic and is closed under fixed-points of definable inflationary operations and hence, by 8.1.2(c), under fixed-points of definable inductive operations. Therefore, inflationary fixed-point logic is sometimes called *inductive fixed-point logic.*

The syntax of FO(IFP): For a vocabulary τ the class of formulas of FO(IFP) of vocabulary τ is given by the calculus

- $\dfrac{}{\varphi}$ φ an atomic second-order formula over τ
- $\dfrac{\varphi}{\neg\varphi}$; $\dfrac{\varphi, \psi}{(\varphi \vee \psi)}$; $\dfrac{\varphi}{\exists x \varphi}$
- $\dfrac{\varphi}{[\mathrm{IFP}_{\overline{x},X}\varphi]\,\overline{t}}$ where the lengths of $\overline{x}$ and $\overline{t}$ are the same and coincide with the arity of X.

Sentences are formulas without free first-order and second-order variables, where the free occurrence of variables is defined in the standard way, adding the clause

$$\mathrm{free}\big([\mathrm{IFP}_{\overline{x},X}\varphi]\,\overline{t}\big) \quad := \quad \mathrm{free}(\overline{t}) \cup (\mathrm{free}(\varphi) \setminus \{\overline{x}, X\}).$$

The semantics is defined inductively w.r.t. the calculus above, the meaning of $[\mathrm{IFP}_{\overline{x},X}\varphi]\,\overline{t}$ being $\overline{t} \in F_\infty^{(X\overline{x} \vee \varphi)}$. More precisely: If X is k-ary and the variables free in $[\mathrm{IFP}_{\overline{x},X}\varphi]\,\overline{t}$ are among $\overline{u}$ and $\overline{Y}$, and $\overline{b}$ and $\overline{S}$ are interpretations in $\mathcal{A}$ of $\overline{u}$ and $\overline{Y}$, respectively, then

$$\mathcal{A} \models [\mathrm{IFP}_{\overline{x},X}\varphi]\,\overline{t}\ [\overline{b}, \overline{S}] \qquad \text{iff} \qquad (t_1[\overline{b}], \ldots, t_k[\overline{b}]) \in F_\infty^{(X\overline{x} \vee \varphi)}.$$

We sometimes denote $F_\infty^{(X\overline{x} \vee \varphi)}$ by $[\mathrm{IFP}_{\overline{x},X}\varphi]$.

As an example, using the formula $\chi(x, y, X) = (x = y \vee \exists z(Xxz \wedge Ezy))$ of the beginning of this section, we see that the FO(IFP)-formula in the language of graphs

$$\psi_0(x, y) \quad := \quad [\mathrm{IFP}_{xy,X}\,(x = y \vee \exists z(Xxz \wedge Ezy))]\,xy$$

expresses "$x = y$ or x, y are connected by a path". Hence, the class of connected graphs is axiomatized by (the graph axioms and) $\forall x \forall y\, \psi_0(x, y)$.

A formula φ of FO(IFP) is said to be first-order, if it does not contain the IFP operator, even though it may contain second-order variables. Call an FO(IFP)-formula φ *positive* (*negative*) in the second-order variable X, if each free occurrence of X in φ is in the scope of an even (odd) number of negation symbols. And call φ *normal*, if for every subformula of φ of the form $[\mathrm{IFP}_{\overline{y},Y}\,\psi]\,\overline{t}$ the formula ψ is positive in Y. A simple induction shows (see the next exercise) that for any normal formula $\varphi(\overline{x}, X, \ldots)$ positive in X the operation F^φ is monotone.

Exercise 8.1.6 (a) Let $\varphi(\overline{x}, X_1, \ldots, X_k, Y_1, \ldots, Y_l)$ be a normal formula of FO(IFP) that is positive in $X_1, \ldots, X_k$ and negative in $Y_1, \ldots, Y_l$. Then $\mathcal{A} \models \varphi[\overline{a}, \overline{R}, \overline{S}]$, $R_1 \subseteq R'_1, \ldots, R_k \subseteq R'_k$, and $S'_1 \subseteq S_1, \ldots, S'_l \subseteq S_l$ imply $\mathcal{A} \models \varphi[\overline{a}, \overline{R}', \overline{S}']$. (Hint: Induction on φ.)

(b) The formula $\varphi(x, X) := [\mathrm{IFP}_{x,Y}\,(Xx \vee (\exists y Yy \wedge \neg Yc))]x$ is positive in X, but F^{φ} is not monotone. Note that φ is not normal. □

Of course, it is decidable whether a formula φ is normal and positive in X, while – even for first-order φ – it is not decidable whether F^{φ} is monotone. This can be seen from Trahtenbrot's Theorem 7.2.1 and the equivalence

$$F^{\forall x(Xx \rightarrow \psi)} \text{ is monotone (in the finite) iff } \psi \text{ is valid in the finite}$$

which holds for every first-order sentence ψ.

We saw in 8.1.2(b) that monotone operations have least fixed-points. This motivates the introduction of a corresponding fixed-point logic, namely *Least Fixed-Point Logic* FO(LFP) that we get by closing first-order logic under least fixed-points of operations definable by positive formulas. From a semantic point of view it would be more natural to consider the closure under arbitrary definable monotone operations. But then, by the preceding remark, we would get a logic with an undecidable syntax. Moreover, the restriction to normal formulas does not lead to a loss of expressive power, as even FO(IFP) is not stronger than FO(LFP) (cf. 8.1.13).

The class of formulas of FO(LFP) of vocabulary τ is the fragment of FO(IFP)$[\tau]$ given by the same calculus as that for FO(IFP), the IFP-rule being restricted to positive formulas:

- $\dfrac{\varphi}{[\mathrm{IFP}_{\overline{x},X}\varphi]\,\overline{t}}$ where φ is positive in X and the lengths of $\overline{x}$ and $\overline{t}$ are the same and coincide with the arity of X.

Thus the FO(LFP)-formulas are the normal FO(IFP)-formulas. Since for normal formulas φ that are positive in X, the operation F^{φ} is monotone, we have, by 8.1.2, that $F_{\infty}^{(X\overline{x} \vee \varphi)} = F_{\infty}^{\varphi}$ and that F_{∞}^{φ} is the least fixed-point of F^{φ}. Hence, in this case, $[\mathrm{IFP}_{\overline{x},X}\varphi]\,\overline{t}$ expresses that $\overline{t}$ is in the least fixed-point of F^{φ}. We therefore write $[\mathrm{LFP}_{\overline{x},X}\varphi]\,\overline{t}$ instead of $[\mathrm{IFP}_{\overline{x},X}\varphi]\,\overline{t}$.

Example 8.1.7 Consider trees as $\{E\}$-structures (cf. 1A.1). The formula of FO(LFP)

$$\varphi_{<}(x, y) \quad := \quad [\mathrm{LFP}_{xy,X}(Exy \vee \exists z(Xxz \wedge Ezy))]\,xy$$

defines a partial ordering (an irreflexive and transitive binary relation), the *induced partial ordering.* The FO(LFP)-formula $\varphi_D(x, y) :=$

$$[\mathrm{LFP}_{xy,X}\,(\forall z(\neg Ezx \wedge \neg Ezy) \vee \exists u \exists v(Xuv \wedge Eux \wedge Evy))]\,xy$$

expresses that x and y have the same depth and therefore, the FO(LFP)-sentence

$$\forall x \forall y (\forall z (\neg Exz \wedge \neg Eyz) \rightarrow \varphi_D(x,y))$$

says that all leaves have the same depth. □

Example 8.1.8 Fix a vocabulary τ and $s \geq 1$. We consider the pebble games introduced in subsection 3.3.1. A simple induction on n shows that for any τ-structure $\mathcal{A}$ we have

$$F_\varphi^{n+1} \quad := \quad \{(a_1, \ldots, a_s, b_1, \ldots, b_s) \mid \text{ the spoiler wins } G_n^s(\mathcal{A}, \overline{a}, \mathcal{A}, \overline{b})\},$$

where

$$\varphi(\overline{x}, \overline{y}, Z) := \bigvee_{\substack{\psi \in \mathrm{FO}^s \\ \psi \text{ atomic}}} (\psi(\overline{x}) \leftrightarrow \neg\psi(\overline{y})) \vee \bigvee_{1 \leq i \leq s} (\exists x_i \forall y_i Z \overline{x}\,\overline{y} \vee \exists y_i \forall x_i Z \overline{x}\,\overline{y}).$$

Thus, for every $\overline{a}, \overline{b} \in A^s$ we have that $\mathcal{A} \models [\mathrm{LFP}_{\overline{x}\,\overline{y}, Z}\, \varphi]\overline{x}\,\overline{y}[\overline{a}, \overline{b}]$ iff the spoiler wins $G_\infty^s(\mathcal{A}, \overline{a}, \mathcal{B}, \overline{b})$. □

Exercise 8.1.9 Given a graph $\mathcal{G}$ and $k \geq 1$, consider the following cops-and-robbers game $\mathrm{CR}^k(\mathcal{G})$ with infinitely many moves: There are k cops and one robber, each of whom after any move stands on a vertex of $\mathcal{G}$. In the first move every cop chooses a position on the graph and afterwards the robber chooses his position. In any subsequent move one of the cops flies in a helicopter to a new vertex. Before the helicopter actually lands the robber will run to a new vertex along a path of the graph; however, he is not permitted to run through a cop. More formally, a position of the game $\mathrm{CR}^k(\mathcal{G})$ is a $(k+1)$-tuple of vertices of G. In the first move the "cops" choose $a_1, \ldots, a_k \in G$ and the "robber" $b \in G$. Then, $(a_1, \ldots, a_k, b)$ is the position after the first move. Suppose $(e_1, \ldots, e_k, f)$ is the position after i moves. In the $(i+1)$th move, first the "cops" choose l with $1 \leq l \leq k$ and $e' \in G$, and then the robber chooses a point f' in the connected component of f in the graph induced on $G \setminus \{e_1, \ldots, e_{l-1}, e_{l+1}, \ldots, e_k\}$. Then $(e_1, \ldots, e_{l-1}, e', e_{l+1}, \ldots, e_k, f')$ is the new position. The "cops" win if eventually a position $(a_1, \ldots, a_k, b)$ with $b \in \{a_1, \ldots, a_k\}$ is reached. We say that a graph $\mathcal{G}$ has *tree-width* $\leq k$ if the "cops" have a winning strategy in $\mathrm{CR}^{k+1}(\mathcal{G})$ (usually, this concept is defined in terms of trees).

Show that the class of graphs of tree-width $\leq k$ is axiomatizable in FO(LFP). Hint: Consider a formula expressing in any graph $\mathcal{G}$ that the "cops" have a winning strategy in $\mathrm{CR}^{k+1}(\mathcal{G})$, e.g.,

$$\exists x_1 \ldots \exists x_{k+1} \forall y [\mathrm{LFP}_{\overline{x}y, Z}(\bigvee_{i=1}^{k+1} x_i = y \vee \bigvee_{l=1}^{k+1} \exists x_l \forall z (\psi_l(\overline{x}, y, z) \rightarrow Z\overline{x}z))]\overline{x}y$$

where

$$\psi_l(\overline{x}, y, z) := [\mathrm{LFP}_{z,U}(z = y \vee \exists u (Uu \wedge Euz \wedge \bigwedge_{i \neq l} z \neq x_i))]z. \qquad \square$$

For convenience we adopt the following convention:

If not explicitly stated otherwise, the notation $\varphi(\overline{x}, \overline{Y})$ only means that the variables in φ that are relevant in the given situation are among $\overline{x}$ and $\overline{Y}$; there may be other variables free in φ.

Exercise 8.1.10 (a) Let $\psi(\overline{x}, X)$ be an FO(LFP)-formula that is positive in X. Show that the greatest fixed-point of F^ψ is defined by the FO(LFP)-formula $\neg[\mathrm{LFP}_{\overline{x},X}\, \neg\psi(\overline{x}, \neg X)]\,\overline{x}$ (here $\psi(\overline{x}, \neg X)$ is obtained from ψ by replacing subformulas $X\overline{t}$ by $\neg X\overline{t}$; hence, $\neg\psi(\overline{x}, \neg X)$ is positive in X), a formula which sometimes is abbreviated by $[\mathrm{GFP}_{\overline{x},X}\, \psi(\overline{x}, X)]\,\overline{x}$. Hint: Use 8.1.4.

(b) We know that with $\varphi(x, X) := \exists y(\neg Xy \wedge Exy)$ we have $F = F^\varphi$ for the antitone function F introduced for the game associated with digraphs (cf. 8.1.5(b)). Hence, for $H := F \circ F$ we have $H = F^\psi$, where

$$\psi(x, X) \quad := \quad \exists y(Exy \wedge \forall z(Eyz \rightarrow Xz))$$

(note that $\psi(x, X)$ is equivalent to the formula $\varphi(x, \varphi(\_, X))$ obtained from φ by replacing subformulas Xy by $\varphi(y, X)$). By Exercise 8.1.5(b) the statement "there are no drawn points" is expressed by

$$\forall x([\mathrm{LFP}_{x,X}\, \psi]\, x \quad \leftrightarrow \quad [\mathrm{GFP}_{x,X}\, \psi]\, x)$$

and the statement "x is won by I" by $[\mathrm{LFP}_{x,X}\, \psi]\, x$. □

Example 8.1.11 Consider orderings as $\{<, S, \min, \max\}$-structures.[1] Fix $k \geq 1$. Identify an ordering $\mathcal{A}$ with its isomorphic copy $(\{0, \ldots, n-1\}, <, S, 0, n-1)$ where $n = \|A\|$. Then every k-tuple $(m_1, \ldots, m_k)$ in A^k can be considered as the n-adic representation of a natural number, namely of

$$[m_1 \ldots m_k] := m_1 \cdot n^{k-1} + \ldots + m_k \cdot n^0.$$

Let S^k be the "k-adic successor relation",

$$S^k x_1 \ldots x_k\, y_1 \ldots y_k \qquad \text{iff} \qquad [y_1 \ldots y_k] = [x_1 \ldots x_k] + 1.$$

S^k is first-order-definable by

$$\bigvee_{1 \leq i \leq k} (x_{i+1} = \ldots = x_k = \max \wedge \neg x_i = \max$$
$$\wedge\, Sx_iy_i \wedge \bigwedge_{i<j\leq k} y_j = \min \wedge \bigwedge_{1\leq j<i} y_j = x_j).$$

Define the relations ADD^k ("addition") and MULT^k ("multiplication") by

$$\mathrm{ADD}^k x_1 \ldots x_k\, y_1 \ldots y_k\, z_1 \ldots z_k \qquad \text{iff} \qquad [\overline{z}] = [\overline{x}] + [\overline{y}]$$

[1] Note that $<$, being the transitive closure of S, is definable in FO(LFP)[$\{S\}$].

and

$$\mathrm{MULT}^k x_1 \dots x_k\, y_1 \dots y_k\, z_1 \dots z_k \quad \text{iff} \quad [\overline{z}] = [\overline{x}] \cdot [\overline{y}].$$

ADD^k and MULT^k are definable in FO(LFP): For ADD^k we imitate the inductive definition

$$\begin{aligned} x + 0 &= x \\ x + (y+1) &= (x+y)+1 \end{aligned}$$

by an application of the LFP operator and obtain:

$$[\mathrm{LFP}_{\overline{x}\,\overline{y}\,\overline{z},X}((\overline{y} = \widetilde{\min} \wedge \overline{z} = \overline{x}) \vee \exists\overline{v}\exists\overline{w}(X\overline{x}\,\overline{v}\,\overline{w} \wedge S^k\overline{v}\,\overline{y} \wedge S^k\overline{w}\,\overline{z}))\,]\,\overline{x}\,\overline{y}\,\overline{z},$$

a formula we denote by $\varphi_+(\overline{x}, \overline{y}, \overline{z})$. For MULT^k we use the inductive definition

$$\begin{aligned} x \cdot 0 &= 0 \\ x \cdot (y+1) &= x \cdot y + x \end{aligned}$$

and obtain

$$[\mathrm{LFP}_{\overline{x}\,\overline{y}\,\overline{z},X}(\overline{y} = \overline{z} = \widetilde{\min} \vee \exists\overline{v}\exists\overline{w}(X\overline{x}\,\overline{v}\,\overline{w} \wedge S^k\overline{v}\,\overline{y} \wedge \varphi_+(\overline{w}, \overline{x}, \overline{z})))\,]\,\overline{x}\,\overline{y}\,\overline{z},$$

a formula with "nested" LFP operators.

Now one easily sees that the graph of any polynomial is definable in FO(LFP) in the following sense: Given $p(x) \in \mathbb{N}[x]$, there is a k and a formula $\varphi_p(x, y_1, \dots, y_k)$ of FO(LFP) such that for all orderings $\mathcal{A}$ containing the coefficients of p we have $\mathcal{A} \models \forall x \exists \overline{y} \varphi_p(x, \overline{y})$, and for all $a, \overline{b} \in A$,

$$\mathcal{A} \models \varphi_p[a, \overline{b}] \;\text{ iff }\; p(a) = [\overline{b}]. \qquad \square$$

The following lemma is immediate. We often use it without mentioning it explicitly. Part (b) shows how parameters can be incorporated into fixed-points.

Lemma 8.1.12 (a) *If X does not occur free in φ then $[\mathrm{IFP}_{\overline{x},X}\varphi(\overline{x})]\,\overline{t}$ and $\varphi(\overline{t})$ are equivalent.*
(b) *The formulas*

$$[\mathrm{IFP}_{\overline{x},X}\, \varphi(\overline{x}, y, X)]\,\overline{t} \quad \textit{and} \quad [\mathrm{IFP}_{\overline{x}u,Y}\, \varphi(\overline{x}, u, Y\_u)]\,\overline{t}y$$

are equivalent.[2] □

By definition, FO(LFP) $\leq$ FO(IFP). As already remarked in the introduction of this chapter we are going to show the converse:

Theorem 8.1.13 *Every* FO(IFP)*-formula is equivalent to an* FO(LFP)*-formula.*

Moreover we shall prove

[2] Here, $\chi(Y\_u)$ is obtained from $\chi(X)$ by replacing each subformula $X\overline{s}$ by $Y\overline{s}u$.

Theorem 8.1.14 *Every* FO(LFP)*-formula is equivalent to an* FO(LFP)*-formula with at most one occurrence of* LFP.

The proofs will be given only in the next section. With the following considerations we point at an essential step.

Let X_0,X_1,... be the sequence of sets obtained when evaluating the outermost LFP operation in a formula of the form

$$[\mathrm{LFP}_{\overline{x},X} \ldots [\mathrm{LFP}_{\overline{y},Y}\varphi(\overline{y}, X, Y)] \ldots] \ldots .$$

In the transition from X_n to X_{n+1} one needs the value $[\mathrm{LFP}_{\overline{y},Y}\varphi(\overline{y}, X_n, Y)]$. We show how this kind of nested fixed-points can be expressed by "simultaneous fixed-points" and how simultaneous fixed-points can be expressed by a single fixed-point. We give the results in a general, more abstract form that will also be useful in later sections.

First, we introduce the notion of simultaneous fixed-point. Let $m \geq 0$ and suppose that $M_0, \ldots, M_m$ are finite sets and that

$$(*)\quad \begin{array}{c} F^0 : \mathrm{Pow}(M_0) \times \ldots \times \mathrm{Pow}(M_m) \to \mathrm{Pow}(M_0) \\ F^1 : \mathrm{Pow}(M_0) \times \ldots \times \mathrm{Pow}(M_m) \to \mathrm{Pow}(M_1) \\ \vdots \\ F^m : \mathrm{Pow}(M_0) \times \ldots \times \mathrm{Pow}(M_m) \to \mathrm{Pow}(M_m). \end{array}$$

We define the sequence $(F^0_{(n)}, \ldots, F^m_{(n)})_{n\geq 0}$ by

$$F^i_{(0)} := \emptyset, \qquad F^i_{(n+1)} := F^i(F^0_{(n)}, \ldots, F^m_{(n)}).$$

If we have for some n that $(F^0_{(n)}, \ldots, F^m_{(n)}) = (F^0_{(n+1)}, \ldots, F^m_{(n+1)})$, we set

$$(F^0_{(\infty)}, \ldots, F^m_{(\infty)}) \quad := \quad (F^0_{(n)}, \ldots, F^m_{(n)})$$

and say that the *simultaneous fixed-point* $(F^0_{(\infty)}, \ldots, F^m_{(\infty)})$ of $(F^0, \ldots, F^m)$ exists. Note that for $i = 0, \ldots, m$,

$$F^i(F^0_{(\infty)}, \ldots, F^m_{(\infty)}) = F^i_{(\infty)}.$$

If the simultaneous fixed-point does not exist, we set $F^i_{(\infty)} := \emptyset$ for $i \leq m$.

Example 8.1.15 (a) Let $F, G : \mathrm{Pow}(M) \times \mathrm{Pow}(M) \to \mathrm{Pow}(M)$ be given by

$$F(X,Y) := M, \qquad G(X,Y) := (M \setminus Y) \cap X.$$

Then $(F_{(n)})_{n\geq 0}$ is the sequence $\emptyset, M, M, \ldots$ and $(G_{(n)})_{n\geq 0}$ is the sequence $\emptyset, \emptyset, M, \emptyset, M, \emptyset, \ldots$. Therefore, the simultaneous fixed-point of (F, G) does not exist and hence, $F_{(\infty)} = G_{(\infty)} = \emptyset$, even though $F_{(1)} = F_{(2)}$ and $G_{(0)} = G_{(1)}$.

(b) Suppose that the fixed-points F_∞ and G_∞ of $F : \mathrm{Pow}(M) \to \mathrm{Pow}(M)$

and $G : \mathrm{Pow}(N) \to \mathrm{Pow}(N)$ exist and define the functions $\hat{F} : \mathrm{Pow}(M) \times \mathrm{Pow}(N) \to \mathrm{Pow}(M)$ and $\hat{G} : \mathrm{Pow}(M) \times \mathrm{Pow}(N) \to \mathrm{Pow}(N)$ by $\hat{F}(X,Y) = F(X)$ and $\hat{G}(X,Y) = G(Y)$. Then the simultaneous fixed-point $(\hat{F}_{(\infty)}, \hat{G}_{(\infty)})$ exists and $(\hat{F}_{(\infty)}, \hat{G}_{(\infty)}) = (F_\infty, G_\infty)$. □

Definition 8.1.16 Let $F^0, \dots, F^m$ be as in $(*)$.

(a) $(F^0, \dots, F^m)$ is *inductive* if for $i \leq m$

$$F^i_{(0)} \subseteq F^i_{(1)} \subseteq F^i_{(2)} \subseteq \dots.$$

(b) $(F^0, \dots, F^m)$ is *inflationary* if for arbitrary $X_0, \dots, X_m$ and all $i \leq m$,

$$X_i \subseteq F^i(X_0, \dots, X_m).$$

(c) $(F^0, \dots, F^m)$ is *monotone* if

$$X_0 \subseteq Y_0, \dots, X_m \subseteq Y_m \text{ imply } F^i(X_0, \dots, X_m) \subseteq F^i(Y_0, \dots, Y_m)$$

for arbitrary $X_0, \dots, X_m, Y_0, \dots, Y_m$ and all $i \leq m$. □

The proof of the following lemma is a straightforward generalization of the proof of the corresponding parts in 8.1.1 and 8.1.2.

Lemma 8.1.17 *Let $F^0, \dots, F^m$ be as in $(*)$.*

(a) *If $(F^0, \dots, F^m)$ is inductive then the simultaneous fixed-point exists.*

(b) *If $(F^0, \dots, F^m)$ is inflationary or monotone then $(F^0, \dots, F^m)$ is inductive.*

(c) *If $(F^0, \dots, F^m)$ is monotone then $(F^0_{(\infty)}, \dots, F^m_{(\infty)})$ is the* simultaneous least fixed-point *of $(F^0, \dots, F^m)$. Even $F^i(X_0, \dots, X_m) \subseteq X_i$ for $i \leq m$ implies $F^0_{(\infty)} \subseteq X_0, \dots, F^m_{(\infty)} \subseteq X_m$.*

(d) *If for $i \leq m$, $G^i : \mathrm{Pow}(M_0) \times \dots \times \mathrm{Pow}(M_m) \to \mathrm{Pow}(M_i)$ is given by*

$$G^i(X_0, \dots, X_m) \quad := \quad X_i \cup F^i(X_0, \dots, X_m)$$

then $(G^0, \dots, G^m)$ is inflationary. In case $(F^0, \dots, F^m)$ is inductive we have $(G^0_{(n)}, \dots, G^m_{(n)}) = (F^0_{(n)}, \dots, F^m_{(n)})$ for all $n \geq 0$ and hence, $(G^0_{(\infty)}, \dots, G^m_{(\infty)}) = (F^0_{(\infty)}, \dots, F^m_{(\infty)})$. □

We introduce extensions of FO(IFP) and FO(LFP) which allow to speak about simultaneous fixed-points. On the one hand these extensions are quite useful when formalizing statements which involve fixed-point operations. On the other hand we shall show that they do not increase the expressive power.

The formulas of *Simultaneous Inflationary Fixed-Point Logic* FO(S-IFP) are obtained by replacing the clause corresponding to IFP in the definition of FO(IFP) by

$$(+) \qquad \frac{\varphi_0, \ldots, \varphi_m}{[\text{S-IFP}_{\overline{x}_0, X_0, \ldots, \overline{x}_m, X_m}\, \varphi_0, \ldots, \varphi_m]\, \overline{t}}$$

where $m \geq 0$, where for $i = 0, \ldots, m$ the arity of X_i equals the length of $\overline{x}_i$, and where $\overline{t}$ has the same length as $\overline{x}_0$. As usual, the variables in each sequence $\overline{x}_i$ are distinct, but the same variable may occur in distinct $\overline{x}_i$'s. A variable x is free in $[\text{S-IFP}_{\overline{x}_0, X_0, \ldots, \overline{x}_m, X_m}\, \varphi_0, \ldots, \varphi_m]\, \overline{t}$, if x is in $\overline{t}$ or if for at least one i, x is free in φ_i and not in $\overline{x}_i$. We define the corresponding inflationary system $(F^0, \ldots, F^m)$ by

$$F^i(X_0, \ldots, X_m) \quad := \quad X_i \cup \{\overline{x}_i \mid \varphi_i(\overline{x}_i, X_0, \ldots, X_m)\}$$

and let $(F^0_{(\infty)}, \ldots, F^m_{(\infty)})$ be their simultaneous fixed-point. Then, by definition, $[\text{S-IFP}_{\overline{x}_0, X_0, \ldots, \overline{x}_m, X_m}\, \varphi_0, \ldots, \varphi_m]\, \overline{t}$ means $\overline{t} \in F^0_{(\infty)}$.

Simultaneous Least Fixed-Point Logic FO(S-LFP) is defined analogously, clause (+) being restricted to formulas $\varphi_0, \ldots, \varphi_m$ which are positive in all the variables $X_0, \ldots, X_m$. The system $(F^0, \ldots, F^m)$ of corresponding functions is now defined by

$$F^i(X_0, \ldots, X_m) \quad := \quad \{\overline{x}_i \mid \varphi_i(\overline{x}_i, X_0, \ldots, X_m)\}.$$

By positivity, the system $(F^0, \ldots, F^m)$ is monotone; hence, by 8.1.17(b),(d), $[\text{S-IFP}_{\overline{x}_0, X_0, \ldots, \overline{x}_m, X_m}\, \varphi_0, \ldots, \varphi_m]\, \overline{t}$ expresses $\overline{t} \in F^0_{(\infty)}$. Since, by 8.1.17(c), $(F^0_{(\infty)}, \ldots, F^m_{(\infty)})$ is the least fixed-point of $(F^0, \ldots, F^m)$, we denote the formula in (+) by $[\text{S-LFP}_{\overline{x}_0, X_0, \ldots, \overline{x}_m, X_m}\, \varphi_0, \ldots, \varphi_m]\, \overline{t}$.[3]

Example 8.1.18 We consider digraphs. Let F and G be the functions corresponding to the FO(S-LFP)-formulas φ_0 and φ_1 that are positive in X (unary) and Y (binary),

$$\begin{aligned} \varphi_0(x, X, Y) \quad &:= \quad Yxx \\ \varphi_1(x, y, X, Y) \quad &:= \quad Exy \vee \exists z(Yxz \wedge Ezy). \end{aligned}$$

Then $G_{(\infty)}$ is the transitive closure of the edge relation and $F_{(\infty)}$ the set of vertices z such that there is a cycle through z. Hence,

$$[\text{S-LFP}_{x, X, xy, Y}\, \varphi_0, \varphi_1]z$$

expresses that there is a cycle through z. Clearly, $[\text{LFP}_{xy,X}\, (Exy \vee \exists z(Yxz \wedge Ezy))]zz$ is equivalent to this formula. Note that there is no first-order formula $\psi(x, X)$ positive in X with free variables among x, X such that $[\text{LFP}_{x,X}\, \psi]z$ expresses that there is a cycle through z. Otherwise, the formula $\exists z[\text{LFP}_{x,X}\, \psi]z$ expresses that there is a cycle. Now, $\exists z[\text{LFP}_{x,X}\, \psi]z$ is

[3] Even though the system $(F^0, \ldots, F^m)$ corresponding to an S-IFP operator and to an S-LFP operator have different definitions, they coincide on the parts relevant for our discussion, namely on the stages and on the fixed-points.

equivalent to $F_1^{\psi} \neq \emptyset$ and hence, to the first-order formula $\exists x \psi(x, \emptyset)$.[4] But we know that there is no first-order formula expressing that there is a cycle (cf. 2.3.9). □

Notes 8.1.19 Inflationary fixed-point logic goes back to Gurevich [70]. For least fixed-point logic cf. the notes to the next section. Theorem 8.1.13 is due to Gurevich and Shelah [76] and Theorem 8.1.14 to Immerman [91]. A further reference is [18].

8.2 Simultaneous Induction and Transitivity

In the following, A will always denote a finite set with at least two elements. For $b \in A$ and $i \geq 0$ let

$$b^i := \underbrace{b \ldots b}_{i \text{ times}}$$

and let $\tilde{b}$ be b^i for some i. The length of $\tilde{b}$ will be irrelevant or clear from the context. If Z is an $(k+l)$-ary relation on A and $\overline{a} \in A^l$, then the *$\overline{a}$-section* $Z\_\overline{a}$ of Z is given by

$$Z\_\overline{a} \quad := \quad \{\overline{b} \in A^k \mid Z\overline{b}\overline{a}\}.$$

Let $m \geq 0$ and

$$(1) \qquad \begin{array}{c} F^0 : \mathrm{Pow}(A^{k_0}) \times \ldots \times \mathrm{Pow}(A^{k_m}) \to \mathrm{Pow}(A^{k_0}) \\ F^1 : \mathrm{Pow}(A^{k_0}) \times \ldots \times \mathrm{Pow}(A^{k_m}) \to \mathrm{Pow}(A^{k_1}) \\ \vdots \\ F^m : \mathrm{Pow}(A^{k_0}) \times \ldots \times \mathrm{Pow}(A^{k_m}) \to \mathrm{Pow}(A^{k_m}). \end{array}$$

We are going to code these operations as sections of a single operation J of higher arity. Setting

$$(2) \qquad k := \max\{k_0, \ldots, k_m\} + (m+1)$$

we define

$$J : \mathrm{Pow}(A^k) \to \mathrm{Pow}(A^k)$$

in such a way that if $J_0, J_1, \ldots$ are the stages of J then for any $a, b \in A$ with $a \neq b$, the $\tilde{a}b^i$-section of J_n codes $F^i_{(n)}$, that is,

$$F^i_{(n)} = J_n\,\_\tilde{a}b^i.$$[5]

---

[4] The formula $\psi(x, \emptyset)$ is obtained from $\psi(x, X)$ by replacing each occurrence of an atomic formula Xt by $\neg t = t$.

[5] Hence, $\tilde{a}$ has length $k - i - k_i$.

Lemma 8.2.1 (Simultaneous Induction) *Let $F^0, \ldots, F^m$ and k be as in (1) and (2) above. Define the function $J : \mathrm{Pow}(A^k) \to \mathrm{Pow}(A^k)$, the* simultaneous join *of $(F^0, \ldots, F^m)$, by*

$$J(Z) := \bigcup_{a,b \in A, a \neq b} ((F^0(Z\_\tilde{a}b^0, \ldots, Z\_\tilde{a}b^m) \times \{\tilde{a}b^0\}) \cup \ldots \cup (F^m(Z\_\tilde{a}b^0, \ldots, Z\_\tilde{a}b^m) \times \{\tilde{a}b^m\})).$$

(a) *For all $n \geq 0$,*

$$J_n = \bigcup_{a,b \in A, a \neq b} ((F^0_{(n)} \times \{\tilde{a}b^0\}) \cup \ldots \cup (F^m_{(n)} \times \{\tilde{a}b^m\})).$$

Therefore, the fixed-point J_∞ of J exists iff the simultaneous fixed-point $(F^0_{(\infty)}, \ldots, F^m_{(\infty)})$ exists. Moreover,

$$J_\infty = \bigcup_{a,b \in A, a \neq b} ((F^0_{(\infty)} \times \{\tilde{a}b^0\}) \cup \ldots \cup (F^m_{(\infty)} \times \{\tilde{a}b^m\})).$$

Thus, for $i = 0, \ldots, m$ and for all $a, b \in A$ with $a \neq b$,

$$F^i_{(n)} = J_n\,\_\tilde{a}b^i \quad \textit{and} \quad F^i_{(\infty)} = J_\infty\,\_\tilde{a}b^i;$$

hence, for $\bar{c} \in A^{k_i}$,

$$F^i_{(\infty)}\bar{c} \quad \textit{iff} \quad J_\infty \bar{c}\tilde{a}b^i;$$

in particular, for $\bar{c} \in A^{k_0}$,

$$(*) \qquad F^0_{(\infty)}\bar{c} \quad \textit{iff} \quad J_\infty \bar{c}\tilde{a}.$$

(b) *If $(F^0, \ldots, F^m)$ is inductive then J is inductive.*

Proof. We prove by induction on n that

$$J_n = \bigcup_{a,b \in A, a \neq b} ((F^0_{(n)} \times \{\tilde{a}b^0\}) \cup \ldots \cup (F^m_{(n)} \times \{\tilde{a}b^m\}))$$

(all the other claims in (a) and (b) are then immediate). The case $n = 0$ is clear. In the induction step we have J_{n+1}

$$\begin{aligned}
&= J(J_n) = J(\textstyle\bigcup_{a \neq b}((F^0_{(n)} \times \{\tilde{a}b^0\}) \cup \ldots \cup (F^m_{(n)} \times \{\tilde{a}b^m\}))) \\
&= \textstyle\bigcup_{a \neq b}((F^0(F^0_{(n)}, \ldots, F^m_{(n)}) \times \{\tilde{a}b^0\}) \\
&\qquad \cup \ldots \cup (F^m(F^0_{(n)}, \ldots, F^m_{(n)}) \times \{\tilde{a}b^m\})) \\
&= \textstyle\bigcup_{a \neq b}((F^0_{(n+1)} \times \{\tilde{a}b^0\}) \cup \ldots \cup (F^m_{(n+1)} \times \{\tilde{a}b^m\})).
\end{aligned}$$

□

We come back to formulas and use the preceding lemma to prove that the extensions of FO(IFP) and FO(LFP) to the corresponding simultaneous fixed-point logics do not increase the expressive power.

To simplify the presentation we consider only structures with at least two elements; in particular, we say that φ and ψ are equivalent and write

$$\models_{\text{fin}} \varphi \leftrightarrow \psi$$

if φ and ψ are equivalent in all structures with at least two elements. After 8.2.11 we illustrate how this restriction can be removed.

First some notations that will be used frequently in the next chapters. We write

$$\models_{\text{fin}} \varphi \leftrightarrow \exists(\forall)x\psi(x),$$

if φ is equivalent to both $\exists x\psi(x)$ and $\forall x\psi(x)$. Then, for any term t, φ is equivalent to $\psi(t)$, a fact we often shall use tacitly.

To speak about distinct sections of a simultaneous fixed-point process, we use, for $l \geq 1$ and $i = 0, \ldots, l$, the formulas $\delta_i^l(x_1, \ldots, x_l, v, w)$ given by

$$\delta_0^l(x_1, \ldots, x_l, v, w) \quad := \quad \neg v = w \land x_1 = \ldots = x_l = v$$

and for $i = 1, \ldots, l$ by

$$\begin{aligned} \delta_i^l(x_1, \ldots, x_l, v, w) \quad := \quad & \neg v = w \land \\ & x_1 = \ldots = x_{l-i} = v \land x_{l-i+1} = \ldots = x_l = w. \end{aligned}$$

We often omit the superscript l and write $\delta_i(\overline{x}, u, v)$, l then being determined by the length of $\overline{x}$. Note that for distinct a, b in a structure $\mathcal{A}$, we have

$$\mathcal{A} \models \delta_i[\tilde{a}b^j ab] \quad \text{iff} \quad i = j.$$

Furthermore, we use some more or less self-explanatory notations. For instance, given a formula $\varphi(X)$, we write $\varphi(Y\_\overline{y})$ for the formula obtained from φ by replacing each subformula $X\overline{t}$ by $Y\overline{t}\overline{y}$; of course, this presupposes that the arities of X and Y and the length of $\overline{y}$ match in the right way.

Theorem 8.2.2 FO(S-LFP) $\equiv$ FO(LFP) *and* FO(S-IFP) $\equiv$ FO(IFP).

Clearly, it suffices to show that FO(S-LFP) $\leq$ FO(LFP) and FO(S-IFP) $\leq$ FO(IFP). The proof is by induction on formulas, the only nontrivial case being handled by the next lemma.

Lemma 8.2.3 (Simultaneous Induction for LFP **and** IFP**)** *Let*

$$\varphi_0(\overline{x}_0, X_0, \ldots, X_m), \ldots, \varphi_m(\overline{x}_m, X_0, \ldots, X_m)$$

be formulas of FO(LFP), *where X_i is k_i-ary and the length of $\overline{x}_i$ is k_i. Furthermore, assume that $\varphi_0, \ldots, \varphi_m$ are all positive in $X_0, \ldots, X_m$. Set*

$k := \max\{k_0, \dots, k_m\} + (m+1)$ *and, using new variables* v *and* w, *define the* FO(LFP)*-formula* $\chi_J(z_1, \dots, z_k, Z)$ *by*

$$\begin{aligned}
\chi_J(z_1,\dots,z_k,Z) \;&:=\; \exists v \exists w(\neg v = w \wedge \\
&((\varphi_0(z_1,\dots,z_{k_0}, Z_\tilde{v}w^0,\dots,Z_\tilde{v}w^m) \wedge \delta_0(z_{k_0+1},\dots,z_k,v,w)) \\
&\vee (\varphi_1(z_1,\dots,z_{k_1}, Z_\tilde{v}w^0,\dots,Z_\tilde{v}w^m) \wedge \delta_1(z_{k_1+1},\dots,z_k,v,w)) \\
&\vdots \\
&\vee (\varphi_m(z_1,\dots,z_{k_m}, Z_\tilde{v}w^0,\dots,Z_\tilde{v}w^m) \wedge \delta_m(z_{k_m+1},\dots,z_k,v,w)))).
\end{aligned}$$

Then, for any new variable u,

$$\models_{\text{fin}} [\text{S-LFP}_{\overline{x}_0, X_0, \dots, \overline{x}_m, X_m}\, \varphi_0, \dots, \varphi_m]\, \overline{t} \leftrightarrow \exists(\forall) u [\text{LFP}_{\overline{z}, Z}\, \chi_J]\, \overline{t}\tilde{u}.$$

If $\varphi_0, \dots, \varphi_m$ *are first-order (existential) formulas then* χ_J *is first-order (existential), too. Moreover,*

$$\text{free}(\chi_J) \subseteq \{\overline{z}\} \cup (\text{free}(\varphi_0) \setminus \{\overline{x}_0\}) \cup \dots \cup (\text{free}(\varphi_m) \setminus \{\overline{x}_m\}).$$

Without the assumption that $\varphi_0, \dots, \varphi_m$ *are all positive in* $X_0, \dots, X_m$ *the claims are true if* LFP *is replaced by* IFP *everywhere.*

Proof. First, we turn to the LFP-case. For $i = 0, \dots, m$ define

$$F^i(X_0, \dots, X_m) \;:=\; \{\overline{x}_i \mid \varphi_i(\overline{x}_i, X_0, \dots, X_m)\}.$$

Then, χ_J (which is positive in Z) defines the simultaneous join of $(F^0, \dots, F^m)$ as introduced in Lemma 8.2.1. Since $[\text{S-LFP}_{\overline{x}_0, X_0, \dots, \overline{x}_m, X_m}\, \varphi_0, \dots, \varphi_m]\, \overline{t}$ expresses $F^0_{(\infty)}\overline{t}$ and $[\text{LFP}_{\overline{z}, Z}\, \chi_J]\overline{t}\tilde{u}$ expresses $J_\infty \overline{t}\tilde{u}$, the claim follows immediately from $(*)$ in part (a) of 8.2.1.

In the the IFP-case, for $i = 0, \dots, m$ we define F^i by

$$F^i(X_0, \dots, X_m) \;:=\; X_i \cup \{\overline{x}_i \mid \varphi_i(\overline{x}_i, X_0, \dots, X_m)\}.$$

Let J be the simultaneous join of $(F^0, \dots, F^m)$ (cf. 8.2.1(a)). One easily verifies that $F^{(Z\overline{z} \vee \chi_J)}$ defines J' where $J'(Z) := Z \cup J(Z)$. Recall that $[\text{IFP}_{\overline{z}, Z}\, \chi_J]\overline{t}\tilde{u}$ means $J'_\infty \overline{t}\tilde{u}$. Since $(F^0, \dots, F^m)$ is inflationary and hence, inductive, J is inductive (cf. 8.2.1(b)). Therefore, $J'_\infty = J_\infty$ (cf. 8.1.2(c)). Thus $[\text{IFP}_{\overline{z}, Z}\, \chi_J]\, \overline{t}\tilde{u}$ expresses $J_\infty \overline{t}\tilde{u}$ which, by $(*)$ in 8.2.1(a), is equivalent to $F^0_{(\infty)}\overline{t}$, that is, to

$$[\text{S-IFP}_{\overline{x}_0, X_0, \dots, \overline{x}_m, X_m}\, \varphi_0, \dots, \varphi_m]\, \overline{t}. \qquad \square$$

As a first application we present a lemma which will frequently be used in the following.

Lemma 8.2.4 *Every* FO(LFP)*-formula* $[\mathrm{LFP}_{\overline{y},Y}\,\varphi]\,\overline{t}$ *with first-order* φ *is equivalent to a formula of the form*

$$\exists(\forall)u[\mathrm{LFP}_{\overline{z},Z}\,\psi]\tilde{u}$$

with first-order ψ. *If* φ *is existential then* ψ *can be chosen existential, too. The claim and the following proof remain true if* LFP *is replaced by* IFP *everywhere.*

As the proof will show, we may require in addition – and will often do so tacitly – that u does not occur free in ψ and that the free variables of $\exists(\forall)u[\mathrm{LFP}_{\overline{z},Z}\,\psi]\tilde{u}$ are among the free variables in $[\mathrm{LFP}_{\overline{y},Y}\,\varphi]\,\overline{t}$.

Proof. Set $\varphi_0(x, X, Y) := Y\overline{t}$ with dummy variables x and X. We show, for a new variable v,

$$(+) \qquad \models_{\mathrm{fin}} [\mathrm{LFP}_{\overline{y},Y}\,\varphi]\,\overline{t} \leftrightarrow [\text{S-LFP}_{x,X,\overline{y},Y}\,\varphi_0, \varphi]\,v.$$

This yields the claim: By the Lemma on Simultaneous Induction for LFP the formula on the right side is equivalent to one of the form $\exists(\forall)u[\mathrm{LFP}_{\overline{z},Z}\,\psi]v\tilde{u}$ with first-order ψ and $v \notin \mathrm{free}(\psi)$ (moreover, ψ is existential if φ is existential). Hence,

$$\models_{\mathrm{fin}} [\mathrm{LFP}_{\overline{y},Y}\,\varphi]\,\overline{t} \leftrightarrow \exists(\forall)u[\mathrm{LFP}_{\overline{z},Z}\,\psi]v\tilde{u}$$

and therefore, as $v \notin \mathrm{free}([\mathrm{LFP}_{\overline{y},Y}\,\varphi]\,\overline{t})$,

$$\models_{\mathrm{fin}} [\mathrm{LFP}_{\overline{y},Y}\,\varphi]\,\overline{t} \leftrightarrow \exists(\forall)u[\mathrm{LFP}_{\overline{z},Z}\,\psi]\tilde{u}.$$

To prove (+), consider the functions F and G corresponding to φ_0 and φ, respectively,

$$F(X,Y) = \{x \mid \varphi_0(x, X, Y)\} \quad \text{and} \quad G(X,Y) = \{\overline{y} \mid \varphi(\overline{y}, Y)\}.$$

For the simultaneous fixed-point $(F_{(\infty)}, G_{(\infty)})$ we immediately get

$$G_{(\infty)} = F_{\infty}^{\varphi} \quad \text{and} \quad F_{(\infty)} = \begin{cases} \text{“universe”} & \text{if } G_{(\infty)}\overline{t} \\ \emptyset & \text{otherwise.} \end{cases}$$

Hence for any v,

$$\overline{t} \in F_{\infty}^{\varphi} \qquad \text{iff} \qquad v \in F_{(\infty)},$$

which is a reformulation of (+). □

One of our aims is to prove that any formula of FO(LFP) is equivalent to a formula containing only one LFP operation. As an essential step we have to express nested fixed-points by single ones. To do so we need a further result, the so-called Transitivity Lemma. It shows how the fixed-point of nested *monotone* operations can be expressed by a simultaneous fixed-point.

Lemma 8.2.5 (Transitivity) *Let (F, G) be monotone, where*

$$F : \text{Pow}(A^k) \times \text{Pow}(A^l) \to \text{Pow}(A^k)$$
$$G : \text{Pow}(A^k) \times \text{Pow}(A^l) \to \text{Pow}(A^l)$$

and let $E : \text{Pow}(A^k) \to \text{Pow}(A^k)$ be given by

$$E(X) \quad := \quad F(X, G(X, \_)_\infty),$$

where $G(X, \_) : \text{Pow}(A^l) \to \text{Pow}(A^l)$ denotes the monotone operation $Y \mapsto G(X, Y)$ and $G(X, \_)_\infty$ denotes its least fixed-point. Then E is monotone and $E_\infty = F_{(\infty)}$.

Proof. The monotonicity of E is clear. To prove $E_\infty = F_{(\infty)}$ we show

(1) $\quad E(F_{(\infty)}) \subseteq F_{(\infty)}$ (hence, $E_\infty \subseteq F_{(\infty)}$ by part (b) of 8.1.2).

(2) $\quad$ For all n, $F_{(n)} \subseteq E_\infty$ and $G_{(n)} \subseteq G(E_\infty, \_)_\infty$ (hence, $F_{(\infty)} \subseteq E_\infty$).

For (1) we note

$$\begin{aligned} E(F_{(\infty)}) &= F(F_{(\infty)}, G(F_{(\infty)}, \_)_\infty) \subseteq F(F_{(\infty)}, G_{(\infty)}) \\ &\qquad \text{(since } G(F_{(\infty)}, G_{(\infty)}) = G_{(\infty)} \text{ we have } G(F_{(\infty)}, \_)_\infty \subseteq G_{(\infty)}) \\ &= F_{(\infty)}. \end{aligned}$$

(2) is proved by induction: Clearly, the inclusions hold for $n = 0$. In the induction step we have

$$\begin{aligned} F_{(n+1)} &= F(F_{(n)}, G_{(n)}) \subseteq F(E_\infty, G(E_\infty, \_)_\infty) \quad \text{(by induction hypothesis)} \\ &= E(E_\infty) = E_\infty \end{aligned}$$

and

$$G_{(n+1)} = G(F_{(n)}, G_{(n)}) \subseteq G(E_\infty, G(E_\infty, \_)_\infty) = G(E_\infty, \_)_\infty. \qquad \square$$

Using the Lemma on Simultaneous Induction for LFP, the preceding lemma gives

Lemma 8.2.6 (Transitivity for LFP) *Let $\varphi(\overline{x}, X, Y)$ and $\psi(\overline{y}, X, Y)$ be first-order formulas that are positive in X and Y. Moreover, assume that no individual variable free in $[\text{LFP}_{\overline{y},Y}\psi(\overline{y}, X, Y)]$, that is, free in ψ and distinct from $\overline{y}$, gets into the scope of a corresponding quantifier or* LFP *operator in* $(*)$. *Then*

$(*)$
$$[\text{LFP}_{\overline{x},X}\varphi(\overline{x}, X, [\text{LFP}_{\overline{y},Y}\psi(\overline{y}, X, Y)])]\,\overline{t}$$

is equivalent to a formula of the form

$$\exists(\forall)u[\text{LFP}_{\overline{z},Z}\,\chi(\overline{z}, Z)]\,\tilde{u}$$

with first-order χ.

Proof. Define F and G by

$$F(X,Y) := \{\overline{x} \mid \varphi(\overline{x},X,Y)\} \quad \text{and} \quad G(X,Y) := \{\overline{y} \mid \psi(\overline{y},X,Y)\}.$$

Then the corresponding E in the Transitivity Lemma is given by

$$\rho_E(\overline{x},X) \ := \ \varphi(\overline{x},X,[\mathrm{LFP}_{\overline{y},Y}\psi(\overline{y},X,Y)]).$$

Since $E_\infty = F_{(\infty)}$, we have

$$\models_{\mathrm{fin}} [\mathrm{LFP}_{\overline{x},X}\varphi(\overline{x},X,[\mathrm{LFP}_{\overline{y},Y}\psi(\overline{y},X,Y)]\,)]\,\overline{t} \ \leftrightarrow \ [\text{S-LFP}_{\overline{x},X,\overline{y},Y}\,\varphi,\psi]\,\overline{t}.$$

By 8.2.3, the formula on the right side is equivalent to $[\mathrm{LFP}_{\overline{z},Z}\,\chi_0]\,\overline{t}\tilde{v}$ for suitable first-order χ_0 and hence, by 8.2.4, to one of the desired form. □

In a first step towards a proof of 8.1.14 we use the Transitivity Lemma to settle the case of FO(LFP)-formulas with only positive occurrences of the LFP operator.

Lemma 8.2.7 *Every* FO(LFP)*-formula* φ *containing only positive occurrences of the* LFP *operator (i.e., each* LFP *operator in* φ *is in the scope of an even number of negation symbols) is equivalent to a formula of the form*

$$\exists(\forall)u[\mathrm{LFP}_{\overline{z},Z}\,\psi]\,\tilde{u}$$

where ψ *is first-order.*

Proof. By 8.2.4 it suffices to give a representation of φ in the form

$$[\mathrm{LFP}_{\overline{z},Z}\,\psi]\,\overline{t}$$

with first-order ψ. We use the connectives $\neg, \wedge, \vee$ and the quantifiers $\forall$ and $\exists$. Then, by hypothesis, we can assume that all negation symbols in φ are in front of atomic formulas. We proceed by induction on φ.

If φ is atomic or the negation of an atomic formula then

$$\models_{\mathrm{fin}} \varphi \leftrightarrow [\mathrm{LFP}_{x,X}\,\varphi]\,x$$

for any dummy variables x and X.

Suppose $\varphi = (\varphi_1 \wedge \varphi_2)$. By induction hypothesis we can assume that

$$\varphi_1 = [\mathrm{LFP}_{\overline{x},X}\,\psi_1(\overline{x},X)]\,\overline{s} \qquad \text{and} \qquad \varphi_2 = [\mathrm{LFP}_{\overline{y},Y}\,\psi_2(\overline{y},Y)]\,\overline{t}$$

with first-order ψ_1 and ψ_2. Moreover, we can assume that no variable in $\overline{x}$ occurs in ψ_2. Then one easily verifies that

$$\models_{\mathrm{fin}} (\varphi_1 \wedge \varphi_2) \leftrightarrow [\mathrm{LFP}_{\overline{x},X}(\psi_1(\overline{x},X) \wedge [\mathrm{LFP}_{\overline{y},Y}\psi_2(\overline{y},Y)]\overline{t}\,)]\,\overline{s}$$

which, by the Transitivity Lemma 8.2.6 (for $\varphi(\overline{x}, X, Y) := \psi_1(\overline{x}, X) \wedge Y\overline{t}$ and for $\psi(\overline{y}, X, Y) := \psi_2(\overline{y}, Y)$), is equivalent to a formula of the desired form.

The proof for $\varphi = (\varphi_1 \vee \varphi_2)$ is completely analogous.

Suppose that $\varphi = \forall v \psi_1$. By induction hypothesis we can assume that $\varphi = \forall v[\text{LFP}_{\overline{y},Y}\, \psi_1]\,\overline{t}$ with a first-order formula ψ_1 and, by 8.1.12(b), that the variables free in ψ_1 are among $\overline{y}$. By 8.1.12(a), for dummy x and X,

$$\models_{\text{fin}} \varphi \leftrightarrow [\text{LFP}_{x,X}\, \forall v[\text{LFP}_{\overline{y},Y}\, \psi_1]\,\overline{t}\,]\, v.$$

By the Transitivity Lemma (for $\varphi(x, X, Y) := \forall v Y\overline{t}$ and $\psi(\overline{y}, X, Y) := \psi_1$), this formula is equivalent to a formula of the form claimed.

The proof for $\varphi = \exists v \psi_1$ is completely analogous.

Finally suppose that

$$\varphi = [\text{LFP}_{\overline{x},X}\, \varphi_1]\,\overline{t}$$

where, by induction hypothesis, we may assume that $\varphi_1 = [\text{LFP}_{\overline{y},Y}\, \psi_1]\,\overline{s}$ with first-order ψ_1 and, by 8.1.12(b), that no variable in $\overline{x}$ is free in $[\text{LFP}_{\overline{y},Y}\, \psi_1]$. Hence,

$$\varphi = [\text{LFP}_{\overline{x},X}\, [\text{LFP}_{\overline{y},Y}\, \psi_1]\,\overline{s}]\,\overline{t}.$$

Once more, the Transitivity Lemma (for the formulas $\varphi(\overline{x}, X, Y) := Y\overline{s}$ and $\psi(\overline{y}, X, Y) := \psi_1$) yields the desired form. □

We now want to extend the preceding lemma to arbitrary formulas of FO(LFP). This will easily be obtained by the methods we are going to develop in order to show that every FO(IFP)-formula is equivalent to an FO(LFP)-formula. So we turn to this problem next.

Let $\varphi(\overline{x}, X)$ be an FO(LFP)-formula. Recall that $[\text{IFP}_{\overline{x},X}\varphi(\overline{x}, X)]\,\overline{t}$ expresses that $\overline{t} \in F_\infty^{(X\overline{x} \vee \varphi(\overline{x},X))}$. Note that X may occur positively *and* negatively in $(X\overline{x} \vee \varphi)$. Replace all (free) negative occurrences of X by $\neg Y$ (where Y is a new variable), thus getting an FO(LFP)-formula $\psi(\overline{x}, X, Y)$, which is positive in X and Y, such that

$$(*) \qquad (X\overline{x} \vee \varphi(\overline{x}, X)) \quad \text{is (equivalent to)} \quad \psi(\overline{x}, X, \neg X)$$

(for example, if $\varphi(x, y, X) = \neg\exists u(\neg Xxu \vee Xuy)$, then $\psi(x, y, X, Y) = (Xxy \vee \neg\exists u(\neg Xxu \vee \neg Yuy)))$. Since ψ is positive in X and Y, the function

$$L(X, Y) \quad := \quad \{\overline{x} \mid \psi(\overline{x}, X, Y)\}$$

is monotone (cf. 8.1.6). We set $H(X) := L(X, X^c)$.[6] Then $H = F^{\psi(\overline{x},X,\neg X)} = F^{(X\overline{x} \vee \varphi(\overline{x},X))}$ and therefore, H is inflationary. We are going to show how the fixed-point of H can be expressed in terms of two relations, the "stage

[6] For $X \subseteq M$, X^c denotes the complement $M \setminus X$.

comparison relations". These relations can be obtained as simultaneous fixed-points of positive formulas and thus, by Simultaneous Induction for FO(LFP), as a fixed-point of a positive formula, i.e., by an FO(LFP)-formula. Once more we study parts of the problem on a more abstract level.

First we introduce the *rank function* $|\,|_H$ and the *stage comparison relations* $\leq_H$ and $\prec_H$ associated with an arbitrary inductive $H : \mathrm{Pow}(M) \to \mathrm{Pow}(M)$ (with finite M). We define $|\,|_H : M \to \mathbb{N} \cup \{\infty\}$ by

$$|a|_H := \begin{cases} n & \text{if} \quad a \in H_n \setminus H_{n-1} \\ \infty & \text{if} \quad a \notin H_\infty \end{cases}$$

and the binary relations $\leq_H$ and $\prec_H$ on M by

$$\begin{array}{lll} a \leq_H b & \text{iff} & a, b \in H_\infty \text{ and } |a|_H \leq |b|_H \\ a \prec_H b & \text{iff} & a \in H_\infty \text{ and } |a|_H < |b|_H, \end{array}$$

where, by definition, $n < \infty$ for $n \in \mathbb{N}$.

Parts (a) – (e) of the next lemma are immediate from the definitions.

Lemma 8.2.8 (a) $\quad a \in H_\infty \quad$ *iff* $\quad a \leq_H a$.

(b) *For* $a \in H_\infty$,

$$\{u \mid u \leq_H a\}^c = \{u \mid a \prec_H u\}.$$

(c) $a \leq_H b$ *iff* $a, b \in H_1 \vee \exists e\,(e \prec_H b \wedge a, b \in H(\{u \mid u \leq_H e\}))$
(*in words:* $a \leq_H b$ *iff both* a, b *are contained in the first stage* H_1 *or there is a stage not containing* b *such that* a *and* b *are members of the next stage*).

(d) $a \prec_H b$ *iff* $(a \in H_1 \wedge b \notin H_1) \vee \exists e\,(e \leq_H e \wedge$
$a \in H(\{u \mid u \leq_H e\}) \wedge b \notin H(\{u \mid u \leq_H e\}))$.

(e) *If* $n_0 \geq 1$ *and* $H_\infty = H_{n_0} \neq H_{n_0-1}$ *then*

$$a \notin H_\infty \quad \text{iff} \quad \exists b\,(b \in H_{n_0} \setminus H_{n_0-1} \wedge b \prec_H a).$$

(f) *If* $n_0 \geq 1$ *and* $H_\infty = H_{n_0} \neq H_{n_0-1}$ *then*

$$b \in H_{n_0} \setminus H_{n_0-1} \quad \text{iff} \quad \forall e\,(e \leq_H b \vee e \notin H(\{u \mid u \leq_H b\})).$$

Proof. (f) First assume that b satisfies the right side. Then, $b \in H_\infty$, as otherwise, for all e, $e \notin H(\{u \mid u \leq_H b\}) = H(\emptyset) = H_1$, hence, $H_1 = \emptyset = H_\infty$, a contradiction. If $b \in H_n \setminus H_{n-1}$ for some $n < n_0$, then no $e \in H_{n+1} \setminus H_n$ would satisfy $(e \leq_H b \vee e \notin H(\{u \mid u \leq_H b\}))$, a contradiction. Conversely, suppose $b \in H_{n_0} \setminus H_{n_0-1}$ and let e be arbitrary: if $e \in H_\infty$ then $e \leq_H b$; if $e \notin H_\infty$ then $e \notin H(\{u \mid u \leq_H b\})$. $\square$

Now suppose that for the inductive function H there is a monotone function $L : \text{Pow}(M) \times \text{Pow}(M) \to \text{Pow}(M)$ such that

$$\text{for all } X \subseteq M, \quad H(X) = L(X, X^c).$$

Then the operation $\widehat{L} : \text{Pow}(M) \times \text{Pow}(M) \to \text{Pow}(M)$ which is given by

$$\text{for all } X, Y \subseteq M, \quad \widehat{L}(X, Y) := (L(X^c, Y^c))^c$$

is monotone, too: If $X_1 \subseteq X_2$ and $Y_1 \subseteq Y_2$ we obtain stepwise

$$X_2^c \subseteq X_1^c,\ Y_2^c \subseteq Y_1^c,\ L(X_2^c, Y_2^c) \subseteq L(X_1^c, Y_1^c),\ \widehat{L}(X_1, Y_1) \subseteq \widehat{L}(X_2, Y_2).$$

Using part (b) of the preceding lemma we can rewrite parts (c) and (d) as

(c′) $a \preceq_H b$ iff $a, b \in H_1 \ \vee \ \exists e\, (e \prec_H b \wedge$
$a, b \in L(\{u \mid u \preceq_H e\}, \{u \mid e \prec_H u\}))$

(d′) $a \prec_H b$ iff $(a \in H_1 \ \wedge \ b \notin H_1) \ \vee \exists e\, (e \preceq_H e \wedge$
$a \in L(\{u \mid u \preceq_H e\}, \{u \mid e \prec_H u\}) \wedge$
$b \in \widehat{L}(\{u \mid e \prec_H u\}, \{u \mid u \preceq_H e\}))$.

We now define operations $F, G : \text{Pow}(M^2) \times \text{Pow}(M^2) \to \text{Pow}(M^2)$ in such a way that

$$F(\preceq_H, \prec_H) = \preceq_H \qquad \text{and} \qquad G(\preceq_H, \prec_H) = \prec_H .$$

(c′) and (d′) show that this is achieved by setting (we use notations such as $a\, F(U,V)\, b$ for $(a, b) \in F(U, V)$)

(1) $a\, F(U,V)\, b$ iff $a, b \in H_1 \ \vee \ \exists e\, (eVb \ \wedge \ a, b \in L(\_Ue,\ eV\_))$

(2) $a\, G(U,V)\, b$ iff $(a \in H_1 \ \wedge \ b \notin H_1) \ \vee \ \exists e\, (eUe \wedge$
$a \in L(\_Ue,\ eV\_) \ \wedge \ b \in \widehat{L}(eV\_,\ \_Ue))$.

By monotonicity of L and $\widehat{L}$, the functions F and G are monotone, too. Once more, we define $F_{(n)}$ and $G_{(n)}$ by

$$F_{(0)} := \emptyset,\ G_{(0)} := \emptyset,\ \ F_{(n+1)} := F(F_{(n)}, G_{(n)}),\ \ G_{(n+1)} := G(F_{(n)}, G_{(n)}).$$

Then we have

(3) $a\, F_{(n)}\, b$ iff $a, b \in H_n \wedge a \preceq_H b$

(4) $a\, G_{(n)}\, b$ iff $a \in H_n \wedge a \prec_H b$.

The proof is by simultaneous induction, the induction step for (3) being:

$$
\begin{aligned}
a\, F_{(n+1)}\, b \quad &\text{iff} \quad a\, F(F_{(n)}, G_{(n)})\, b \\
&\text{iff} \quad a, b \in H_1 \;\vee\; \exists e\, (e\, G_{(n)}\, b \wedge a, b \in L(\_F_{(n)}e,\, eG_{(n)}\_)) \\
&\text{iff} \quad a, b \in H_1 \;\vee\; \exists e\, (e \in H_n \wedge e \prec_H b \wedge \\
&\qquad a, b \in L(\{u \mid u \leq_H e\}, \{u \mid e \prec_H u\})) \text{ (by ind. hyp.)} \\
&\text{iff} \quad a, b \in H_1 \;\vee\; \exists e\, (e \in H_n \wedge e \prec_H b \wedge \\
&\qquad a, b \in H(\{u \mid u \leq_H e\}) \quad \text{(by choice of } L) \\
&\text{iff} \quad a, b \in H_{n+1} \text{ and } a \leq_H b \text{ (by definition of } H_{n+1}).
\end{aligned}
$$

From (3) and (4) we get

$$(5) \qquad (F_{(\infty)}, G_{(\infty)}) = (\leq_H, \prec_H).$$

Thus $(\leq_H, \prec_H)$ is the simultaneous fixed-point of the monotone system (F, G).

We come back to our original aim of expressing the result of an IFP operation by an application of the LFP operator.

Suppose that $\varphi(\overline{x}, X)$ is an FO(LFP)-formula. Choose an FO(LFP)-formula $\psi(\overline{x}, X, Y)$ positive in X and Y such that

$$(X\overline{x} \vee \varphi(\overline{x}, X)) \qquad \text{and} \qquad \psi(\overline{x}, X, \neg X)$$

are equivalent. As (the inductive) H and (the monotone) L of the preceding discussion take the operations given by

$$H(X) := \{\overline{x} \mid X\overline{x} \vee \varphi(\overline{x}, X)\} \quad \text{and} \quad L(X, Y) := \{\overline{x} \mid \psi(\overline{x}, X, Y)\},$$

respectively. Then $H(X) = L(X, X^c)$, $\widehat{L}(X, Y) = \{\overline{x} \mid \neg\psi(\overline{x}, \neg X, \neg Y)\}$, and $\neg\psi(\overline{x}, \neg X, \neg Y)$ is positive in X and Y. The first stage H_1 of H is defined by $\varphi(\overline{x}, \emptyset)$, where $\varphi(\overline{x}, \emptyset)$ is obtained from $\varphi(\overline{x}, X)$ by replacing each occurrence of an atomic part of the form $X\overline{t}$ by $\neg t_1 = t_1$. By (1) and (2), the corresponding operations F and G can be defined by the FO(LFP)-formulas

$$
\begin{aligned}
\epsilon(\overline{x}, \overline{y}, U, V) \;&:=\; (\varphi(\overline{x}, \emptyset) \wedge \varphi(\overline{y}, \emptyset)) \vee \exists \overline{z}\, (\overline{z} V \overline{y} \wedge \\
&\qquad \psi(\overline{x}, \_\, U\overline{z}, \overline{z}V\_) \wedge \psi(\overline{y}, \_\, U\overline{z}, \overline{z}V\_)) \\
\delta(\overline{x}, \overline{y}, U, V) \;&:=\; (\varphi(\overline{x}, \emptyset) \wedge \neg\varphi(\overline{y}, \emptyset)) \vee \exists \overline{z}\, (\overline{z} U \overline{z} \wedge \\
&\qquad \psi(\overline{x}, \_\, U\overline{z}, \overline{z}V\_) \wedge \neg\psi(\overline{y}, \neg\, \overline{z}V\_, \neg\_U\overline{z})),
\end{aligned}
$$

respectively. Both are positive in U, V. As $(\leq_H, \prec_H) = (F_{(\infty)}, G_{(\infty)})$ by (5), the relations $\leq_H$ and $\prec_H$ are definable by

$$
\begin{aligned}
\varphi_{\leq_H}(\overline{x}, \overline{y}) \;&:=\; [\text{S-LFP}_{\overline{x}\,\overline{y}, U, \overline{x}\,\overline{y}, V}\, \epsilon, \delta]\, \overline{x}\,\overline{y}, \\
\varphi_{\prec_H}(\overline{x}, \overline{y}) \;&:=\; [\text{S-LFP}_{\overline{x}\,\overline{y}, V, \overline{x}\,\overline{y}, U}\, \delta, \epsilon]\, \overline{x}\,\overline{y},
\end{aligned}
$$

respectively. By 8.2.3 there are FO(LFP)-formulas $\psi_{\leq_H}$ and $\psi_{\prec_H}$ such that

$$\models_{\text{fin}} \varphi_{\leq_H}(\overline{x},\overline{y}) \leftrightarrow \exists(\forall)u[\text{LFP}_{\overline{z},Z}\,\psi_{\leq_H}]\,\tilde{u},$$
$$\models_{\text{fin}} \varphi_{\prec_H}(\overline{x},\overline{y}) \leftrightarrow \exists(\forall)u[\text{LFP}_{\overline{z},Z}\,\psi_{\prec_H}]\,\tilde{u}.$$

Finally, using part (a) of 8.2.8, we get that

$$\models_{\text{fin}} [\text{IFP}_{\overline{x},X}\,\varphi]\,\overline{t} \leftrightarrow \varphi_{\leq_H}(\overline{t},\overline{t}).$$

Altogether, we can replace an IFP operation by an equivalent LFP operation. This immediately allows us to handle the only nontrivial step of an inductive proof of Theorem 8.1.13.

With the following considerations we turn to the proof of Theorem 8.1.14. First, we show that every formula $\neg[\text{LFP}_{\overline{x},X}\,\varphi(\overline{x},X)]\,\overline{t}$ with first-order φ is equivalent to an FO(LFP)-formula containing only positive occurrences of the LFP operator. So let $\varphi(\overline{x},X)$ be first-order. Then, the formulas ϵ and δ are first-order and, by 8.2.3, $\psi_{\leq_H}$ and $\psi_{\prec_H}$ can be chosen first-order. If $n_0 \geq 1$ and $H_\infty = H_{n_0} \neq H_{n_0-1}$ then, by 8.2.8(f) and by the definition of $\widehat{L}$, we have

$$b \in H_{n_0} \setminus H_{n_0-1} \text{ iff } \forall e\,(e \leq_H b \vee e \in \hat{L}(\{u \mid b \prec_H u\}, \{u \mid u \leq_H b\})).$$

Thus, $\overline{z} \in H_{n_0} \setminus H_{n_0-1}$ can be expressed by

$$\varphi_{\max}(\overline{z}) \quad := \quad \forall\overline{v}(\varphi_{\leq_H}(\overline{v},\overline{z}) \vee \neg\psi(\overline{v}, \neg\varphi_{\prec_H}(\overline{z},\_\,), \neg\varphi_{\leq_H}(\_\,,\overline{z})).$$

Together with 8.2.8(e) we get the equivalence of $\neg[\text{LFP}_{\overline{x},X}\,\varphi(\overline{x},X)]\,\overline{t}$ and

$$(*) \qquad \forall\overline{x}\neg\varphi(\overline{x},\emptyset) \vee (\exists\overline{x}\varphi(\overline{x},\emptyset) \wedge \exists\overline{z}(\varphi_{\max}(\overline{z}) \wedge \varphi_{\prec_H}(\overline{z},\overline{t}))).$$

Since $\varphi_{\leq_H}$ and $\varphi_{\prec_H}$ only contain positive occurrences of the LFP operator, the same applies to the formula in $(*)$.

Thus, every FO(LFP)-formula is equivalent to one with only positive occurrences of LFP. In view of Lemma 8.2.7 this finishes the proof of 8.1.14 and, in fact, gives the following strengthening of it:

Theorem 8.2.9 *For every* FO(LFP)*-formula φ there is a first-order formula ψ such that*

$$\models_{\text{fin}} \varphi \leftrightarrow \exists(\forall)u[\text{LFP}_{\overline{z},Z}\,\psi]\,\tilde{u}. \qquad \square$$

Corollary 8.2.10 *For every* FO(LFP)*-formula φ with free variables among $\overline{x} = x_1 \ldots x_n$ with $n \geq 1$ there are first-order formulas ψ_0, ψ_1 with free individual variables among $\overline{x}$ and $\overline{y}$, respectively, such that*

$$\models_{\text{fin}} \varphi(\overline{x}) \leftrightarrow [\text{S-LFP}_{\overline{x},X,\overline{y},Y}\,\psi_0,\psi_1]\,\overline{x}.$$

Proof. With the preceding theorem choose a first-order formula $\psi(\overline{x}, \overline{z}, Z)$ such that φ and $\exists(\forall)u[\text{LFP}_{\overline{z},Z}\,\psi]\,\tilde{u}$ are equivalent. We can assume that $\{\overline{z}\} \cap \{\overline{x}\} = \emptyset$ and that the free individual variables of ψ are among $\overline{x}\,\overline{z}$. Let Y be of arity length($\overline{x}$)+length($\overline{z}$). Then

$$\begin{aligned}\psi_0(\overline{x}, X, Y) &:= Y\overline{x}\tilde{x}_1\\ \psi_1(\overline{x}, \overline{z}, X, Y) &:= \psi(\overline{x}, \overline{z}, Y\overline{x}\_)\end{aligned}$$

satisfy the claim, where $\overline{y} = \overline{x}\,\overline{z}$. □

We saw in 8.1.18 that in the preceding corollary we cannot replace the formula $[\text{S-LFP}_{\overline{x},X,\overline{y},Y}\,\psi_0, \psi_1]\,\overline{x}$ by a formula of the form $[\text{LFP}_{\overline{x},X}\,\psi]\,\overline{x}$: a counterexample is given by a formula $\varphi(x)$ expressing in digraphs that there is a cycle through x.

Exercise 8.2.11 Consider a formula $\chi = \exists(\forall)u[\text{IFP}_{\overline{x},X}\varphi]\tilde{u}$ where φ is first-order and Σ_1 (cf. page 7). Show that χ is equivalent to a formula $\exists(\forall)u[\text{LFP}_{\overline{z},Z}\rho]\,\tilde{u}$ where ρ is first-order and Δ_2 (that is, equivalent to a Σ_2-formula and to a Π_2-formula). Hint: Assume that φ is a Σ_1-formula and let ψ, ϵ, δ be as in the exposition following 8.2.8. Then ϵ is equivalent to a Σ_1-formula and δ to a Σ_2-formula; hence, the corresponding $\psi_{\leq_H}$ is equivalent to a Σ_2-formula. To obtain a Π_2 representation, note that 8.2.8(d) can be replaced by

$$\begin{aligned} a \prec_H b \quad \text{iff} \quad & (a \leq_H a \wedge b \notin H_1) \wedge \\ & \forall e(a \leq_H e \vee a \prec_H e \vee b \notin H(\{u \mid u \leq_H e\})). \end{aligned}$$

Therefore, as $\delta(\overline{x}, \overline{y}, U, V)$ we can use the Π_1-formula

$$\overline{x}U\overline{x} \wedge \neg\varphi(\overline{y}, \emptyset) \wedge \forall\overline{z}(\overline{x}U\overline{z} \vee \overline{x}V\overline{z} \vee \neg\psi(\overline{y}, \neg\overline{z}V\_, \neg\_U\overline{z})).$$

Then the corresponding χ_J is equivalent to a Π_2-formula (note that χ_J has the form $\exists v \exists w(\neg v = w \wedge \theta)$, but is equivalent to $\forall v \forall w(\neg v = w \rightarrow \theta)$. An inspection of the proof of 8.2.4 shows that we get a Δ_2-formula if we start from a Δ_2-formula. □

As promised, we now show how one can remove the (tacitly assumed) restriction to structures with at least two elements. As an example we take the preceding theorem (Theorem 8.2.9). First of all we note that in structures with only one element the formulas

$$[\text{LFP}_{\overline{v},V}\,\chi(\overline{v}, V)]\,\overline{t} \quad \text{and} \quad \chi(\overline{t}, \emptyset)$$

are equivalent. Hence, for every FO(LFP)-formula ρ there is a first-order formula ρ^* equivalent to it in these structures. Now, given $\varphi \in$ FO(LFP), we use the preceding theorem to choose a first-order ψ such that φ and $\exists(\forall)u[\text{LFP}_{\overline{z},Z}\,\psi]\,\tilde{u}$ are equivalent in structures with at least two elements.

…r new x and y, the formulas φ and $\exists(\forall)u[\text{LFP}_{\bar{z},Z}\,(\forall x\forall y x = y \wedge \varphi^*) \vee$ …$\exists y \neg x = y \wedge \psi)]\,\tilde{u}$ are equivalent in any finite structure.

We close this section by giving a further result demonstrating the value of simultaneous fixed-point logic: We show that, on ordered structures, it captures the class of polynomial time computable functions in terms of the stages of simultaneous fixed-points. First we introduce the corresponding terminology. Let

$$\varphi_0(\overline{x}_0, X_0, \ldots, X_m), \ldots, \varphi_m(\overline{x}_m, X_0, \ldots, X_m)$$

be first-order formulas that are positive in $X_0, \ldots, X_m$, where $\text{arity}(X_i) = \text{length}(\overline{x}_i)$ for $i \leq m$. Given a structure $\mathcal{A}$, let $F^0, \ldots, F^m$ be the functions corresponding to $\varphi_0, \ldots, \varphi_m$,

$$F^i(X_0, \ldots, X_m) \quad = \quad \{\overline{x}_i \mid \varphi_i(\overline{x}_i, X_0, \ldots, X_m)\}.$$

For $i \leq m$, let $|\varphi_i|(\mathcal{A})$ be the number of different stages of F^i, more precisely,

$$|\varphi_i|(\mathcal{A}) \quad := \quad \|\{n \mid F^i_{(n)} \neq F^i_{(n+1)}\}\|.$$

Clearly, the function $\mathcal{A} \mapsto |\varphi_i|(\mathcal{A})$ is a function computable in polynomial time. A converse is also true:

Theorem 8.2.12 *Let $< \in \tau$ and let K be a class of ordered τ-structures in* PTIME. *Then for every function $f : K \to \mathbb{N}$ computable in polynomial time*[7] *there are first-order formulas $\varphi_0(\overline{x}_0, X_0, \ldots, X_m), \ldots, \varphi_m(\overline{x}_m, X_0, \ldots, X_m)$ that are positive in $X_0, \ldots, X_m$ such that for all $\mathcal{A} \in K$ with at least two elements,*

$$f(\mathcal{A}) = |\varphi_0|(\mathcal{A}).$$

Proof. Since f is computable in polynomial time, there is an r such that $f(\mathcal{A}) < \|A\|^r$ for all $\mathcal{A} \in K$ with at least two elements. For $j < \|A\|^r$ let $|j|_r$ be the j-th r-tuple in the lexicographic ordering of A^r. By assumption, the class

$$\{(\mathcal{A}, a_1, \ldots, a_r) \mid \mathcal{A} \in K,\ |f(\mathcal{A})|_r = (a_1, \ldots, a_r)\}$$

is in PTIME. Hence, by 7.3.4 and 8.1.13, there is a formula $\varphi(x_1, \ldots, x_r)$ of FO(LFP) such that for any ordered structure $\mathcal{B}$ and $\overline{b} \in B^r$,

$$\mathcal{B} \models \varphi[\overline{b}] \qquad \text{iff} \qquad \mathcal{B} \in K \text{ and } \overline{b} = |f(\mathcal{B})|_r.$$

By 8.2.10, we obtain first-order formulas $\psi_0(\overline{x}, X, Y), \psi_1(\overline{y}, X, Y)$ that are positive in X, Y such that $\varphi(\overline{x})$ and $[\text{S-LFP}_{\overline{x},X,\overline{y},Y}\,\psi_0, \psi_1]\,\overline{x}$ are equivalent. To get the desired representation of $f(\mathcal{A})$, we construct a simultaneous fixed-point process, where the stages of the first component Z (of arity r) successively take up the r-tuples in lexicographic order that are smaller than $|f(\mathcal{A})|_r$ as soon as $|f(\mathcal{A})|_r$ gets into X.

[7] We assume that $\mathcal{A} \cong \mathcal{B}$ implies $f(\mathcal{A}) = f(\mathcal{B})$.

We set

$$\psi(\overline{z}, Z, X, Y) \; := \; \exists \overline{x}(X\overline{x} \wedge \overline{z} < \overline{x} \wedge (\overline{z} = \widetilde{\min} \vee \exists \overline{u}(Z\overline{u} \wedge \overline{z} = \overline{u} + 1))).$$

For the system $\psi(\overline{z}, Z, X, Y)$, $\psi_0(\overline{x}, Z, X, Y)$, $\psi_1(\overline{y}, Z, X, Y)$ one easily verifies that $f(\mathcal{A}) = |\psi|(\mathcal{A})$. □

In the preceding theorem we cannot replace $\varphi_0, \ldots, \varphi_m$ by a single formula. To give a counterexample, we consider the function f defined on the class of ordered graphs by

$$f(\mathcal{G}) := \begin{cases} 1 & \text{if } \mathcal{G} \text{ is connected} \\ 0 & \text{otherwise.} \end{cases}$$

Suppose that, for some first-order formula $\varphi(\overline{x}, X)$ positive in X, we have $f(\mathcal{G}) = |\varphi|(\mathcal{G})$. Then, $\mathcal{G}$ is connected iff $F_\infty^\varphi \neq \emptyset$. But $F_\infty^\varphi \neq \emptyset$ is equivalent to $F_1^\varphi \neq \emptyset$. Hence, $\mathcal{G}$ is connected iff $\mathcal{G}$ is a model of $\exists \overline{x} \varphi(\overline{x}, \emptyset)$. But we saw in 2.3.7 that the class of finite connected ordered graphs is not first-order axiomatizable.

Notes 8.2.13 The book [122] contains the first systematic study of least fixed-point operators (in the context of infinite structures); in particular, in this book the Lemma on Simultaneous Induction and the Transitivity Lemma are proven and the stage comparison relations are introduced. The result of Exercise 8.2.11 is due to Dahlhaus [23], Theorem 8.2.12 to Kolaitis and Thakur [103].

8.3 Partial Fixed-Point Logic

In Chapter 7 a further fixed-point logic turned out to be relevant, namely partial fixed-point logic, its importance being documented by the fact that, in ordered structures, exactly the queries computable in polynomial space are expressible in partial fixed-point logic. In this section we recall the definition of partial fixed-point logic, discuss the scope of totally defined fixed-points, and show that every formula of partial fixed-point logic is equivalent to a formula containing exactly one occurrence of the PFP operator.

Partial Fixed-Point Logic FO(PFP) contains first-order logic and "F_∞ for any definable F": For a vocabulary τ the class of formulas of FO(PFP) of vocabulary τ is given by the calculus

- $\dfrac{}{\varphi}$ where φ is an atomic second-order formula over τ
- $\dfrac{\varphi}{\neg\varphi}, \quad \dfrac{\varphi, \psi}{(\varphi \vee \psi)}, \quad \dfrac{\varphi}{\exists x \varphi}$

- $\dfrac{\varphi}{[\mathrm{PFP}_{\overline{x},X}\varphi]\,\overline{t}}$ where the lengths of $\overline{x}$ and $\overline{t}$ are the same and coincide with the arity of X.

The semantics is defined inductively w.r.t. this calculus, the meaning of $[\mathrm{PFP}_{\overline{x},X}\varphi]\,\overline{t}$ being $\overline{t} \in F_\infty^\varphi$. In particular, $[\mathrm{PFP}_{\overline{x},X}\varphi]\,\overline{t}$ is false if the fixed-point of F^φ does not exist. All other definitions and conventions are as for FO(LFP).

Clearly, we have

Proposition 8.3.1 FO(LFP) $\leq$ FO(PFP). □

Consider, for example, orderings as $\{<, S, \min, \max\}$-structures. Then the sentence $\exists x\,[\mathrm{PFP}_{x,X}\,\psi(x,X)]\,x$ where

$$\psi(x,X) = (\forall y\,\neg Xy \wedge x = \min) \vee (X\max \wedge x = \max) \vee \exists y(Xy \wedge \exists u(Syu \wedge Sux))$$

just holds in orderings of odd length.

As a further example consider the formula

$$[\mathrm{PFP}_{x,X}\,\exists y(\neg Xy \wedge Exy)]\,x.$$

By 8.1.5(b), in digraphs it expresses, at the same time, that x is won for player I and that there are no drawn points. By 8.1.10(b) we see that this formula is equivalent to an FO(LFP)-formula. This result is generalized in the following (cf. 8.1.5(a))

Exercise 8.3.2 Assume that $\varphi(\overline{x}, X)$ is an FO(LFP)-formula negative in X and that X does not occur in the scope of any LFP operator. Then

- $\psi(\overline{x}, X) := \varphi(\overline{x}, \varphi(\_, X))$ is an FO(LFP)-formula positive in X.
- F^φ is antitone and $F^\psi = F^\varphi \circ F^\varphi$.
- The "FO(LFP,PFP)"-formula $[\mathrm{PFP}_{\overline{x},X}\,\varphi(\overline{x}, X)]\,\overline{t}$ is equivalent to the formula of FO(LFP)

$$\forall\overline{x}([\mathrm{LFP}_{\overline{x},X}\,\psi]\,\overline{x} \leftrightarrow [\mathrm{GFP}_{\overline{x},X}\,\psi]\,\overline{x}) \wedge [\mathrm{LFP}_{\overline{x},X}\,\psi]\,\overline{t}.$$ □

We aim at showing

Theorem 8.3.3 *Every* FO(PFP)*-formula is equivalent to an* FO(PFP)*-formula with at most one occurrence of* PFP.

We shall point out similarities and differences to the proof of the corresponding result for FO(LFP). As there, we start by introducing *Simultaneous Partial Fixed-Point Logic* FO(S-PFP) and by showing that it has not more expressive power than FO(PFP), even though it will turn out to be a valuable tool in many considerations.

The formulas of FO(S-PFP) are obtained by replacing the PFP rule above by

$$\frac{\varphi_0, \ldots, \varphi_m}{[\text{S-PFP}_{\overline{x}_0, X_0, \ldots, \overline{x}_m, X_m}\, \varphi_0, \ldots, \varphi_m]\, \overline{t}}$$

where $m \geq 0$, where for $i = 0, \ldots, m$ the arity of X_i equals the length of $\overline{x}_i$, and where $\overline{t}$ has the same length as $\overline{x}_0$. As usual, the variables in each sequence $\overline{x}_i$ are distinct, but the same variable may occur in distinct $\overline{x}_i$'s. A variable x is free in $[\text{S-PFP}_{\overline{x}_0, X_0, \ldots, \overline{x}_m, X_m}\, \varphi_0, \ldots, \varphi_m]\, \overline{t}$, if x occurs in $\overline{t}$ or if for at least one i, x is free in φ_i and not in $\overline{x}_i$. Introducing $F^0, \ldots, F^m$ by

$$F^i(X_0, \ldots, X_m) \ := \ \{\overline{x}_i \mid \varphi_i(\overline{x}_i, X_0, \ldots, X_m)\},$$

the meaning of $[\text{S-PFP}_{\overline{x}_0, X_0, \ldots, \overline{x}_m, X_m}\, \varphi_0, \ldots, \varphi_m]\, \overline{t}$ is defined to be $\overline{t} \in F^0_{(\infty)}$.

Once more, for simplicity, we restrict ourselves to structures with at least two elements.

Theorem 8.3.4 FO(PFP) $\equiv$ FO(S-PFP).

The proof of FO(S-PFP) $\leq$ FO(PFP) is by induction on formulas, the main step being taken care of by the following lemma.

Lemma 8.3.5 (Simultaneous Induction for PFP**)** *Let*

$$\varphi_0(\overline{x}_0, X_0, \ldots, X_m), \ldots, \varphi_m(\overline{x}_m, X_0, \ldots, X_m)$$

be formulas of FO(PFP)*, where X_i is k_i-ary and the length of $\overline{x}_i$ is k_i. Set $k := \max\{k_0, \ldots, k_m\} + (m+1)$ and define the* FO(PFP)*-formula $\chi_J(z_1, \ldots, z_k, Z)$ by* [8]

$$\begin{aligned}
\chi_J(z_1, \ldots, z_k, Z) \ := \ & \exists v \exists w (\neg v = w \land \\
& ((\varphi_0(z_1, \ldots, z_{k_0}, Z_\tilde{v}w^0, \ldots, Z_\tilde{v}w^m) \land \delta_0(z_{k_0+1}, \ldots, z_k, v, w)) \\
& \lor (\varphi_1(z_1, \ldots, z_{k_1}, Z_\tilde{v}w^0, \ldots, Z_\tilde{v}w^m) \land \delta_1(z_{k_1+1}, \ldots, z_k, v, w)) \\
& \vdots \\
& \lor (\varphi_m(z_1, \ldots, z_{k_m}, Z_\tilde{v}w^0, \ldots, Z_\tilde{v}w^m) \land \delta_m(z_{k_m+1}, \ldots, z_k, v, w)))).
\end{aligned}$$

Then, for any new variable u,

$$\models_{\text{fin}} [\text{S-PFP}_{\overline{x}_0, X_0, \ldots, \overline{x}_m, X_m}\, \varphi_0, \ldots, \varphi_m]\, \overline{t} \leftrightarrow \exists(\forall) u [\text{PFP}_{\overline{z}, Z}\, \chi_J]\, \overline{t}\tilde{u}.$$

If $\varphi_0, \ldots, \varphi_m$ are first-order formulas then χ_J is first-order. Moreover, given a structure and an assignment to variables, the simultaneous fixed-point of the PFP *operation of the formula on the right side in the equivalence above exists iff the fixed-point of the* PFP *operation on the left side exists.*

The lemma is a literal translation of the Lemma on Simultaneous Induction for LFP (cf. 8.2.3), and also the proof can be translated; of course, here we do not have the assumption that the formulas are positive in the second-order variables and hence, the same applies to χ_J. The additional claim on the existence of the fixed-point immediately follows from 8.2.1(a). □

Also the proof of 8.2.4 can be literally translated and gives

[8] For the definition of δ_i see page 179.

Lemma 8.3.6 *Every* FO(PFP)*-formula* $[\text{PFP}_{\overline{y},Y}\,\varphi]\,\overline{t}$ *with first-order* φ *is equivalent to a formula of the form*

$$\exists(\forall)u[\text{PFP}_{\overline{z},Z}\,\psi]\,\tilde{u}$$

with first-order ψ. *If* φ *is existential then* ψ *can be chosen existential, too. For a given structure and assignment, if* F_∞^φ *exists then so does* F_∞^ψ. *Furthermore, we may require that* u *does not occur free in* ψ *and that the free variables of* $\exists(\forall)u[\text{PFP}_{\overline{z},Z}\,\psi]\,\tilde{u}$ *are among the free variables of* $[\text{PFP}_{\overline{y},Y}\,\varphi]\,\overline{t}$. □

In the proof of the Transitivity Lemma for LFP we have made essential use of the monotonicity of the operations in question, so it does not translate to the present case. However, it is easy to derive a version of the Transitivity Lemma for PFP (see 8.3.10 below), if we restrict ourselves to formulas whose fixed-points always exist. For this purpose call a formula of FO(PFP) *totally defined*, if for all its subformulas of the form $[\text{PFP}_{\overline{x},X}\,\psi]\,\overline{t}$ the fixed-point F_∞^ψ exists in every structure and for all assignments.

Exercise 8.3.7 Show, using Trahtenbrot's theorem, that it is not decidable whether an FO(PFP)-formula is totally defined. □

Nevertheless we can show:

Proposition 8.3.8 *For every* FO(PFP)*-formula* φ *there is an equivalent totally defined* FO(PFP)*-formula* ψ. *If* φ *has the form* $[\text{PFP}_{\overline{x},X}\,\varphi']\,\overline{t}$ *with first-order* φ' *then* ψ *can be chosen of the form* $[\text{PFP}_{\overline{x},X}\,\psi']\,\overline{s}$ *with first-order* ψ'.

For the proof we need the following

Lemma 8.3.9 *Assume* M *is finite. Let* $F:\text{Pow}(M)\to\text{Pow}(M)$. *Then there is an* $l\geq 1$ *such that* $F_l=F_{2\cdot l}$.

Proof. By finiteness of M, the sequence $F_0, F_1, \ldots$ must become periodic (cf. 7.1.1(a)). So there are n_0 and $l_0\geq 1$ such that $F_m=F_{m+l_0}$ holds for all $m>n_0$. Choose s such that $l:=s\cdot l_0>n_0$. Then $F_l=F_{s\cdot l_0}=F_{s\cdot l_0+s\cdot l_0}=F_{2\cdot l}$. □

Proof (of 8.3.8). The proof proceeds by induction on FO(PFP)-formulas, the main step being the PFP operator. So let $[\text{PFP}_{\overline{x},X}\,\chi]\,\overline{t}$ be given and suppose, by induction hypothesis, that χ is totally defined. Consider the function F^χ for a given structure $\mathcal{A}$ and an assignment of values to the parameters. Set $X_n:=F_n^\chi$ and $X_\infty:=F_\infty^\chi$. Choose the smallest $l\geq 1$ such that $X_{2\cdot l}=X_l$. We consider the sequences $(Y_n)_{n\geq 0}$, $(Z_n)_{n\geq 0}$, and $(W_n)_{n\geq 0}$, whose values are shown by the following table:

X	X_0	X_1	X_2	$\ldots$	$\ldots$	X_l	X_{l+1}	X_{l+2}	$\ldots$
Y	X_0	X_1	X_2	$\ldots$	$\ldots$	X_l	X_∞	X_∞	$\ldots$
Z	X_0	X_2	X_4	$\ldots$	$\ldots$	X_{2l}	X_∞	X_∞	$\ldots$
W	$\emptyset$	A	A	$\ldots$	$\ldots$	A	A	A	$\ldots$

In the first l steps (Y_n) proceeds as (X_n) and (Z_n) as $(X_{2\cdot n})$, i.e., with twice the velocity of (X_n). When $n > 0$ (which is signaled by $W_n \neq \emptyset$) and $Y_n = Z_n$ then $Y_m = Z_m = X_\infty$ for $m > n$. So, all these new sequences have fixed-points (this guarantees that the formula obtained below is totally defined). The fixed-points are the simultaneous fixed-point of the operations given by the formulas φ_Y, φ_Z, and φ_W, where

$$\begin{aligned}
\varphi_Y(\overline{y}, Y, Z, W) &:= ((W = \emptyset \vee Y \neq Z) \wedge \chi(\overline{y}, Y)) \vee (Y = \chi(_, Y) \wedge Y\overline{y})^9 \\
\varphi_Z(\overline{z}, Y, Z, W) &:= ((W = \emptyset \vee Y \neq Z) \wedge \chi(\overline{z}, \chi(_, Z))) \\
&\quad\ \vee (Z = \chi(_, Z) \wedge Z\overline{z}) \\
\varphi_W(w, Y, Z, W) &:= w = w.
\end{aligned}$$

By the first two lines of the table we know that

$$\models_{\text{fin}} [\text{PFP}_{\overline{x},X}\, \chi]\,\overline{t} \leftrightarrow [\text{S-PFP}_{\overline{y},Y,\overline{z},Z,w,W}\, \varphi_Y, \varphi_Z, \varphi_W]\,\overline{t}$$

and therefore, by the Lemma on Simultaneous Induction for PFP, we have for the corresponding χ_J that

$$\models_{\text{fin}} [\text{PFP}_{\overline{x},X}\, \chi]\,\overline{t} \leftrightarrow [\text{PFP}_{\overline{v},V}\, \chi_J]\,\overline{t}\tilde{u}$$

and that the formula on the right side is totally defined. □

Next we show an FO(PFP)-version of the Transitivity Lemma. It will be used in the proof of 8.3.3.

Lemma 8.3.10 (Transitivity for PFP) *Let $\varphi(\overline{x}, X, Y)$ and $\psi(\overline{y}, X, Y)$ be first-order and let $[\text{PFP}_{\overline{y},Y}\, \psi(\overline{y}, X, Y)]\,\overline{y}$ be totally defined. (Moreover, assume that no individual variable free in $[\text{PFP}_{\overline{y},Y}\, \psi(\overline{y}, X, Y)]$ gets into the scope of a corresponding quantifier or PFP operator in (*).) Then the formula*

$$(*) \qquad [\text{PFP}_{\overline{x},X}\, \varphi(\overline{x}, X, [\text{PFP}_{\overline{y},Y}\, \psi(\overline{y}, X, Y)])]\,\overline{t}$$

is equivalent to a totally defined formula

$$\exists(\forall)u[\text{PFP}_{\overline{z},Z}\, \chi(\overline{z}, Z)]\,\tilde{u}$$

with first-order χ.

Proof. Set $\rho := [\text{PFP}_{\overline{x},X}\, \varphi(\overline{x}, X, [\text{PFP}_{\overline{y},Y}\, \psi(\overline{y}, X, Y)])]\,\overline{t}$. We want to write the nested fixed-point as a simultaneous fixed-point. For this purpose let

$$\varphi'(\overline{x}, X, Y) := (Y \neq \psi(\_, X, Y) \wedge X\overline{x}) \vee (Y = \psi(\_, X, Y) \wedge \varphi(\overline{x}, X, Y))$$

("if Y is not a fixed-point then X does not change, otherwise X goes one step further"),

[9] $Y = \chi(\_, Y)$ is an abbreviation for $\forall\overline{x}(Y\overline{x} \leftrightarrow \chi(\overline{x}, Y))$.

$$\begin{aligned}\psi'(\overline{y}, X, Y) \;:=\; & (Y \neq \psi(\_, X, Y) \wedge \psi(\overline{y}, X, Y)) \vee \\ & (Y = \psi(\_, X, Y) \wedge (X = \varphi(\_, X, Y) \wedge Y\overline{y}))\end{aligned}$$

("if Y is not a fixed-point then Y goes one step further; otherwise Y is set back to $\emptyset$ except X happens to be a fixed-point; then Y does not change").
Since $[\mathrm{PFP}_{\overline{y},Y}\, \psi(\overline{y}, X, Y)]\,\overline{t}$ is totally defined, we immediately have

$$\models_{\mathrm{fin}} \rho \leftrightarrow [\text{S-PFP}_{\overline{x},X,\overline{y},Y}\, \varphi', \psi']\,\overline{t}.$$

For the corresponding χ_J, which is first-order (cf. 8.3.5), we therefore get

$$\models_{\mathrm{fin}} \rho \leftrightarrow [\mathrm{PFP}_{\overline{z},Z}\, \chi_J]\,\overline{t}\tilde{u}.$$

To get a totally defined formula of the desired form, we first apply 8.3.8 to $[\mathrm{PFP}_{\overline{z},Z}\, \chi_J]\,\overline{t}\tilde{u}$ and then 8.3.6. □

Exercise 8.3.11 Let $\varphi(x, X, Y) := (x = c \vee Yx)$ and $\psi(y, X, Y) := \neg Yy$. Show that $[\mathrm{PFP}_{x,X}\, \varphi(x, X, [\mathrm{PFP}_{y,Y}\, \psi(y, X, Y)])]\, x$ is equivalent to the formula $x = c$, while the simultaneous fixed-point of φ' and ψ' defined according to the previous proof does not exist. Note that $[\mathrm{PFP}_{y,Y}\, \psi(y, X, Y)]\, y$ is not totally defined. □

We now get Theorem 8.3.3 in the following sharper form.

Theorem 8.3.12 *For every* FO(PFP)*-formula φ there is a first-order formula ψ such that*

$$\models_{\mathrm{fin}} \varphi \leftrightarrow \exists(\forall)u[\mathrm{PFP}_{\overline{z},Z}\, \psi]\,\tilde{u}$$

and $\exists(\forall)u[\mathrm{PFP}_{\overline{z},Z}\, \psi]\,\tilde{u}$ is totally defined.

Proof. We proceed by induction on φ. The atomic case and the cases for $\vee$, $\exists$, and PFP are handled as in 8.2.7, now using the preceding lemma instead of 8.2.6. Concerning the negation step, suppose $\varphi = \neg\varphi_1$ and, by induction hypothesis, assume that

$$\varphi_1 = [\mathrm{PFP}_{\overline{v},V}\, \psi_1(\overline{v}, V)]\,\overline{t}$$

is totally defined and ψ_1 is first-order. Set $\rho(\overline{x}, X, Z) := \neg Z\overline{t}$ for dummy variables $\overline{x}$ and X. Then φ $(= \neg[\mathrm{PFP}_{\overline{v},V}\, \psi_1(\overline{v}, V)]\,\overline{t})$ is equivalent to

$$[\mathrm{PFP}_{\overline{x},X}\, \rho(\overline{x}, X, [\mathrm{PFP}_{\overline{v},V}\, \psi_1(\overline{v}, V)])]\,\overline{t}$$

which, by the preceding lemma, is equivalent to a formula of the desired form. □

We close this section by showing that in the normal form given in the preceding theorem the formula ψ on the right hand side can be chosen existential, a result we shall also obtain by different methods in Chapter 9. The following exercise gives the idea underlying the procedure to eliminate universal quantifiers.

Exercise 8.3.13 Show that in models of $\neg c = d$ the formulas $\forall x Rx$ and

$$[\mathrm{PFP}_{x,X}\,(x = c \vee (x = d \wedge \exists z \neg Rz)) \wedge ((Xc \wedge Xd) \rightarrow \neg x = x)]\,c$$

are equivalent. □

Theorem 8.3.14 *For every* FO(PFP)*-formula* φ *there is an existential first-order formula* ρ *such that* $\exists(\forall)u[\mathrm{PFP}_{\overline{y},Y}\,\rho]\,\tilde{u}$ *is totally defined and*

$$\models_{\mathrm{fin}} \varphi \leftrightarrow \exists(\forall)u[\mathrm{PFP}_{\overline{y},Y}\,\rho]\,\tilde{u}.$$

Proof. By the preceding theorem we can assume that $\varphi = [\mathrm{PFP}_{\overline{y},Y}\,\psi]\,\tilde{u}$,[10] where $[\mathrm{PFP}_{\overline{y},Y}\,\psi]\,\tilde{u}$ is totally defined and where ψ is first-order and in prenex normal form. It suffices to show that if ψ is a Σ_{k+1}-formula (compare 1.B for the definition of Σ_n-formulas) with $k > 0$ then ψ can be replaced by a Σ_k-formula (and hence, by induction, by a Σ_1-formula; the result then follows from 8.3.6). So assume that $\psi(\overline{y}, Y) = \exists \overline{z} \forall \overline{u} \psi'$ with $\psi' = \psi'(\overline{y}, \overline{z}, \overline{u}, Y) \in \Sigma_{k-1}$, and let $Y_n := F_n^{\psi}$. Define the sequences $(U_{(n)}), (V_{(n)}), (W_{(n)})$ simultaneously, using the formulas χ_U, χ_V, χ_W, respectively, where

$$\begin{aligned}
\chi_U(\overline{y}, U, V, W) &:= W \neq \emptyset \wedge \exists \overline{z} \neg V \overline{y}\,\overline{z} \\
\chi_V(\overline{y}\,\overline{z}, U, V, W) &:= \neg \forall \overline{u} \psi'(\overline{y}, \overline{z}, \overline{u}, U) \\
\chi_W(x, U, V, W) &:= x = x.
\end{aligned}$$

Note that $W_{(0)} = \emptyset$ and $W_{(n)} =$ universe for $n > 0$. Hence, we get $U_{(1)} = \emptyset$ and for $n \geq 2$:

$$\begin{aligned}
U_{(n)} &= \{\overline{y} \mid \exists \overline{z} \neg V_{(n-1)} \overline{y}\,\overline{z}\} \\
&= \{\overline{y} \mid \exists \overline{z} \forall \overline{u} \psi'(\overline{y}, \overline{z}, \overline{u}, U_{(n-2)})\}.
\end{aligned}$$

Since $Y_n = \{\overline{y} \mid \exists \overline{z} \forall \overline{u} \psi'(\overline{y}, \overline{z}, \overline{u}, Y_{n-1})\}$, a simple induction shows that $U_{(2n)} = Y_n$ and $U_{(2n+1)} = Y_n$; in particular, $(U_{(n)})$ eventually becomes constant. As $V_{(n)} = \{\overline{y}\,\overline{z} \mid \neg \forall \overline{u} \psi'(\overline{y}, \overline{z}, \overline{u}, U_{(n-1)})\}$, the sequence $(V_{(n)})$ becomes constant, too. Thus, the simultaneous fixed-point $(U_{(\infty)}, V_{(\infty)}, W_{(\infty)})$ exists and $U_{(\infty)} = Y_\infty$. Hence φ, that is, $[\mathrm{PFP}_{\overline{y},Y}\,\psi]\,\tilde{u}$, is equivalent to $[\text{S-PFP}_{\overline{y},U,\overline{y}\,\overline{z},V,x,W}\,\chi_U, \chi_V, \chi_W]\,\tilde{u}$. Since χ_U, χ_V, χ_W are equivalent to a Σ_1, Σ_k, Σ_1-formula, respectively, a closer look at the corresponding formula χ_J in 8.3.5 shows that it is equivalent to a Σ_k-formula and that, for a new variable v, we have

$$\models_{\mathrm{fin}} \varphi \leftrightarrow \exists(\forall)v[\mathrm{PFP}_{\overline{z},Z}\,\chi_J]\,\tilde{u}\tilde{v},$$

and hence, $\models_{\mathrm{fin}} \varphi \leftrightarrow \exists(\forall)u[\mathrm{PFP}_{\overline{z},Z}\,\chi_J]\,\tilde{u}\tilde{u}$. □

Exercise 8.3.15 (A version of the game of life) Let $\mathcal{G} = (G, E^G, C^G)$ where (G, E^G) is a graph and $C^G \subseteq G$. Interpret C^G as the set of vertices hosting a live cell. Set $C_0 = \emptyset, C_1 = C^G$, and for $n > 1$, define C_n, the "n-th generation", by the following "reproduction rule": For $a \in G$, $a \in C_n$ iff (i) or (ii) holds, where

[10] Recall that a formula $\exists(\forall)x\chi(x)$ is equivalent to $\chi(t)$ for any term t.

(i) $a \in C_{n-1}$ and, in C_{n-1}, a has exactly two neighbours hosting a live cell (that is, there are exactly two $b \in C_{n-1}$ such that $E^G ab$);
(ii) in C_{n-1}, a has exactly three neighbours hosting a live cell.

Show that there is a sentence φ of FO(PFP) such that φ holds in a structure $\mathcal{G}$ as above iff for some $n \geq 1$, $C_n = \emptyset$ (the species dies out). Give an example of a structure $\mathcal{G}$ such that $(C_n)_{n\geq 0}$ has no fixed-point. □

Exercise 8.3.16 (a) Assume that M is finite and $F : \text{Pow}(M) \to \text{Pow}(M)$. Define

$$\begin{aligned} F_{\text{true}} &:= \{a \in M \mid \exists n_0 \forall m \geq n_0 : a \in F_m\}, \\ F_{\text{false}} &:= \{a \in M \mid \exists n_0 \forall m \geq n_0 : a \notin F_m\}, \end{aligned}$$

and $F_{\text{und}} := M \setminus (F_{\text{true}} \cup F_{\text{false}})$, the set of *undefined* points. Introduce the class of formulas of the logic FO(3-PFP) as that of FO(PFP) at the beginning of this section replacing, however, the last rule of that calculus by

$$\frac{\varphi}{[t\text{-PFP}_{\overline{x},X}\varphi]\,\overline{t}}, \quad \frac{\varphi}{[f\text{-PFP}_{\overline{x},X}\varphi]\,\overline{t}}, \quad \frac{\varphi}{[u\text{-PFP}_{\overline{x},X}\varphi]\,\overline{t}},$$

the meaning of these formulas being

$$\text{“}\overline{t} \in F^{\varphi}_{\text{true}}\text{”}, \quad \text{“}\overline{t} \in F^{\varphi}_{\text{false}}\text{”}, \text{ and “}\overline{t} \in F^{\varphi}_{\text{und}}\text{”},$$

respectively. Show that FO(3-PFP) $\equiv$ FO(PFP). Hint: Argue as in the proof of 8.3.8, using, e.g., for the truth set, an additional sequence $(V_n)_{n\geq 0}$ given by $V_0 = X_0, V_1 = X_1, \ldots, V_l = X_l, V_{l+1} = V_l \cap X_{l+1}, V_{l+2} = V_{l+1} \cap X_{l+2}, \ldots$.

(b) For the game associated with digraphs and $\varphi(x, X) := \exists y(\neg Xy \wedge Exy)$ the formulas $[t\text{-PFP}_{x,X}\varphi]\,x$, $[f\text{-PFP}_{x,X}\varphi]x$, and $[u\text{-PFP}_{x,X}\varphi]\,x$ express “x is won for I”, “x is won for II”, and “x is drawn”, respectively.

Notes 8.3.17 Partial fixed-point logic goes back to [2]. The presentation of this section owes much to Grohe [61].

8.4 Fixed-Point Logics and $L^{\omega}_{\infty\omega}$

In the first part of this section we show that the fixed-point logics introduced so far are contained in $L^{\omega}_{\infty\omega}$. Therefore, when exploring the expressive power of fixed-point logics, we can prove nonaxiomatizability results by using nonaxiomatizability results for $L^{\omega}_{\infty\omega}$. The second part is concerned with the question whether FO(LFP) $\equiv$ FO(PFP). We shall see that, when studying this problem, we can restrict ourselves to ordered structures. Thus, by the results of Chapter 7, the problem FO(LFP) $\equiv$ FO(PFP) is equivalent to the question whether the complexity classes PTIME and PSPACE coincide.

Throughout this section, for simplicity, we only consider relational vocabularies. The following lemma shows a way how to define the stages of a fixed-point construction in first-order logic.

Proposition 8.4.1 *Let $\varphi(\overline{x}, X)$ be a first-order formula where all variables are among $v_1, \ldots, v_k$ and X. Suppose X is s-ary and $\overline{x} = x_1 \ldots x_s$ (with $x_1, \ldots, x_s$ among $v_1, \ldots, v_k$). Then, for every n, there is a formula $\varphi^n(\overline{x})$ in FO^{k+s} defining the stage F^{φ}_n.*

Proof. Let $y_1 = v_{k+1}, \ldots, y_s = v_{k+s}$. Then $\varphi^n(\overline{x})$ can be defined inductively by

$$\begin{aligned}\varphi^0(\overline{x}) &:= \neg x_1 = x_1,\\ \varphi^{n+1}(\overline{x}) &:= \varphi(\overline{x}, X)\frac{\exists \overline{y}(\overline{y} =\_ \wedge \exists \overline{x}(\overline{x} = \overline{y} \wedge \varphi^n(\overline{x})))}{X\_}\end{aligned}$$

(i.e., we replace in $\varphi(\overline{x}, X)$ each occurrence of an atomic subformula of the form $X\overline{t}$ by $\exists \overline{y}(\overline{y} = \overline{t} \wedge \exists \overline{x}(\overline{x} = \overline{y} \wedge \varphi^n(\overline{x})))$; note that some variables of $\overline{x}$ may occur in $\overline{t}$). □

As a corollary we get:

Theorem 8.4.2 $\mathrm{FO(PFP)} \leq L^{\omega}_{\infty\omega}$.

Proof. By 8.3.12, it suffices to show for first-order φ that $[\mathrm{PFP}_{\overline{x},X}\,\varphi]\overline{t}$ is equivalent to a formula of $L^{\omega}_{\infty\omega}$.

So suppose that φ, k, s, X, $\overline{x}$, $\overline{y}$ are as in the preceding proposition and that the variables in $\overline{t}$ are among $v_1, \ldots, v_k$. Then $[\mathrm{PFP}_{\overline{x},X}\,\varphi]\overline{t}$ is equivalent to the $L^{k+s}_{\infty\omega}$-formula

$$\bigvee_{n \geq 0} (\forall \overline{x}(\varphi^n(\overline{x}) \leftrightarrow \varphi^{n+1}(\overline{x})) \wedge \varphi^n(\overline{t}))$$

where, to stay within $L^{k+s}_{\infty\omega}$, we take $\exists \overline{y}(\overline{y} = \overline{t} \wedge \exists \overline{x}(\overline{x} = \overline{y} \wedge \varphi^n(\overline{x})))$ for $\varphi^n(\overline{t})$. □

Exercise 8.4.3 Show that $\forall x \forall y[\mathrm{LFP}_{y,Y}(y = x \vee \exists x(Yx \wedge Exy))]y$ is not equivalent to a formula of $L^2_{\infty\omega}$. (Hint: Use part (b) of 3.3.6). □

The preceding theorem allows us to translate some results from $L^{\omega}_{\infty\omega}$ to FO(PFP). We give two examples. By 3.3.25(b) we get

Proposition 8.4.4 *For a relational vocabulary containing only unary relation symbols the expressive power of* FO(PFP) *and* FO *coincide; in particular, the class* EVEN *of sets of even cardinality is not axiomatizable in* FO(PFP). □

The last statement of the proposition shows that neither FO(LFP) strongly captures PTIME nor FO(PFP) strongly captures PSPACE (cf. page 157). From the 0–1 laws for $L^{\omega}_{\infty\omega}$ obtained in Chapter 4 we get for example:

Theorem 8.4.5 FO(PFP) *satisfies the labeled and the unlabeled* 0–1 *law.* □

Now we turn to the question whether FO(LFP) $\equiv$ FO(PFP). We first recall some definitions: In subsection 3.3.2, to every τ-structure $\mathcal{A}$ and to every $s \geq 1$ we have assigned a τ/s-structure $\mathcal{A}/s$ which captures the $\mathrm{L}^s_{\infty\omega}$-theory of $\mathcal{A}$. Recall that A/s consists of the classes of the equivalence relation $\sim$ on A^s given by

$$\overline{a} \sim \overline{b} \quad \text{iff} \quad \overline{a} \text{ and } \overline{b} \text{ satisfy the same } \mathrm{L}^s_{\infty\omega}\text{-formulas in } \mathcal{A}$$

and recall that the vocabulary τ/s and the structures $\mathcal{A}/s$ are given as follows:

– for every k-ary $R \in \tau \cup \{=\}$ and any $i_1, \ldots, i_k$ with $1 \leq i_1, \ldots, i_k \leq s$ the vocabulary τ/s contains a unary relation symbol $R_{i_1 \ldots i_k}$ with interpretation

$$R^{A/s}_{i_1 \ldots i_k} \quad := \quad \{[\overline{a}] \mid \overline{a} \in A^s, R^{\mathcal{A}} a_{i_1} \ldots a_{i_k}\};$$

– for $i = 1, \ldots, s$ the vocabulary τ/s contains a binary relation symbol S_i with

$$S^{A/s}_i \quad := \quad \{\, ([\overline{a}], [\overline{a}']) \;\mid\; \overline{a}, \overline{a}' \in \; A^s, \text{there is } a \in A \text{ such that } [\overline{a}'] = [\overline{a}\tfrac{a}{i}]\}.$$

Since FO(PFP) $\leq \mathrm{L}^\omega_{\infty\omega}$, for any $\varphi \in$ FO(PFP)$[\tau]$ we can choose an s such that $\varphi = \varphi(v_1, \ldots, v_s)$ is equivalent to an $\mathrm{L}^s_{\infty\omega}$-formula, thus

$$\mathcal{A} \models \varphi[\overline{a}] \text{ and } \overline{a} \sim \overline{b} \text{ imply } \mathcal{A} \models \varphi[\overline{b}].$$

Therefore, any FO(PFP) definable relation on $\mathcal{A}$ is the union of some equivalence classes. This will enable us to show that the FO(LFP) (and the FO(PFP)) definable relations on $\mathcal{A}$ and on the $\mathcal{A}/s$ correspond to each other. Since $\mathcal{A}/s$ has an FO(LFP) definable ordering, we thus shall be able to translate the question whether FO(LFP) $\equiv$ FO(PFP) from arbitrary structures to ordered ones.

Fix s. We start by showing that $\sim$ is definable in FO(LFP) (cf. 8.1.8). Let $\overline{a}$ and $\overline{b}$ range over A^s. For $j \geq 0$ define $\sim_j$ on A^s by induction:

$$\overline{a} \sim_0 \overline{b} \quad \text{iff} \quad \overline{a} \text{ and } \overline{b} \text{ satisfy the same atomic formulas in } \mathcal{A}$$

$$\overline{a} \sim_{j+1} \overline{b} \quad \text{iff} \quad \overline{a} \sim_0 \overline{b} \text{ and for all } i = 1, \ldots, s \text{ and all } a \in A \ (b \in A) \text{ there is } b \in A \ (a \in A) \text{ such that } \overline{a}\tfrac{a}{i} \sim_j \overline{b}\tfrac{b}{i}$$

(in the terminology of subsection 3.3.1 we have $\overline{a} \sim_j \overline{b}$ iff the duplicator has a winning strategy in the pebble game $\mathrm{G}^s_j(\mathcal{A}, \overline{a}, \mathcal{A}, \overline{b})$ with s pebbles and j moves). Clearly, $\sim_0 \supseteq \sim_1 \supseteq \ldots$, so that $\sim_l = \sim_{l+1}$ for some l.[11] For such an l we have $\sim_l = \sim$. For the complements $\not\sim_j$ of $\sim_j$ we have $\not\sim_0 \subseteq \not\sim_1 \subseteq \ldots$.

[11] In subsection 3.3.3 the least such l was called the s-rank $r(\mathcal{A})$.

They are the stages $F_1^{\varphi_{\not\sim}}, F_2^{\varphi_{\not\sim}}, \ldots$, where $\varphi_{\not\sim}(x_1,\ldots,x_s,y_1,\ldots,y_s,Z)$ is the following formula positive in Z:

$$\bigvee_{\substack{\psi\in\mathrm{FO}^s\\ \psi \text{ atomic}}} (\psi(\overline{x}) \leftrightarrow \neg\psi(\overline{y})) \vee \bigvee_{1\le i\le s} (\exists x_i \forall y_i Z\overline{x}\,\overline{y} \vee \exists y_i \forall x_i Z\overline{x}\,\overline{y}).$$

Altogether we get

Lemma 8.4.6 *There is a τ-formula $\varphi_\sim(\overline{x},\overline{y})$ in* FO(LFP) *with variables among $x_1,\ldots,x_s,y_1,\ldots,y_s$ such that for all τ-structures $\mathcal{A}$ and $\overline{a}, \overline{b} \in A^s$,*

$$\overline{a} \sim \overline{b} \qquad \textit{iff} \qquad \mathcal{A} \models \varphi_\sim[\overline{a},\overline{b}].$$

Proof. Take as $\varphi_\sim(\overline{x},\overline{y})$ the formula $\neg[\mathrm{LFP}_{\overline{x}\,\overline{y},Z}\,\varphi_{\not\sim}(\overline{x},\overline{y},Z)]\,\overline{x}\,\overline{y}$. □

By this lemma there is an FO(LFP) definition of $\mathcal{A}/s$ in $\mathcal{A}$ and hence, any FO(LFP) statement on $\mathcal{A}/s$ can be translated to an FO(LFP) statement on $\mathcal{A}$.

Proposition 8.4.7 *For any $\varphi(x_1,\ldots,x_k)$ in* FO(LFP)$[\tau/s]$ *whose free variables are among $x_1,\ldots,x_k$ there is a formula $\tilde{\varphi}(x_{11},\ldots,x_{1s},\ldots,x_{k1},\ldots,x_{ks})$ in* FO(LFP)$[\tau]$ *such that for all τ-structures $\mathcal{A}$ and $\overline{a}_1,\ldots,\overline{a}_k$ in A^s,*

$$\mathcal{A}/s \models \varphi[[\overline{a}_1],\ldots,[\overline{a}_k]] \qquad \textit{iff} \qquad \mathcal{A} \models \tilde{\varphi}[\overline{a}_1,\ldots,\overline{a}_k].$$

The same is true if FO(LFP) *is replaced by* FO(PFP) *everywhere.*

Proof. For a variable y let $\overline{y}$ be $y_1\ldots y_s$. The definition of $\tilde{\varphi}$ is by induction on φ. We give the main steps:

φ	$\tilde{\varphi}$
$y = z$	$\varphi_\sim(\overline{y},\overline{z})$
$R_{i_1\ldots i_l}y$	$Ry_{i_1}\ldots y_{i_l}$
$S_i yz$	$\exists y_i \varphi_\sim(\overline{y},\overline{z})$
$Xy_1\ldots y_r$	$X'\overline{y}_1\ldots\overline{y}_r$
$\exists y\psi$	$\exists y_1\ldots\exists y_s\tilde{\psi}$
$[\mathrm{LFP}_{y_1\ldots y_r,X}\,\psi]\,z_1\ldots z_r$	$[\mathrm{LFP}_{\overline{y}_1\ldots\overline{y}_r,X'}\,\tilde{\psi}]\,\overline{z}_1\ldots\overline{z}_r$

where X' has s-times the arity of X. □

We come to a translation in the opposite direction:

Proposition 8.4.8 *For any* FO(LFP)$[\tau]$*-formula φ without free second-order variables there is an $s \ge 1$ and an* FO(LFP)$[\tau/s]$*-formula $\varphi^*(v)$ with at most the free variable v such that the individual variables in φ are among $v_1,\ldots,v_s$ and such that for all τ-structures $\mathcal{A}$ and $\overline{a} \in A^s$,*

$$\mathcal{A} \models \varphi[a_1,\ldots,a_s] \qquad \textit{iff} \qquad \mathcal{A}/s \models \varphi^*[[\overline{a}]].$$

If φ is a sentence then φ^* can be chosen to be a sentence.

The same is true for FO(PFP) *instead of* FO(LFP).

Proof. By Theorem 8.1.14, we may assume that the LFP operator occurs only once in φ, w.l.o.g. in the form $[\mathrm{LFP}_{v_1\ldots v_r,X}\,\psi]\,\overline{t}$. Moreover, we may assume that the occurrences of X have the form $Xv_1\ldots v_r$ (if $Xz_1\ldots z_r$ occurs, replace it by $\exists\overline{y}(\overline{y}=\overline{z}\wedge\exists\overline{v}(\overline{v}=\overline{y}\wedge X\overline{v}))$ where $\overline{y}$ are new variables and $\overline{v}=v_1\ldots v_r$). Treating $\overline{t}$ similarly, we may assume that the LFP operator in φ has the form $[\mathrm{LFP}_{v_1\ldots v_r,X}\,\psi]\,v_1\ldots v_r$. Let the variables in φ be among $v_1,\ldots,v_s$. Finally, we may assume that $r=s$; otherwise, we replace X by an s-ary variable Y and $[\mathrm{LFP}_{v_1\ldots v_r,X}\,\psi(X)]\,v_1\ldots v_r$ by $[\mathrm{LFP}_{v_1\ldots v_s,Y}\,\psi(Y\_v_{r+1}\ldots v_s)]\,v_1\ldots v_s$.

The definition of φ^* is by induction on φ, the main steps being (we use a unary relation variable V):

φ	φ^*
$v_i=v_j$	$=_{ij}\,v$
$Rv_{i_1}\ldots v_{i_k}$	$R_{i_1\ldots i_k}v$
$Xv_1\ldots v_s$	Vv
$\exists v_i\psi$	$\exists w(S_ivw\wedge\psi^*(w))$
$[\mathrm{LFP}_{v_1\ldots v_s,X}\psi]v_1\ldots v_s$	$[\mathrm{LFP}_{v,V}\psi^*]v.$

For a sentence φ pass to the sentence $\exists v\varphi^*$. □

The next proposition shows that from the point of view of FO(LFP) the structures $\mathcal{A}/s$ can be regarded as ordered structures.

Proposition 8.4.9 *There is an* FO(LFP)$[\tau/s]$*-formula* $\psi_<(u,v)$ *such that for all* τ*-structures* $\mathcal{A}$,

$$\psi_<^{\mathcal{A}/s} \;:=\; \{([\overline{a}],[\overline{b}])\mid \mathcal{A}/s\models\psi_<[[\overline{a}],[\overline{b}]]\}$$

is an ordering on $\mathcal{A}/s$.

Proof. Let $\mathcal{A}$ be a τ-structure and let $\sim_j$ be the equivalence relations introduced before 8.4.6. Assume $\sim_l=\sim$. For $j\geq 0$ we define partial orderings (that means, irreflexive and transitive relations) $<_j$ on $\mathcal{A}/s$ such that

(1) $$<_0\subseteq<_1\subseteq<_2\ldots$$

(2) $$([\overline{a}]<_j[\overline{b}]\text{ or }[\overline{b}]<_j[\overline{a}])\quad\text{iff}\quad\overline{a}\not\sim_j\overline{b}$$

(thus, $<_j$ induces an ordering on the equivalence classes of $\sim_j$). Then $<_l$ will be the intended ordering.

Let $\chi_1,\ldots,\chi_m$ be an enumeration – say in lexicographic order – of the atomic τ-formulas in the variables $v_1,\ldots,v_s$ and let χ_i^* be the τ/s-formula associated to χ_i in Proposition 8.4.8. As $<_0$ we take $<_0:=\psi_0^{\mathcal{A}/s}(\_,\_)$ where

$$\psi_0(x,y)\;:=\;\bigvee_{1\leq i\leq m}(\chi_i^*(x)\wedge\neg\chi_i^*(y)\wedge\bigwedge_{1\leq j<i}(\chi_j^*(x)\leftrightarrow\chi_j^*(y))).$$

Clearly, $<_0$ is a partial ordering that satisfies (2) for $j = 0$.

Now suppose that $<_j$ has already been defined. To extend $<_j$ to $<_{j+1}$, assume that $\overline{a} \sim_j \overline{b}$ and $\overline{a} \not\sim_{j+1} \overline{b}$. For $1 \leq i \leq s$ set $M(\overline{a}, i) := \{[\overline{a}\frac{a}{i}]_j \mid a \in A\}$ and $M(\overline{b}, i) := \{[\overline{b}\frac{b}{i}]_j \mid b \in A\}$, where $[\overline{c}]_j$ denotes the equivalence class of $\overline{c}$ modulo $\sim_j$. Since $\overline{a} \not\sim_{j+1} \overline{b}$, there is a smallest number i such that $M(\overline{a}, i) \neq M(\overline{b}, i)$. Set $[\overline{a}] <_{j+1} [\overline{b}]$ if the $<_j$-minimal element of $(M(\overline{a}, i) \setminus M(\overline{b}, i)) \cup (M(\overline{b}, i) \setminus M(\overline{a}, i))$ belongs to $M(\overline{a}, i)$; otherwise, set $[\overline{b}] <_{j+1} [\overline{a}]$. One easily verifies that $<_{j+1}$ is a partial ordering (extending $<_j$) and that (2) holds for $j + 1$.

Clearly, $< := <_l$ is a total ordering on $\mathcal{A}/s$. It is the fixed-point of the increasing relations $<_j$ and is thus definable in $\mathcal{A}/s$ by $[\text{IFP}_{uv,X}\, \chi]\, uv$ where, with $x \equiv y$ for $\neg Xxy \wedge \neg Xyx$,

$$\begin{aligned} \chi(u, v, X) \;\; := \;\; & \psi_0(u, v) \vee \bigvee_{1 \leq i \leq s} (\exists u'(S_i uu' \wedge \forall v'(S_i vv' \rightarrow \\ & (Xu'v' \vee (Xv'u' \wedge \exists u''(S_i uu'' \wedge u'' \equiv v'))))) \\ & \wedge \bigwedge_{1 \leq k < i} (\forall u'(S_k uu' \rightarrow \exists v'(S_k vv' \wedge u' \equiv v')) \\ & \wedge \forall v'(S_k vv' \rightarrow \exists u'(S_k uu' \wedge u' \equiv v')))). \end{aligned}$$

□

We are now able to show the main result of this section.

Theorem 8.4.10 *The following are equivalent:*
(i) FO(PFP) $\equiv$ FO(LFP).
(ii) FO(PFP) $\equiv$ FO(LFP) *on ordered structures.*

Proof. The direction from (i) to (ii) is clear. Assume FO(PFP) $\equiv$ FO(LFP) on ordered structures. Let τ be a vocabulary and suppose φ is an FO(PFP)$[\tau]$-sentence. We must show that there is an FO(LFP)$[\tau]$-sentence with the same finite models. Choose $s \geq 1$ and $\varphi^* \in$ FO(PFP)$[\tau/s]$ according to 8.4.8. By (ii) there is an FO(LFP)$[\tau/s\dot{\cup}\{<\}]$-sentence ψ such that

$$\text{“} < \text{ is an ordering”} \models \varphi^* \leftrightarrow \psi.$$

Let $\chi := \psi\frac{\psi_<(\ldots,\_)}{\ldots <\_}$, i.e., χ is obtained from ψ by replacing any atomic part of the form $x < y$ by $\psi_<(x, y)$, where $\psi_<(u, v)$ is the formula of the preceding proposition. Hence, $\chi \in$ FO(LFP)$[\tau/s]$. Choose $\tilde{\chi} \in$ FO(LFP)$[\tau]$ according to 8.4.7. Then, for any τ-structure $\mathcal{A}$ we have

$$\begin{array}{lll} \mathcal{A} \models \varphi & \text{iff} & \mathcal{A}/s \models \varphi^* \\ & \text{iff} & (\mathcal{A}/s, \psi_<^{\mathcal{A}/s}) \models \psi \\ & \text{iff} & \mathcal{A}/s \models \chi \\ & \text{iff} & \mathcal{A} \models \tilde{\chi}. \end{array}$$

□

With 7.5.7 we now get

Corollary 8.4.11 FO(LFP) $\equiv$ FO(PFP) *iff* PTIME = PSPACE. □

The ordered structure $(\mathcal{A}/s, <)$ is isomorphic to exactly one structure of the form $(\{0, \ldots, n\}, \ldots, <)$, the *numerical s-invariant* of $\mathcal{A}$. By the results above, a query on the class of τ-structures is definable in FO(LFP) iff for all sufficiently large s, the corresponding query on the numerical s-invariants is definable in FO(LFP). Therefore, by 7.5.2, a query on the class of τ-structures is definable in FO(LFP) iff the corresponding query on the numerical invariants is in PTIME. That is, on arbitrary finite structures exactly those queries are definable in FO(LFP) which correspond to PTIME queries on the numerical invariants. The same applies to FO(PFP) and PSPACE.

Exercise 8.4.12 A τ-structure $\mathcal{A}$ is *s-rigid* if $\tilde{a} \not\sim \tilde{b}$ for all $a, b \in A$, $a \neq b$ (where $\sim$ is the equivalence relation leading to $\mathcal{A}/s$ and $\tilde{a}, \tilde{b}$ are the constant sequences of length s). A class K of structures is *s-rigid*, if every structure in K is s-rigid. Show:

(a) $\mathcal{A}$ is rigid iff $\mathcal{A}$ is s-rigid for some s.
(b) The following are equivalent:
 (i) K is s-rigid for some s.
 (ii) There is a formula in FO(LFP) defining an ordering on K (i.e., on every structure in K).
 (iii) There is a formula in $L^{\omega}_{\infty\omega}$ defining an ordering on K.

Hint: (i) $\Rightarrow$ (ii): If the τ-structure $\mathcal{A}$ is s-rigid, then $\chi^{\mathcal{A}}(\_,\_)$ with $\chi(x, y) = \tilde{\psi}_<(x, \ldots, x, y, \ldots, y)$ is an ordering, where $\tilde{\psi}_<(\overline{u}, \overline{v})$ is the formula associated with the τ/s-formula $\psi_<(u, v)$ of 8.4.9 according to 8.4.7. – (iii) $\Rightarrow$ (i): If $\varphi(x, y) \in L^{\omega}_{\infty\omega}$ defines an ordering, use formulas as defined in part (a) of 3.3.1, here built up from $\varphi(x, y)$ instead of $x < y$, to show that K is s-rigid for some s.

Conclude: If K is an s-rigid class of structures, then the PTIME-computable queries on K are exactly the FO(LFP) definable ones.

In contrast to (a) there is a first-order axiomatizable class of rigid structures that is not s-rigid for any s (see [77]). □

Exercise 8.4.13 Let K be a class of structures. We say that K is *fixed-point bounded*, if for any first-order formula $\varphi(\overline{x}, X)$ positive in X with free variables among $\overline{x}, X$, there is an m_0 such that

$$K \models \forall \overline{x}(\varphi^{m_0+1}(\overline{x}) \rightarrow \varphi^{m_0}(\overline{x})),$$

where $\varphi^m(\overline{x})$ is a first-order formula defining F^{φ}_m (see 8.4.1). Show the equivalence of (i) – (iii):

(i) K is fixed-point bounded.
(ii) K is bounded (in the sense of Theorem 3.3.24).
(iii) On K, every $L^{\omega}_{\infty\omega}$-formula is equivalent to an FO-formula.

Hint: For (i) $\Rightarrow$ (ii) apply the hypothesis to the formula $\varphi_{\not\prec}(\overline{x},\overline{y},X)$ introduced before 8.4.6. The equivalence of (ii) and (iii) was shown in 3.3.24. For the direction from (iii) to (i) suppose that $\varphi(\overline{x},X)$ is as above and witnesses that there is no m_0 as claimed. For $M \subseteq \mathbb{N}$ set $\varphi_M := \bigvee_{m\in M}(\varphi^{m+1}(\overline{x}) \wedge \neg\varphi^m(\overline{x}))$. Then, φ_M and φ_L are not equivalent on K for $L \neq M$.

In (iii), one cannot replace $L^{\omega}_{\infty\omega}$ by FO(LFP) (see [74]). □

8.4.1 The Logic FO(PFP$_{\text{PTIME}}$)

While least fixed-point operations reach their fixed-point in polynomially many steps, partial fixed-point operations may need exponentially many steps. The intermediate logic FO(PFP$_{\text{PTIME}}$) is the fragment of FO(PFP) consisting of those formulas for which every occurrence of the PFP operator reaches a fixed-point in polynomially many steps in all structures. Clearly,

$$\text{FO(LFP)} \leq \text{FO(PFP}_{\text{PTIME}}).$$

Exercise 8.4.16 shows that FO(LFP) $\equiv$ FO(PFP$_{\text{PTIME}}$) for arbitrary structures is quite unlikely. On the other hand we have

Theorem 8.4.14 *On ordered structures,* FO(LFP) $\equiv$ FO(PFP$_{\text{PTIME}}$).

Proof. The claim is clear by the results of Chapter 7: By definition, every query definable in FO(PFP$_{\text{PTIME}}$) can be evaluated in polynomial time and is thus definable in FO(LFP). Nevertheless, for later purposes we sketch a direct proof. Let $[\text{PFP}_{\overline{x},X}\varphi(\overline{x},X)]\,\overline{t}$ be given and suppose that, for some polynomial p, the sequence $(X_l)_{l\geq 0}$ (with $X_l := F^{\varphi}_l$) gets constant after at most p(cardinality of the universe) many steps. The idea of how to express this fixed-point in FO(LFP) is simple: We code the stages $(X_l)_{l\leq p(\dots)}$ by a relation variable Y with fixed-point Y_∞ where

$$Y_\infty m\overline{x} \qquad \text{iff} \qquad m \leq p(\dots) \wedge X_m\overline{x}.$$

The stages of $Y_0, Y_1, Y_2, \dots$ will form the increasing sequence

$$\begin{aligned}
&\{(0,\overline{x}) \mid \overline{x} \in X_0\} \quad (= \emptyset),\\
&\{(0,\overline{x}) \mid \overline{x} \in X_0\} \cup \{(1,\overline{x}) \mid \overline{x} \in X_1\},\\
&\{(0,\overline{x}) \mid \overline{x} \in X_0\} \cup \{(1,\overline{x}) \mid \overline{x} \in X_1\} \cup \{(2,\overline{x}) \mid \overline{x} \in X_2\}, \dots .
\end{aligned}$$

The arithmetics needed for this encoding is definable in FO(LFP). In fact, by 8.1.11 there are $k \geq 1$ and formulas $S^k\overline{u}\,\overline{v}$ and $\overline{v} = p(x+1)$ in FO(LFP) expressing (in the corresponding number representation) that $\overline{v}$ is the successor of $\overline{u}$ and that $\overline{v}$ equals $p(x+1)$, respectively. Therefore, as a formula equivalent to $[\text{PFP}_{\overline{x},X}\varphi(\overline{x},X)]\,\overline{t}$ we can take

$$\begin{aligned}
\exists\overline{w}(\overline{w} = p(\max +1) \wedge [\text{IFP}_{\overline{u}\,\overline{x},Y}((S^k\widetilde{\min}\,\overline{u} \wedge \varphi(\overline{x},\emptyset))\\
\vee\exists\overline{v}\,\overline{y}\,(Y\overline{v}\,\overline{y} \wedge \overline{v} < \overline{w} \wedge S^k\overline{v}\,\overline{u} \wedge \varphi(\overline{x},Y\overline{v}_)))\,]\,\overline{w}\overline{t}).
\end{aligned}$$

□

Exercise 8.4.15 (FO(LFP) $<$ $L^{\omega}_{\infty\omega} \cap$ PTIME) Let $\tau = \{E, U\}$ and let $\tau' = \{<, U\}$ with unary U and binary $<, E$. Denote by STG the class of (binary) "strings", i.e., the class of *ordered* τ'-structures. A finite τ-structure $\mathcal{T} = (T, E^T, U^T)$ is a *labeled complete binary tree* if

- (T, E^T) is a tree, where every element has out-degree 0 or 2 and all leaves have the same depth (cf. 8.1.7).
- For any $a, b \in T$ of the same depth we have $U^A a$ iff $U^A b$.

Denote by LCT the class of labeled complete binary trees. Show:

- LCT is axiomatizable in FO(LFP) (use 8.1.7).
- If $\mathcal{T}_1, \mathcal{T}_2 \in$ LCT and $\mathcal{T}_1 \equiv^{L^2_{\infty\omega}} \mathcal{T}_2$ then $\mathcal{T}_1 \cong \mathcal{T}_2$.
- Any class K with $K \subseteq$ LCT is axiomatizable in $L^{\omega}_{\infty\omega}$.

There is a natural correspondence between strings and labeled complete binary trees obtained by encoding the same U-information with the i-th bit of the string as with the i-th level of the tree. More formally, if $\varphi_i(x) \in \mathrm{FO}[\tau']$ expresses that x is the i-th element of $<$ and $\psi_i(x) \in \mathrm{FO}[\tau]$ expresses that x has depth i, then for $\mathcal{A} \in$ STG and $\mathcal{T} \in$ LCT set $\mathcal{A} \approx \mathcal{T}$ iff for all $i \geq 0$,

- $\mathcal{A} \models \exists x \varphi_i(x)$ iff $\mathcal{T} \models \exists x \psi_i(x)$
- $\mathcal{A} \models \forall x(\varphi_i(x) \to Ux)$ iff $\mathcal{T} \models \forall x(\psi_i(x) \to Ux)$.

Show

- If $H \subseteq$ LCT is axiomatizable in FO(LFP) then $\mathrm{STG}(H) := \{\mathcal{A} \mid \mathcal{A} \approx \mathcal{T}$ for some $\mathcal{T} \in H\}$ is axiomatizable in FO(LFP). Hint: Show for $\mathcal{T} \in$ LCT and $\bar{a}, \bar{b} \in T^s$ that

$$(\mathcal{T}, \bar{a}) \cong (\mathcal{T}, \bar{b}) \qquad \text{iff} \qquad (d(a_i \wedge a_j))_{1 \leq i \leq j \leq s} = (d(b_i \wedge b_j))_{1 \leq i \leq j \leq s},$$

  where $d(c)$ is the depth of c and where $a \wedge b$ denotes the element c of least depth with $c \leq a$ and $c \leq b$ in the induced partial ordering (cf. 8.1.7). Thus s-tuples of elements of the tree can be encoded by $\frac{s(s+1)}{2}$-tuples in the corresponding string. The set of the latter tuples being definable in FO, this can be used to translate the FO(LFP)-axiomatization of H into an FO(LFP)-axiomatization of STG(H).
- If $K \subseteq$ STG is in TIME(2^n) (that is, K is accepted by a 2^n time-bounded machine), then $\mathrm{LCT}(K)_<$ is in PTIME, where $\mathrm{LCT}(K) := \{\mathcal{T} \mid \mathcal{A} \approx \mathcal{T}$ for some $\mathcal{A} \in K\}$.

It is a well-known result of complexity theory that there are classes $K \subseteq$ STG that are in TIME(2^n) but not in PTIME. Conclude for such a K that LCT(K) is in PTIME and is $L^{\omega}_{\infty\omega}$-axiomatizable, but not FO(LFP)-axiomatizable. □

Exercise 8.4.16 (FO($\mathrm{PFP_{PTIME}}$) $\equiv$ FO(LFP) $\Rightarrow$ PSPACE = PTIME) Let $\tau_1 = \{E, U\}$ and $\tau_2 = \{E, U, V\}$, where U and V are unary and E is binary. Fix $k \geq 1$. Recall the definition of strings (= structures in STG) and of labeled complete binary trees of the preceding exercise. For the present purpose we encode strings of length n in a suitable complete binary tree of depth n^k. The class TREE_k consists of τ_2-structures $\mathcal{T} = (T, E^T, U^T, V^T)$ that satisfy

(T, E^T, U^T) and (T, E^T, V^T) are labeled complete binary trees;
$\forall x \forall y((Exy \wedge Vy) \rightarrow Vx)$
$\forall x(Ux \rightarrow Vx)$
"The depth of $\mathcal{T}$ is m^k, where m is the number of levels labeled by V".

TREE_k is axiomatizable in FO(LFP); in particular, the last condition can be expressed in FO(LFP) by developing the arithmetic of Example 8.1.11 for complete binary trees instead of orderings, the j-th level taking over the role of the j-th element of the ordering.

Consider the relation $\asymp$ between strings and structures in TREE_k defined by using $\approx$ of the preceding exercise as follows:

$$\mathcal{A} \asymp \mathcal{T} \quad \text{iff} \quad \mathcal{A} \approx (V^T, E^T \cap (V^T \times V^T), U^T).$$

Show:

(a) If $K \subseteq \text{STG}$ is axiomatizable in FO(PFP), then for some k, $\{\mathcal{T} \in \text{TREE}_k \mid \mathcal{A} \asymp \mathcal{T}$ for some $\mathcal{A} \in K\}$ is axiomatizable in $\text{FO(PFP}_{\text{PTIME}})$.
(b) If $H \subseteq \text{TREE}_k$ is axiomatizable in FO(LFP), then $\{\mathcal{A} \in \text{STG} \mid \mathcal{A} \asymp \mathcal{T}$ for some $\mathcal{T} \in H\}$ is axiomatizable in FO(LFP).

Let $K \subseteq \text{STG}$ be in PSPACE and therefore, axiomatizable in FO(PFP). Using (a), choose k such that $\{\mathcal{T} \in \text{TREE}_k \mid \mathcal{A} \asymp \mathcal{T}$ for some $\mathcal{A} \in K\}$ is axiomatizable in $\text{FO(PFP}_{\text{PTIME}})$ and hence, in FO(LFP), if we assume $\text{FO(PFP}_{\text{PTIME}}) \equiv \text{FO(LFP)}$. Then, by (b), K is axiomatizable in FO(LFP) and therefore, K is in PTIME. Hence, PSPACE = PTIME by the remark following 7.5.7. □

8.4.2 Fixed-Point Logic with Counting

We have seen that FO(LFP) does not strongly capture PTIME and remarked that it is still open whether there exists a logic strongly capturing PTIME (cf. the remarks after 7.5.19). While pursuing this problem, various extensions of FO(LFP) have been studied as candidates for strongly capturing PTIME. As an example we present fixed-point logic with counting.

Recall that, when proving Theorem 8.4.14, we used the fact that, on ordered structures, FO(LFP) can describe the arithmetics needed to code polynomially many stages of a fixed-point operation in one relation. But arithmetics is not all orderings allows to do: Using orderings we can enumerate the elements of a structure and thus successively look at each single element. For instance, by enumeration we can calculate in FO(LFP) the sizes of definable subsets. In fact, let $\varphi(x)$ be an FO(LFP)-formula. Then for any ordered structure $\mathcal{A}$ (with $\mathcal{A} \not\models \forall x\varphi(x)$),

$$a \text{ is the } \|\varphi^{\mathcal{A}}(\_)\|\text{-th element of } <^{\mathcal{A}} \quad \text{iff} \quad \mathcal{A} \models [\text{LFP}_{xu,X}\chi] \max u[a],$$

where

$$\begin{aligned}\chi(x,u,X) := \ & (x = \min \wedge \neg\varphi(x) \wedge u = \min) \vee (x = \min \wedge \varphi(x) \wedge S \min u) \\ & \vee \exists x'u'(Xx'u' \wedge Sx'x \wedge ((\neg\varphi(x) \wedge u = u') \vee (\varphi(x) \wedge Su'u))).\end{aligned}$$

Fixed-point logic with counting, FO(IFP, #), is an extension of fixed-point logic that contains arithmetic (in the size of the structure) and allows to calculate cardinalities. If we only consider ordered structures, every query definable in fixed-point logic with counting is in PTIME; so in this case the logic FO(IFP, #) coincides with least fixed-point logic and hence, captures PTIME. On the other hand, FO(IFP, #) does not strongly capture PTIME. In fact, we shall see that FO(IFP, #) is (equivalent to) a fragment of $\mathrm{C}^{\omega}_{\infty\omega}$, infinitary logic with counting quantifiers and finitely many variables, and that there is a graph property in PTIME which is not even expressible in this logic (cf. section 12.2). Nevertheless, FO(IFP, #) has turned out to be quite an important logic, since it "captures PTIME on many concrete classes" (cf. Chapter 11). Since FO(IFP, #) contains arithmetics (in the size of the structure), those results for FO(LFP) and ordered structures where we used the ordering only for arithmetical operations or size calculations translate to fixed-point logic with counting and arbitrary finite structures (compare 8.4.19 below and 9.2.3 in the next chapter).

In *Fixed-Point Logic with Counting*, FO(IFP, #), any τ-structure $\mathcal{A}$ is thought of to be accompanied by its *number part* consisting of a disjoint virtual universe $\{0, 1, \ldots, \|A\|\}$ together with the natural ordering, the successor relation, and constants for the first and the last element, respectively. We assume that $\tau \cap \{<, S, \min, \max\} = \emptyset$. $\mathcal{A}$ is viewed as a two-sorted structure, the elements of $\mathcal{A}$ itself being of the first sort, the *point sort*, those of the virtual universe being of the second sort, the *number sort.* On the syntactic side we have variables $x, y, z, \ldots$ for the first sort (*point variables*) and variables $\mu, \nu, \ldots$ for the second sort (*number variables*). The formulas of FO(IFP, #) are obtained from atomic formulas by closing under the connectives, first-order quantification of both sorts, and inflationary fixed-points of mixed sorted relations; moreover, for any formula φ and variable x there is a term $\#_x\varphi$ of the second sort denoting the cardinality of the set of elements x satisfying φ. More precisely, terms of both sorts and formulas are given simultaneously by the following clauses:

- All first-order τ-terms (with variables $x, y, \ldots$) are terms of the first sort; all first-order terms of vocabulary $\{<, S, \min, \max\}$ (with variables $\mu, \nu, \ldots$) are terms of the second sort.
- All atomic τ-formulas and all atomic $\{<, S, \min, \max\}$-formulas (with terms of the corresponding sort) are formulas.

- If X is a second-order variable of mixed arity n_1, n_2 then $X\bar{t}\bar{\rho}$ is a formula, where $\bar{t}$ and $\bar{\rho}$ are sequences of terms of the first and second sort of lengths n_1 and n_2, respectively.
- If φ and ψ are formulas then so are $\neg\varphi$ and $(\varphi \vee \psi)$.
- If φ is a formula then $\exists x\varphi$ and $\exists \nu\varphi$ are formulas.
- If φ is a formula then $[\text{IFP}_{\bar{x}\,\bar{\mu},X}\varphi]\,\bar{t}\bar{\rho}$ is a formula (the lengths of $\bar{x}, \bar{\mu}, \bar{t}, \bar{\rho}$ are according to the mixed arity of X).
- If φ is a formula then $\#_x\varphi$ is a term of the second sort.

The variable x does not occur free in $\#_x\varphi$. The semantics should be clear: If $\varphi = \varphi(x, \bar{y})$ and $\bar{a} \in \mathcal{A}$, the value $\#_x\varphi(x, \bar{a})^{\mathcal{A}}$ is given by the number $\|\{a \in A \mid \mathcal{A} \models \varphi[a, \bar{a}]\}\|$.

Examples 8.4.17 (a) Let $\tau = \{P\}$ with unary P. Then $\#_x Px = \#_x \neg Px$ is a sentence of FO(IFP, #) not equivalent to any $L^{\omega}_{\infty\omega}$-sentence (cf. 3.3.25(b) and 2.3.12).

(b) The class EVEN of structures of even cardinality is axiomatizable in FO(IFP, #) by

$$\neg[\text{IFP}_{\mu,X}\,(\mu = \min \vee \exists\nu\,(X\nu \wedge \text{“}\mu = \nu + 2\text{”}))]\,\max\,.$$

(c) For equivalence relations E_1 and E_2 the following sentence of FO(IFP, #) expresses that, for all cardinalities l, E_1 and E_2 have the same number of equivalence classes of cardinality l:

$$\forall\mu(\#_x(\#_y E_1xy = \mu) = \#_x(\#_y E_2xy = \mu)).$$

(d) For graphs the following formula of FO(IFP, #) says that the distance between x and y is λ:

$$[\text{IFP}_{z\mu,U}\,(z = x \wedge \mu = 0) \vee (\neg\exists\nu\, Uz\nu \wedge \exists z'\exists\mu'(Uz'\mu' \wedge Ez'z \wedge S\mu'\mu))]\,y\lambda.$$

(e) Note that the negation symbol is superfluous in FO(IFP, #). In fact, if x is a variable not occurring in φ then

$$\models_{\text{fin}} \neg\varphi \leftrightarrow \#_x\varphi = \min\,. \qquad \Box$$

It should be clear how FO(PFP, #), *Partial Fixed-Point Logic with Counting*, is defined.

Proposition 8.4.18 FO(PFP, #) $\leq C^{\omega}_{\infty\omega}$.

Proof. Given a sentence φ of FO(PFP, #) we show that it is equivalent to a sentence of the form

$$\bigvee_{n\geq 1} (\exists^{=n} x\, x = x \wedge \varphi_n)$$

where φ_n is a sentence of first-order logic with counting quantifiers that captures the meaning of φ in models of cardinality n. The translation from

φ to φ_n, though easy in principle, is involved if one wants to give a clean definition (the reader is encouraged to translate the formulas in the preceding examples to $C^{\omega}_{\infty\omega}$). We give a translation in two steps: First, we pass from φ to a sentence φ'_n eliminating # and PFP, but using atomic formulas and first-order quantifications of the second sort (they will be eliminated in the second step). Let φ'_n result from φ by replacing any subformula $\#_x\varphi = \mu$ by

$$\bigvee_{i \leq n} (\exists^{=i} x\varphi \wedge \text{“}\mu \text{ is the i-th element of } <\text{”})$$

and by treating PFP operations as in the proof of 8.4.2, taking into account that we can replace the infinite disjunctions used there by finite ones, since we only consider structures of cardinality n. In the second step we pass to an FO(C)-sentence φ_n. More generally, for every formula $\psi(\overline{x}, \mu_1, \ldots, \mu_k)$, where atomic subformulas and quantifications of the second sort are still allowed, and for every $\overline{m} = m_1 \ldots m_k$ with $m_1, \ldots, m_k \leq n$ we define an FO(C)-formula $\psi_{\overline{m}}(\overline{x})$ such that, in structures of cardinality n, $\psi_{\overline{m}}(\overline{x})$ says the same as $\psi(\overline{x}, m_1, \ldots, m_k)$. We then set $\varphi_n := (\varphi'_n)_{\overline{m}}$ with $\overline{m} = \emptyset$ (note that φ'_n is a sentence). The main steps in an inductive definition of $\psi_{\overline{m}}(\overline{x})$ are

– If $\psi(\overline{x}, \overline{\mu}) = \mu_i < \mu_j$ then $\psi_{\overline{m}}(\overline{x}) := \begin{cases} \text{T} & \text{if } m_i < m_j \\ \text{F} & \text{otherwise} \end{cases}$

– If $\psi(\overline{x}, \overline{\mu}) = \exists\mu\chi(\overline{x}, \overline{\mu}, \mu)$ then $\psi_{\overline{m}}(\overline{x}) := \bigvee_{m \leq n} \chi_{\overline{m}m}(\overline{x})$. □

Since in the proof of 8.4.14 we only used arithmetical properties of the ordering, in essentially the same way we get:

Proposition 8.4.19 FO(IFP, #) $\equiv$ FO($\text{PFP}_{\text{PTIME}}$, #). □

Here, FO($\text{PFP}_{\text{PTIME}}$, #) denotes the fragment of FO(PFP, #) consisting of those formulas for which each PFP operation closes in polynomially many steps.

Notes 8.4.20 Theorem 8.4.5 goes back to [12]. The main result 8.4.10 is due to Abiteboul and Vianu [4]. The proof presented here is based on [30]. Fixed-point logic with counting originates in ideas of Immerman [92]; in the form described here it has been shaped by Grädel and Otto [57, 124]. Otto has extended Theorem 8.4.10 to fixed-point logic with counting; cf. [125].

8.5 Fixed-Point Logics and Second-Order Logic

The results of Chapter 7 give us some information about the relationship between fixed-point logics and second-order logic SO: Let $K = \text{Mod}(\varphi)$ for some sentence φ of FO(LFP). Then, $K_< \in$ PTIME $\subseteq$ NPTIME and hence, $K = \text{Mod}(\psi)$ for some Σ^1_1-sentence ψ (by 7.5.14). Therefore, we have

(1) $$\mathrm{FO(LFP)} \le \Sigma_1^1.$$

As the class EVEN of finite sets of even cardinality is not axiomatizable in FO(LFP) (see 8.4.4) but in Σ_1^1 (see the beginning of section 3.1), we have $\Sigma_1^1 \not\le \mathrm{FO(LFP)}$. On ordered structures the statement $\Sigma_1^1 \not\le \mathrm{FO(LFP)}$ is equivalent to NPTIME $\neq$ PTIME (cf. 7.5.7) and hence, is still open.

Since every class of ordered structures axiomatizable in SO is in PSPACE (cf. 7.4.3(b)), the Main Theorem (cf. 7.5.2) yields

(2) $$\mathrm{SO} \le \mathrm{FO(PFP)} \quad \text{on ordered structures.}$$

In this section we aim at purely modeltheoretic proofs of (1) and (2). The proof of (1) leads to an Ehrenfeucht-Fraïssé method for FO(LFP) that can also be used to get nonaxiomatizability results for FO(LFP). (However, in most cases it is easier to prove nonaxiomatizability even for $\mathrm{L}^{\omega}_{\infty\omega}$, as we did in the last section for EVEN.)

Theorem 8.5.1 *On ordered structures,* SO $\le$ FO(PFP) *and* SO(PFP) $\equiv$ FO(PFP).[12]

Proof. Let τ be a vocabulary with $\{<, S, \min, \max\} \subseteq \tau$. We restrict our attention to finite structures whose universe is $\{0, \dots, n\}$ for some n and whose $\{<, S, \min, \max\}$-reduct is the natural ordering on $\{0, \dots, n\}$. For every formula φ of SO (or, of SO(PFP)) we inductively define an equivalent FO(PFP)-formula, the only nontrivial case being second-order quantification. So let $\varphi = \forall Z \psi(Z)$ and, for simplicity, assume that Z is unary. Below we show the existence of a first-order formula $\chi(x, y, X)$ with binary X such that in any ordered structure $\mathcal{A}$ the fixed-point F_∞^χ exists and is nonempty, and the sequence of sets $(\{x \mid \exists y\, F_k^\chi xy\})_{k \ge 0}$ runs through all subsets of A. Then $\forall Z \psi(Z)$ is equivalent to

$$\exists x \exists y\, [\mathrm{PFP}_{xy,X}\, (\chi(x, y, X) \wedge \psi(\exists y X\_y))]\, xy.$$

To define χ, we code a sequence $(a_0, \dots, a_n)$ of elements of A by the set $X := \{(a_i, i) \mid i \le n\}$. It is easy to write down a formula $\chi(x, y, X)$ such that the stages $F_1^\chi, F_2^\chi, \dots, F_{(n+1)^{n+1}}^\chi$ are codes of the sequences

$$(0, \dots, 0),\ (0, \dots, 0, 1), \dots, (n, \dots, n),$$

respectively, and $F_\infty^\chi = F_{(n+1)^{n+1}}^\chi$. □

We now show that $\mathrm{FO(LFP)} \le \Sigma_1^1$. For the least fixed-point F_∞ of a monotone operation $F : \mathrm{Pow}(A^k) \to \mathrm{Pow}(A^k)$ we have (see part (b) of 8.1.2)

$$F_\infty = \bigcap \{X \subseteq A^k | F(X) \subseteq X\}.$$

[12] Cf. 7.5.19

Thus

$$\begin{array}{lll} \overline{a} \in F_\infty & \text{iff} & \text{for all } X \subseteq A^k, \text{ if } F(X) \subseteq X \text{ then } \overline{a} \in X \\ & \text{iff} & \text{for all } X \subseteq A^k, \text{ if } \overline{a} \notin X \text{ then there is some} \\ & & \overline{b} \text{ with } \overline{b} \notin X \text{ and } \overline{b} \in F(X). \end{array}$$

Hence, for an arbitrary formula φ of FO(LFP), we have

$$(*) \qquad \models_{\text{fin}} [\text{LFP}_{\overline{x},X}\, \varphi]\, \overline{t} \leftrightarrow \forall X(\neg X\overline{t} \rightarrow \exists \overline{x}(\neg X\overline{x} \wedge \varphi)).$$

Since, by 8.2.9, every formula of FO(LFP) is equivalent to a formula of the form $\forall v[\text{LFP}_{\overline{x},X}\, \varphi]\, \tilde{v}$ with first-order φ, $(*)$ shows that FO(LFP) $\leq \Pi^1_1$. Since FO(LFP) is closed under negation, we get

Theorem 8.5.2 FO(LFP) $\leq \Delta^1_1$. □

The equivalence $(*)$ can be used to obtain a characterization in the Ehrenfeucht-Fraïssé style of classes axiomatizable in FO(LFP). Essentially, the corresponding Ehrenfeucht-Fraïssé game is obtained by taking into account relation moves that correspond to the right side of $(*)$. Suppose that in a play the spoiler aims at showing

$$\mathcal{A} \models [\text{LFP}_{x,X}\, \varphi(x, X)]\, y\, [a] \;\text{ and }\; \mathcal{B} \not\models [\text{LFP}_{x,X}\, \varphi(x, X)]\, y\, [b]$$

(a and b being the interpretations of y), that is, by $(*)$,

$$(+) \qquad \begin{array}{rrcl} & \mathcal{A} & \models & \forall X(\neg Xy \rightarrow \exists x(\neg Xx \wedge \varphi))\, [a] \\ \text{and} & \mathcal{B} & \not\models & \forall X(\neg Xy \rightarrow \exists x(\neg Xx \wedge \varphi))\, [b]. \end{array}$$

For this purpose he starts by choosing a subset B_0 of B such that $b \notin B_0$ and $\mathcal{B} \not\models \exists x(\neg Xx \wedge \varphi(x, X))\, [B_0]$. The duplicator then selects a subset A_0 of A with $a \notin A_0$ claiming that it behaves like B_0. However, by $(+)$,

$$\mathcal{A} \models \exists x(\neg Xx \wedge \varphi(x, X))\, [A_0].$$

Thus the spoiler can choose an element $a' \notin A_0$ such that $\mathcal{A} \models \varphi[a', A_0]$, while for any $b' \notin B_0$ chosen by the duplicator we have $\mathcal{B} \not\models \varphi[b', B_0]$. Now the spoiler has a "simpler" task, namely to show that $\mathcal{A} \models \varphi[a', A_0]$ and $\mathcal{B} \not\models \varphi[b', B_0]$.

The order and type of choices of the spoiler and the duplicator in this example will be the defining clauses of a so-called LFP move in the game we are going to introduce. To simplify the notation and the presentation we consider only relational vocabularies and *monadic least fixed-point logic* FO(M-LFP), which is obtained by restricting FO(LFP) to formulas in which only unary relation variables (set variables) are allowed. By $(*)$, we have FO(M-LFP) $\leq$ MSO. Furthermore, we remark that FO(M-LFP) like monadic second-order logic MSO has turned out to be an interesting logic in the realm of infinite structures (cf. [14, 41]).

We write $\mathcal{A} \equiv_m^{\mathrm{M}} \mathcal{B}$ iff $\mathcal{A}$ and $\mathcal{B}$ satisfy the same sentences of FO(M-LFP) of quantifier rank $\leq m$, where the quantifier rank is defined inductively as for first-order logic, adding the clause

$$\mathrm{qr}([\mathrm{LFP}_{x,X}\,\varphi]\,y) \quad := \quad 1 + \mathrm{qr}(\varphi).$$

In order to determine the winner of a play we need the notion of a generalized partial isomorphism. It takes into consideration that in $[\mathrm{LFP}_{x,X}\,\varphi]\,y$ the formula φ is positive in X.

Definition 8.5.3 Let $\mathcal{A}$ and $\mathcal{B}$ be τ-structures (where τ is relational). Suppose that $\overline{a} = a_1 \dots a_k \in A$, $\overline{b} = b_1 \dots b_k \in B$, that $\overline{U} = U_1 \dots U_l$ and $\overline{V} = V_1 \dots V_n$ are sequences of subsets of A, and that $\overline{S} = S_1 \dots S_l$ and $\overline{T} = T_1 \dots T_n$ are sequences of subsets of B. Then $(\overline{a} \mapsto \overline{b}, \overline{U} \mapsto \overline{S}, \overline{V} \leftarrow \overline{T})$ is a *generalized partial isomorphism* from $\mathcal{A}$ to $\mathcal{B}$ iff $\overline{a} \mapsto \overline{b}$ is a partial isomorphism from $\mathcal{A}$ to $\mathcal{B}$ and for $i = 1, \dots, k$ and $j = 1, \dots, l$ we have:

- $a_i \in U_j$ implies $b_i \in S_j$,
- $b_i \in T_j$ implies $a_i \in V_j$. □

As the Ehrenfeucht-Fraïssé game $\mathrm{G}_m(\mathcal{A}, \mathcal{B})$ for first-order logic, the game MLFP-$\mathrm{G}_m(\mathcal{A}, \mathcal{B})$ is played by two players, the spoiler and the duplicator, and consists of m moves. In the i-th move $(1 \leq i \leq m)$ an element a_i of A and an element b_i of B are chosen, and in some moves subsets will be chosen, too. More precisely: There are two types of moves, *point moves* and LFP *moves.* In each move of a play the spoiler first decides what type of move he wants. Point moves are as in the first-order case. Now suppose that the spoiler decides that the i-th move should be an LFP move. Then he selects a $j < i$ and a structure, $\mathcal{A}$ or $\mathcal{B}$. We speak of a *positive* LFP move if he selects $\mathcal{A}$, and of a *negative* LFP move if he selects $\mathcal{B}$.

In the positive case the spoiler chooses a subset S of B with $b_j \notin S$, and the duplicator answers by a subset U of A with $a_j \notin U$. Then the spoiler chooses an element $a_i \in A \setminus U$, and the move is finished by the selection of an element $b_i \in B \setminus S$ by the duplicator.

In the negative case the roles of A and B are interchanged, that is: The spoiler chooses a $V \subseteq A$ with $a_j \notin V$, and the duplicator answers by a $T \subseteq B$ with $b_j \notin T$; then the spoiler chooses an element $b_i \in B \setminus T$, and finally the duplicator selects an element $a_i \in A \setminus V$.

In the course of a play, besides the elements $a_1 \dots a_m$ and $b_1 \dots b_m$, subsets

$$U_1 \dots U_l \text{ of } A \qquad \text{and} \qquad S_1 \dots S_l \text{ of } B$$

have been chosen in positive LFP moves, and subsets

$$V_1 \dots V_n \text{ of } A \qquad \text{and} \qquad T_1 \dots T_n \text{ of } B$$

in negative ones. By definition, the duplicator wins the play if $(\overline{a} \mapsto \overline{b}, \overline{U} \mapsto \overline{S}, \overline{V} \leftarrow \overline{T})$ is a generalized partial isomorphism.

The main theorem now reads as follows:

Theorem 8.5.4 *Let $\mathcal{A}$ and $\mathcal{B}$ be structures and $m \in \mathbb{N}$. Then the following are equivalent:*

(i) $\mathcal{A} \equiv^{\mathrm{M}}_m \mathcal{B}$, *i.e., $\mathcal{A}$ and $\mathcal{B}$ satisfy the same sentences of* FO(M-LFP) *of quantifier rank $\leq m$.*

(ii) *The duplicator has a winning strategy in the game* MLFP-G$_m(\mathcal{A}, \mathcal{B})$.

Before presenting the proof we give an example.

Example 8.5.5 Let $\mathcal{G}_l$ and $\mathcal{G}_l \dot{\cup} \mathcal{G}_l$ be graphs consisting of one and two cycles of length $l+1$, respectively. To be definite, set

$$G_l := \{0, \ldots, l\}, \quad E^{G_l} := \{(i, i+1) \mid i < l\} \cup \{(i+1, i) \mid i < l\} \cup \{(0, l), (l, 0)\}$$

and $G_l \dot{\cup} G_l := \{0, \ldots, l\} \times \{0, 1\}$. – The spoiler has a winning strategy in the game MLFP-G$_4(\mathcal{G}_l, \mathcal{G}_l \dot{\cup} \mathcal{G}_l)$, which is exemplified by the following table where we have underlined his selections. He plays according to the fact that

$$\begin{array}{rcl} \mathcal{G}_l & \models & \exists x \forall y [\mathrm{LFP}_{z,Z}\, z = x \vee \exists u (Zu \wedge Euz)]\, y \\ \mathcal{G}_l \dot{\cup} \mathcal{G}_l & \not\models & \exists x \forall y [\mathrm{LFP}_{z,Z}\, z = x \vee \exists u (Zu \wedge Euz)]\, y. \end{array}$$

move	$\mathcal{G}_l$	$\mathcal{G}_l \dot{\cup} \mathcal{G}_l$	$j < i$
1	$\underline{0}$	(b, k)	
2	a	$\underline{(b, 1-k)}$	
3	U (with $a \notin U$)	$\underline{\{(c, k) \mid c \leq l\}}$	2
	$\underline{a'} := \min\{u \mid u \notin U\}$	$(d, 1-k)$	
4	$\underline{a' - 1}$	?	

No fourth move of the duplicator will lead to a generalized partial isomorphism, because $a' - 1$ is connected with a' by an edge and $a' - 1 \in U$, but there is no element in $\mathcal{G}_l \dot{\cup} \mathcal{G}_l$ related to $(d, 1-k)$ and $\{(c, k) \mid c \leq l\}$ in this way. (If $a' = 0$ then already in the third move there is no answer for the duplicator leading to a generalized partial isomorphism.) □

Proof (of 8.5.4). First suppose that the duplicator has a winning strategy for MLFP-G$_m(\mathcal{A}, \mathcal{B})$. We have to show that ($\mathcal{A} \models \varphi$ iff $\mathcal{B} \models \varphi$) holds for all FO(M-LFP)-sentences of quantifier rank $\leq m$. Essentially, the proof runs in the standard way: As usual, point moves take care of first-order quantifiers and LFP moves of LFP operators as outlined before introducing the game. More precisely: Suppose that in a play, where the duplicator has followed his winning strategy, $\overline{a} \mapsto \overline{b}$, $\overline{U} \mapsto \overline{S}$ and $\overline{V} \leftarrow\!\!\dashv \overline{T}$ have been the choices in the first i moves; then show that

$$\mathcal{A} \models \psi(\overline{x}, \overline{X}, \overline{Y})\,[\overline{a}, \overline{U}, \overline{V}] \quad \text{implies} \quad \mathcal{B} \models \psi(\overline{x}, \overline{X}, \overline{Y})\,[\overline{b}, \overline{S}, \overline{T}]$$

for any FO(M-LFP)-formula ψ of quantifier rank $\leq m-i$ which is positive in $\overline{X}$ and negative in $\overline{Y}$ (a corresponding inductive proof runs on $m-i$).

For the other direction we introduce an analogue of j-isomorphism types (cf. 2.2.5). Fix a structure $\mathcal{A}$. For $\overline{a} = a_1 \dots a_k \in A$, "positive" subsets $\overline{U} = U_1 \dots U_l$ of A and "negative" subsets $\overline{V} = V_1 \dots V_n$ of A let

$$\chi^0_{\overline{a},\overline{U},\overline{V}} \;:=\; \varphi^0_{\overline{a}} \wedge \bigwedge_{1\leq i\leq k} \Big(\bigwedge_{1\leq j\leq l;\; a_i \in U_j} X_j v_i \wedge \bigwedge_{1\leq j\leq n;\; a_i \notin V_j} \neg Y_j v_i \Big),$$

where $\varphi^0_{\overline{a}}$ is the (formula denoting the) 0-isomorphism type of $\overline{a}$ in $\mathcal{A}$. Then, for arbitrary $\mathcal{B}, \overline{b}, \overline{S}$, and $\overline{T}$, we have

$$\mathcal{B} \models \chi^0_{\overline{a},\overline{U},\overline{V}}[\overline{b}, \overline{S}, \overline{T}] \quad \text{iff} \quad (\overline{a} \mapsto \overline{b}, \overline{U} \mapsto \overline{S}, \overline{V} \leftarrow\!\!\dashv \overline{T}) \text{ is a generalized partial isomorphism.}$$

For $j > 0$ we define the formula $\chi^j_{\overline{a},\overline{U},\overline{V}}$ such that $\mathcal{B} \models \chi^j_{\overline{a},\overline{U},\overline{V}}[\overline{b}, \overline{S}, \overline{T}]$ whenever in a play of the FO(M-LFP) game, in which $\overline{a} \mapsto \overline{b}$, $\overline{U} \mapsto \overline{S}$, and $\overline{V} \leftarrow\!\!\dashv \overline{T}$ have been selected so far, there is a strategy for the duplicator which allows him to answer j further moves correctly. Thus we set (writing v, X, and Y for v_{k+1}, X_{l+1}, and Y_{n+1}, respectively):

$$\begin{aligned}\chi^j_{\overline{a},\overline{U},\overline{V}} \;:=\; & \bigwedge_{a\in A} \exists v \chi^{j-1}_{\overline{a}a,\overline{U},\overline{V}} \wedge \forall v \bigvee_{a \in A} \chi^{j-1}_{\overline{a}a,\overline{U},\overline{V}} \wedge \\ & \bigwedge_{1\leq i\leq k} \Big(\forall X \big(\neg X v_i \to \bigvee_{U\subseteq A;\, a_i\notin U} \bigwedge_{a\notin U} \exists v(\neg Xv \wedge \chi^{j-1}_{\overline{a}a,\overline{U}U,\overline{V}})\big) \\ & \wedge \bigwedge_{V\subseteq A;\, a_i \notin V} \exists Y\big(\neg Y v_i \wedge \forall v(\neg Y v \to \bigvee_{a\notin V} \chi^{j-1}_{\overline{a}a,\overline{U},\overline{V}V})\big)\Big).\end{aligned}$$

If $\overline{a}, \overline{U}, \overline{V}$ are empty, we write $\chi^j_{\mathcal{A}}$ for the sentence $\chi^j_{\overline{a},\overline{U},\overline{V}}$. We show that, up to equivalence,

(1) $\quad \chi^j_{\overline{a},\overline{U},\overline{V}}$ is an FO(M-LFP)-formula of quantifier rank $\leq j$.

This proves the claim: In fact, since $\mathcal{A} \models \chi^m_{\mathcal{A}}$, we get $\mathcal{B} \models \chi^m_{\mathcal{A}}$ from $\mathcal{A} \equiv^{\mathrm{M}}_m \mathcal{B}$ and hence, the duplicator has a winning strategy in MLFP-$\mathrm{G}_m(\mathcal{A}, \mathcal{B})$.

To obtain (1) we first note that a simple induction on j shows that, given the lengths of the sequences which appear as subscripts, there are only finitely many formulas of the form $\chi^j_{\overline{a},\overline{U},\overline{V}}$. Thus the disjunctions and conjunctions in the definition of $\chi^j_{\overline{a},\overline{U},\overline{V}}$ are finite.

The proof of (1) also proceeds by induction on j. In $\chi^j_{\overline{a},\overline{U},\overline{V}}$ the conjunct

$$\exists Y(\neg Y v_i \wedge \forall v(\neg Y v \to \bigvee_{a \notin V} \chi^{j-1}_{\overline{a}a,\overline{U},\overline{V}V}))$$

is equivalent to

$$\neg \forall Y(\neg Y v_i \to \exists v(\neg Y v \wedge \neg \bigvee_{a \notin V} \chi^{j-1}_{\overline{a}a,\overline{U},\overline{V}V})),$$

which, by (*) before Theorem 8.5.2, is equivalent to

$$\neg[\mathrm{LFP}_{v,Y} \neg \bigvee_{a \notin V} \chi^{j-1}_{\overline{a}a,\overline{U},\overline{V}V}]\, v_i$$

and hence, by induction hypothesis, to an FO(M-LFP)-formula of quantifier rank $\leq j$. Finally, we show how to rewrite

$$\forall X(\neg X v_i \to \bigvee_{U \subseteq A;\, a_i \notin U} \bigwedge_{a \notin U} \exists v(\neg X v \wedge \chi^{j-1}_{\overline{a}a,\overline{U}U,\overline{V}}))$$

as an FO(M-LFP)-formula of quantifier rank $\leq j$. This formula has the form

$$\forall X(\neg X v_i \to \bigvee_{\gamma \in \Gamma} \bigwedge_{\delta \in \Delta_\gamma} \exists v(\neg X v \wedge \varphi_{\gamma\delta}))$$

for suitable finite sets Γ and Δ_γ. Let Π be the set of functions h with domain Γ such that $h(\gamma) \in \Delta_\gamma$ for $\gamma \in \Gamma$. Then, as usual, we can interchange the order of $\bigvee$ and $\bigwedge$, thus obtaining the equivalent formula

$$\forall X(\neg X v_i \to \bigwedge_{h \in \Pi} \bigvee_{\gamma \in \Gamma} \exists v(\neg X v \wedge \varphi_{\gamma h(\gamma)}))$$

which in turn is equivalent to

$$\bigwedge_{h \in \Pi} \forall X(\neg X v_i \to \exists v(\neg X v \wedge \bigvee_{\gamma \in \Gamma} \varphi_{\gamma h(\gamma)})).$$

But, using (*) again, this formula can be rewritten as

$$\bigwedge_{h \in \Pi} [\mathrm{LFP}_{v,X} \bigvee_{\gamma \in \Gamma} \varphi_{\gamma h(\gamma)}]\, v_i.$$

Note that the $\varphi_{\gamma h(\gamma)}$ have quantifier rank $\leq j-1$. □

The following corollaries are now proven as in the case of first-order logic.

Corollary 8.5.6 *For every* FO(M-LFP)*-sentence* φ *of quantifier rank* m,

$$\models_{\mathrm{fin}} \varphi \leftrightarrow \bigvee\{\chi^m_{\mathcal{A}} \mid \mathcal{A} \text{ is a finite model of } \varphi\}.$$ □

Corollary 8.5.7 *For a class K of finite structures the following are equivalent:*

(i) *K is not axiomatizable in* FO(M-LFP).
(ii) *For each $m \geq 1$ there are finite structures $\mathcal{A}$ and $\mathcal{B}$ such that*

$$\mathcal{A} \in K, \; \mathcal{B} \notin K, \quad \textit{and} \quad \mathcal{A} \cong^{\mathrm{M}}_{m} \mathcal{B}$$

(where $\mathcal{A} \cong^{\mathrm{M}}_{m} \mathcal{B}$ means that the duplicator has a winning strategy in MLFP-$\mathrm{G}_m(\mathcal{A}, \mathcal{B})$). □

8.5.1 Digression: Implicit Definability

Let $\mathcal{L}$ be a logic. An n-ary query $\mathcal{Q}$ on the class of τ-structures is *implicitly $\mathcal{L}$-definable* if – with R a new n-ary relation symbol – there is an $\mathcal{L}[\tau \cup \{R\}]$-sentence $\psi(R)$ such that

(1) $\models_{\text{fin}} \exists^{=1} X \psi(X)$;
(2) for any τ-structure $\mathcal{A}$, $\quad (\mathcal{A}, \mathcal{Q}(\mathcal{A})) \models \psi(R)$.

We then say that $\psi(R)$ (or, $\psi(X)$) *implicitly* $\mathcal{L}$-defines $\mathcal{Q}$. In section 3.5 we used the notion of implicit definability in a slightly different way; essentially, (1) was replaced by $\models_{\text{fin}} \exists^{\leq 1} X \psi(X)$. In order to distinguish both notions, one sometimes speaks of weak (in the present case) and strong implicit definability.

Clearly, any $\mathcal{L}[\tau \cup \{R\}]$-sentence $\psi(R)$ such that

$$\models_{\text{fin}} \exists^{=1} X \psi(X)$$

implicitly $\mathcal{L}$-defines a query. Note that for first-order $\psi(X)$ the formula $\exists^{=1} X \psi(X)$ is a second-order formula.

We show that any formula of FO(LFP) is equivalent to a first-order formula that contains relations implicitly definable in FO:

Theorem 8.5.8 *Given an* FO(LFP)*-formula $\varphi(\overline{x})$, there are a first-order formula $\psi_0(X)$ and a first-order formula $\psi_1(\overline{x}, X)$ (where the free variables always are among the displayed ones) such that*

(a) $\models_{\text{fin}} \exists^{=1} X \psi_0(X)$
(b) $\models_{\text{fin}} \forall X (\psi_0(X) \rightarrow (\varphi(\overline{x}) \leftrightarrow \psi_1(\overline{x}, X)))$.

Proof. For notational simplicity let φ be a sentence. By 8.2.9, we can assume that $\varphi = \exists u [\mathrm{LFP}_{\overline{z}, Z}\, \chi(\overline{z}, Z)]\, \tilde{u}$ with first-order χ. Let Z be k-ary. Recall the definition of the stage comparison relation $\leq_\chi$ associated with χ (denoted by $\leq_{F^\chi}$ in section 8.2):

$$\overline{y} \leq_\chi \overline{z} \qquad \text{iff} \qquad \text{for some } n \geq 1,\; \overline{y} \in F^\chi_n \text{ and } \overline{z} \in F^\chi_n \setminus F^\chi_{n-1}$$

($\overline{y}$ and $\overline{z}$ range over k-tuples).

We show that $\leq_\chi$ is implicitly FO-definable by a formula $\psi_0(\leq)$. Then we are done, since $\bar{t} \in F_\infty^\chi$ is equivalent to $\bar{t} \leq_\chi \bar{t}$ and hence, we have

$$\models_{\text{fin}} \forall \leq (\psi_0(\leq) \to (\varphi \leftrightarrow \exists u\, \tilde{u} \leq \tilde{u})).$$

When defining $\psi_0(\leq)$ we use the abbreviations:

$$\begin{array}{rll} \bar{y} < \bar{z} & \text{for} & \bar{y} \leq \bar{z} \wedge \neg \bar{z} \leq \bar{y} \\ \text{pred}(\bar{z}) & \text{for} & \{\bar{y} \mid \bar{y} < \bar{z}\} \\ \text{field}(\leq) & \text{for} & \{\bar{z} \mid \exists \bar{y}(\bar{y} \leq \bar{z} \vee \bar{z} \leq \bar{y})\}. \end{array}$$

As $\psi_0(\leq)$ we take the conjunction of (1)-(4) where

(1) " $\leq$ is a reflexive and transitive relation on field($\leq$)" $\wedge$
$\forall \bar{y} \in \text{field}(\leq) \forall \bar{z} \in \text{field}(\leq)\, (\bar{y} \leq \bar{z} \vee \bar{z} \leq \bar{y})$

(2) $\forall \bar{z}(\bar{z} \leq \bar{z} \leftrightarrow \chi(\bar{z}, \text{field}(\leq)))$

(3) $\forall \bar{z}((\bar{z} \leq \bar{z} \wedge \neg \exists \bar{y}\, \bar{y} < \bar{z}) \leftrightarrow \chi(\bar{z}, \emptyset))$

(4) $\forall \bar{y} \forall \bar{z} \in \text{field}(\leq)(\bar{y} \leq \bar{z} \leftrightarrow \chi(\bar{y}, \text{pred}(\bar{z})))$.

We prove that $\psi_0(\leq)$ implicitly defines $\leq_\chi$: In a model of $\psi_0(\leq)$ let $\preceq$ be the ordering induced by $\leq$ on the classes of the equivalence relation $\sim$ given by

$$\bar{y} \sim \bar{z} \quad \text{iff} \quad \bar{y} \leq \bar{z} \text{ and } \bar{z} \leq \bar{y}.$$

Then, an induction on $n \geq 1$ using (3) and (4), shows that for $\bar{y} \in \text{field}(\leq)$ the class of $\bar{y}$ is the n-th element of $\preceq$ iff $\bar{y} \in F_n \setminus F_{n-1}$.

Together with (2) this yields that $\psi_0(\leq)$ implicitly defines $\leq_\chi$. □

Given a logic $\mathcal{L}$, let IMP($\mathcal{L}$) be the logic that allows to define exactly those queries that are expressible in $\mathcal{L}$ using implicitly $\mathcal{L}$-definable queries. More precisely: An IMP($\mathcal{L}$)[τ]-formula $\varphi(\bar{x})$ with free variables among $\bar{x}$ is a tuple

$$(\psi_1(R_1), \ldots, \psi_m(R_m), \psi(\bar{x}, R_1, \ldots, R_m))$$

where $m \geq 0$, each $\psi_i(R_i)$ is an $\mathcal{L}[\tau \dot{\cup} \{R_i\}]$-sentence, and $\psi(\bar{x}, R_1, \ldots, R_m) \in \mathcal{L}[\tau \dot{\cup} \{R_1, \ldots, R_m\}]$ has free variables among $\bar{x}$ and

$$\models_{\text{fin}} \exists^{=1} X_1 \psi_1(X_1) \wedge \ldots \wedge \exists^{=1} X_m \psi_m(X_m).$$

The meaning of $\varphi(\bar{x})$ is that of

$$\forall X_1 \ldots \forall X_m((\psi_1(X_1) \wedge \ldots \wedge \psi_m(X_m)) \to \forall \bar{x}(\varphi(\bar{x}) \leftrightarrow \psi(\bar{x}, X_1, \ldots, X_m))),$$

i.e., for any τ-structure $\mathcal{A}$ and $\bar{a} \in A$,

$$\mathcal{A} \models \varphi[\bar{a}] \qquad \text{iff} \qquad \mathcal{A} \models \psi[\bar{a}, R_1, \ldots, R_m]$$

where R_i is uniquely determined by $\mathcal{A} \models \psi_i[R_i]$. The preceding theorem shows

Corollary 8.5.9 FO(LFP) $\leq$ IMP(FO). □

It has been shown (cf. [29]) that IMP(FO) $\not\leq$ FO(LPF), even IMP(FO) $\not\leq \mathrm{L}^{\omega}_{\infty\omega}$.

Since the formula

$$(\psi_1(R_1), \ldots, \psi_m(R_m), \psi(\overline{x}, R_1, \ldots, R_m))$$

in the definition of IMP($\mathcal{L}$) is equivalent to

$$\forall X_1 \ldots \forall X_m((\psi_1(X_1) \wedge \ldots \wedge \psi_m(X_m)) \rightarrow \psi(\overline{x}, X_1, \ldots, X_m))$$

and to

$$\exists X_1 \ldots \exists X_m(\psi_1(X_1) \wedge \ldots \wedge \psi_m(X_m) \wedge \psi(\overline{x}, X_1, \ldots, X_m))$$

we get

Proposition 8.5.10 IMP(FO) $\leq \Delta^1_1$. □

Exercise 8.5.11 Show: IMP(IMP(FO)) $\equiv$ IMP(FO). □

Exercise 8.5.12 Show that IMP(FO(PFP)) $\equiv$ FO(PFP) on ordered structures (use Theorem 8.5.1). □

Exercise 8.5.13 Let $\mathcal{A}$ be a τ-structure. For an ordered representation $(\mathcal{A}, <)$ let $w_{(\mathcal{A},<)}$ be the $\{0,1\}$-word obtained by concatenating the inscriptions of all input tapes of a Turing machine started with $(\mathcal{A}, <)$. For fixed $\mathcal{A}$, all orderings $<$ lead to words of the same lengths, say l.

(a) Show: $w_{(\mathcal{A},<_1)} = w_{(\mathcal{A},<_2)}$ iff $(\mathcal{A}, <_1) \cong (\mathcal{A}, <_2)$.

An ordering $<$ on $\mathcal{A}$ is *minimal*, if for all orderings $<'$ on $\mathcal{A}$ the word $w_{(\mathcal{A},<')}$ is not smaller than $w_{(\mathcal{A},<)}$ in the lexicographic ordering of $\{0,1\}^l$. Hence, by (a), there is a unique minimal ordering iff $\mathcal{A}$ is rigid. Let $\mathcal{Q}$ be the query that assigns the minimal ordering to a rigid structure and the empty binary relation to a nonrigid one.

(b) Show that $\mathcal{Q}$ is expressible in IMP(FO(PFP)) (and thus, IMP(FO(PFP)) $\not\leq \mathrm{L}^{\omega}_{\infty\omega}$ by 8.4.12). Hint: A formula $\psi(X)$ implicitly defines $\mathcal{Q}$ if it expresses in any $\mathcal{A}$ that "X is an ordering, and for all orderings Y different from X we have: If π is the map with $\pi : (A, X) \cong (A, Y)$, then not $\pi : (\mathcal{A}, X) \cong (\mathcal{A}, Y)$, and if R is the first relation and $\overline{a}$ the first X-tuple such that

$$R^{\mathcal{A}}\overline{a} \leftrightarrow R^{\mathcal{A}}\pi(\overline{a}) \text{ does not hold,}$$

then

$$\text{not } R^{\mathcal{A}}\overline{a} \text{ and } R^{\mathcal{A}}\pi(\overline{a}),$$

or there is no such ordering and $X = \emptyset$". To get $\psi(X)$ as an FO(PFP)-formula note that in the presence of X and Y, the map π is definable in FO(LFP) and that, in the presence of the ordering X, the second-order quantification "for all Y" can be expressed in FO(PFP) (cf. the proof of 8.5.1). □

Notes 8.5.14 The game-theoretic characterization of FO(LFP)-equivalence is due to Bosse [14]. Theorem 8.5.8 is due to Kolaitis [101], Exercise 8.5.13 is taken from [81].

8.6 Transitive Closure Logic

One of the most prominent examples of a global relation expressible in least fixed-point logic but not in first-order logic is the transitive closure of a given relation. *Transitive Closure Logic* FO(TC) is obtained by adding to first-order logic the operation of taking the transitive closure of definable relations. We have introduced FO(TC) already in Chapter 7, where we showed that FO(TC) captures NLOGSPACE. This section contains a more detailed analysis of the expressive power of FO(TC); it can be read independently of Chapter 7.

First we recall some definitions. Let $k \geq 1$ and X be a $2k$-ary relation on a set A. The *transitive closure* $\mathrm{TC}(X)$ of X is defined by

$$\begin{aligned}\mathrm{TC}(X) := \{(\overline{a},\overline{b}) \in A^{2k} \mid{} & \text{there exist } n > 0 \text{ and } \overline{e}_0,\dots,\overline{e}_n \in A^k \text{ such} \\ & \text{that } \overline{a} = \overline{e}_0, \overline{b} = \overline{e}_n, \text{ and for all } i < n,\ (\overline{e}_i,\overline{e}_{i+1}) \in X\},\end{aligned}$$

and the *deterministic transitive closure* $\mathrm{DTC}(X)$ by

$$\begin{aligned}\mathrm{DTC}(X) := \{(\overline{a},\overline{b}) \in A^{2k} \mid{} & \text{there exist } n > 0 \text{ and } \overline{e}_0,\dots,\overline{e}_n \in A^k \text{ such} \\ & \text{that } \overline{a} = \overline{e}_0, \overline{b} = \overline{e}_n, \text{ and for all } i < n,\ \overline{e}_{i+1} \text{ is the} \\ & \text{unique } \overline{e} \text{ for which } (\overline{e}_i,\overline{e}) \in X\}.\end{aligned}$$

For a vocabulary τ the class of formulas of FO(TC) of vocabulary τ is given by the calculus with the following rules:

(i) $\dfrac{}{\varphi}$ where φ is an atomic first-order formula over τ

(ii) $\dfrac{\varphi}{\neg\varphi}, \quad \dfrac{\varphi,\psi}{(\varphi \vee \psi)}, \quad \dfrac{\varphi}{\exists x \varphi}$

(iii) $\dfrac{\varphi}{[\mathrm{TC}_{\overline{x},\overline{y}}\,\varphi]\,\overline{s}\overline{t}}$ where the variables in $\overline{x},\overline{y}$ are pairwise distinct and where the tuples $\overline{x},\overline{y},\overline{s}$, and $\overline{t}$ are all of the same length, $\overline{s}$ and $\overline{t}$ being tuples of terms.

For FO(DTC), rule (iii) is replaced by $\dfrac{\varphi}{[\mathrm{DTC}_{\overline{x},\overline{y}}\,\varphi]\,\overline{s}\overline{t}}$ with the same side conditions.

Sentences are formulas without free variables, where the free occurrence of a variable is defined in the standard way with the additional clause

$$\mathrm{free}([\mathrm{TC}_{\overline{x},\overline{y}}\,\varphi]\,\overline{s}\overline{t}) \quad := \quad \mathrm{free}(\overline{s}) \cup \mathrm{free}(\overline{t}) \cup (\mathrm{free}(\varphi) \setminus \{\overline{x},\overline{y}\}),$$

and similarly for DTC.

The semantics is defined inductively w.r.t. the calculus above, the meaning of $[\mathrm{TC}_{\overline{x},\overline{y}}\,\varphi(\overline{x},\overline{y})]\,\overline{s}\overline{t}$ being $(\overline{s},\overline{t}) \in \mathrm{TC}(\{(\overline{x},\overline{y}) \mid \varphi(\overline{x},\overline{y})\})$; and $[\mathrm{DTC}_{\overline{x},\overline{y}}\,\varphi(\overline{x},\overline{y})]\,\overline{s}\overline{t}$ expresses that $(\overline{s},\overline{t}) \in \mathrm{DTC}(\{(\overline{x},\overline{y}) \mid \varphi(\overline{x},\overline{y})\})$.[13]

Examples and Remarks 8.6.1 (a) FO(DTC) $\leq$ FO(TC) since

$$\models_{\text{fin}} [\mathrm{DTC}_{\overline{x},\overline{y}}\,\varphi(\overline{x},\overline{y})]\,\overline{s}\overline{t} \leftrightarrow [\mathrm{TC}_{\overline{x},\overline{y}}\,(\varphi(\overline{x},\overline{y}) \wedge \forall\overline{z}(\varphi(\overline{x},\overline{z}) \to \overline{z} = \overline{y}))]\,\overline{s}\overline{t}.$$

(b) FO $\not\equiv$ FO(TC) since

$$\forall x \forall y (\neg x = y \to [\mathrm{TC}_{x,y} Exy]\, xy)$$

expresses "connectivity" in the class of graphs (cf. 2.3.8).

(c) FO $\not\equiv$ FO(DTC) since

$$\forall x \forall y [\mathrm{DTC}_{x,y} Exy]\, xy$$

is an FO(DTC)-sentence which, for every $l \geq 1$, holds in the digraph $\mathcal{D}_l$ consisting of a cycle of length $l+1$, but not in the union $\mathcal{D}_l \dot{\cup} \mathcal{D}_l$. Moreover,

$$\forall u \forall v \exists w \exists z [\mathrm{DTC}_{xy,x'y'}(Exy \wedge Ex'y' \wedge x' = y \wedge \neg y' = x)]\, uwvz$$

separates the "undirected" cycles $\mathcal{G}_l$ from $\mathcal{G}_l \dot{\cup} \mathcal{G}_l$ for $l \geq 2$.

(d) FO(TC) $\leq$ FO(LFP) since

$$\models_{\text{fin}} [\mathrm{TC}_{\overline{x},\overline{y}}\,\varphi(\overline{x},\overline{y})]\,\overline{s}\overline{t} \leftrightarrow [\mathrm{LFP}_{\overline{x}\,\overline{y},X}\,(\varphi(\overline{x},\overline{y}) \vee \exists\overline{u}(X\overline{x}\,\overline{u} \wedge \varphi(\overline{u},\overline{y})))]\,\overline{s}\overline{t}. \quad \Box$$

Exercise 8.6.2 Show that the class of bipartite graphs is axiomatizable in FO(TC). Hint: Use 6.3.6. □

Exercise 8.6.3 For $r \geq 1$ let FO(TC^r) be the fragment of FO(TC) consisting of those formulas, where for each subformula of the form $[\mathrm{TC}_{\overline{x},\overline{y}}\,\varphi]\,\overline{s}\overline{t}$ we have length$(\overline{x}) = r$. Clearly, FO(TC^1) $\leq$ FO(TC^2) $\leq \ldots$ Show that FO(TC^1) $\equiv$ MSO on the class of word models (cf. section 6.2). (Hint: FO(TC^1) $\leq$ MSO is true in general; for MSO $\leq$ FO(TC^1) give a proof by induction on regular expressions (cf. section 6.1).)

Define FO(DTC^r) similarly and show that FO(DTC^2) $\not\leq$ MSO on the class of word models. (Hint: Use the fact that the set $\{a^n b^n \mid n \geq 1\}$ is not regular.) □

8.6.1 FO(DTC) < FO(TC)

By the results of Chapter 7 we know that

---

[13] Recall that the notation $\varphi(\overline{x},\overline{y})$ exhibits only the variables relevant in this situation. So φ may contain free variables that are not among $\overline{x},\overline{y}$.

- FO(DTC) $\not\equiv$ FO(TC) on ordered structures iff LOGSPACE $\neq$ NLOGSPACE
- FO(TC) $\not\equiv$ FO(LFP) on ordered structures iff NLOGSPACE $\neq$ PTIME.

The inequalities on the right hand side are prominent open problems of complexity theory. The following remarks will lead to a proof of FO(DTC) $\not\equiv$ FO(TC) for *arbitrary* finite structures; the corresponding inequality FO(TC) $\not\equiv$ FO(LFP) will be shown in subsection 8.6.3.

Definition 8.6.4 A subset C of a structure $\mathcal{A}$ is said to be *closed* if for all $a \in A \setminus C$ there is an automorphism π of $\mathcal{A}$ pointwise fixing C with $\pi(a) \neq a$. □

Every DTC-path starting in a closed set must stay in it:

Lemma 8.6.5 *Let C be a closed subset of a structure $\mathcal{A}$. For every formula $\varphi(\overline{x}, \overline{y}, \overline{z})$ with free variables among $\overline{x}, \overline{y}, \overline{z}$ and for every $\overline{c} \in C$,*

$$\mathcal{A} \models [\mathrm{DTC}_{\overline{x},\overline{y}}\, \varphi(\overline{x}, \overline{y}, \overline{c})]\, \overline{x}\, \overline{y}\, [\overline{a}\overline{b}] \quad \textit{and} \quad \overline{a} \in C \quad \textit{imply} \quad \overline{b} \in C.$$

Proof. Since $(\overline{a}, \overline{b}) \in \mathrm{DTC}(\varphi^{\mathcal{A}}(\_,\_, \overline{c}))$, there is a sequence $\overline{e}_0, \ldots, \overline{e}_m$ such that

$$(*) \quad \overline{a} = \overline{e}_0,\, \overline{b} = \overline{e}_m, \text{ and for } i < m, \ \ \mathcal{A} \models \forall \overline{y}\big(\varphi(\overline{e}_i, \overline{y}, \overline{c}) \leftrightarrow \overline{y} = \overline{e}_{i+1}\big).$$

If $\overline{b} \notin C$ then there is a $j < m$ such that $\overline{e}_j \in C$ and $\overline{e}_{j+1} \notin C$, say, $\overline{e}_{j+1} = e_1 \ldots e_k$ and $e_i \notin C$. Choose an automorphism π of $\mathcal{A}$ pointwise fixing C with $\pi(e_i) \neq e_i$. Then

$$\mathcal{A} \models \varphi[\overline{e}_j, \overline{e}_{j+1}, \overline{c}], \quad \mathcal{A} \models \varphi[\overline{e}_j, \pi(\overline{e}_{j+1}), \overline{c}], \quad \text{and} \quad \overline{e}_{j+1} \neq \pi(\overline{e}_{j+1}),$$

contradicting $(*)$. □

By the lemma there is a bound for the length of DTC-paths without repetitions that start in a closed set of a given cardinality. This is the reason why in classes of structures with many closed sets of controllable cardinality, so-called indiscernible classes, every DTC operation is first-order definable. We thus are led to a proof of FO(DTC) $\not\equiv$ FO(TC) by showing

- FO(DTC) $\equiv$ FO on every indiscernible class.
- FO(TC) $\not\equiv$ FO on some indiscernible class.

Definition 8.6.6 Let $f : \mathbb{N} \to \mathbb{N}$ be monotone (that is, $m \leq n$ implies $f(m) \leq f(n)$). A structure $\mathcal{A}$ is said to be *f-indiscernible*, if for every subset X of A there is a closed set Y such that $X \subseteq Y \subseteq A$ and $\|Y\| \leq f(\|X\|)$. A class K of structures is *indiscernible* if, for some monotone function f, every structure in K is f-indiscernible. □

Proposition 8.6.7 *Let K be an indiscernible class. Then, on K, every formula of* FO(DTC) *is equivalent to a first-order formula.*

Proof. Let K be indiscernible via f and $\varphi(\overline{x},\overline{y},\overline{z})$ with $\overline{x} = x_1 \dots x_k$, $\overline{y} = y_1 \dots y_k$, and $\overline{z} = z_1 \dots z_l$ be an FO-formula. Suppose that $\mathcal{A} \in K$, $\overline{c} \in A$, and $(\overline{a},\overline{b}) \in \text{DTC}(\varphi^{\mathcal{A}}(\_,\_,\overline{c}))$. Choose a closed set $C \subseteq A$ such that $\{\overline{a},\overline{c}\} \subseteq C$ and $\|C\| \leq m := f(k+l)$. By the preceding lemma, every deterministic path connecting $\overline{a}$ and $\overline{b}$ stays within C; hence, the shortest such path has length $\leq m^k$. Therefore, $[\text{DTC}_{\overline{x},\overline{y}}\,\varphi(\overline{x},\overline{y},\overline{z})]\,\overline{s}\overline{t}$ is equivalent to a first-order formula expressing "there exists a deterministic φ-path of length $\leq m^k$ from $\overline{s}$ to $\overline{t}$ ". □

For any structure $\mathcal{A}$ in a relational vocabulary τ we denote, as in section 5.2, by $\mathcal{A}\times 2$ the structure we get by duplicating each element. To be precise, let $\mathcal{A} \times 2$ be the τ-structure with universe $A \times \{0,1\}$ such that for any n-ary $R \in \tau$,

$$R^{A\times 2} := \{((a_1,i_1),\dots,(a_n,i_n)) \mid R^{A}a_1\dots a_n \text{ and } i_1,\dots,i_n \in \{0,1\}\}.$$

Lemma 8.6.8 *$\mathcal{A} \times 2$ is f_0-indiscernible, where $f_0(m) = 2{\cdot}m$.*

Proof. Clearly, for $X \subseteq A \times 2$ the set

$$C(X) \quad := \quad X \cup \{(a,i) \mid (a,1{-}i) \in X\}$$

is closed: for any $(b,i) \in (A \times 2) \setminus C(X)$, the function $\pi_b : A \times 2 \to A \times 2$ given by

$$\pi_b((a,j)) := \begin{cases} (a,j) & \text{if } a \neq b \\ (b,1{-}j) & \text{if } a = b \end{cases}$$

is the desired automorphism. □

Theorem 8.6.9 FO(DTC) $\not\equiv$ FO(TC).

Proof. First of all note that

$$(*) \qquad \mathcal{A} \cong_m \mathcal{B} \text{ implies } \mathcal{A} \times 2 \cong_m \mathcal{B} \times 2.$$

Let $K := \{\mathcal{G}_l \mid l \geq 2\}$ be the set of cycles and let $H := \{\mathcal{G}_l\dot{\cup}\mathcal{G}_l \mid l \geq 2\}$. By the preceding lemma the sets $K \times 2$ $(:= \{\mathcal{A} \times 2 \mid \mathcal{A} \in K\})$ and $H \times 2$ are indiscernible. Since $\mathcal{G}_l \cong_m \mathcal{G}_l\dot{\cup}\mathcal{G}_l$ for $l \geq 2^m$ (cf. 2.3.8), $(*)$ shows that there is no first-order sentence separating $K \times 2$ and $H \times 2$, and hence by 8.6.7 and 8.6.8, no FO(DTC)-sentence. But the structures in $K \times 2$ are connected graphs and those in $H \times 2$ are not; thus there is a separating FO(TC)-sentence (cf. 8.6.1(b)). □

Exercise 8.6.10 Show that the class of bipartite graphs is not axiomatizable in FO(DTC). □

8.6.2 FO(posTC) and Normal Forms

Let FO(posDTC) and FO(posTC) be the logics consisting of the formulas of FO(DTC) and FO(TC) in which DTC and TC, respectively, only occur positively, i.e., within the scope of an even number of negation signs (we assume that the formulas of FO(DTC) are built up from atomic formulas using $\neg, \wedge, \vee, \forall, \exists$, and DTC; similarly for FO(TC)).

Proposition 8.6.11 FO(posDTC) $\equiv$ FO(DTC).

Proof. The proof of the nontrivial part proceeds by induction on the number of occurrences of logical operations ($\neg, \wedge, \vee, \forall, \exists$, DTC) in a formula of FO(DTC). In the main step of the induction we have to show that an FO(DTC)-formula of the form

$$(*) \qquad \neg[\mathrm{DTC}_{\overline{x},\overline{y}}\varphi(\overline{x},\overline{y})]\,\overline{s}\overline{t}$$

is equivalent to an FO(posDTC)-formula where, by induction hypothesis, we may assume that φ and $\neg\varphi$ are equivalent to FO(posDTC)-formulas. Obviously, $(*)$ just says that the deterministic φ-path starting at $\overline{s}$ reaches, without passing through $\overline{t}$, a point $\overline{z}$, where it ends or where a φ-cycle starts that does not contain $\overline{t}$. Noting that the deterministic φ-path ends at $\overline{z}$ if $\overline{z}$ has no or more than one φ-neighbour, we thus can express $\neg[\mathrm{DTC}_{\overline{x},\overline{y}}\varphi(\overline{x},\overline{y})]\,\overline{s}\overline{t}$ by

$$\begin{aligned}&\exists\overline{z}((\overline{s}=\overline{z} \vee [\mathrm{DTC}_{\overline{x},\overline{y}}\,(\varphi(\overline{x},\overline{y}) \wedge \neg\overline{y}=\overline{t})]\,\overline{s}\,\overline{z}) \wedge (\forall\overline{x}\neg\varphi(\overline{z},\overline{x})\\ &\vee \exists\overline{x}\exists\overline{y}\,(\varphi(\overline{z},\overline{x}) \wedge \varphi(\overline{z},\overline{y}) \wedge \neg\overline{x}=\overline{y}) \vee [\mathrm{DTC}_{\overline{x},\overline{y}}(\varphi(\overline{x},\overline{y}) \wedge \neg\overline{y}=\overline{t})]\,\overline{z}\,\overline{z})).\end{aligned}$$

Applying the induction hypothesis to φ and $\neg\varphi$ we see that this formula is equivalent to an FO(posDTC)-formula. □

Note that the maximal number of *nested* occurrences of the DTC operator in an FO(DTC)-formula and in the FO(posDTC)-formula assigned to it in the preceding proof are the same. We use this fact when showing

Proposition 8.6.12 FO(DTC) $\leq$ FO(posTC).

Proof. By the preceding proposition it suffices to show that FO(posDTC) $\leq$ FO(posTC). The proof is by induction on the number of nested DTC operators, the only nontrivial step being the case of an FO(DTC)-formula of the form

$$[\mathrm{DTC}_{\overline{x},\overline{y}}\,\varphi(\overline{x},\overline{y})]\overline{s}\overline{t}.$$

By the remark preceding the proposition, $\neg\varphi$ is equivalent to an FO(posDTC)-formula with the same maximal number of nested occurrences of the DTC operator as φ. Thus, by induction hypothesis, both φ and $\neg\varphi$ are equivalent to formulas of FO(posTC); therefore the same holds for $[\mathrm{DTC}_{\overline{x},\overline{y}}\,\varphi(\overline{x},\overline{y})]\overline{s}\overline{t}$, since this formula is equivalent to $[\mathrm{TC}_{\overline{x},\overline{y}}\,(\varphi(\overline{x},\overline{y}) \wedge \forall\overline{z}(\neg\varphi(\overline{x},\overline{z}) \vee \overline{z}=\overline{y}))]\,\overline{s}\overline{t}$. □

We have already shown in Chapter 7 that FO(posTC) $\equiv$ FO(TC) on ordered structures. Below, we give a purely modeltheoretic proof of this fact. On arbitrary structures, FO(posTC) has less expressive power than FO(TC). The next proposition contains a normal form for a fragment of FO(posTC) that will be useful for our purposes.

Proposition 8.6.13 *Suppose that τ contains two constants c and d. Let φ be an* existential FO(posTC)*-formula, that is a formula built up from atomic and negated atomic formulas with the help of $\wedge$, $\vee$, TC, and $\exists$. Then, in models of $\neg c = d$, φ is equivalent to a formula of the form*

$$[\mathrm{TC}_{\overline{x},\overline{x}'}\psi(\overline{x},\overline{x}')]\,\tilde{c}\tilde{d}$$

where ψ is quantifier-free.

Proof. The proof proceeds by induction on φ. Of course, any first-order formula φ is equivalent to $[\mathrm{TC}_{\overline{x},\overline{x}'}\varphi]\,\tilde{c}\tilde{d}$, where $\overline{x},\overline{x}'$ are variables not occurring in φ. Let $\varphi = (\varphi_1 \vee \varphi_2)$ and suppose that φ_i is equivalent to $[\mathrm{TC}_{\overline{x},\overline{x}'}\psi_i(\overline{x},\overline{x}')]\tilde{c}\tilde{d}$ with quantifier-free ψ_i (add dummy variables to obtain sequences $\overline{x},\overline{x}'$ of the same length for φ_1 and φ_2). Then φ will be equivalent to a formula $[\mathrm{TC}_{\overline{x}x,\overline{x}'x'}\psi(\overline{x}x,\overline{x}'x')]\tilde{c}c\tilde{d}d$, where the additional arguments x and x' in ψ are used as a switch as indicated by the figure

$$\begin{array}{ccc} \tilde{c}c & \ldots \xrightarrow{\psi_1} \ldots & \tilde{d}c \\ \downarrow & & \downarrow \\ \tilde{c}d & \ldots \xrightarrow{\psi_2} \ldots & \tilde{d}d \end{array}$$

In the formula ψ below, the first line corresponds to the upper path in the figure leading from $\tilde{c}c$ to $\tilde{d}d$, and the second line to the lower path:

$$\begin{aligned} \psi(\overline{x}x,\overline{x}'x') \quad := \quad & (x = x' = c \wedge \psi_1) \vee (\overline{x} = \overline{x}' = \tilde{d} \wedge x = c \wedge x' = d) \\ & \vee (\overline{x} = \overline{x}' = \tilde{c} \wedge x = c \wedge x' = d) \vee (x = x' = d \wedge \psi_2). \end{aligned}$$

Similarly, for the conjunction $(\varphi_1 \wedge \varphi_2)$ we take as $\psi(\overline{x}x,\overline{x}'x')$ a formula expressing the existence of a path as indicated by

$$\tilde{c}c \ldots \xrightarrow{\psi_1} \ldots \tilde{d}c \longrightarrow \tilde{c}d \ldots \xrightarrow{\psi_2} \ldots \tilde{d}d.$$

Now suppose that $\varphi = [\mathrm{TC}_{\overline{u},\overline{v}}\varphi']\,\overline{s}\overline{t}$ where, by induction hypothesis, we can assume that φ' has the form $[\mathrm{TC}_{\overline{x},\overline{x}'}\,\psi'(\overline{x},\overline{x}',\overline{u},\overline{v})]\,\tilde{c}\tilde{d}$ with quantifier-free ψ'. Let $\overline{s} = \overline{e}_0,\ldots,\overline{e}_k = \overline{t}$ be a path witnessing that $(\overline{s},\overline{t}) \in \mathrm{TC}(\varphi'(\_,\_))$. Then, for $i = 0,\ldots,k-1$ there is a ψ'-path witnessing that $(\tilde{c},\tilde{d}) \in \mathrm{TC}(\psi'(\_,\_,\overline{e}_i,\overline{e}_{i+1}))$. Therefore, φ is equivalent to

$$[\mathrm{TC}_{\overline{u}\,\overline{v}\,\overline{x},\overline{u}'\overline{v}'\overline{x}'}\psi(\overline{u},\overline{v},\overline{x},\overline{u}',\overline{v}',\overline{x}')]\,\tilde{c}\tilde{c}\tilde{c}\tilde{d}\tilde{d}\tilde{d}$$

where in the formula ψ given below the first line sets the corresponding starting values, the second line takes care of witnessing that $(\tilde{c},\tilde{d}) \in$

$\mathrm{TC}(\psi'(\_,\_,\overline{e}_i,\overline{e}_{i+1}))$, the next line allows to pass from the tuple $\overline{e}_i,\overline{e}_{i+1}$ to $\overline{e}_{i+1},\overline{e}_{i+2}$, and the last line realizes that $\overline{e}_k=\overline{t}$:

$$\begin{aligned}\psi(\overline{u},\overline{v},\overline{x},\overline{u}',\overline{v}',\overline{x}') \;:=\; & (\overline{u}=\tilde{c}\wedge\overline{v}=\tilde{c}\wedge\overline{x}=\tilde{c}\wedge\overline{u}'=\overline{s}\wedge\overline{x}'=\tilde{c})\\ & \vee(\neg\overline{x}=\tilde{d}\wedge\overline{u}'=\overline{u}\wedge\overline{v}'=\overline{v}\wedge\psi'(\overline{x},\overline{x}',\overline{u},\overline{v}))\\ & \vee(\overline{x}=\tilde{d}\wedge\neg\overline{v}=\overline{t}\wedge\overline{u}'=\overline{v}\wedge\overline{x}'=\tilde{c})\\ & \vee(\overline{x}=\tilde{d}\wedge\overline{v}=\overline{t}\wedge\overline{u}'=\tilde{d}\wedge\overline{v}'=\tilde{d}\wedge\overline{x}'=\tilde{d}).\end{aligned}$$

For the existential quantifier note that

$$\models_{\text{fin}} \exists z\varphi(z)\leftrightarrow[\mathrm{TC}_{x,y}((x=c\wedge\varphi(y))\vee(\varphi(x)\wedge y=d))]\,cd.$$

So we can apply what we already know to obtain the desired normal form also for this case. □

Corollary 8.6.14 *Let $\tau\supseteq\{<,S,\min,\max\}$. On ordered τ-structures with at least two elements, every* FO(posTC)*-formula is equivalent to a formula of the form*

$$[\mathrm{TC}_{\overline{x},\overline{x}'}\,\psi(\overline{x},\overline{x}')]\,\widetilde{\min}\widetilde{\max}$$

with quantifier-free ψ.

Proof. In ordered models, $\forall x\varphi(x)$ is equivalent to

$$[\mathrm{TC}_{x,y}\,(\varphi(x)\,\wedge\,Sxy\,\wedge\,\varphi(y))]\,\min\max\,.$$

Hence, the universal quantifier can be eliminated in FO(posTC)-formulas. Thus the claim follows from the preceding proposition. □

We use this normal form for FO(posTC)-formulas when proving now that FO(posTC) $\equiv$ FO(TC) on ordered structures.

Fix a vocabulary τ for ordered structures, for simplicity, $\tau\supseteq\{<,S,\min,\max\}$. We saw in 8.1.11 that for $k\geq 1$ the successor relation S^k of the lexicographic ordering of k-tuples is first-order definable. For any ordered τ-structure $\mathcal{A}$ with $A=\{0,\ldots,n-1\}$ and for any $m_1,\ldots,m_k\in A$ we set

$$[m_1,\ldots,m_k]\;:=\;m_1\cdot n^{k-1}+\ldots+m_{k-1}\cdot n+m_k.$$

In the following we use self-explanatory notations such as $\overline{l}=\overline{m}+1$ for $S^k\overline{m}\overline{l}$ or, equivalently, for $[\overline{l}]=[\overline{m}]+1$. Often we write 0 for min.

Theorem 8.6.15 *On ordered structures,* FO(posTC) $\equiv$ FO(TC).

Proof. The proof – a purely modeltheoretic version of the proof given in 7.5.20 – proceeds by induction on FO(TC)-formulas. The only nontrivial case is the negation step. By induction hypothesis and the corollary above it suffices to show for first-order ψ that

$$\neg[\mathrm{TC}_{\overline{x},\overline{y}}\, \psi(\overline{x},\overline{y})]\, \overline{s}\overline{t}$$

is equivalent to an FO(posTC)-formula.[14]

Suppose $\overline{x} = x_1 \ldots x_r$. Given a structure $\mathcal{A}$ and $\overline{a}, \overline{b} \in A^r$, let $d_\psi(\overline{a},\overline{b})$ be the length of the shortest ψ-path connecting $\overline{a}$ and $\overline{b}$,

$$d_\psi(\overline{a},\overline{b}) := \min\{k > 0 \mid \text{there exist } \overline{a}_0 = \overline{a}, \overline{a}_1, \ldots, \overline{a}_k = \overline{b} \text{ such that } \mathcal{A} \models \psi[\overline{a}_i, \overline{a}_{i+1}] \text{ for } i < k\},$$

where $d_\psi(\overline{a},\overline{b}) := \infty$ in case the set on the right side is empty. Obviously:

(1) If $d_\psi(\overline{a},\overline{b}) < \infty$ then $0 < d_\psi(\overline{a},\overline{b}) \le \|A\|^r$.

Moreover, $\neg\,[\mathrm{TC}_{\overline{x},\overline{y}}\, \psi(\overline{x},\overline{y})]\, \overline{s}\overline{t}$ is equivalent to

$$\|\{\overline{v} \mid d_\psi(\overline{s},\overline{v}) < \infty\}\| = \|\{\overline{v} \mid d_{(\psi(\overline{x},\overline{y})\,\wedge\,\neg\overline{y}=\overline{t})}(\overline{s},\overline{v}) < \infty\}\|.$$

We show stepwise that the ingredients of this equality are definable in FO(posTC). Note that the number $\|A\|^r$ has the string $1\tilde{0}$ of length $r+1$ as its $\|A\|$-adic representation.

(2) For $\varphi(x_1, \ldots, x_r, y_1, \ldots, y_r) \in$ FO(posTC) there is an FO(posTC)-formula $\chi(\overline{x}, \overline{y}, z_1, \ldots, z_{r+1})$ expressing $d_\varphi(\overline{x},\overline{y}) \le \overline{z}$; more precisely: for any ordered $\mathcal{A}$ and $\overline{a}, \overline{b}, \overline{m} \in A$,

$$\mathcal{A} \models \chi[\overline{a},\overline{b},\overline{m}] \quad \text{iff} \quad d_\varphi(\overline{a},\overline{b}) \le [\overline{m}].$$

To obtain such a formula χ, we use the TC operator to go through all the tuples of a path leading from $\overline{x}$ to $\overline{y}$:

$$\chi(\overline{x},\overline{y},\overline{z}) \;:=\; [\mathrm{TC}_{\overline{w}\,\overline{z},\overline{w}'\,\overline{z}'}(\varphi(\overline{w},\overline{w}') \wedge \overline{z} < \overline{z}')]\overline{x}\, 0\tilde{0}\, \overline{y}\, \overline{z}.$$

(3) For first-order $\varphi(\overline{x},\overline{y})$ there is an FO(posTC)-formula $\rho_\varphi(\overline{x},\overline{z})$ expressing $\|\{\overline{y} \mid d_\varphi(\overline{x},\overline{y}) < \infty\}\| = \overline{z}$.

For a proof we first note that by (1),

$$\|\{\overline{y} \mid d_\varphi(\overline{x},\overline{y}) < \infty\}\| = \|\{\overline{y} \mid d_\varphi(\overline{x},\overline{y}) \le \|A\|^r\}\|.$$

To obtain a formula $\rho_\varphi(\overline{x},\overline{z})$ we use an inductive definition of the function

$$g(\overline{u})\,(= g(\overline{x},\overline{u})) \;:=\; \|\{\overline{y} \mid d_\varphi(\overline{x},\overline{y}) \le \overline{u}\}\|,$$

$g(1\tilde{0})$ $(= \|\{\overline{y} \mid d_\varphi(\overline{x},\overline{y}) \le \|A\|^r\}\|)$ being the value we are interested in.

By (1), $g(0\tilde{0}) = 0\tilde{0}$. Suppose $g(\overline{u}) = \|\{\overline{y} \mid d_\varphi(\overline{x},\overline{y}) \le \overline{u}\}\| = \overline{z}$ where $\overline{u} = u_1 \ldots u_{r+1}$. Fix $\overline{y}$. If $\overline{u} = 0\tilde{0}$ then $d_\varphi(\overline{x},\overline{y}) \not\le \overline{u}+1$ is equivalent to $\neg\varphi(\overline{x},\overline{y})$. If $\overline{u} \ne 0\tilde{0}$ then $d_\varphi(\overline{x},\overline{y}) \not\le \overline{u}+1$ iff there are $\overline{z}$ many $\overline{w}$ such that

[14] The restriction to structures with at least two elements is removed as for FO(LFP) after 8.2.11.

$\overline{w} \neq \overline{y}$, $d_\varphi(\overline{x},\overline{w}) \leq \overline{u}$, and $\neg\varphi(\overline{w},\overline{y})$. In the following formula $\chi_\varphi(\overline{x},\overline{u},\overline{z},\overline{z}')$ the inner TC operator serves to count these $\overline{w}$ ($\overline{q} = q_1 \ldots q_{r+1}$ being the counting variables) and thus to check whether there are $\overline{z}$ many. If $g(\overline{u}) = \overline{z}$ then $\chi_\varphi(\overline{x},\overline{u},\overline{z},\overline{z}')$ expresses that $g(\overline{u}+1) = \overline{z}'$. Set with $\overline{v} = v_1 \ldots v_{r+1}$

$$\begin{aligned}\chi_\varphi(\overline{x},\overline{u},\overline{z},\overline{z}') := \ & [\mathrm{TC}_{\overline{y}\,\overline{v},\overline{y}'\,\overline{v}'}\,((\overline{y}' = \overline{y}+1 \land \overline{v}' = \overline{v}+1 \land d_\varphi(\overline{x},\overline{y}) \leq \overline{u}+1)\\ &\lor (\overline{y}' = \overline{y}+1 \land \overline{v}' = \overline{v} \land \overline{u} = \tilde{0} \land \neg\varphi(\overline{x},\overline{y}))\\ &\lor (\overline{y}' = \overline{y}+1 \land \overline{v}' = \overline{v} \land \neg\overline{u} = \tilde{0}\\ &\land [\mathrm{TC}_{\overline{w}\,\overline{q},\overline{w}'\,\overline{q}'}\,((\overline{w}' = \overline{w}+1 \land \overline{q}' = \overline{q}) \lor (\overline{w}' = \overline{w}+1 \land \overline{q}' = \overline{q}+1\\ &\land \neg\overline{w} = \overline{y} \land d_\varphi(\overline{x},\overline{w}) \leq \overline{u} \land \neg\varphi(\overline{w},\overline{y})))]\tilde{0}\,\tilde{0}\tilde{0}\,\tilde{1}\,\overline{z}))]\tilde{0}\,\tilde{0}\tilde{0}\,\tilde{1}\,\overline{z}'.\end{aligned}$$

By (2), we can view χ_φ as an FO(posTC)-formula. Now

$$\rho_\varphi(\overline{x},\overline{z}) \ := \ [\mathrm{TC}_{\overline{u}\,\overline{z},\overline{u}'\,\overline{z}'}\,(\overline{u}' = \overline{u}+1 \land \chi_\varphi(\overline{x},\overline{u},\overline{z},\overline{z}'))]0\tilde{0}\,0\tilde{0}\,1\tilde{0}\,\overline{z}$$

expresses that $g(1\tilde{0}) = \overline{z}$ and hence, that $\|\{\overline{y} \mid d_\varphi(\overline{x},\overline{y}) < \infty\}\| = \overline{z}$. We can regard ρ_φ as an FO(posTC)-formula. This finishes the proof of (3).

As remarked above, $\neg[\mathrm{TC}_{\overline{x},\overline{y}}\,\psi(\overline{x},\overline{y})]\,\overline{s}\overline{t}$ is equivalent to

$$\|\{\overline{y}' \mid d_\psi(\overline{s},\overline{y}') < \infty\}\| = \|\{\overline{y}' \mid d_{(\psi(\overline{x},\overline{y})\land\neg\overline{y}=\overline{t})}\,(\overline{s},\overline{y}') < \infty\}\|$$

and thus, by (3), to the FO(posTC)-formula

$$\exists\overline{z}(\rho_\psi(\overline{s},\overline{z}) \land \rho_{(\psi(\overline{x},\overline{y})\land\neg\overline{y}=\overline{t})}(\overline{s},\overline{z})). \qquad \square$$

In view of 8.6.14 we now have

Corollary 8.6.16 *Let* $\tau \supseteq \{<, S, \min, \max\}$. *On ordered structures with at least two elements, every* FO(TC)*-formula is equivalent to a formula of the form*

$$[\mathrm{TC}_{\overline{x},\overline{x}'}\,\psi(\overline{x},\overline{x}')]\widetilde{\min}\widetilde{\max}$$

where ψ *is quantifier-free.* $\square$

Exercise 8.6.17 Let τ be arbitrary. Show that for any existential formula φ of FO(posTC) there is a quantifier-free ψ such that in structures with at least two elements, φ and $\exists(\forall)u[\mathrm{TC}_{\overline{x},\overline{x}'}\,\psi(\overline{x},\overline{x}')]\,\tilde{u}$ are equivalent. Conclude that, on ordered structures, the same results hold for FO(TC). Hint: Note that in structures with at least two elements any φ of the form

$$(*) \qquad [\mathrm{TC}_{\overline{x},\overline{x}'}\,\chi(\overline{x},\overline{x}')]\,\overline{s}\overline{t}$$

is equivalent to $\exists(\forall)u[\mathrm{TC}_{\overline{x}vw,\overline{x}'v'w'}\,\psi(\overline{x}vw,\overline{x}'v'w')]\,\tilde{u}\,\tilde{u}$, where

$$\begin{aligned}\psi(\overline{x}vw,\overline{x}'v'w') \ := \ & (\overline{x} = \tilde{v} = \tilde{w} \land \overline{x}' = \overline{s} \land \neg v' = w')\\ \lor\ & (\neg v = w \land \chi(\overline{x},\overline{x}') \land \neg v' = w')\\ \lor\ & (\overline{x} = \overline{t} \land \overline{x}' = \tilde{v} = \tilde{w}).\end{aligned}$$

To show that any existential FO(posTC)-formula φ is equivalent to a formula of the form (*) with quantifier-free χ, get rid of the constants c and d in the proof of 8.6.13: the role of c is taken over by any (all) pairs vw with $v = w$, that of d by any (all) vw with $v \neq w$ (similarly as in ψ). $\square$

The next example shows that the class of formulas having the normal form of Proposition 8.6.13 does not contain all FO(posTC)-formulas.

Example 8.6.18 Let $\tau = \{E, c, d\}$ where E is binary. There is no sentence of the form

$$(*) \qquad [\mathrm{TC}_{\overline{x},\overline{x}'}\psi(\overline{x},\overline{x}')]\,\tilde{c}\tilde{d},$$

where ψ is first-order, that is equivalent to $\forall x \forall y [\mathrm{TC}_{x,y} Exy]\, xy$. By contradiction, suppose that (*) is such a sentence. Let $\overline{x} = x_1 \dots x_m$ and set $k := \mathrm{qr}(\psi)$. Choose l big enough compared with k and m. Since $(\mathcal{G}_l, 0, 1) \models \forall x \forall y [\mathrm{TC}_{x,y} Exy]\, xy$ (for $\mathcal{G}_l$ cf. 2.3.8), we have $(\mathcal{G}_l, 0, 1) \models [\mathrm{TC}_{\overline{x},\overline{x}'}\psi(\overline{x},\overline{x}')]\,\tilde{c}\tilde{d}$. Hence, there are $\tilde{0} = \overline{e}_0, \overline{e}_1, \dots, \overline{e}_n = \tilde{1}$ with

$$(\mathcal{G}_l, 0, 1) \models \psi[\overline{e}_i, \overline{e}_{i+1}]$$

for $i < n$. In view of 2.3.8 and the choice of l, we have

$$(\mathcal{G}_l, 0, 1)\,\dot{\cup}\,\mathcal{G}_l \models \psi[\overline{e}_i, \overline{e}_{i+1}],$$

where $\overline{e}_i, \overline{e}_{i+1}$ denote the corresponding elements in the first copy. But then

$$(\mathcal{G}_l, 0, 1)\,\dot{\cup}\,\mathcal{G}_l \models [\mathrm{TC}_{\overline{x},\overline{x}'}\psi(\overline{x},\overline{x}')]\,\tilde{c}\tilde{d},$$

even though $(\mathcal{G}_l, 0, 1)\,\dot{\cup}\,\mathcal{G}_l \not\models \forall x \forall y [\mathrm{TC}_{x,y} Exy]\, xy$. $\square$

8.6.3 FO(TC) < FO(LFP)

The preceding example motivates how to handle the TC operation in a gametheoretic characterization of the FO(TC)-equivalence of structures. Given $m, r \geq 1$ and structures $\mathcal{A}$ and $\mathcal{B}$, the game $\text{TC-G}^r_m(\mathcal{A}, \mathcal{B})$ consists of m moves and is played, as usual, by two players, the spoiler and the duplicator (we shall see below that the number r is related to the arity of the TC operators). In every move there are chosen some elements (possibly more than one) both in $\mathcal{A}$ and $\mathcal{B}$. Let $a_1, \dots, a_s \in A$ and $b_1, \dots, b_s \in B$ be the elements chosen in the first i moves of a play and denote by p_i the assignment $a_1 \dots a_s \mapsto b_1 \dots b_s$. The duplicator wins the play if p_i is a partial isomorphism for all $i \leq m$ (or, equivalently, if p_m is a partial isomorphism). In the i-th move the spoiler first decides whether the move should be a point move or a TC move; moreover, he chooses one of the structures, say $\mathcal{A}$. The point moves are as in the first-order case. In a TC move the spoiler selects a j with $1 \leq j \leq r$ and, for some $k \geq 1$, a sequence $\overline{a}_0, \dots, \overline{a}_k$ of j-tuples in $\mathcal{A}$ with

$\overline{a}_0$ and $\overline{a}_k$ in the domain of p_{i-1} (i.e., $\overline{a}_0$ and $\overline{a}_k$ consists of elements already chosen or of constants). The duplicator answers by selecting, for some $l \geq 1$, a sequence $\overline{b}_0, \ldots, \overline{b}_l$ of j-tuples in $\mathcal{B}$ with $\overline{b}_0 = p_{i-1}(\overline{a}_0)$ and $\overline{b}_l = p_{i-1}(\overline{a}_k)$. Then the spoiler selects some $l' < l$ and the duplicator answers with a $k' < k$. This finishes the i-th move, the elements chosen in this move being $\overline{a}_{k'}, \overline{a}_{k'+1}$ and $\overline{b}_{l'}, \overline{b}_{l'+1}$.

TC moves in which the spoiler chooses the structure $\mathcal{A}$ are called *positive*, those in which he chooses $\mathcal{B}$, *negative.*

Define the quantifier rank as in the case of FO with the additional clause

$$\mathrm{qr}([\mathrm{TC}_{\overline{x},\overline{y}}\, \varphi]\, \overline{s}\overline{t}) \quad := \quad \mathrm{qr}(\varphi) + 1.$$

Then one can show:

Theorem 8.6.19 *The following are equivalent:*

(i) *The duplicator has a winning strategy for* TC-$\mathrm{G}^r_m(\mathcal{A}, \mathcal{B})$.
(ii) *$\mathcal{A}$ and $\mathcal{B}$ satisfy the same* FO(TC)*-sentences of quantifier rank $\leq m$ containing the* TC *operator only in the form* $[\mathrm{TC}_{\overline{x},\overline{y}}\, \varphi]\, \overline{s}\overline{t}$ *with* $\mathrm{length}(\overline{x}) \leq r$.

Proof. The proof proceeds along the usual lines; for the implication (ii) $\Rightarrow$ (i) we give a direct proof of the only nontrivial step.

We assume (ii) and explain a winning strategy for the duplicator in TC-$\mathrm{G}^r_m(\mathcal{A}, \mathcal{B})$. Suppose that in the first s moves of a play elements $\overline{a}$ in $\mathcal{A}$ and $\overline{b}$ in $\mathcal{B}$ have been chosen and that

$$(*) \qquad \mathcal{A} \models \varphi[\overline{a}] \quad \text{iff} \quad \mathcal{B} \models \varphi[\overline{b}]$$

holds for all FO(TC)-formulas $\varphi(\overline{x})$ of quantifier rank $\leq m - s$ ($s = 0$ corresponds to the assumption (ii)). Let $p := \overline{a} \mapsto \overline{b}$. Assume that the spoiler decides that the (s+1)-th move should be a TC-move, and let j with $1 \leq j \leq r$ and the sequence $\overline{a}_0, \ldots, \overline{a}_k$ of j-tuples be his selections. For $i < k$ set

$$\Psi_i \quad := \quad \{\psi(\overline{x}, \overline{y}, \overline{z}) \mid \mathrm{qr}(\psi) \leq m - s - 1,\ \mathcal{A} \models \psi[\overline{a}, \overline{a}_i, \overline{a}_{i+1}]\}$$

and let $\chi(\overline{x}, \overline{y}, \overline{z}) := \bigvee_{i<k} \bigwedge \Psi_i$. (Of course, $\bigwedge \Psi_i$ is equivalent to a finite conjunction.) Clearly,

$$\mathcal{A} \models [\mathrm{TC}_{\overline{y},\overline{z}}\, \chi(\overline{a}, \overline{y}, \overline{z})]\, \overline{y}\, \overline{z}\, [\overline{a}_0, \overline{a}_k].$$

By $(*)$,

$$(+) \qquad \mathcal{B} \models [\mathrm{TC}_{\overline{y},\overline{z}}\, \chi(\overline{b}, \overline{y}, \overline{z})]\, \overline{y}\, \overline{z}\, [p(\overline{a}_0), p(\overline{a}_k)].$$

The duplicator selects a χ-path $p(\overline{a}_0) = \overline{b}_0, \overline{b}_1, \ldots, \overline{b}_l = p(\overline{a}_k)$ witnessing (+). Suppose the spoiler now chooses $\overline{b}_{l'}, \overline{b}_{l'+1}$. Then, for some $k' < k$, $\mathcal{B} \models \Psi_{k'}[\overline{b}, \overline{b}_{l'}, \overline{b}_{l'+1}]$, and the duplicator answers by $\overline{a}_{k'}, \overline{a}_{k'+1}$. By definition of $\Psi_{k'}$,

$$\mathcal{A} \models \psi[\overline{a}, \overline{a}_{k'}, \overline{a}_{k'+1}] \quad \text{iff} \quad \mathcal{B} \models \psi[\overline{b}, \overline{b}_{l'}, \overline{b}_{l'+1}]$$

holds for all FO(TC)-formulas of quantifier rank $\leq m - s - 1$. □

We now aim at a proof of FO(TC) $<$ FO(LFP). The following classes of FO(TC)-formulas will be useful. For $m \geq 0$ and $r \geq 1$ define P^r_m by induction on m as follows:

- P^r_0 is the set of quantifier-free formulas.
- P^r_{2m+1} is the closure under conjunctions and disjunctions of the set $P^r_{2m} \cup \{[\mathrm{TC}_{\overline{x},\overline{y}}\, \varphi]\, \overline{s}\overline{t} \mid \varphi \in P^r_{2m},\ \mathrm{length}(\overline{x}) \leq r\}$.
- P^r_{2m+2} is the closure under conjunctions and disjunctions of the set $P^r_{2m+1} \cup \{\neg[\mathrm{TC}_{\overline{x},\overline{y}}\, \neg\varphi]\, \overline{s}\overline{t} \mid \varphi \in P^r_{2m+1},\ \mathrm{length}(\overline{x}) \leq r\}$.

Since $\exists x \varphi(x)$ and $[\mathrm{TC}_{x,y}((x = c \land \varphi(y)) \lor (\varphi(x) \land y = c))]\, cc$ are equivalent, one gets by a simple induction on formulas:

Lemma 8.6.20 *If the vocabulary τ contains at least one constant then every formula of* FO(TC) *is equivalent to a formula in* $\bigcup_{m \geq 0, r \geq 1} P^r_m$. □

We show FO(TC) $<$ FO(LFP) by exhibiting a class of structures axiomatizable in FO(LFP) but not in FO(TC). The underlying idea is already present in the following example.

Example 8.6.21 Let $r \geq 1$ and $\tau = \{E, F, c\}$ with E binary and F ("full") unary. Let $\mathcal{A}$ and $\mathcal{B}$ be the structures

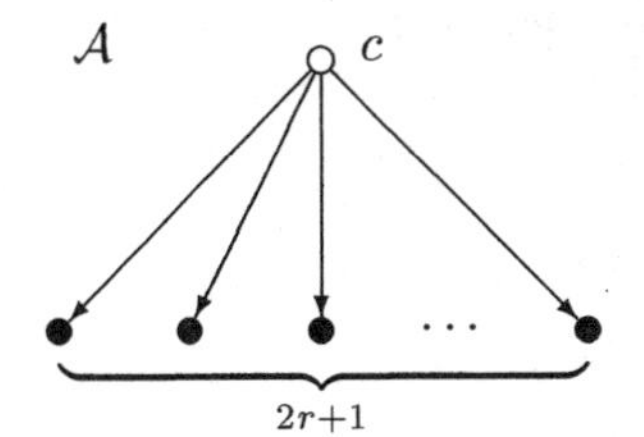

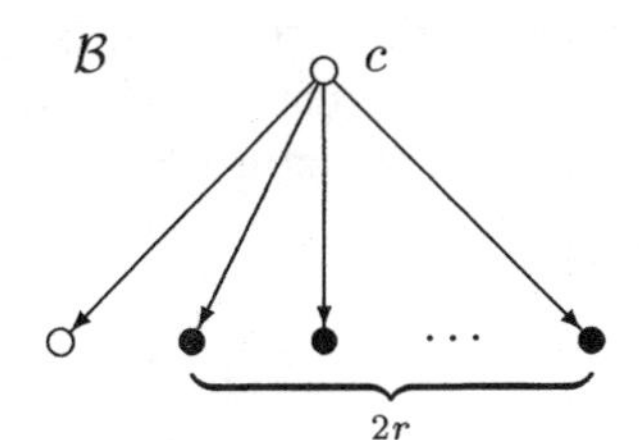

Points for which F holds are represented by full dots, $\overset{a}{\underset{b}{\downarrow}}$ means that $Ea\,b$. We show for quantifier-free φ and $\overline{x}$ with $\mathrm{length}(\overline{x}) \leq r$ that

$$\mathcal{A} \models [\mathrm{TC}_{\overline{x},\overline{y}}\, \varphi]\, \tilde{c}\tilde{c} \quad \text{implies} \quad \mathcal{B} \models [\mathrm{TC}_{\overline{x},\overline{y}}\, \varphi]\, \tilde{c}\tilde{c}.$$

In fact, suppose that $\tilde{c} = \overline{a}_0, \ldots, \overline{a}_l = \tilde{c}$ witnesses $(\tilde{c}, \tilde{c}) \in \mathrm{TC}(\varphi^{\mathcal{A}}(\_,\_))$. Successively choose $\overline{b}_0 = \tilde{c}, \overline{b}_1, \ldots, \overline{b}_l = \tilde{c}$ in $\mathcal{B}$ such that $\overline{a}_i\, \overline{a}_{i+1} \mapsto \overline{b}_i\, \overline{b}_{i+1}$ is a partial isomorphism (such elements exist, since $\mathcal{B}$ contains $2r$ full elements). As φ is quantifier-free, $\mathcal{B} \models \varphi[\overline{b}_i, \overline{b}_{i+1}]$ and hence, $(\tilde{c}, \tilde{c}) \in \mathrm{TC}(\varphi^{\mathcal{B}}(\_,\_))$. □

Using suitable refinements of $\mathcal{A}$, $\mathcal{B}$ one can handle nested TC operators and show:

Theorem 8.6.22 FO(TC) $<$ FO(LFP).

Proof. Fix an $r \geq 1$ and let $\tau_0 := \{E, F\}$, where E and F are as in the preceding example. For $m \geq 0$ define τ_0-structures $\mathcal{A}_{r,m}$ and $\mathcal{B}_{r,m}$ and their *roots* by induction on m as follows:

$$\mathcal{A}_{r,0}: \quad \circ \qquad\qquad \mathcal{B}_{r,0}: \quad \bullet$$

and the respective element is the root. Suppose that $\mathcal{A}_{r,m}$ and $\mathcal{B}_{r,m}$ and their roots have already been defined. Set

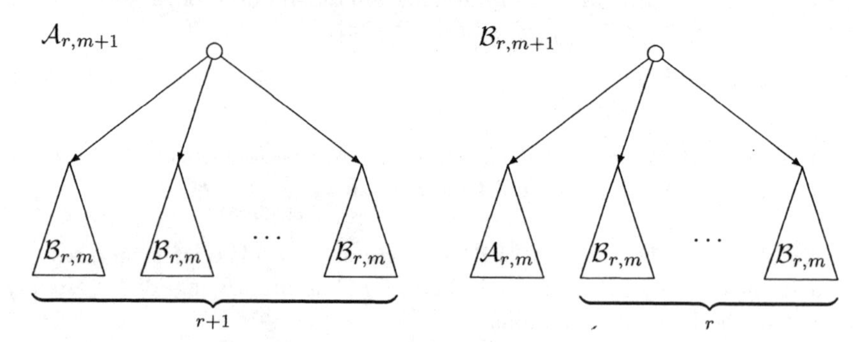

that is, $\mathcal{A}_{r,m+1}$ is the disjoint union of $r+1$ copies of $\mathcal{B}_{r,m}$ together with a new element, the root of $\mathcal{A}_{r,m+1}$, that is connected by the relation E to the roots of the distinct copies of $\mathcal{B}_{r,m}$. The structure $\mathcal{B}_{r,m+1}$ is defined in the same way, but one copy of $\mathcal{B}_{r,m}$ is replaced by $\mathcal{A}_{r,m}$. Note that only leaves can be full. The structure $\mathcal{B}_{2,2}$ looks like

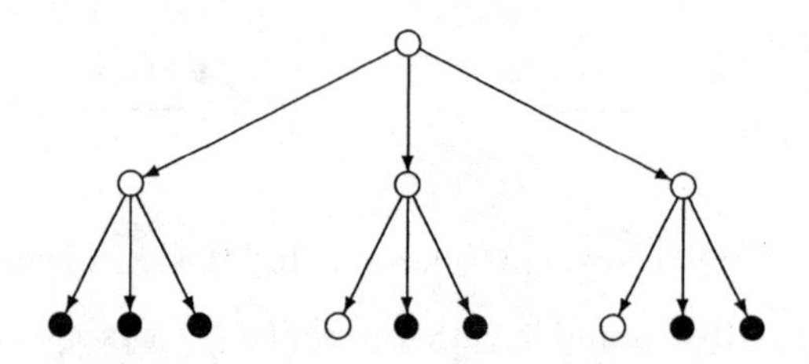

Let $\tau = \tau_0 \cup \{c\}$ and let $\mathcal{A}'_{r,m}$ and $\mathcal{B}'_{r,m}$ be the τ-structures ($\mathcal{A}_{r,m}$, "root of $\mathcal{A}_{r,m}$") and ($\mathcal{B}_{r,m}$, "root of $\mathcal{B}_{r,m}$"), respectively.

We recall or introduce some terminology. Let $\mathcal{D}$ be one of the structures $\mathcal{A}_{r,m}$ or $\mathcal{B}_{r,m}$. For $d, d' \in D$ we say that d' is a *successor* of d if $E^D dd'$, and that d' is a descendant of d if for some $l \geq 0$ there are $d_0 = d, d_1, \ldots, d_l = d'$ such that $E^D d_i d_{i+1}$ for $i < l$. The *height* $h(d)$ of d is the length of the largest E-path starting at d, i.e.,

$$h(d) := \max\{l \geq 0 \mid \text{there are } d_0 = d, d_1, \ldots, d_l \text{ with } E^D d_i d_{i+1} \text{ for } i < l\}.$$

The substructure $\mathcal{T}(d)$ of $\mathcal{D}$ whose universe $T(d)$ is the set of descendants of d is called the *tree generated by* d.

Define the function colour: $D \to \{\text{black, white}\}$ by

$$(*) \qquad \text{colour}(d) := \begin{cases} \text{black} & \text{if } \mathcal{T}(d) \cong \mathcal{B}_{r,h(d)} \\ \text{white} & \text{otherwise.} \end{cases}$$

The inductive definition of $\mathcal{A}_{r,m}$ and $\mathcal{B}_{r,m}$ leads to the following definition of colour by induction on the height: The point d is of black colour if it is full (i.e., if $F^D d$ holds) or if it has a white successor. More precisely, colour(d) = black iff

$$\mathcal{D} \models [\text{IFP}_{x,X}(Fx \vee \exists y \exists z(Exy \wedge Exz \wedge Xy \wedge \neg Xz))]\, x\, [d].$$

Using ◍ for a black point and ○ for a white point, the structure $\mathcal{B}_{2,2}$ looks as follows:

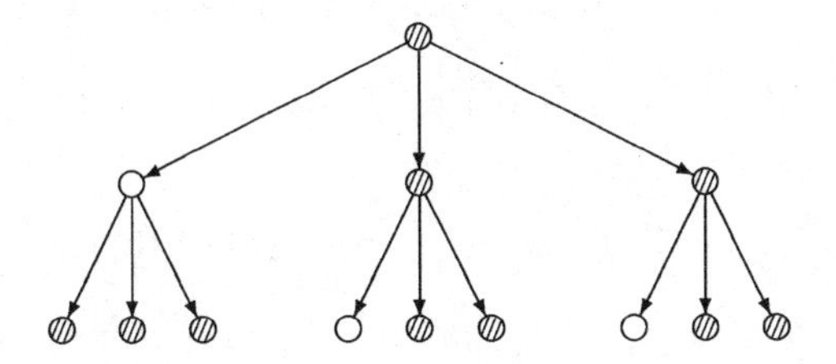

Let φ_0 be a sentence of FO(LFP)$[\tau]$ saying that the root is black, that is, an FO(LFP)-sentence equivalent to $[\text{IFP}_{x,X}(Fx \vee \exists y \exists z(Exy \wedge Exz \wedge Xy \wedge \neg Xz))]\, c$. Then, in view of $(*)$, we have

(1) For all $m \geq 0, r \geq 1$: $\quad \mathcal{A}'_{r,m} \models \neg\varphi_0$ and $\mathcal{B}'_{r,m} \models \varphi_0$.

Therefore, by Lemma 8.6.20, we are done if we show:

(2) Let $k \geq 0$ and let φ be an FO(TC)-sentence in P^r_k. Then for $m > 2 \cdot k$,

$$\mathcal{A}'_{2r,m} \models \varphi \qquad \text{iff} \qquad \mathcal{B}'_{2r,m} \models \varphi.$$

We prove (2) by induction on k, showing a corresponding statement for formulas and assignments to their variables. As in the example preceding this theorem one carefully has to choose the elements in the right copies. To make precise the notion of "right copy" we have in mind, we introduce the concept of an *amenable* function. A function $\pi : A_{r,m} \to B_{r,m}$ is *amenable* if

- $\pi : \mathcal{A}_{r,m}|\{E\} \cong \mathcal{B}_{r,m}|\{E\}$
- For every $l \leq m$ there exists exactly one $a \in A_{r,m}$ such that $h(a) = l$ and colour$(\pi(a)) \neq$ colour(a). The element a is called the π-*critical* point of height l.

Note that if π is amenable, a its critical point of height l, a' its critical point of height l', and $l' < l$, then $a' \in T(a)$. There is an amenable function from

$\mathcal{A}_{r,m}$ to $\mathcal{B}_{r,m}$; this is obtained by induction on m : If $\pi_0 : \mathcal{A}_{r,m} \to \mathcal{B}_{r,m}$ is amenable, then $\pi : \mathcal{A}_{r,m+1} \to \mathcal{B}_{r,m+1}$ is amenable, where (compare the figures defining $\mathcal{A}_{r,m+1}$ and $\mathcal{B}_{r,m+1}$) π maps the root of $\mathcal{A}_{r,m+1}$ to the root of $\mathcal{B}_{r,m+1}$, π equals π_0^{-1} on the leftmost copy of $\mathcal{B}_{r,m}$ that is attached to the root of $\mathcal{A}_{r,m+1}$, and is the identity on the remaining copies.

In the example preceding this theorem, when choosing the elements $\overline{b}_i$, we avoided the element of $\mathcal{B}$, which is not full and distinct from the root. Here we have to avoid elements below critical points of suitable height. More precisely, one can show by induction on k:

(3) Suppose that $s, k \geq 0$ and $m > 2 \cdot k$. Let $\overline{a}_1, \overline{a}'_1, \dots, \overline{a}_s, \overline{a}'_s$ be r-tuples in $A_{2r,m}$. Furthermore, let $\pi : A_{2r,m} \to \mathcal{B}_{2r,m}$ be amenable and let a be its critical point of height $2k$. Finally, assume that

$$\{\overline{a}_1, \overline{a}'_1, \dots, \overline{a}_s, \overline{a}'_s\} \cap T(a) = \emptyset.$$

Then, for every $\varphi(\overline{x}_1, \overline{x}'_1, \dots, \overline{x}_s, \overline{x}'_s) \in P^r_k$ with free variables among the displayed ones,

$$\mathcal{A}'_{2r,m} \models \varphi[\overline{a}_1, \overline{a}'_1, \dots, \overline{a}_s, \overline{a}'_s] \text{ iff } \mathcal{B}'_{2r,m} \models \varphi[\pi(\overline{a}_1), \pi(\overline{a}'_1), \dots, \pi(\overline{a}_s), \pi(\overline{a}'_s)].$$

Once this is shown, we get (2) by taking $s := 0$ and as π any amenable function from $\mathcal{A}_{2r,m}$ to $\mathcal{B}_{2r,m}$.

In the inductive proof of (3) the case $k = 0$ is obvious. To prove (3) for $k \geq 1$ under the assumption that it is already proved for $k-1$, it suffices to show for $m > 2 \cdot k$ (writing $\overline{a}$ for $\overline{a}_1 \overline{a}'_1 \dots \overline{a}_s \overline{a}'_s$)

$$(+) \qquad \mathcal{A}'_{2r,m} \models [\mathrm{TC}_{\overline{x},\overline{y}}\, \psi]\, \overline{s}\overline{t}\, [\overline{a}] \text{ iff } \mathcal{B}'_{2r,m} \models [\mathrm{TC}_{\overline{x},\overline{y}}\, \psi]\, \overline{s}\overline{t}\, [\pi(\overline{a})]$$

where $\text{length}(\overline{x}) = r$ and $\psi \in P^r_{k-1}$.

First we assume that the left side of (+) holds, $\overline{c}_0, \dots, \overline{c}_l$ being a witnessing sequence. Our aim is to construct a witnessing sequence for the right side. As $\overline{c}_1$ contains less than $2r+1$ elements, not all of the $2r+1$ copies of $\mathcal{B}_{2r,2k-1}$ attached to a contain one of these elements. Hence, we can find $\overline{d}_1$ in $\mathcal{B}_{2r,m}$, but outside the copy of $\mathcal{A}_{2r,2k-1}$ attached to $\pi(a)$, and an amenable π' : $\mathcal{A}'_{2r,m} \to \mathcal{B}'_{2r,m}$ that coincides with π outside of $T(a) \setminus \{a\}$ such that

$$\pi'(\overline{c}_1) = \overline{d}_1$$

and such that $\overline{a}, \overline{c}_0, \overline{c}_1$ and π' satisfy (3) for $k-1$. Hence, by induction hypothesis,

$$\mathcal{B}'_{2r,m} \models \psi[\pi'(\overline{a}), \pi'(\overline{c}_0), \pi'(\overline{c}_1)].$$

Treating now the pair $\overline{c}_1, \overline{c}_2$ in the same way ($\overline{c}_1, \overline{c}_2$ contains less than $2r+1$ elements!), where the counterpart $\overline{d}_1$ of $\overline{c}_1$ is already defined, and continuing in this way up to the pair $\overline{c}_{l-1}, \overline{c}_l$, we get a witnessing sequence $\overline{d}_0, \overline{d}_1, \dots, \overline{d}_l$ for the right side of (+).

For the other direction we argue similarly. Concerning elements of the witnessing sequence in the copy of $\mathcal{A}_{2r,2k-1}$ attached to $\pi(a)$, we proceed with this copy as with $T(a)$ above, thus, altogether, decreasing the height of the critical point of the induction hypothesis by two. □

Notes 8.6.23 Theorem 8.6.9 is due to Grädel and McColm [56]. Subsection 8.6.2 is based on [93, 94]. The trees in subsection 8.6.3 go back to [18].

8.7 Bounded Fixed-Point Logic

In the preceding section we already mentioned that an application of the TC operator can be replaced by an application of the LFP operator. In fact,

$$\models_{\text{fin}} [\text{TC}_{\overline{x},\overline{y}}\varphi(\overline{x},\overline{y})]\,\overline{s}\overline{t} \leftrightarrow [\text{LFP}_{\overline{y},Y}\,(\varphi(\overline{s},\overline{y}) \vee \exists\overline{x}(Y\overline{x} \wedge \varphi(\overline{x},\overline{y})))]\,\overline{t}.$$

Conversely, every FO(LFP)-formula of the form

$$[\text{LFP}_{\overline{y},Y}(\psi(\overline{y}) \vee \exists\overline{x}(Y\overline{x} \wedge \varphi(\overline{x},\overline{y})))]\,\overline{t}$$

where Y does not occur in ψ and φ, is equivalent to

$$\psi(\overline{t}) \vee \exists\overline{x}\big(\psi(\overline{x}) \wedge [\text{TC}_{\overline{x},\overline{y}}\,\varphi(\overline{x},\overline{y})]\,\overline{x}\overline{t}\big).$$

In this sense, transitive closures correspond to those least fixed-points, where each tuple in a new stage is already witnessed by a *single* tuple of the preceding stage. We generalize transitive closure logic to *Bounded Fixed-Point Logic* FO(BFP), which allows the LFP operator only if there is a bound $r \geq 1$ such that each tuple in a new stage is already witnessed by a set of at most r many tuples of the preceding stage. More precisely: For a vocabulary τ the class FO(BFP)[τ] of formulas of FO(BFP) of vocabulary τ is given by

- the rules of first-order logic starting from first-order atomic formulas (i.e., no relation variables are allowed)
- the "BFP-rule"

$$\frac{\varphi_0(\overline{y}),\quad \varphi_1(\overline{y},\overline{x}_1,\ldots,\overline{x}_r)}{[\text{LFP}_{\overline{y},Y}(\varphi_0(\overline{y}) \vee \exists\overline{x}_1\ldots\exists\overline{x}_r\,(Y\overline{x}_1 \wedge \ldots \wedge Y\overline{x}_r \wedge \varphi_1(\overline{y},\overline{x}_1,\ldots,\overline{x}_r)))]\,\overline{t}}$$

  where Y is an s-ary relation variable for some $s \geq 1$, and where $r \geq 0$ and $\overline{x}_1,\ldots,\overline{x}_r$ are s-tuples of variables, all variables in $\overline{x}_1,\ldots,\overline{x}_r$ being pairwise distinct. (Note that Y does not occur free in φ_0 or φ_1.)

We abbreviate the formula above by

$$[\text{LFP}_{\overline{y},Y}\,(\varphi_0(\overline{y}) \vee \exists\overline{x}_1 \in Y \ldots \exists\overline{x}_r \in Y\varphi_1(\overline{y},\overline{x}_1,\ldots,\overline{x}_r))]\,\overline{t}.$$

Sometimes, bounded fixed-point logic is called *stratified fixed-point logic.* This name will become clear in the next chapter.

Denote by $\mathrm{FO(BFP}_{r_0})$ the logic obtained by restricting the BFP-rule to r with $r \le r_0$. The introductory remarks show part (c) of the following proposition, parts (a) and (b) being trivial.

Proposition 8.7.1 (a) $\mathrm{FO(BFP)} \le \mathrm{FO(LFP)}$
(b) $\mathrm{FO(BFP_0)} \equiv \mathrm{FO}$
(c) $\mathrm{FO(BFP_1)} \equiv \mathrm{FO(TC)}$
(d) $\mathrm{FO(BFP_2)} \equiv \mathrm{FO(BFP)}$.

Proof. (d) The proof proceeds by induction on FO(BFP)-formulas. In the main step we show that

$$(*) \qquad [\mathrm{LFP}_{\overline{y},Y}\,(\varphi_0(\overline{y}) \vee \exists \overline{x}_1 \in Y \ldots \exists \overline{x}_r \in Y \varphi_1(\overline{y}, \overline{x}_1, \ldots, \overline{x}_r))]\,\overline{t}$$

is equivalent to a formula of $\mathrm{FO(BFP_2)}$ under the assumption that φ_0 and φ_1 are already in $\mathrm{FO(BFP_2)}$. For simplicity we assume that Y is unary and hence, $\overline{y} = y, \overline{x}_1 = x_1, \ldots, \overline{x}_r = x_r$. Denote by $Y_0, Y_1, \ldots$ the stages corresponding to the LFP operation in $(*)$. Let Z be a new r-ary relation symbol. The fixed-point operation of the following formula ψ stepwise incorporates the elements of a new stage Y_n into the different components of the tuples in Z.

$$\begin{aligned}\psi(y_1, \ldots, y_r) \quad := \quad & [\mathrm{LFP}_{\overline{z},Z}\,((\varphi_0(z_1) \wedge \ldots \wedge \varphi_0(z_r)) \vee \exists \overline{x} \in Z \exists \overline{y} \in Z \exists y \\ & (\varphi_1(y, \overline{x}) \wedge \bigvee_{i=1,\ldots,r} (z_i = y \wedge \bigwedge_{j \ne i} z_j = y_j)))]\,\overline{y}.\end{aligned}$$

If $Z_0, Z_1, \ldots$ are the stages of the LFP operation in ψ, then one easily shows by induction on n:

$$(1) \qquad \text{for all } n, \qquad Z_n \subseteq Y_n^r$$

$$(2) \qquad \text{for all } n \text{ there is an } m \text{ such that} \quad Y_n^r \subseteq Z_m.$$

Hence, $Y_\infty^r = Z_\infty$, and therefore,

$$\models_{\mathrm{fin}} [\mathrm{LFP}_{y,Y}\,(\varphi_0 \vee \exists x_1 \in Y \ldots \exists x_r \in Y \varphi_1)]\,t \leftrightarrow \psi(t, \ldots, t). \qquad \square$$

We come back to the trees $\mathcal{A}_{r,m}$ and $\mathcal{B}_{r,m}$ introduced in the proof of 8.6.22. A point of one of these structures got the colour "black", if it was (a leaf and) full or if at least (= exactly) one successor was white. Thus the colour of a point was determined by the colour of its $r+1$ successors. It is therefore not surprising that we can find a sentence in $\mathrm{FO(BFP}_{r+1})$ which separates the classes $\{\mathcal{A}'_{r,m} \mid m \ge 0\}$ and $\{\mathcal{B}'_{r,m} \mid m \ge 0\}$. In fact, we have (recall that c denotes the root of the structures in question)

$$F_\infty^{\psi_r} = \{(c, y) \mid \text{ colour}(y) = \text{ black}\} \cup \{(x, y) \mid x \ne c, \text{ colour}(y) = \text{ white}\}$$

for the formula $\psi_r(x, y, Y) :=$

$$((x = c \wedge Fy) \vee (\neg x = c \wedge \forall z \neg Eyz \wedge \neg Fy)) \vee \exists x_0 y_0 \in Y \dots \exists x_r y_r \in Y$$
$$(\bigwedge_{0 \le i < j \le r} y_i \ne y_j \wedge \bigwedge_{i \le r} Eyy_i \wedge$$
$$((x = c \wedge \bigvee_{i \le r} \neg x_i = c) \vee (\neg x = c \wedge \bigwedge_{i \le r} x_i = c))).$$

Hence for all $m > 0$,

$$\mathcal{A}'_{r,m} \models \neg[\text{LFP}_{xy,Y}\psi_r]\,cc \quad \text{and} \quad \mathcal{B}'_{r,m} \models [\text{LFP}_{xy,Y}\psi_r]\,cc.$$

It can be shown that (2) in the proof of 8.6.22 can be strengthened to

for any FO(TC)-sentence φ there is an m_0 such that for all $m \ge m_0$,

$$\mathcal{A}'_{2,m} \models \varphi \;\text{ iff }\; \mathcal{B}'_{2,m} \models \varphi.$$

Therefore, the FO(BFP)-sentence $[\text{LFP}_{xy,Y}\,\psi_2]cc$ is not equivalent to an FO(TC)-sentence. This gives the first inequality in the next proposition. To obtain the second inequality one shows that the FO(LFP)-sentence φ_0 of (1) in the proof of 8.6.22 is not equivalent to an FO(BFP)-sentence; that is, there is no FO(BFP)-sentence separating the sets $\{\mathcal{A}'_{r,m} \mid r \ge 1, m \ge 0\}$ and $\{\mathcal{B}'_{r,m} \mid r \ge 1, m \ge 0\}$.

Theorem 8.7.2 FO(TC) $<$ FO(BFP) $<$ FO(LFP). □

We shall see in Theorem 9.1.6 that FO(BFP) $\equiv$ FO(LFP) on ordered structures. In the BFP-rule the fixed-point operation is applied to formulas $(\varphi_0 \vee \exists \overline{x}_1 \in Y \dots \exists \overline{x}_r \in Y\,\varphi_1)$ which do not contain free relation variables besides Y, and where Y only occurs positively. In fact, this is the only essential restriction, as is shown by the following result that we need in the next chapter.

Proposition 8.7.3 *Let Y be a second-order variable and let ψ be a formula built up from* FO(BFP$_r$)*-formulas and atomic formulas of the form $Y\overline{s}$ with $\neg, \wedge, \vee$, and $\exists$. Furthermore, suppose that Y is not in the scope of any negation sign. If Y occurs at most r times in ψ, then $[\text{LFP}_{\overline{y},Y}\psi]\,\overline{t}$ is equivalent to a formula of* FO(BFP$_r$).

Proof. To each such ψ with exactly s occurrences of Y we associate a formula ψ^* of the form $\exists \overline{x}_1 \in Y \dots \exists \overline{x}_n \in Y\,\psi_1$, where ψ_1 is an FO(BFP$_r$)-formula that does not contain Y and is equivalent to ψ whenever a nonempty relation is assigned to Y. Then, $[\text{LFP}_{\overline{y},Y}\psi(\overline{y}, Y)]\,\overline{t}$ is equivalent to

$$[\text{LFP}_{\overline{y},Y}(\psi(\overline{y}, \emptyset) \vee \exists \overline{x}_1 \in Y \dots \exists \overline{x}_n \in Y\psi_1)]\,\overline{t},$$

a formula of FO(BFP$_r$). We define ψ^* by induction on ψ: If $\psi = Y\overline{s}$ then $\psi^* := \exists \overline{x} \in Y\,\overline{x} = \overline{s}$. Set $\psi^* := \psi$ for any formula not containing Y; in particular, for $\psi = \neg\varphi$ we have $\psi^* = \neg\varphi$ since, by hypothesis, φ does not contain

Y. The definition of ψ^* for the other cases is straightforward; it uses that $(\varphi \vee \exists \overline{x}_1 \in Y \ldots \exists \overline{x}_n \in Y \, \psi)$ and $\exists \overline{x}_1 \in Y \ldots \exists \overline{x}_n \in Y \, (\varphi \vee \psi)$ are equivalent in case no variable in $\overline{x}_1, \ldots, \overline{x}_n$ is free in φ and provided a nonempty relation is assigned to Y (and a similar equivalence for the conjunction). □

Notes 8.7.4 Stratified logic was introduced and studied first by Chandra and Harel [19]. References for the results given are [23, 102, 61].

9. Logic Programs

In the present chapter we introduce a group of languages that originate in database theory. Our aim is to exhibit a close relationship to fixed-point logics, thus demonstrating that the two groups of logical systems are two sides of one and the same coin. We therefore may use methods and results from one group when investigating the other. Examples will illustrate the methodological means we thus have gained.

9.1 DATALOG

Fix a vocabulary τ. A *general logic program* is a finite set Π of *clauses* of the form

$$\gamma \leftarrow \gamma_1, \ldots, \gamma_l$$

where $l \geq 0$ and where $\gamma, \gamma_1, \ldots, \gamma_l$ are atomic or negated atomic first-order formulas. The formula γ must be an atomic formula of the form $R\bar{t}$. It is called the *head* of the clause. The sequence $\gamma_1, \ldots, \gamma_l$ constitutes the *body* of the clause. Every relation symbol occurring in the head of some clause of Π is *intentional*, all the other symbols in τ are *extensional*. Denote by $(\tau, \Pi)_{\text{int}}$ and $(\tau, \Pi)_{\text{ext}}$ the set of intentional and extensional symbols, respectively. Hence, $(\tau, \Pi)_{\text{ext}} = \tau \setminus (\tau, \Pi)_{\text{int}}$. We often write τ_{int} and τ_{ext} if no confusion is possible.

A DATALOG *program* is a general logic program, in which no intentional symbol occurs negated in the body of any clause. Examples of DATALOG programs are

$$\begin{array}{llllll}
\Pi_0: & Txy \leftarrow Exy & \Pi_1: & P\min \leftarrow & \Pi_2: & Px \leftarrow \neg x = c \\
 & Txz \leftarrow Txy, Eyz & & Qy \leftarrow Px, Sxy & & Px \leftarrow \neg Rx \\
 & & & Py \leftarrow Qx, Sxy & &
\end{array}$$

The intended meaning of a DATALOG program Π is obtained by interpreting its clauses in a dynamic way as rules, the clause $\gamma \leftarrow \gamma_1, \ldots, \gamma_l$ as the rule

$$(*) \qquad \text{whenever } \gamma_1, \ldots, \gamma_l \text{ then } \gamma.$$

The values of the extensional symbols are given by a τ_{ext}-structure. All the rules of Π will be applied simultaneously to generate consecutive stages of the intentional symbols. For Π_0 for example, the final value of T in a graph $\mathcal{G} = (G, E^G)$ will be the transitive closure of E^G. We come to the exact definition. To simplify the presentation of the semantics we suppose that for any intentional symbol P of Π, say, of arity r, there are distinct variables $\overline{x}_P = x_{P,1} \ldots x_{P,r}$ such that any clause in Π with head symbol P has the form

$$P\overline{x}_P \leftarrow \gamma_1, \ldots, \gamma_l$$

(otherwise, carry out suitable changes having in mind the intended interpretation; for example, replace the clause $P\min \leftarrow$ in Π_1 by $Px \leftarrow x = \min$ (where $\overline{x}_P = x$), and a clause $Rxyx \leftarrow Py, Sx\min$ by $Rxyz \leftarrow Py, Sx\min, x = z$ (where $\overline{x}_R = xyz$)). With every intentional P we associate the formula

$$\varphi_P(\overline{x}_P) \quad := \quad \bigvee \{\exists \overline{y}\, (\gamma_1 \wedge \ldots \wedge \gamma_l) \mid P\overline{x}_P \leftarrow \gamma_1, \ldots, \gamma_l \text{ in } \Pi\}$$

where $\overline{y}$ contains the variables in $\gamma_1, \ldots, \gamma_l$ that are distinct from the variables in $\overline{x}_P$. Assume that $P^1, \ldots, P^k$ are the intentional symbols of Π. By definition of a DATALOG program they do not occur negated in the body of any clause. Therefore $\varphi_{P^1}, \ldots, \varphi_{P^k}$ are positive in $P^1, \ldots, P^k$ and hence, their simultaneous fixed-point $(P^1_{(\infty)}, \ldots, P^k_{(\infty)})$ given by $P^i_{(\infty)} := \bigcup_{n \geq 0} P^i_{(n)}$ exists, where

$$P^i_{(0)} := \emptyset, \qquad P^i_{(n+1)} := \{\overline{x}_{P^i} \mid \varphi_{P^i}(\overline{x}_{P^i}, P^1_{(n)}, \ldots, P^k_{(n)})\}.$$

For a τ_{ext}-structure $\mathcal{A}$ we can rewrite the definition of $P^i_{(n)}$ in such a way that the procedural character of the DATALOG program Π is emphasized:

$$P^i_{(0)} = \emptyset$$

$$(+) \qquad \begin{aligned} P^i_{(n+1)} \quad = \quad & \{\overline{a} \mid \text{there are } P^i\overline{x}_{P^i} \leftarrow \gamma_1, \ldots, \gamma_l \text{ in } \Pi \text{ and } \overline{b} \in A \\ & \text{such that } (\mathcal{A}, P^1_{(n)}, \ldots, P^k_{(n)}) \models \gamma_1 \wedge \ldots \wedge \gamma_l\, [\overline{a}, \overline{b}]\} \end{aligned}$$

(where $\overline{a}$ interprets the variables in $\overline{x}_{P^i}$ and $\overline{b}$ the remaining ones).

Thus, a DATALOG program Π and a τ_{ext}-structure $\mathcal{A}$ give rise to the τ-structure

$$\mathcal{A}[\Pi] \quad := \quad (\mathcal{A}, P^1_{(\infty)}, \ldots, P^k_{(\infty)}).$$

And every intentional symbol P of Π defines a query on the class of τ_{ext}-structures given by $\mathcal{A} \overset{\Pi}{\longmapsto} P^{\mathcal{A}[\Pi]}$.

For the DATALOG program Π_1 at the beginning of this section and any ordering $\mathcal{B}$ as $\{<, S, \min, \max\}$-structure, we have $\mathcal{B}[\Pi_1] = (\mathcal{B}, P^B, Q^B)$, where P^B and Q^B consist of the elements at odd and even positions, respectively.

A DATALOG *formula* has the form $(\Pi, P)\overline{t}$ where P is an intentional symbol of Π, say of arity r, and $\overline{t} = t_1 \ldots t_r$ are terms. It is a formula of vocabulary

τ_{ext}, its free variables being those occurring in some t_i. For a τ_{ext}-structure $\mathcal{A}$ and $\overline{a} \in A$,

$$\mathcal{A} \models (\Pi, P)\overline{t}\,[\overline{a}] \qquad \text{iff} \qquad (t_1[\overline{a}], \ldots, t_r[\overline{a}]) \in P^{\mathcal{A}[\Pi]}.$$

In order to get sentences and thus to be able to compare the expressive power of DATALOG programs with other logics, it is desirable to allow zero-ary relation symbols. If $P \in \tau$ is 0-ary then, in a structure $\mathcal{A}$, $P^{\mathcal{A}}$ is the boolean value TRUE or FALSE. Denoting the empty sequence by $\emptyset$, for zero-ary P^i the inductive step (+) above reads as follows:

$$\begin{aligned} P^i_{(n+1)} \quad := \quad & \{\emptyset \mid \text{there are a clause } P^i \leftarrow \gamma_1, \ldots, \gamma_l \text{ in } \Pi \text{ and } \overline{b} \in A \\ & \quad \text{such that } (\mathcal{A}, P^1_{(n)}, \ldots, P^k_{(n)}) \models \gamma_1 \wedge \ldots \wedge \gamma_l\,[\emptyset, \overline{b}]\}. \end{aligned}$$

So, "$P^{\mathcal{A}} = \emptyset$" means "$P^{\mathcal{A}} =$ FALSE", "$P^{\mathcal{A}} = \{\emptyset\}$" means "$P^{\mathcal{A}} =$ TRUE".

To give an example, consider the program

$$\begin{aligned} \Pi_4 : \qquad Txy &\leftarrow Exy \\ Txz &\leftarrow Txy, Tyz \\ P &\leftarrow Tcd \end{aligned}$$

with a 0-ary relation symbol P. Then, for any graph $\mathcal{G} := (G, E^G, c^G, d^G)$ with distinguished points c^G and d^G, $P^{\mathcal{G}[\Pi_4]}$ is true just in case c^G and d^G are connected by a path; in other words, $\mathcal{G} \models (\Pi_4, P)$ iff c^G and d^G are connected by a path.

We now come to a generalization of DATALOG programs. Let Π be a DATALOG program and $\mathcal{A}$ a $(\tau, \Pi)_{\text{ext}}$-structure. Then $\mathcal{A}[\Pi]$ is a τ-structure. To $\mathcal{A}[\Pi]$ one can apply a further DATALOG program. In such a program all symbols in τ are extensional and thus may occur negated in the bodies of clauses. An iteration of this process leads to the notion of a stratified DATALOG program. A *stratified* DATALOG program (by short: S-DATALOG program) Σ consists of a sequence $\Pi_0, \ldots, \Pi_n$ of DATALOG programs of vocabularies $\tau_0, \ldots, \tau_n$, where $(\tau_{m+1}, \Pi_{m+1})_{\text{ext}} = \tau_m$ for $m < n$. Given a $(\tau_0, \Pi_0)_{\text{ext}}$-structure $\mathcal{A}$, we set

$$\mathcal{A}[\Sigma] \quad := \quad (\ldots(\mathcal{A}[\Pi_0])[\Pi_1]\ldots)[\Pi_n].$$

As with DATALOG, $(\Sigma, P)\overline{t}$ is an S-DATALOG *formula* if P is an intentional symbol of one of the constituents Π_i. For example, $\Sigma := \Pi_0, \Pi_1$ with

$$\Pi_0 : \quad Px \leftarrow Exy \qquad \Pi_1 : \quad Lx \leftarrow \neg Px.$$

is an S-DATALOG program. In any tree, the S-DATALOG formula $(\Sigma, L)x$ defines the set of leaves.

The following result explains why bounded fixed-point logic is sometimes called stratified fixed-point logic.

Theorem 9.1.1 S-DATALOG $\equiv$ FO(BFP).

We give the proof in such a way that it yields a stronger result: The *breadth* of a DATALOG program is the number of occurrences of intentional symbols in the bodies. Thus the program

$$\begin{array}{rcl} Txy & \leftarrow & Exy \\ Txz & \leftarrow & Txy, Tyz \\ P & \leftarrow & Tcd \end{array}$$

has breadth 3. The *breadth* of an S-DATALOG program $\Pi_0, \Pi_1, \ldots, \Pi_k$ is, by definition, given by $\max_{i \leq k}$ breadth(Π_i). Denote by S-DATALOG$_r$ the class of S-DATALOG formulas $(\Sigma, Q)\overline{t}$ where Σ is of breadth $\leq r$. Then the preceding theorem can be strengthened to:

Lemma 9.1.2 *For all* $r \geq 0$, S-DATALOG$_r \equiv$ FO(BFP$_r$).

Proof. We show FO(BFP$_r$) $\leq$ S-DATALOG$_r$ by assigning to every formula φ of FO(BFP$_r$) with free variables among $\overline{y}$ an equivalent S-DATALOG$_r$ formula $(\Sigma_\varphi, Q_\varphi)\overline{y}$. We do this by induction on φ, tacitly assuming that the intentional symbols of the programs Π_ψ, $\Pi_\chi, \ldots$ we refer to are distinct.

- If $\varphi(\overline{y})$ is atomic, set $\Sigma_\varphi := \{Q_\varphi \overline{y} \leftarrow \varphi\}$.
- If $\varphi(\overline{y}) = \neg\psi$, set $\Sigma_\varphi := \Sigma_\psi, \{Q_\varphi \overline{y} \leftarrow \neg Q_\psi \overline{y}\}$.
- If $\varphi(\overline{y}) = (\psi \vee \chi)$, set $\Sigma_\varphi := \Sigma_\psi, \Sigma_\chi, \{Q_\varphi \overline{y} \leftarrow Q_\psi \overline{y},\ Q_\varphi \overline{y} \leftarrow Q_\chi \overline{y}\}$.
- If $\varphi(\overline{y}) = \exists x \psi(\overline{y}, x)$, set $\Sigma_\varphi := \Sigma_\psi, \{Q_\varphi \overline{y} \leftarrow Q_\psi \overline{y} x\}$.
- For $\varphi(\overline{y}) = [\text{LFP}_{\overline{x}, X}(\psi(\overline{y}, \overline{x}) \vee \exists \overline{x}_1 \in X \ldots \exists \overline{x}_s \in X\, \chi(\overline{y}, \overline{x}, \overline{x}_1, \ldots, \overline{x}_s))]\,\overline{t}$ with $s \leq r$ let

$$\begin{array}{llcl} \Pi: & Q\overline{y}\,\overline{x} & \leftarrow & Q_\psi \overline{y}\,\overline{x} \\ & Q\overline{y}\,\overline{x} & \leftarrow & Q\overline{y}\,\overline{x}_1, \ldots, Q\overline{y}\,\overline{x}_s, Q_\chi \overline{y}\,\overline{x}\,\overline{x}_1 \ldots \overline{x}_s \end{array}$$

  and set $\Sigma_\varphi := \Sigma_\psi, \Sigma_\chi, \Pi, \{Q_\varphi \overline{y} \leftarrow Q\overline{y}\overline{t}\}$.

For the direction S-DATALOG$_r \leq$ FO(BFP$_r$) we first study the case of a (plain) DATALOG formula $(\Pi, P)\overline{t}$ where Π is of breadth $\leq r$. Let $P^1, \ldots, P^k$ with $P = P^1$ be its intentional symbols. By previous remarks we may assume that the clauses have the form

$$Q\overline{x}_Q \leftarrow \gamma_1, \ldots, \gamma_l$$

where $\overline{x}_Q$ is a sequence of distinct variables. Let

$$\varphi_Q(\overline{x}_Q) \quad := \quad \bigvee \{\exists \overline{y}(\gamma_1 \wedge \ldots \wedge \gamma_l) \mid Q\overline{x}_Q \leftarrow \gamma_1 \ldots, \gamma_l \text{ in } \Pi\}$$

where $\overline{y}$ contains all the variables in the clause that are not in $\overline{x}_Q$. By definition of the semantics of DATALOG, the formula $(\Pi, P)\overline{t}$ is equivalent to

$$[\text{S-LFP}_{\overline{x}_{P^1},P^1,\ldots,\overline{x}_{P^k},P^k}\,\varphi_{P^1},\ldots,\varphi_{P^k}]\,\overline{t}.$$

Since $\varphi_{P^1},\ldots,\varphi_{P^k}$ are existential and altogether, $P^1,\ldots,P^k$ occur at most r times, the corresponding formula $\chi_J(\overline{z},Z)$ in 8.2.3 is existential, too, and contains Z at most r times, and $(\Pi,P^i)\overline{t}$ is equivalent to

$$(*) \qquad \exists u[\text{LFP}_{\overline{z},Z}\,\chi_J]\,\overline{t}\tilde{u}.[1]$$

By 8.7.3 this formula is equivalent to a formula in FO(BFP_r).

Now in any S-DATALOG$_r$ program of breadth $\leq r$ of the form $\Sigma := \Pi,\Pi_1$ we can deal with Π_1 as with Π, where for the intentional symbols of Π occurring in Π_1 (as extensional symbols!) we use the defining formulas in FO(BFP_r) just gotten. For an arbitrary S-DATALOG program $\Sigma := \Pi_0,\ldots,\Pi_n$ of breadth $\leq r$ the result follows by induction on n. □

Using 8.7.1 we get:

Corollary 9.1.3 S-DATALOG$_0$ $\equiv$ FO, S-DATALOG$_1$ $\equiv$ FO(TC), *and* S-DATALOG$_2$ $\equiv$ S-DATALOG. □

What is the logic corresponding to plain DATALOG? We get a hint from the observation that in the formula in $(*)$ of the preceding proof the least fixed-point operator is applied to an existential formula. In fact, we have:

Theorem 9.1.4 DATALOG *and* Existential Fixed-Point Logic E(LFP) *have the same expressive power.*

Here the class of formulas of E(LFP) is given by

- $\dfrac{}{\varphi}$ where φ is an atomic second-order formula
- $\dfrac{}{\neg\varphi}$ where φ is an atomic first-order formula
- $\dfrac{\varphi,\psi}{(\varphi\wedge\psi)}$, $\dfrac{\varphi,\psi}{(\varphi\vee\psi)}$, $\dfrac{\varphi}{\exists x\varphi}$
- $\dfrac{\varphi}{[\text{LFP}_{\overline{y},Y}\varphi]\,\overline{t}}$

Proof (of 9.1.4). DATALOG $\leq$ E(LFP): We saw in the proof of the preceding theorem that any DATALOG formula $(\Pi,P^i)\overline{t}$ is equivalent to the formula in $(*)$. But this formula is a formula of E(LFP).

E(LFP) $\leq$ DATALOG: The proof parallels that of FO(BFP) $\leq$ S-DATALOG, but now in every step of the induction the S-DATALOG program given there has to be joined to a single DATALOG program. Proceed as follows:

[1] This formula is only equivalent to the DATALOG program in models with at least two elements. To remove this restriction one can proceed as follows: On structures of cardinality 1, the k-th stage is already the final one. But the k-th stage of P^1 can be expressed by an existential first-order formula ψ (compare 9.3.1). Then, for arbitrary structures, we have to replace the formula χ_J in $(*)$ by $(\psi\vee\chi_J)$.

- For $\varphi(\overline{y})$ atomic set $\Pi_\varphi := \{Q_\varphi \overline{y} \leftarrow \varphi\}$.
- For $\varphi(\overline{y}) = \neg\psi$ with atomic ψ set $\Pi_\varphi := \{Q_\varphi \overline{y} \leftarrow \neg\psi\}$.
- For $\varphi(\overline{y}) = (\psi \wedge \chi)$ set $\Pi_\varphi := \Pi_\psi \cup \Pi_\chi \cup \{Q_\varphi \overline{y} \leftarrow Q_\psi \overline{y}, Q_\chi \overline{y}\}$.
- For $\varphi(\overline{y}) = (\psi \vee \chi)$ set $\Pi_\varphi := \Pi_\psi \cup \Pi_\chi \cup \{Q_\varphi \overline{y} \leftarrow Q_\psi \overline{y},\ Q_\varphi \overline{y} \leftarrow Q_\chi \overline{y}\}$.
- For $\varphi(\overline{y}) = \exists x \psi(\overline{y}, x)$ set $\Pi_\varphi := \Pi_\psi \cup \{Q_\varphi \overline{y} \leftarrow Q_\psi \overline{y} x\}$.
- Suppose $\varphi(\overline{y}) = [\mathrm{LFP}_{\overline{x},X}\psi]\,\overline{t}$. Assume w.l.o.g. that the individual variables free in ψ are among $\overline{x}$. Consider X in ψ as a relation symbol (and not as a relation variable) and let Π_ψ be the DATALOG program corresponding to ψ by induction hypothesis and having X as an extensional symbol; X does not occur negated in the body of a clause, since ψ is positive in X. Then as Π_φ we can take the following program (that contains X as an intentional symbol): $\Pi_\varphi := \Pi_\psi \cup \{X\overline{x} \leftarrow Q_\psi \overline{x},\ Q_\varphi \overline{y} \leftarrow X\overline{t}\}$. □

Since in graphs the E(LFP) formula

$$[\mathrm{LFP}_{xy,X}\,(Exy \vee \exists z(Xxz \wedge Ezy))]\,cd$$

expresses that there is a path from c to d, we see that E(LFP) $\not\leq$ FO. On the other hand, we also have FO $\not\leq$ E(LFP). An example of an FO-formula not equivalent to a formula of E(LFP) is given by $\forall x Px$, since a simple proof by induction shows that E(LFP) has the following monotonicity property:

Proposition 9.1.5 *If $\varphi(x_1, \dots, x_n)$ is a formula of* E(LFP) *and $a_1, \dots, a_n \in A$, then*

$$\mathcal{A} \models \varphi[a_1, \dots, a_n] \text{ and } \mathcal{A} \subseteq \mathcal{B} \quad \text{imply} \quad \mathcal{B} \models \varphi[a_1, \dots, a_n]. \qquad \Box$$

Let $\tau \supseteq \{<, S, \min, \max\}$. In presence of $S, \min, \max$ we have for ordered structures:

$$\mathcal{A} \subseteq \mathcal{B} \quad \text{implies} \quad \mathcal{A} = \mathcal{B}.$$

Thus, on ordered structures, the monotonicity property of the preceding proposition holds for all logics.

Theorem 9.1.6 *Let $\tau \supseteq \{<, S, \min, \max\}$. On ordered structures, we have* E(LFP) $\equiv$ FO(LFP), *that is, the expressive power of* DATALOG, S-DATALOG, E(LFP), FO(BFP), *and* FO(LFP) *coincide.*

Proof. To show that FO(LFP) $\leq$ E(LFP) recall (cf. 8.2.9) that every formula of FO(LFP) is equivalent to a formula of the form

$$\exists x\,[\mathrm{LFP}_{\overline{y},Y}\psi]\,\overline{t}$$

where ψ is first-order. Thus it suffices to show that on ordered structures universal quantification can be replaced by an LFP operation (according to E(LFP)). For this note that $\forall x \varphi(x)$ is equivalent to

$$[\mathrm{LFP}_{x,X}\,(\varphi(\min) \wedge x = \min) \vee \exists y(Xy \wedge Syx \wedge \varphi(x))]\,\max. \qquad \Box$$

Exercise 9.1.7 A DATALOG formula $(\Pi, P)\bar{t}$ is *positive* if no negated atomic formula at all occurs in the body of any clause. Introduce positive existential fixed-point logic and show that it has the same expressive power as positive DATALOG. □

Exercise 9.1.8 Let $\Sigma := \Pi_0, \ldots, \Pi_k$ and $\Sigma' := \Pi'_0, \ldots, \Pi'_l$ be S-DATALOG programs with $\Pi_0 \cup \ldots \cup \Pi_k = \Pi'_0 \cup \ldots \cup \Pi'_l$. Show that $\mathcal{A}[\Sigma] = \mathcal{A}[\Sigma']$ for any structure $\mathcal{A}$. Hint: Consider a finest subdivision of $\Pi_0 \cup \ldots \cup \Pi_k$ leading to an S-DATALOG program and show that Σ and Σ' are equivalent to it. □

Exercise 9.1.9 Show that on ordered structures the successor relation S is not definable in E(LFP) in terms of $<, \min, \max$. (Hint: Use 9.1.5.) Show that $<$ is definable in E(LFP) using S. □

Exercise 9.1.10 Extend the semantics of E(LFP) and DATALOG to infinite structures and show the equivalence of E(LFP) and DATALOG for arbitrary structures. Conclude

- If $\varphi(\overline{x})$ is a formula of E(LFP), $\mathcal{A}$ is an arbitrary structure, $\overline{a} \in A$, and $\mathcal{A} \models \varphi[\overline{a}]$, then there is a finite $A_0 \subseteq A$ such that $\overline{a} \in A_0$, and for all $\mathcal{B}$ with $A_0 \subseteq B$, if A_0 generates the same substructure in $\mathcal{B}$ as in $\mathcal{A}$ then $\mathcal{B} \models \varphi[\overline{a}]$.
- If an E(LFP)-formula has a model then it has a finite model. □

9.2 I-DATALOG and P-DATALOG

Until now we have introduced some natural classes of general logic programs, and we have seen that they correspond to certain fixed-point logics. We now ask the other way round: What classes of programs correspond to, say, inflationary (= least) fixed-point logic and to partial fixed-point logic?

First we show that inflationary fixed-point logic FO(IFP) can be captured by general logic programs, if we equip them with an "inflationary" semantics. We speak of I-DATALOG, the "I" standing for the inflationary aspect of its semantics.

Being a general logic program, an I-DATALOG program Π may have clauses whose body also contains negated intentional symbols. The semantics is fixed in a way that consistently extends the conventions for DATALOG. Similarly as with DATALOG, we may suppose that for any intentional symbol P of Π there are distinct variables $\overline{x}_P$ such that any clause in Π with head symbol P has the form

$$P\overline{x}_P \leftarrow \gamma_1, \ldots, \gamma_l.$$

We set

$$\varphi_P \quad := \quad \bigvee \{\exists \overline{y}\, (\gamma_1 \wedge \ldots \wedge \gamma_l) \mid P\overline{x}_P \leftarrow \gamma_1, \ldots, \gamma_l \text{ in } \Pi\}$$

where $\overline{y}$ contains the variables of the corresponding clause that are distinct from the variables in $\overline{x}_P$. Assume that $P^1, \ldots, P^k$ are the intentional symbols of Π. Then the simultaneous inflationary operations corresponding to $\varphi_{P^1}, \ldots, \varphi_{P^k}$ give rise to the sequences

$$P^i_{(0)} := \emptyset, \quad P^i_{(n+1)} := P^i_{(n)} \cup \{\overline{x}_{P^i} \mid \varphi_{P^i}(\overline{x}_{P^i}, P^1_{(n)}, \ldots, P^k_{(n)})\}$$

with the simultaneous fixed-point given by $P^i_{(\infty)} := \bigcup_{n \geq 0} P^i_{(n)}$.

Let $\mathcal{A}$ be a τ_{ext}-structure. Then Π leads to the τ-structure

$$\mathcal{A}[\Pi] \quad := \quad (\mathcal{A}, P^1_{(\infty)}, \ldots, P^k_{(\infty)}).$$

If P is an intentional symbol of Π, $(\Pi, P)\overline{t}$ is an I-DATALOG *formula*, its meaning in a τ_{ext}-structure $\mathcal{A}$ being $\overline{t} \in P^{\mathcal{A}[\Pi]}$.

Exercise 9.2.1 Let $\Pi = \{Px \leftarrow Rx, \quad Qx \leftarrow \neg Px\}$ be an I-DATALOG program and Σ be the S-DATALOG program $\{Px \leftarrow Rx\}$, $\{Qx \leftarrow \neg Px\}$. Show for $\mathcal{A} = (A, R^A)$ that $Q^{\mathcal{A}[\Pi]} = A$ and $Q^{\mathcal{A}[\Sigma]} = A \setminus R^A$. In this sense the semantics of I-DATALOG and S-DATALOG programs are not compatible. □

Theorem 9.2.2 I-DATALOG $\equiv$ FO(IFP).

Proof. By the definitions just given, an I-DATALOG formula $(\Pi, P)\overline{t}$ with intentional symbols $(P =)P^1, P^2, \ldots, P^k$ is equivalent to

$$[\text{S-IFP}_{\overline{x}_{P^1}, P^1, \ldots, \overline{x}_{P^k}, P^k}\, \varphi_{P^1}, \ldots, \varphi_{P^k}]\,\overline{t}$$

and hence, is definable in FO(IFP). Thus, I-DATALOG $\leq$ FO(IFP).

To show FO(IFP) $\leq$ I-DATALOG, to every formula φ of FO(IFP) with free first-order variables among $\overline{y}$ we assign an equivalent I-DATALOG-formula $(\Pi_\varphi, Q_\varphi)\overline{y}$. Free relation variables in φ are treated as relation symbols, hence, they are extensional symbols of Π_φ. Π_φ contains an intentional 0-ary symbol DONE_φ. DONE_φ becomes true at the last step of the evaluation of Π_φ (more precisely, when DONE_φ is true, all intentional symbols have reached their final value). We proceed by induction on φ:

$\varphi(\overline{y})$ atomic: Set $\Pi_\varphi := \left\{ \begin{array}{rcl} Q_\varphi \overline{y} & \leftarrow & \varphi \\ \text{DONE}_\varphi & \leftarrow & \end{array} \right\}$

$\varphi(\overline{y}) := \neg\psi$: Let Π_φ consist of the clauses of Π_ψ and

$$\begin{array}{rcl} Q_\varphi \overline{y} & \leftarrow & \text{DONE}_\psi, \neg Q_\psi \overline{y} \\ \text{DONE}_\varphi & \leftarrow & \text{DONE}_\psi \end{array}$$

$\varphi(\overline{y}) := (\psi \vee \chi)$: Let Π_φ consist of the clauses of $\Pi_\psi \cup \Pi_\chi$ and the clauses

$$\begin{array}{rcl} Q_\varphi \overline{y} & \leftarrow & Q_\psi \overline{y} \\ Q_\varphi \overline{y} & \leftarrow & Q_\chi \overline{y} \\ \text{DONE}_\varphi & \leftarrow & \text{DONE}_\psi, \text{DONE}_\chi \end{array}$$

$\varphi(\overline{y}) := \exists x \psi(\overline{y}, x)$: Let Π_φ consist of the clauses of Π_ψ and the clauses

$$\begin{array}{rcl} Q_\varphi \overline{y} & \leftarrow & Q_\psi \overline{y} x \\ \text{DONE}_\varphi & \leftarrow & \text{DONE}_\psi \end{array}$$

$\varphi(\overline{y}) := [\text{IFP}_{\overline{x},X} \psi]\, \overline{t}$: Assume w.l.o.g. that all first-order variables free in ψ are among $\overline{x} = x_1 \dots x_k$. By induction hypothesis there is a program Π_ψ corresponding to ψ (with X as an extensional symbol). Thus, for a given interpretation of X, the program Π_ψ returns as Q_ψ the set $\{\overline{x} \mid \psi(\overline{x}, X)\}$. For $\varphi(\overline{y})$ the program Π_ψ must be repeated several times until the fixed-point of the inflationary operation is reached. But note that in order to get the stages of the IFP operation, after each iteration step the intentional symbols of Π_ψ have to get empty again before the next iteration step starts. However, such a set back is excluded by the "inflationary" semantics of I-DATALOG. To overcome this difficulty we increase the arity of the intentional symbols, using the additional components for timestamps, the timestamps for the $(n+1)$-th iteration step being given by the new elements of X obtained in the n-th iteration step.

The program Π_φ given below contains as intentional symbols the intentional symbols of Π_ψ and X, Q_φ, DONE_φ, the 0-ary symbols NOTEMPTY, DELAY, DONE, the k-ary symbols OLD, I, and for every intentional symbol R of Π_ψ a new symbol R^* with $\text{arity}(R^*) = \text{arity}(R) + k$. We use the following notation: For new variables $\overline{z} = z_1 \dots z_k$, the program $\Pi^*_\psi[\overline{z}]$ is obtained by replacing in every clause of Π_ψ each (possibly negated) atomic formula of the form $R\overline{s}$ with an intentional R by $R^* \overline{s}\, \overline{z}$. And if Π is any program then $\Pi \,||\, \gamma_1, \dots, \gamma_m$ is obtained by appending $\gamma_1, \dots, \gamma_m$ to the bodies of all clauses in Π.

The following remarks should help to read Π_φ: The clauses in (1) take care of the first iteration of ψ, (2) - (4) stop the program in case the first iteration returns an empty result, and (5) updates the value of X. The subsequent iterations of ψ are simulated by the clauses in (6) timestamped with the tuples added at the previous stage. By (7) the value of X is updated and the old value is stored in OLD by use of (8). (9) checks whether no new value has been added to X. In this case (10) and (12) stop the program. By (11) the goal symbol Q_φ gets its right value.

(1) $\Pi_\psi \,||\, \neg\text{DONE}_\psi$

(2) $\text{NOTEMPTY} \leftarrow \text{DONE}_\psi, Q_\psi \overline{x}$

(3) $\text{DELAY} \leftarrow \text{DONE}_\psi$

(4) $\text{DONE}_\varphi \leftarrow \text{DELAY}, \neg\text{NOTEMPTY}$

(5) $X\overline{x} \leftarrow \text{DELAY}, Q_\psi \overline{x}$

(6) $\Pi^*_\psi[\overline{z}] \,||\, X\overline{z}, \neg\text{OLD}\overline{z}$
(7) $X\overline{x} \leftarrow \text{DONE}^*_\psi\overline{z}, Q^*_\psi\overline{x}\,\overline{z},$
(8) $\text{OLD}\overline{z} \leftarrow \text{DONE}^*_\psi\overline{z}$
(9) $I\overline{z} \leftarrow \text{DONE}^*_\psi\overline{z}, Q^*_\psi\overline{x}\,\overline{z}, \neg X\overline{x}$
(10) $\text{DONE} \leftarrow \text{OLD}\overline{z}, \neg I\overline{z}$
(11) $Q_\varphi\overline{y} \leftarrow \text{DONE}, X\overline{t}$
(12) $\text{DONE}_\varphi \leftarrow \text{DONE}$ □

Remark 9.2.3 Recall the definition of FO(IFP, #), fixed-point logic with counting, given in subsection 8.4.1. DATALOG(#), *Datalog with Counting*, extends DATALOG by allowing two sorts (the point sort and the number sort) and admitting two-sorted intentional relation symbols and certain terms of the number sort in its clauses. More precisely: Call a term t of the number sort *basic* if it is either a variable μ or min, max or of the form $\#_x R\overline{y}\,\overline{\mu}$ for an intentional R. Now, a DATALOG(#) program is a finite set of clauses

$$\gamma \leftarrow \gamma_1, \ldots, \gamma_l$$

where γ is atomic and relational and $\gamma_1, \ldots, \gamma_l$ are atomic or negated atomic formulas, all of whose terms of the second sort are basic; moreover, intentional symbols do not occur in negated atomic formulas. This last condition is not a real restriction, since we already remarked in part (e) of 8.4.17 that $\neg R\overline{y}\,\overline{\mu}$ is equivalent to $\#_x R\overline{y}\,\overline{\mu} = \min$. So negated intentional relation symbols are available via unnegated atomic formulas containing counting terms. We equip DATALOG(#) with the inflationary semantics. Now it is not difficult to adapt the proof of the preceding theorem in order to show

$$\text{DATALOG}(\#) \equiv \text{FO(IFP, \#)}.$$ □

We turn to partial fixed-point logic and introduce P-DATALOG that will correspond to FO(PFP). The syntax of P-DATALOG is the same as that of I-DATALOG, that is, every general logic program is a P-DATALOG program. The semantics is given by the following conventions: Let Π be a P-DATALOG program. As in the preceding cases we assume that every clause with the intentional symbol P in its head has the form

$$P\overline{x}_P \leftarrow \gamma_1, \ldots, \gamma_l.$$

Again, set

$$\varphi_P(\overline{x}_P) \quad := \quad \bigvee\{\exists\overline{y}\,(\gamma_1 \wedge \ldots \wedge \gamma_l) \mid P\overline{x}_P \leftarrow \gamma_1, \ldots, \gamma_l \text{ in } \Pi\}.$$

Assume that $P^1, \ldots, P^k$ are the intentional symbols of Π. Set

$$P^i_{(0)} := \emptyset, \quad P^i_{(n+1)} := \{\overline{x}_{P^i} \mid \varphi_{P^i}(\overline{x}_{P^i}, P^1_{(n)}, \ldots, P^k_{(n)})\}$$

and let $(P^1_{(\infty)}, \ldots, P^k_{(\infty)})$ be the simultaneous fixed-point. Thus, in case there is an n_0 such that $P^i_{(n_0+1)} = P^i_{(n_0)}$ for $i = 1, \ldots, k$, we have $P^1_{(\infty)} = P^1_{(n_0)}, \ldots, P^k_{(\infty)} = P^k_{(n_0)}$; otherwise, $P^i_{(\infty)} = \emptyset$ for $i = 1, \ldots, k$. Let $\mathcal{A}$ be a τ_{ext}-structure. Then Π leads to the τ-structure

$$\mathcal{A}[\Pi] \quad := \quad (\mathcal{A}, P^1_{(\infty)}, \ldots, P^k_{(\infty)}).$$

For any intentional symbol Q of Π and terms $\bar{t}$, $(\Pi, Q)\bar{t}$ is a P-DATALOG formula, its meaning in a τ_{ext}-structure $\mathcal{A}$ being $\bar{t} \in Q^{\mathcal{A}[\Pi]}$. [2]

Theorem 9.2.4 P-DATALOG $\equiv$ FO(PFP).

Proof. By the definitions just given, a P-DATALOG formula $(\Pi, Q)\bar{t}$, where Π has the intentional symbols $(Q =)P^1, P^2, \ldots, P^k$, is equivalent to

$$[\text{S-PFP}_{\bar{x}_{P^1}, P^1, \ldots, \bar{x}_{P^k}, P^k}\, \varphi_{P^1}, \ldots, \varphi_{P^k}]\,\bar{t}$$

and hence, to a formula of FO(PFP).

For the direction FO(PFP) $\leq$ P-DATALOG we proceed as in the case of FO(IFP) $\leq$ I-DATALOG, assigning to every formula φ of FO(PFP) with free first-order variables among $\bar{y}$ an equivalent P-DATALOG formula $(\Pi_\varphi, Q_\varphi)\bar{y}$, where Π_φ contains a 0-ary symbol DONE_φ taking the value TRUE in the last stage of the evaluation. Again, free second-order variables of φ are treated as extensional symbols in Π_φ. The atomic case and the cases $\varphi = \neg\psi$, $\varphi = (\psi \vee \chi)$ and $\varphi = \exists x \psi$ are handled as in 9.2.2. For $\varphi(\bar{y}) := [\text{PFP}_{\bar{x}, X}\psi]\,\bar{t}$, where w.o.l.g. the first-order variables free in ψ are among $\bar{x}$, the program Π_φ given below is simpler than the corresponding one for I-DATALOG, since in the semantics of P-DATALOG an intentional symbol Z gets $\emptyset$ in an evaluation step in case no positive Z-information is obtained. Π_φ contains as intentional symbols the intentional symbols of Π_ψ, X, DONE_φ, Q_φ, and the 0-ary symbols UPDATE, NOTFIXPOINT.

The clauses in (1) take care of an application of Π_ψ. Clauses (3) and (4) check whether a fixed-point has been reached. In the positive case, clauses (7)–(10) stop the program. Clauses (5) and (6) serve to update the values of the intentional symbols for the next application of Π_ψ.

(1) $\begin{array}{c}\Pi_\psi \\ X\bar{x} \leftarrow X\bar{x}\end{array} \; \| \; \neg\text{DONE}_\psi, \neg\text{UPDATE}$

(2) $\text{UPDATE} \leftarrow \text{DONE}_\psi$

(3) $\text{NOTFIXPOINT} \leftarrow \text{DONE}_\psi,\ Q_\psi\bar{x},\ \neg X\bar{x}$

(4) $\text{NOTFIXPOINT} \leftarrow \text{DONE}_\psi,\ X\bar{x},\ \neg Q_\psi\bar{x}$

(5) $Q_\psi\bar{x} \leftarrow \text{DONE}_\psi, Q_\psi\bar{x}$

(6) $X\bar{x} \leftarrow \text{UPDATE},\ \text{NOTFIXPOINT},\ Q_\psi\bar{x}$

[2] One easily verifies that the semantics extends the conventions for DATALOG.

(7) $\mathrm{DONE}_\varphi \leftarrow \mathrm{UPDATE}, \neg\mathrm{NOTFIXPOINT}$
(8) $Q_\varphi \overline{y} \leftarrow \mathrm{UPDATE}, \neg\mathrm{NOTFIXPOINT}, Q_\psi \overline{t}$
(9) $\mathrm{UPDATE} \leftarrow \mathrm{UPDATE}, \neg\mathrm{NOTFIXPOINT}$
(10) $Q_\psi \overline{x} \leftarrow \mathrm{UPDATE}, \neg\mathrm{NOTFIXPOINT}, Q_\psi \overline{x}$ □

9.3 A Preservation Theorem

A DATALOG program is *positive* if no negated atomic formula at all occurs in the body of any clause. Recall that, given a DATALOG program Π, an intentional symbol P, and $n \geq 0$, we denote by $P_{(n)}$ the n-th stage of P in the evaluation of Π.

Proposition 9.3.1 *Let Π be a DATALOG program. Then, for any intentional symbol P and $n \geq 0$, the relation $P_{(n)}$ is expressible by an existential first-order formula φ_P^n. If Π is positive then φ_P^n can be chosen existential positive (cf. 2.5.3).*

Proof. The proof proceeds by induction on n. If $n = 0$ then $\varphi_P^0 := \mathrm{F}$. For $n > 0$, let

$$\varphi_P(\overline{x}_P) = \bigvee \{\exists \overline{y}(\gamma_1 \wedge \ldots \wedge \gamma_l) \mid P\overline{x}_P \leftarrow \gamma_1, \ldots, \gamma_l \text{ in } \Pi\},$$

where $\overline{y}$ contains the variables in $\gamma_1, \ldots, \gamma_l$ that are not in $\overline{x}_P$. To obtain $\varphi_P^n(\overline{x}_P)$, replace in $\varphi_P(\overline{x}_P)$ any subformula $Q\overline{t}$ with intentional Q by $\varphi_Q^{n-1}(\overline{t})$. □

A DATALOG formula $(\Pi, Q)\overline{t}$ is *bounded* if there is an $n \geq 0$ such that $Q_{(n)} = Q_{(\infty)}$ holds in all structures. Thus a bounded (positive) DATALOG formula is equivalent to an existential (positive) first-order formula. On the other hand, an existential (positive) first-order formula is equivalent to a bounded (positive) DATALOG formula. In fact, let $\varphi(\overline{x})$ be existential (positive), w.l.o.g. of the form

$$\bigvee_{i=1,\ldots,s} \exists y_1 \ldots \exists y_{m_i} (\varphi_{i1} \wedge \ldots \wedge \varphi_{ik_i}),$$

where the φ_{ij} are atomic or negated atomic (atomic) formulas. Then $\varphi(\overline{x})$ is equivalent to the DATALOG formula $(\Pi, Q)\overline{x}$, where Π has the clauses

$$Q\overline{x} \leftarrow \varphi_{i1}, \ldots, \varphi_{ik_i}$$

for $i = 1, \ldots, s$ and where Q is a symbol not in φ. As $Q_{(1)} = Q_{(\infty)}$, $(\Pi, Q)\overline{x}$ is bounded.

Thus, by the proposition, the *bounded* positive DATALOG formulas correspond to the existential positive first-order formulas. The following preservation theorem shows that they are the only positive DATALOG formulas equivalent to first-order formulas, thus documenting once more that first-order logic lacks the ability to express recursion.

Theorem 9.3.2 *For a positive* DATALOG *formula* $(\Pi, Q)\overline{t}$ *the following are equivalent:*

(i) $(\Pi, Q)\overline{t}$ *is equivalent to a first-order formula.*

(ii) $(\Pi, Q)\overline{t}$ *is bounded.*

At the end of this section we sketch a proof of this result. First we draw some consequences.

Corollary 9.3.3 *Suppose that the positive* DATALOG *formulas* $(\Pi_1, Q_1)\overline{s}$ *and* $(\Pi_2, Q_2)\overline{t}$ *are equivalent. If* $(\Pi_1, Q_1)\overline{s}$ *is bounded, then* $(\Pi_2, Q_2)\overline{t}$ *is bounded, too.* □

Proposition 9.3.4 *For a positive* DATALOG *formula* $(\Pi, Q)\overline{x}$ *the following are equivalent:*

(i) $(\Pi, Q)\overline{x}$ *is bounded.*

(ii) $(\Pi, Q)\overline{x}$ *is equivalent to an existential positive first-order formula.*

(iii) $(\Pi, Q)\overline{x}$ *is equivalent to a first-order formula.*

(iv) *There is an* $n \geq 1$ *such that for all* $\mathcal{A}$ *and* $\overline{a} \in Q^{\mathcal{A}[\Pi]}$ *there is a substructure* $\mathcal{B}$ *of* $\mathcal{A}$ *with* $\|B\| \leq n$ *and* $\overline{a} \in Q^{\mathcal{B}[\Pi]}$.

Proof. The equivalence of (i) and (ii) has already been shown, and that of (i) and (iii) is given by the theorem. Next we show that (ii) implies (iv): Assume (ii) and let $(\Pi, Q)\overline{x}$ be equivalent to the existential formula $\varphi(\overline{x})$, say $\varphi(\overline{x}) = \exists \overline{y} \psi(\overline{x}, \overline{y})$ with quantifier-free ψ. Given $\mathcal{A}$ and $\overline{a} \in Q^{\mathcal{A}[\Pi]}$, that is, $\mathcal{A} \models \varphi[\overline{a}]$, choose $\overline{b}$ such that $\mathcal{A} \models \psi[\overline{a}, \overline{b}]$. Then, for the substructure $\mathcal{B}$ generated by $\{\overline{a}, \overline{b}\}$, we have $\mathcal{B} \models \varphi[\overline{a}]$ and hence, $\overline{a} \in Q^{\mathcal{B}[\Pi]}$. Thus, as n we can take the sum of length($\overline{x}$), length($\overline{y}$), and of the number of constant symbols.

Finally we show that (iv) implies (i). Clearly, given any n and any DATALOG formula $(\Pi, Q)\overline{x}$, there is an s such that $Q_{(s)} = Q_{(\infty)}$ in all structures of cardinality $\leq n$. Suppose that the positive DATALOG formula $(\Pi, Q)\overline{x}$ satisfies (iv). Choose n according to (iv) and choose s corresponding to $(\Pi, Q)\overline{x}$ and n. We show that $(\Pi, Q)\overline{x}$ is bounded by s. If $\mathcal{A}$ and $\overline{a} \in Q^{\mathcal{A}[\Pi]}$ are given, (iv) yields a substructure $\mathcal{B}$ of $\mathcal{A}$ with $\|B\| \leq n$ such that $\overline{a} \in Q^{\mathcal{B}[\Pi]}$; therefore, $\overline{a} \in Q_{(s)}^{\mathcal{B}[\Pi]}$ and hence, $\overline{a} \in Q_{(s)}^{\mathcal{A}[\Pi]}$. □

We give a further application of 9.3.2. A *universal Horn sentence* is a conjunction of sentences of the form

(1) $\forall \overline{x} \psi$

(2) $\forall \overline{x}((\psi_0 \wedge \ldots \wedge \psi_m) \rightarrow \psi)$

(3) $\forall \overline{x}(\neg\psi_0 \vee \ldots \vee \neg\psi_m)$,

where the ψ's and the ψ_i's are atomic. A class K of τ-structures is *projective Horn* if there is a relational vocabulary σ disjoint from τ and a universal Horn sentence φ of vocabulary $\tau \cup \sigma$ such that the following holds:

– in all subformulas of φ of type (1) or (2) the formula ψ is relational, its relation symbol being from σ
– K is the class of τ-reducts of models of φ (that is, K is the class of models of the second-order sentence $\exists P^1 \ldots \exists P^s \varphi$ where $\sigma = \{P^1, \ldots, P^s\}$).

Since the first-order formulas involved are universal, a projective Horn class is closed under substructures. We have the following "compactness type" converse:

Theorem 9.3.5 *Let K be a projective Horn class. If K is axiomatizable in first-order logic then there is an $n \geq 1$ such that an arbitrary structure belongs to K if all substructures of cardinality at most n belong to K.*

Proof. Suppose σ, τ, and φ are as above and K is the class of τ-reducts of the models of φ. Choose a new 0-ary relation symbol Q and let Π be the positive existential DATALOG program obtained from φ by replacing each conjunct of the form (1) by the clause

(1*) $$\psi \leftarrow$$

each conjunct of the form (2) by the clause

(2*) $$\psi \leftarrow \psi_0, \ldots, \psi_m$$

and each conjunct of the form (3) by the clause

(3*) $$Q \leftarrow \psi_0, \ldots, \psi_m$$

together with

$$R\overline{x} \leftarrow R\overline{x}$$

for $R \in \sigma$. Hence, $\sigma \cup \{Q\}$ is the set of intentional symbols of Π. We show that the class of models of (Π, Q) is the complement K^c of K, that is, for every τ-structure $\mathcal{A}$,

(+) $$\mathcal{A} \in K^c \text{ iff } \mathcal{A} \models (\Pi, Q).$$

Then, K^c is axiomatizable by the positive DATALOG formula (Π, Q) and, as K, is first-order axiomatizable. Hence, K^c satisfies (iv) in 9.3.4, that is, there is an $n \geq 1$ such that all τ-structures $\mathcal{A} \in K^c$ contain a substructure of cardinality $\leq n$ belonging to K^c. Rephrasing this for K we get our claim.

To prove (+), assume first that $\mathcal{A} \in K^c$. Then, in particular, $\mathcal{B} := (\mathcal{A}, (P^{\mathcal{A}[\Pi]})_{P \in \sigma}) \not\models \varphi$. By its definition, $\mathcal{B}$ satisfies the subformulas of type (1) and (2) in φ. Hence, there is a subformula of type (3), say $\forall \overline{x}(\neg\psi_0 \vee \ldots \vee \neg\psi_m)$, such that $\mathcal{B} \models \neg\forall\overline{x}(\neg\psi_0 \vee \ldots \vee \neg\psi_m)$. Choose $\overline{a} \in A$ such that $\mathcal{B} \models (\psi_0 \wedge \ldots \wedge \psi_m)[\overline{a}]$. By (3*), $\mathcal{A} \models (\Pi, Q)$.

Conversely, suppose $\mathcal{A} \in K$, say $(\mathcal{A}, (P^A)_{P \in \sigma}) \models \varphi$. A simple induction on n shows that $P_{(n)} \subseteq P^A$ for $P \in \sigma$; hence, $P^{\mathcal{A}[\Pi]} \subseteq P^A$. Now, if we had $\mathcal{A} \models (\Pi, Q)$, then for some clause of the form (3*) and some $\overline{a} \in A$, we would have $(\mathcal{A}, (P^{\mathcal{A}(\Pi)})_{P \in \sigma}) \models \psi_0 \wedge \ldots \wedge \psi_m[\overline{a}]$ and hence, $(\mathcal{A}, (P^A)_{P \in \sigma}) \models \psi_0 \wedge \ldots \wedge \psi_m[\overline{a}]$ contrary to $(\mathcal{A}, (P^A)_{P \in \sigma}) \models \varphi$. $\Box$

Theorem 9.3.5 can be used to prove nonaxiomatizability results. For example, the class K of acyclic digraphs is projective Horn as can be seen via the axiom

$$\exists P(\forall x \forall y(Exy \to Pxy) \wedge \forall x \forall y((Pxy \wedge Pyz) \to Pxz) \wedge \forall x \neg Pxx)$$

(take as P the transitive closure of E). Since K does not satisfy the conclusion in Theorem 9.3.5, acyclicity is not expressible in first-order logic.

We close by sketching the main idea of a proof of Theorem 9.3.2 using concepts and results from section 2.5 and refer to [7] for the full proof. Let (Π, Q) be a positive DATALOG formula where, for simplicity, we assume that Q is 0-ary. Suppose that (Π, Q) is equivalent to the first-order sentence φ. We have to show that (Π, Q) is bounded. Otherwise, (Π, Q) and, hence, φ have arbitrarily large minimal models. On the other hand, by positivity, (Π, Q) and, hence, φ are preserved under strict homomorphisms. By 2.5.5, for some l and m, no minimal model of φ has an l-scattered subset of cardinality $\geq m$. Finally, one obtains a contradiction by showing that in large minimal models there are such subsets.

9.4 Normal Forms for Fixed-Point Logics

The equivalence of I-DATALOG with FO(IFP) and of P-DATALOG with FO(PFP) (together with 8.2.3, the Lemma on Simultaneous Induction) provides an alternative way to derive normal forms for FO(IFP) and FO(PFP). We shall exemplify this first and then further improve these results, thus obtaining normal forms that, in particular, yield some kind of measure to compare FO(IFP) and FO(PFP) with transitive closure logic and bounded fixed-point logic. Moreover, they will enable us in Chapter 12 to get representations of FO(IFP) and FO(PFP) as logics with generalized quantifiers.

Once more, for simplicity, we consider only structures with at least two elements. In particular, we say that φ and ψ are equivalent if they are equivalent in all finite structures with at least two elements.[3] And we say that φ and $\exists(\forall)x\psi$ are equivalent if φ is equivalent (in this sense) to both $\exists x\psi$ and $\forall x\psi$.

Theorem 9.4.1 *Let φ be an* FO(IFP)*-formula. Then there is an existential first-order formula ψ such that φ is equivalent to*

$$\exists(\forall)u[\mathrm{IFP}_{\overline{z},Z}\,\psi]\,\tilde{u}.$$

Proof. Let $\varphi = \varphi(\overline{v})$ with free first-order variables among $\overline{v}$. Choose an I-DATALOG formula $(\Pi, Q)\overline{v}$ equivalent to φ and let $P^1, \ldots, P^k$ with $Q = P^1$ be the intentional symbols of Π.

[3] In most cases the restriction to structures with at least two elements can be removed as in similar situations before, or the corresponding statements are trivially false.

We may assume that the clauses of Π have the form

$$P\overline{x}_P \leftarrow \gamma_1, \ldots, \gamma_l$$

where $\overline{x}_P$ is a sequence of distinct variables. Let

$$\varphi_P(\overline{x}_P) := \bigvee\{\exists\overline{y}(\gamma_1 \wedge \ldots \wedge \gamma_l) \mid P\overline{x}_P \leftarrow \gamma_1, \ldots, \gamma_l \text{ in } \Pi\}$$

where $\overline{y}$ contains all the variables of the corresponding clause not in $\overline{x}_P$. Then $(\Pi, Q)\overline{v}$ is equivalent to

$$[\text{S-IFP}_{\overline{x}_{P^1}, P^1, \ldots, \overline{x}_{P^k}, P^k}\, \varphi_{P^1}, \ldots, \varphi_{P^k}]\, \overline{v}.$$

Since $\varphi_{P^1}, \ldots, \varphi_{P^k}$ are existential, the corresponding χ_J in 8.2.3 is existential, too, and

$$\models_{\text{fin}} [\text{S-IFP}_{\overline{x}_{P^1}, P^1, \ldots, \overline{x}_{P^k}, P^k}\, \varphi_{P^1}, \ldots, \varphi_{P^k}]\, \overline{v} \leftrightarrow \exists(\forall)u[\text{IFP}_{\overline{z}, Z}\, \chi_J]\, \overline{v}\tilde{u}.$$

By 8.2.4, the formula on the right side is equivalent to a formula of the form $\exists(\forall)u[\text{IFP}_{\overline{z}, Z}\, \psi]\, \tilde{u}$ with existential ψ. □

The corresponding result for FO(PFP) can be obtained along the same lines, using FO(PFP) $\equiv$ P-DATALOG. Moreover, we already proved it in 8.3.14 by different methods.

For FO(LFP) the kernel of the fixed-point operator in a normal form cannot be chosen existential (cf. the example before 9.1.5); however, we have

Theorem 9.4.2 *Suppose that φ is an* FO(LFP)*-formula. Then there is a Δ_2-formula ψ, that is, a first-order formula ψ equivalent to both a Σ_2-formula and a Π_2-formula, such that φ is equivalent to*

$$\exists(\forall)u[\text{LFP}_{\overline{z}, Z}\, \psi]\, \tilde{u}.$$

Proof. By the previous theorem there is an existential formula χ such that

$$\models_{\text{fin}} \varphi \leftrightarrow \exists(\forall)u[\text{IFP}_{\overline{z}, Z}\, \chi]\, \tilde{u}.$$

By 8.2.11, for every existential χ there is a Δ_2-formula ψ such that

$$\models_{\text{fin}} \exists(\forall)u[\text{IFP}_{\overline{z}, Z}\, \chi]\, \tilde{u} \leftrightarrow \exists(\forall)u[\text{LFP}_{\overline{y}, Y}\, \psi]\, \tilde{u}.$$

Thus, φ is equivalent to $\exists(\forall)u[\text{LFP}_{\overline{y}, Y}\, \psi]\, \tilde{u}$. □

Our next goal is to further simplify the normal forms we have obtained so far with respect to the structure of the kernel, in particular with respect to the form in which quantifiers appear in it. We first treat the case of FO(PFP) and FO(IFP). To state the result we need the notion of a nontrivial formula. An r-ary relation R on a set A is *nontrivial* if $\emptyset \neq R \neq A^r$. Furthermore, a formula of FO(PFP) is said to be *nontrivial* if for any subformula $\psi = [\text{PFP}_{\overline{x}, X}\, \chi]\, \overline{t}$ the stages $F_1^\chi, F_2^\chi, \ldots$ are nontrivial (for any assignment). The definitions for FO(IFP) and FO(LFP) are similar.

Theorem 9.4.3 *Let φ be an* FO(PFP)*-formula. Then φ is equivalent to a nontrivial, totally defined formula of the form*

$$\exists(\forall)u[\text{PFP}_{\overline{z},Z}\,(\psi_0 \vee \exists\overline{x} \in Z \exists\overline{y} \notin Z\psi_1)]\,\tilde{u}$$

where ψ_0 and ψ_1 are quantifier-free and do not contain Z. The same result is true if PFP *is replaced by* IFP *everywhere.*

We have seen in 8.7.1(c) that TC operations correspond to LFP operations in which every element of a new stage is already witnessed by a single element of the preceding stage. Similarly, the fixed-point operations of bounded fixed-point logic FO(BFP) correspond to LFP operations where every element of a new stage is witnessed by two elements (cf. 8.7.1(d)). The preceding theorem tells us that IFP operations and PFP operations can be rewritten in such a way that an element of a new stage is witnessed by two elements, but now one belonging to the preceding stage, the other one belonging to its complement.

The proof of 9.4.3 is given in three steps; each step essentially corresponds to one of the next lemmas. We state the lemmas and the proofs for FO(PFP). They remain literally true for FO(IFP) with a single exception which we point out at the corresponding place.

Lemma 9.4.4 *Let φ be an existential (or quantifier-free) first-order formula containing a second-order variable X. Then, provided X is interpreted by a nontrivial relation, φ is equivalent to a formula of the form*

$$\exists\overline{x}_1 \in X \ldots \exists\overline{x}_k \in X \exists\overline{y}_1 \notin X \ldots \exists\overline{y}_l \notin X\psi$$

where ψ is existential (or quantifier-free) and does not contain X. If φ is positive in X then l can be chosen to be 0.

Lemma 9.4.5 *Suppose that*

$$\varphi = \exists(\forall)u[\text{PFP}_{\overline{x},X}\,(\varphi_0 \vee \exists\overline{x}_1 \in X \ldots \exists\overline{x}_k \in X\exists\overline{y}_1 \notin X \ldots \exists\overline{y}_l \notin X\varphi_1)]\,\tilde{u}$$

is a nontrivial, totally defined formula, where φ_0 is quantifier-free, φ_1 is existential, and X does not occur in φ_0 and φ_1. Then φ is equivalent to a nontrivial, totally defined formula of the form

$$\exists(\forall)u[\text{PFP}_{\overline{y},Y}(\psi_0 \vee \exists\overline{v} \in Y\exists\overline{w} \notin Y\psi_1)]\tilde{u}$$

where ψ_0 is quantifier-free, ψ_1 is existential, and Y does not occur in ψ_0, ψ_1.

Lemma 9.4.6 *If φ is as in the conclusion of the preceding lemma then φ is equivalent to a formula of the same form with the additional property that ψ_1 is quantifier-free.*

Postponing the proof of these lemmas, we give the

Proof (of 9.4.3). Let φ be an FO(PFP)-formula. By 8.3.14 we can assume that

$$\varphi = \exists(\forall)u[\text{PFP}_{\overline{u},U}\chi]\tilde{u},$$

that φ is totally defined, and that χ is existential. Suppose U is s-ary. First, we pass to a nontrivial formula. For this purpose we introduce an $(s+2)$-ary relation symbol Y where we use the additional components to ensure nontriviality by setting

$$\rho(\overline{u}, u_{s+1}, u_{s+2}) \quad := \quad \rho_0(\overline{u}, u_{s+1}, u_{s+2}) \vee \rho_1(\overline{u}, u_{s+1}, u_{s+2}, Y)$$

with

$$\begin{aligned} \rho_0(\overline{u}, u_{s+1}, u_{s+2}) \quad &:= \quad u_s = u_{s+1} \wedge \neg u_{s+1} = u_{s+2} \\ \rho_1(\overline{u}, u_{s+1}, u_{s+2}, Y) \quad &:= \quad u_{s+1} = u_{s+2} \wedge \exists \overline{v} v_{s+1} v_{s+2} Y \overline{v} v_{s+1} v_{s+2} \\ &\qquad \wedge \chi(Y_{-} u_{s+1} u_{s+2}). \end{aligned}$$

The stages $F_1^\rho, F_2^\rho, \ldots$ are nontrivial, since they contain all tuples of the form $\ldots uuv$ with $u \neq v$ and no tuple of the form $\ldots uvu$ with $u \neq v$. Moreover, $F_1^\rho = \{(u_1, \ldots, u_{s+2}) \mid u_s = u_{s+1} \text{ and } u_{s+1} \neq u_{s+2}\}$. An evaluation of the stages shows

$$F_n^\chi = F_{n+1}^\rho{}_{-}uu \quad \text{for } n \geq 0, \text{ and } \; F_\infty^\chi = F_\infty^\rho{}_{-}uu.$$

By 9.4.4, provided a nontrivial relation is assigned to Y, ρ_1 is equivalent to a formula ψ_1 of the form

$$\psi_1 \quad := \quad \exists \overline{x}_1 \in Y \ldots \exists \overline{x}_k \in Y \exists \overline{y}_1 \notin Y \ldots \exists \overline{y}_l \notin Y \psi$$

where ψ is existential and does not contain Y. Using dummy variables if necessary, we may assume that $k \geq 1$. Then, if Y is interpreted by the empty relation, both ρ_1 and ψ_1 are false. Hence, we have $F_n^{\rho_0 \vee \rho_1} = F_n^{\rho_0 \vee \psi_1}$ for all n. Thus, φ is equivalent to the nontrivial, totally defined formula

$$\exists(\forall)u[\text{PFP}_{\overline{y},Y}(\rho_0(\overline{y}) \vee \exists \overline{x}_1 \in Y \ldots \exists \overline{x}_k \in Y \exists \overline{y}_1 \notin Y \ldots \exists \overline{y}_l \notin Y \psi]\, \tilde{u}.$$

As claimed, ρ_0 is quantifier-free, ψ is existential, and Y does not occur in ρ_0 and ψ. The theorem now follows by first applying Lemma 9.4.5 and then Lemma 9.4.6. □

Proof (of 9.4.4). Let φ be an existential (or, quantifier-free) formula containing the variable X. We may assume that all negation symbols in φ are in front of atomic formulas. For such φ we inductively define φ^* of the form claimed. If $\varphi = X\overline{t}$ set $\varphi^* := \exists \overline{x} \in X\, \overline{t} = \overline{x}$. If $\varphi = \neg X\overline{t}$ set $\varphi^* := \exists \overline{y} \notin X\, \overline{t} = \overline{y}$. Let φ^* be φ for any other atomic or negated atomic formula φ. The definition for the remaining cases is straightforward, using the induction hypothesis and equivalences such as that of $(\varphi_1 \vee \exists \overline{x}_1 \in X \ldots \exists \overline{x}_k \in X \exists \overline{y}_1 \notin X \ldots \exists \overline{y}_l \notin X \varphi_2)$ and $\exists \overline{x}_1 \in X \ldots \exists \overline{x}_k \in X \exists \overline{y}_1 \notin X \ldots \exists \overline{y}_l \notin X(\varphi_1 \vee \varphi_2)$, provided a nontrivial relation is assigned to X and the variables in $\overline{x}_1, \ldots, \overline{y}_l$ do not occur in φ_1. □

Proof (of 9.4.5). Suppose X is r-ary. The lemma claims that the number of quantifiers $\exists\overline{x} \in X, \exists\overline{y} \notin X$ in the kernel of the fixed-point operator can be reduced to a single positive and a single negative one. We achieve this by increasing the arity of the second-order variable to $(k+l)\cdot r$, whose first k r-blocks will consist of r-tuples in X and the last l r-blocks of r-tuples in the complement X^c of X. The corresponding r-tuples are incorporated in $(k+l)$ steps into the distinct components, thus making necessary additional components for a counter: One iteration step of X will correspond to $(k+l)$ steps for the new variable.

To simplify the presentation, assume that $k = l = 2$ and that X is unary, hence

$$\varphi = \exists(\forall)u[\mathrm{PFP}_{x,X}\,(\varphi_0 \vee \exists x_1 \in X \exists x_2 \in X \exists y_1 \notin X \exists y_2 \notin X \varphi_1)]\,\tilde{u}$$

where $\varphi_1(x, x_1, x_2, y_1, y_2)$ is existential and $\varphi_0(x)$ and φ_1 do not contain X. Let X_n be the n-th stage of the operation corresponding to the PFP operator in φ. Choose a $(4+4)$-ary new variable Y, where the first 4 components are used as a counter for the 4 steps corresponding to one step of X, the next two ones for elements of X, and the last two ones for elements of X^c. For the counters we introduce formulas $\epsilon_i(z_1, \ldots, z_4)$ for $i = 1, \ldots, 4$ by

$$\begin{aligned}
\epsilon_1(z_1,\ldots,z_4) &:= z_1 = z_2 = z_3 = z_4 \\
\epsilon_2(z_1,\ldots,z_4) &:= \neg z_1 = z_2 \wedge z_1 = z_3 = z_4 \\
\epsilon_3(z_1,\ldots,z_4) &:= \neg z_1 = z_3 \wedge z_1 = z_2 = z_4 \\
\epsilon_4(z_1,\ldots,z_4) &:= \neg z_1 = z_4 \wedge z_1 = z_2 = z_3.
\end{aligned}$$

If Y_n denotes the n-th stage of the operation corresponding to the PFP operator in the formula ψ below, set

$$Y_n i\_ \;:=\; \{\overline{x} \mid \text{ there are } \overline{z} \text{ such that } \epsilon_i(\overline{z}) \text{ and } Y_n\overline{z}\,\overline{x}\}$$

($\overline{x} = x_1 \ldots x_4$ will correspond to $x_1x_2y_1y_2$ in φ_1 above). For all $m \geq 0$ and any structure $\mathcal{A}$ we will have (1) and (2).

$$(1)\qquad \begin{aligned}
Y_{4m+1}1_ &= X_{m+1} \times A \times A \times A \\
Y_{4m+2}2_ &= X_{m+1} \times A \times X^c_{m+1} \times A \\
Y_{4m+3}3_ &= X_{m+1} \times X_{m+1} \times X^c_{m+1} \times A \\
Y_{4m+4}4_ &= X_{m+1} \times X_{m+1} \times X^c_{m+1} \times X^c_{m+1}.
\end{aligned}$$

Here the IFP case differs from the PFP case; the values for IFP are:

$$(1')\qquad \begin{aligned}
Y_{4m+1}1_ &= \textstyle\bigcup_{l\leq m+1} X_l \times A \times A \times A \\
Y_{4m+2}2_ &= \textstyle\bigcup_{l\leq m+1} X_l \times A \times X^c_l \times A \\
Y_{4m+3}3_ &= \textstyle\bigcup_{l\leq m+1} X_l \times X_l \times X^c_l \times A \\
Y_{4m+4}4_ &= \textstyle\bigcup_{l\leq m+1} X_l \times X_l \times X^c_l \times X^c_l.
\end{aligned}$$

$$(2) \qquad Y_{4m+i+j}i_- = Y_{4m+i}i_- \quad \text{for } 1 \leq i \leq 4 \text{ and } 1 \leq j \leq 3.$$

This is achieved if we use the formula $\psi :=$

$$\exists(\forall)u[\mathrm{PFP}_{\overline{w}\,z,Y}\,(\epsilon_1(\overline{w}) \wedge \varphi_0(z_1)) \vee \exists \overline{u}\,\overline{x} \in Y \exists \overline{v}\,\overline{y} \notin Y \bigvee_{i=1}^{4} (\epsilon_i(\overline{w}) \wedge \psi_i)]\,\tilde{u}$$

where the ψ_i together with their meaning (formulated in a way suitable for an inductive proof of (1) and (2)) are given as follows:

$\psi_1 := \epsilon_4(\overline{u}) \wedge \varphi_1(z_1, x_1, x_2, x_3, x_4)$
"$z_1 \in X_{m+1}$ is witnessed by $\overline{x} \in Y_{4m}4_-$ via $\varphi_1(z_1, x_1, x_2, x_3, x_4)$"

$\psi_2 := \epsilon_1(\overline{u}) \wedge \epsilon_1(\overline{v}) \wedge x_2 = y_2 \wedge x_3 = y_3 \wedge x_4 = y_4 \wedge z_1 = x_1 \wedge z_3 = y_1$
"$z_3 \in X^c_{m+1}$ is witnessed by $\overline{x} \in Y_{4m+1}1_-$ and $z_3x_2x_3x_4 \notin Y_{4m+1}1_-$"

$\psi_3 := \epsilon_2(\overline{u}) \wedge \epsilon_2(\overline{v}) \wedge x_1 = y_1 \wedge x_2 = y_2 \wedge x_4 = y_4 \wedge z_1 = x_1 \wedge z_2 = y_3 \wedge z_3 = x_3$
"$z_2 \in X_{m+1}$ is witnessed by $\overline{x} \in Y_{4m+2}2_-$ and $x_1x_2z_2x_4 \notin Y_{4m+2}2_-$"

$\psi_4 := \epsilon_3(\overline{u}) \wedge \epsilon_3(\overline{v}) \wedge x_2 = y_2 \wedge x_3 = y_3 \wedge x_4 = y_4 \wedge z_1 = x_1 \wedge z_2 = x_2 \wedge z_3 = x_3 \wedge z_4 = y_1$
"$z_4 \in X^c_{m+1}$ is witnessed by $\overline{x} \in Y_{4m+3}3_-$ and $z_4x_2x_3x_4 \notin Y_{4m+3}3_-$".

Since X_∞ exists, Y_∞ exists, too, and $Y_\infty 1_- = X_\infty \times A \times A \times A$. As $\epsilon_1(u,u,u,u)$, but not $\epsilon_i(u,u,u,u)$ for $i = 2,3,4$, we thus see that φ and ψ are equivalent. □

Proof (of 9.4.6). When proving this lemma, we again use additional components to simulate the unbounded existential quantifiers. Let

$$\varphi = \exists(\forall)u[\mathrm{PFP}_{\overline{x},X}\,(\varphi_0 \vee \exists \overline{v} \in X \exists \overline{w} \notin X \varphi_1)]\,\tilde{u}$$

with quantifier-free φ_0 and existential φ_1, both not containing X. Assume that X is r-ary and

$$\varphi_1 = \exists y_1 \ldots \exists y_l \chi(\overline{x}, \overline{v}, \overline{w}, \overline{y})$$

with quantifier-free χ. Let Z be $(r+l)$-ary. Given a structure $\mathcal{A}$, let $X_0, X_1, \ldots$ denote the stages of the PFP operator in φ and $Z_0, Z_1, \ldots$ the stages of the PFP operator in

$$\psi = \exists(\forall)u[\mathrm{PFP}_{\overline{x}\,\overline{y},Z}\,(\varphi_0 \vee \exists \overline{v}\,\overline{y}_1 \in Z \exists \overline{w}\,\overline{y}_2 \notin Z \chi(\overline{x}, \overline{v}, \overline{w}, \overline{y}_1))]\,\tilde{u}.$$

Then a simple induction shows that $Z_n = X_n \times A^l$. Thus φ and ψ are equivalent. □

Turning to FO(LFP), the improved normal form reads as follows:

Theorem 9.4.7 *Let φ be an* FO(LFP)*-formula. Then φ is equivalent to a nontrivial formula of the form*

$$(1) \qquad \exists(\forall)u[\mathrm{LFP}_{\overline{z},Z}\,(\psi_0 \vee \forall\overline{x}\exists\overline{y} \in Z\,\psi_1)]\,\tilde{u}$$

with quantifier-free ψ_0 and ψ_1 not containing Z, and φ is also equivalent to a formula of the form

$$(2) \qquad \exists(\forall)u[\mathrm{LFP}_{\overline{z},Z}\,\exists\overline{x}\forall\overline{y} \notin Z\,\psi_1]\,\tilde{u}$$

with quantifier-free ψ_1 not containing Z (note that $\forall\overline{y} \notin Z\psi_1$ abbreviates $\forall\overline{y}(Z\overline{y} \vee \psi_1)$ and hence, is positive in Z).

Moreover, in (1) *we can choose ψ_1 of the form $\psi_{11} \rightarrow \psi_{12}$, where no variable of $\overline{y}$ is free in ψ_{11} and no variable of $\overline{z}$ is free in ψ_{12}.*

Proof. We sketch a proof along the lines of the proof for the corresponding result 9.4.3 for FO(PFP). To obtain (1), we start from the fact (cf. 9.4.2) that φ is equivalent to a formula of the form

$$\exists(\forall)u[\mathrm{LFP}_{\overline{u},U}\,\forall\overline{v}\exists\overline{w}\chi]\,\tilde{u}$$

where χ is quantifier-free (and positive in U). Let s be the arity of U. Using an $(s+2)$-ary Y, we ensure nontriviality as in 9.4.3 by passing to

$$\rho(\overline{u}, u_{s+1}, u_{s+2}) \quad := \quad \rho_0(\overline{u}, u_{s+1}, u_{s+2}) \vee \rho_1(\overline{u}, u_{s+1}, u_{s+2}, Y)$$

with

$$\begin{aligned}
\rho_0(\overline{u}, u_{s+1}, u_{s+2}) \quad &:= \quad u_s = u_{s+1} \wedge \neg u_{s+1} = u_{s+2} \\
\rho_1(\overline{u}, u_{s+1}, u_{s+2}, Y) \quad &:= \quad u_{s+1} = u_{s+2} \wedge \exists\overline{u}'u'_{s+1}u'_{s+2}Y\overline{u}'u'_{s+1}u'_{s+2} \\
&\qquad \wedge \forall\overline{v}\exists\overline{w}\chi(Y_u_{s+1}u_{s+2}).
\end{aligned}$$

The formula ρ_1 is equivalent to

$$\forall\overline{v}\exists\overline{w}(u_{s+1} = u_{s+2} \wedge \exists\overline{u}'u'_{s+1}u'_{s+2}Y\overline{u}'u'_{s+1}u'_{s+2} \wedge \chi(Y\_u_{s+1}u_{s+2})).$$

As $\chi(Y\_u_{s+1}u_{s+2})$ is positive in Y, by 9.4.4 it is equivalent to

$$\exists\overline{y}_1 \in Y \ldots \exists\overline{y}_k \in Y\chi_1$$

with quantifier-free χ_1 not containing Y, provided Y is interpreted by a nontrivial relation. Using dummy variables if necessary, we can assume that $k \geq 1$. Altogether, φ is equivalent to

$$\varphi_1 \quad := \quad \exists(\forall)u[\mathrm{LFP}_{\overline{y},Y}(\rho_0 \vee \forall\overline{v}\exists\overline{w}\exists\overline{y}_1 \in Y \ldots \exists\overline{y}_k \in Y\chi_2)]\tilde{u}$$

where $\rho_0(\overline{y})$ and $\chi_2(\overline{y}, \overline{v}, \overline{w}, \overline{y}_1, \ldots, \overline{y}_k)$ are quantifier-free and do not contain Y. Let Z be $k{\cdot}r$-ary where r is the arity of Y. Then for $n \geq 0$, if Y_n denotes the n-th stage of the LFP operation in φ_1,

$$Z_n := \underbrace{Y_n \times \ldots \times Y_n}_{k \text{ times}}$$

are the stages of the LFP operations in

$$\exists(\forall)u[\mathrm{LFP}_{\overline{z}_1\ldots\overline{z}_k,Z}\,((\rho_0(\overline{z}_1) \wedge \ldots \wedge \rho_0(\overline{z}_k))\vee$$

$$\forall\overline{u}\forall\overline{v}\exists\overline{w}\exists\overline{y}_1\ldots\overline{y}_k \in Z(\textstyle\bigvee_{i=1}^k \overline{u} = \overline{z}_i \rightarrow (\rho_0(\overline{u}) \vee \chi_1(\overline{u},\overline{v},\overline{w},\overline{y}_1,\ldots,\overline{y}_k))))]\,\tilde{u}.$$

Now, as we have already done in previous proofs, we can eliminate the existential quantifiers $\exists\overline{w}$ by replacing Z, thereby using a relation variable of higher arity. We thus obtain a formula equivalent to φ of the form claimed in (1).

To obtain (2), we may assume (again by 9.4.2) that φ has the form

$$\varphi = \exists(\forall)u[\mathrm{LFP}_{\overline{y},Y}\,\exists\overline{v}\forall\overline{w}\psi]\,\tilde{u}$$

with quantifier-free ψ. Suppose that Y is r-ary. A simple induction, using the equivalence of $Y\overline{y}$ and $\forall\overline{z} \notin Y \neg\overline{z} = \overline{y}$, shows that ψ is equivalent to a formula

$$\forall\overline{z}_1 \notin Y \ldots \forall\overline{z}_k \notin Y \rho(\overline{y},\overline{v},\overline{w},\overline{z}_1,\ldots,\overline{z}_k)$$

with quantifier-free ρ not containing Y, provided Y is interpreted by a relation different from the set of all r-tuples.

So we may assume that φ is equivalent to the formula

$$\varphi_1 := \exists(\forall)u[\mathrm{LFP}_{\overline{y},Y}\,\exists\overline{v}\forall\overline{w}\forall\overline{z}_1 \notin Y \ldots \forall\overline{z}_k \notin Y \rho(\overline{y},\overline{v},\overline{w},\overline{z}_1,\ldots,\overline{z}_k)]\,\tilde{u}$$

(note that the proviso "Y different from the set of all r-tuples" is irrelevant here, as the fixed-point operations in φ and φ_1 become stationary in case they arrive at the set of all r-tuples). Now, let Z be $k\cdot r$-ary and let φ_2 be the formula

$$\exists(\forall)u[\mathrm{LFP}_{\overline{y}_1\ldots\overline{y}_k,Z}\,\exists\overline{y}\exists\overline{v}\forall\overline{w}\forall\overline{z}_1\ldots\overline{z}_k \notin Z \bigvee_{i=1}^{k}(\overline{y} = \overline{y}_i \wedge \rho(\overline{y},\overline{v},\overline{w},\overline{z}_1,\ldots,\overline{z}_k))]\,\tilde{u}.$$

Then, for the complements Y_n^c and Z_n^c of the stages of the LFP operations in φ_1 and φ_2, we have

$$Z_n^c = \underbrace{Y_n^c \times \ldots \times Y_n^c}_{k \text{ times}}.$$

Hence, φ_1 and φ_2 are equivalent. To eliminate the universal quantifiers $\forall\overline{w}$ in φ_2, we use a variable X with $\mathrm{arity}(X) = \mathrm{length}(\overline{w}) + \mathrm{arity}(Z)$ and pass to the equivalent formula

$$\exists(\forall)u[\mathrm{LFP}_{\overline{y}'\overline{y}_1\ldots\overline{y}_k,X}\,\exists\overline{y}\exists\overline{v}\forall\overline{z}'\overline{z}_1\ldots\overline{z}_k \notin X \textstyle\bigvee_{i=1}^k(\overline{y} = \overline{y}_i$$

$$\wedge\, \rho(\overline{y},\overline{v},\overline{z}',\overline{z}_1,\ldots,\overline{z}_k))]\,\tilde{u},$$

a formula of the desired form (note that $X_n^c = \mathrm{universe}^{\mathrm{length}(\overline{w})} \times Z_n^c$ for all n).

□

As an application of the preceding theorem we treat a variant of the TC operator, the so-called *alternating transitive closure operator* ATC which turns out to be equivalent to the LFP operator.

We consider structures of the form $\mathcal{A} = (A, E^A, U^A)$ with binary E and unary U. In this context the elements of U^A are called *universal*, the elements of $A \setminus U^A$ *existential*. The *alternating transitive closure* $\mathrm{ATC}(E^A, U^A)$ of (E^A, U^A) consists of those pairs $(a, b) \in A \times A$ for which there is an E^A-path from a to b with the property that whenever a universal point c is passed there must also be an E^A-path with this property from d to b for each $d \neq b$ with $E^A cd$. More precisely, using the LFP operator, we define:

$$\begin{aligned}(a, b) \in \mathrm{ATC}(E^A, U^A) \ \text{ iff } \ & (A, E^A, U^A) \models \\ & [\mathrm{LFP}_{xy,X}(\neg Ux \wedge (Exy \vee \exists z(Exz \wedge Xzy))) \\ & \vee (Ux \wedge \exists z Exz \wedge \forall z(Exz \rightarrow (z = y \vee Xzy)))]xy[a, b].\end{aligned}$$

The logic FO(ATC) is defined as FO(TC), replacing the TC-clause by

$$\frac{\varphi(\overline{x}, \overline{y}), \psi(\overline{x})}{[\mathrm{ATC}_{\overline{x},\overline{y}}\, \varphi(\overline{x}, \overline{y}), \psi(\overline{x})]\, \overline{s}\overline{t}}$$

where $\overline{x}, \overline{y}, \overline{s}, \overline{t}$ all have the same length. The meaning of the new formula is given by

$$(\overline{s}, \overline{t}) \in \mathrm{ATC}(\varphi(\_,\_), \psi(\_)).$$

We now have:

Theorem 9.4.8 FO(ATC) $\equiv$ FO(LFP).

Corollary 9.4.9 FO(ATC) *captures* PTIME. □

Proof (of 9.4.8). As FO(ATC) $\leq$ FO(LFP) is clear from the definition of ATC given above, we need only prove the other direction.

So let φ be an FO(LFP)-sentence. By 9.4.7(1) we may assume that φ is equivalent to the nontrivial sentence

$$\exists u[\mathrm{LFP}_{\overline{x},X}\, (\psi_0(\overline{x}) \vee \forall \overline{y} \exists \overline{z} \in X(\psi_1(\overline{x}, \overline{y}) \rightarrow \psi_2(\overline{y}, \overline{z})))]\, \tilde{u},$$

where X does not appear in the quantifier-free formulas ψ_0, ψ_1, ψ_2 and where the free variables are among the displayed ones. In view of nontriviality, this formula is equivalent to

$$\exists u[\mathrm{LFP}_{\overline{x},X}\, (\psi_0(\overline{x}) \vee \forall \overline{y}(\psi_1(\overline{x}, \overline{y}) \rightarrow \exists \overline{z} \in X \psi_2(\overline{y}, \overline{z})))]\, \tilde{u}.$$

Let F_∞ be the fixed-point of the fixed-point process in the last formula. The alternating transitive closure capturing it will make use of a binary and a unary relation on the set of all tuples $\overline{x}\,\overline{y}v_1 \ldots v_4$. For $i = 1, \ldots, 4$ we write $\overline{x}\,\overline{y}\,i$ for any such tuple $\overline{x}\,\overline{y}\,\overline{v}$ with $\epsilon_i(\overline{v})$, where the ϵ_i are as in the proof of 9.4.5. Then the relations will be defined by FO-formulas φ_E and φ_U in such a way that $\overline{x} \in F_\infty$ iff there is an ATC-path from $\overline{x}\,\overline{x}\,1$ to a point of the form $\overline{y}\,\overline{y}\,4$; more precisely:

$$(*)\quad \overline{x} \in F_\infty \quad \text{iff} \quad \begin{array}{l}\text{some (or equivalently, all) pairs of tuples of the form}\\ (\overline{x}\,\overline{x}1, \overline{y}\,\overline{y}\,4) \text{ are in } \mathrm{ATC}(\varphi_E(\_,\_), \varphi_U(\_)).\end{array}$$

As universal points we take all tuples $\overline{x}\,\overline{y}\,1$, that is, we set

$$\varphi_U(\overline{x}, \overline{y}, \overline{v}) \quad := \quad \epsilon_1(\overline{v}).$$

The binary relation consists of edges that take care of the disjuncts $\psi_0(\overline{x})$ and $\forall \overline{y}(\psi_1(\overline{x}, \overline{y}) \rightarrow \exists \overline{z} \in X \psi_2(\overline{y}, \overline{z}))$ of the fixed-point formula above. Concerning $\psi_0(\overline{x})$, it contains the edges indicated by

$$\overline{x}\,\overline{x}\,1 \xrightarrow{\psi_0(\overline{x})} \overline{x}\,\overline{x}\,2 \longrightarrow \overline{y}\,\overline{y}\,4,$$

and concerning the second disjunct, the edges indicated by

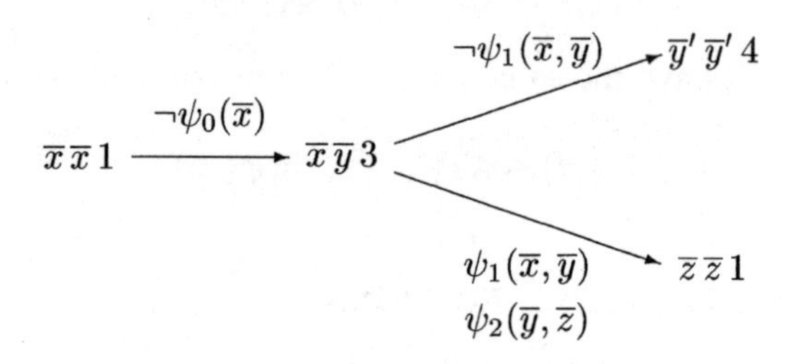

So as φ_E we take

$$\begin{array}{rcl}\varphi_E(\overline{x}\,\overline{y}\,\overline{v}, \overline{x}'\overline{y}'\overline{v}') & := & (\epsilon_1(\overline{v}) \wedge \psi_0(\overline{x}) \wedge \overline{x} = \overline{y} = \overline{x}' = \overline{y}' \wedge \epsilon_2(\overline{v}'))\\ & \vee & (\epsilon_2(\overline{v}) \wedge \overline{x} = \overline{y} \wedge \overline{x}' = \overline{y}' \wedge \epsilon_4(\overline{v}))\\ & \vee & (\epsilon_1(\overline{v}) \wedge \neg\psi_0(\overline{x}) \wedge \overline{x} = \overline{y} \wedge \overline{x}' = \overline{x} \wedge \epsilon_3(\overline{v}'))\\ & \vee & (\epsilon_3(\overline{v}) \wedge \neg\psi_1(\overline{x}, \overline{y}) \wedge \overline{x}' = \overline{y}' \wedge \epsilon_4(\overline{v}'))\\ & \vee & (\epsilon_3(\overline{v}) \wedge \psi_1(\overline{x}, \overline{y}) \wedge \psi_2(\overline{y}, \overline{x}') \wedge \overline{x}' = \overline{y}' \wedge \epsilon_1(\overline{v}')).\end{array}$$

Then $(*)$ is obvious and yields that φ is equivalent to

$$\begin{array}{l}\exists \overline{u} \exists \overline{z} \exists \overline{w} \exists \overline{w}'(\epsilon_1(\overline{w}) \wedge \epsilon_4(\overline{w}') \wedge\\ \quad [\mathrm{ATC}_{\overline{x}\,\overline{y}\,\overline{v}, \overline{x}'\,\overline{y}'\,\overline{v}'}\, \varphi_E(\overline{x}\,\overline{y}\,\overline{v}, \overline{x}'\,\overline{y}'\,\overline{v}'), \varphi_U(\overline{x}, \overline{y}, \overline{v})]\, \tilde{u}\tilde{u}\tilde{w}\overline{z}\,\overline{z}\,\overline{w}').\end{array}$$

□

9.5 An Application of Negative Fixed-Point Logic

The normal form obtained in Theorem 9.4.3 tells us – as already pointed out in the remarks following this theorem – that in FO(PFP) we can restrict ourselves to PFP operations where every element of a new stage is witnessed by two elements, one belonging to the previous stage (a positive witness), the other one belonging to its complement (a negative witness). In this section we analyze the expressive power of the fragment of FO(PFP) containing PFP operators with only negative witnesses and give an application of this fragment to so-called well-founded DATALOG.

For $k, l \in \mathbb{N}$ let PFP(k,l) be the class of FO(PFP)-formulas containing only fixed-point operations where every element of a new stage (besides the first one) is witnessed by k positive and l negative elements, i.e., $\varphi \in$ FO(PFP) is in PFP(k,l) if all its subformulas starting with a PFP operator have the form

$$[\mathrm{PFP}_{\overline{x},X}\,(\varphi_0 \vee \exists \overline{y}_1 \in X \ldots \exists \overline{y}_k \in X \exists \overline{z}_1 \notin X \ldots \exists \overline{z}_l \notin X \varphi_1]\,\overline{t}$$

where φ_0 and φ_1 do not contain X.

We have a complete picture of the expressive power of the different PFP(k,l):

Theorem 9.5.1 (a) *For* $k, l \geq 1$, PFP$(k,l) \equiv$ PFP$(1,1) \equiv$ FO(PFP).
(b) *For* $k \geq 2$, PFP$(k,0) \equiv$ PFP$(2,0) \equiv$ FO(BFP).
(c) PFP$(1,0) \equiv$ FO(TC).
(d) *For* $l \geq 1$, PFP$(0,l) \equiv$ PFP$(0,1) \equiv$ FO(LFP). *Moreover, every* $\varphi \in$ FO(LFP) *is equivalent to a totally defined formula of the form*

$$\exists(\forall) u[\mathrm{PFP}_{\overline{x},X}\, \exists \overline{y} \notin X \psi(\overline{x},\overline{y})]\,\tilde{u}$$

where ψ *is quantifier-free and does not contain* X.

Proof. Part (a) holds by Theorem 9.4.3, parts (b) and (c) hold by Proposition 8.7.1. We turn to (d) and first prove PFP$(0,l) \leq$ FO(LFP) by induction on PFP$(0,l)$-formulas. So, consider a formula $\varphi = [\mathrm{PFP}_{\overline{x},X}\,\psi]\,\overline{t}$ with $\psi = (\varphi_0 \vee \exists \overline{y}_1 \notin X \ldots \exists \overline{y}_l \notin X\, \varphi_1)$ where φ_0 and φ_1 do not contain X. By induction hypothesis, we may assume that φ_0 and φ_1 are FO(LFP)-formulas. Then ψ is equivalent to the FO(LFP)-formula

$$(\varphi_0 \vee \exists \overline{y}_1 \ldots \exists \overline{y}_l(\neg X\overline{y}_1 \wedge \ldots \wedge \neg X\overline{y}_l \wedge \varphi_1))$$

negative in X. Therefore, by 8.3.2, $[\mathrm{PFP}_{\overline{x},X}\,\psi]\,\overline{t}$ is equivalent to an FO(LFP)-formula.

We sketch a proof of FO(LFP) $\leq$ PFP$(0,1)$ and of the normal form claimed in (d) and refer the reader to [112] for details. Let φ be a formula of FO(LFP). We may assume that

$$\varphi = \exists(\forall)u[\mathrm{LFP}_{\overline{z},Z}\,\varphi_0]\,\tilde{u}$$

with first-order φ_0. For simplicity, let φ be a sentence.

One can simulate the semantics of FO(LFP)-formulas by a game for a definable digraph, thereby obtaining a quantifier-free first-order formula $\psi(\overline{x},\overline{y})$ with $\overline{x} = x_1 \dots x_k$ and $\overline{y} = y_1 \dots y_k$ for suitable k such that (1),(2), and (3) hold for all $\mathcal{A}$:

(1) $(A^k, \psi^{\mathcal{A}}(\_,\_))$ is a digraph, i.e., $\mathcal{A} \models \forall\overline{x}\neg\psi(\overline{x},\overline{x})$.
(2) $\mathcal{A} \models [\mathrm{LFP}_{\overline{z},Z}\,\varphi_0]\,\tilde{u}$ iff $\tilde{u}$ is won in the game associated with $(A^k, \psi^{\mathcal{A}}(\_,\_))$ (cf. 8.1.5(b)).
(3) $(A^k, \psi^{\mathcal{A}}(\_,\_))$ has no drawn points.

Therefore (cf. the example previous to 8.3.2), $\exists(\forall)u[\mathrm{PFP}_{\overline{x},X}\,\exists\overline{y} \notin X\psi(\overline{x},\overline{y})]\,\tilde{u}$ is a totally defined formula equivalent to φ. □

We apply the normal form just proven to *well-founded* DATALOG, WF-DATALOG. (The exercise at the end of this section shows how one can obtain some results for FO(LFP) as immediate consequences of this normal form.) Consider the program

$$\begin{aligned} Qx &\leftarrow Qx \\ Qx &\leftarrow \neg Qx \end{aligned}$$

In the semantics of both I-DATALOG and P-DATALOG, for a structure $\mathcal{A}$ we have $Q_\infty = A$, the elements of A getting into Q using the "wrong" information "$\neg Q = A$" in the first step. In WF-DATALOG, intuitively speaking, one requires that the values obtained for the intentional symbols are "re-confirmed by their positive occurrences". More precisely: The semantics of WF-DATALOG, the so-called *well-founded semantics*, treats positive and negative occurrences of intentional symbols in the bodies of clauses in an asymmetric way: Let Π be a WF-DATALOG program, i.e., a general logic program. For every intentional symbol P replace all occurrences of P in the heads of clauses and all positive occurrences of P in the bodies by a new relation symbol P'. Now, the original P does not further occur in the head of any clause of the resulting program Π'; hence, P is an extensional symbol of Π' and thus, Π' is a DATALOG program. In order to simplify the presentation, we assume that Π only contains a single intentional relation symbol P.

In every τ_{ext}-structure $\mathcal{A}$, the program Π gives rise to a sequence $(P_n)_{n\geq 0}$ of relations on A defined by

$P_0 := \emptyset$,
P_{n+1} is the result for P' of the evaluation of the DATALOG program Π' in $(\mathcal{A}, P_n)$, i.e., taking P_n as interpretation of the extensional symbol P of Π'.

For example, for the WF-DATALOG program

$$\begin{array}{llll} \Pi_0: & Pxy & \leftarrow & Rxyz, \neg Pux \\ & Pxy & \leftarrow & \neg Sxy, Rxyz, Pxz, \neg Pyy \end{array}$$

we have

$$\begin{array}{llll} \Pi_0': & P'xy & \leftarrow & Rxyz, \neg Pux \\ & P'xy & \leftarrow & \neg Sxy, Rxyz, P'xz, \neg Pyy \end{array}$$

and the stages P_n of the evaluation of Π_0 as described above are just the stages F_n^φ of F^φ, where

$$\begin{array}{rcl} \varphi(xy, P) & := & [\text{LFP}_{xy,P'} (\exists z \exists u(Rxyz \wedge \neg Pux) \\ & & \vee \exists z(\neg Sxy \wedge Rxyz \wedge P'xz \wedge \neg Pyy))]\, xy. \end{array}$$

We come back to the general case above, i.e., to Π, the τ_{ext}-structure $\mathcal{A}$, and the sequence $(P_n)_{n \geq 0}$. Let P_{true} be the truth set of this sequence, i.e.,

$$P_{\text{true}} \quad = \quad \{a \in A \mid \exists n_0 \forall n \geq n_0 : a \in P_n\}$$

(cf. 8.3.16). Then, by definition, in the WF-semantics, Π leads to the τ-structure $\mathcal{A}(\Pi) := (\mathcal{A}, P_{\text{true}})$. For terms $\bar{t}$, the formula $(\Pi, P)\bar{t}$ is a WF-DATALOG *formula* of vocabulary τ_{ext}, its meaning being "$\bar{t} \in P_{\text{true}}$".

Note that $P_{\text{true}} = P_\infty$ if the fixed-point P_∞ of the sequence $(P_n)_{n \geq 0}$ exists. Call Π a *totally defined* WF-DATALOG program and $(\Pi, P)\bar{t}$ a *totally defined* WF-DATALOG formula, if the fixed-point exists in all structures.

In case Π is a DATALOG program, the WF-DATALOG formula $(\Pi, P)\bar{t}$ is equivalent to the DATALOG formula $(\Pi, P)\bar{t}$; in case Π is a totally WF-DATALOG program and the bodies of the clauses in Π contain only negative occurrences of P, the WF-DATALOG formula $(\Pi, P)\bar{t}$ is equivalent to the P-DATALOG formula $(\Pi, P)\bar{t}$. The same applies to the program in part (b) of the following theorem; so, in a certain sense, we can do without the well-founded semantics.

Theorem 9.5.2 (a) WF-DATALOG $\equiv$ FO(LFP).

(b) *Every* WF-DATALOG *formula is equivalent to a totally defined* WF-DATALOG *formula* $(\Pi, P)\bar{x}$ *with* Π *of the form*

$$\begin{array}{rcl} Z\bar{x}\bar{y} & \leftarrow & \gamma_{11}, \ldots \gamma_{1k_1}, \neg Z\bar{x}\bar{z} \\ \vdots & \leftarrow & \vdots \\ Z\bar{x}\bar{y} & \leftarrow & \gamma_{s1}, \ldots \gamma_{sk_s}, \neg Z\bar{x}\bar{z} \\ P\bar{x} & \leftarrow & Z\bar{x}\tilde{u} \end{array}$$

where the γ_{ij} *contain neither* Z *nor* P.

Proof. (a) WF-DATALOG $\leq$ FO(LFP): Let $(\Pi, P)\bar{t}$ be a formula of WF-DATALOG and again, for simplicity (otherwise, argue with FO(S-LFP)), assume that P is the unique intentional symbol, i.e., Π has the form

$$\begin{array}{ccc} P\overline{x} & \leftarrow & \gamma_{11}, \ldots, \gamma_{1k_1} \\ \vdots & \leftarrow & \vdots \\ P\overline{x} & \leftarrow & \gamma_{s1}, \ldots, \gamma_{sk_s}. \end{array}$$

Then the stages P_n of Π evaluated in a τ_{ext}-structure are just the stages F_n^{φ} of F^{φ}, where

$$\varphi(\overline{x}, P) \quad := \quad [\text{LFP}_{\overline{x},P'} \bigvee_{i=1}^{s} \exists \overline{y}_i(\gamma'_{i1} \wedge \ldots \wedge \gamma'_{ik_i})]\,\overline{x}.$$

where $\overline{y}_i$ are the variables in $(\gamma_{i1} \wedge \ldots \wedge \gamma_{ik_i})$ distinct from $\overline{x}$ and where γ'_{ij} is obtained from γ_{ij} by replacing positive occurrences of P by P'. Then, $\varphi(\overline{x}, P)$ is negative in P. Hence, F^{φ} is antitone, i.e.,

$$X \subseteq Y \quad \text{implies} \quad F^{\varphi}(Y) \subseteq F^{\varphi}(X)$$

(see Exercise 8.1.6). Therefore,

$$F_0^{\varphi} \subseteq F_2^{\varphi} \subseteq F_4^{\varphi} \subseteq \ldots F_5^{\varphi} \subseteq F_3^{\varphi} \subseteq F_1^{\varphi}$$

(cf. 8.1.5(a)). Clearly, $G := F^{\varphi} \circ F^{\varphi}$ is monotone, and $G = F^{\psi}$ holds for the formula $\psi(\overline{x}, P) := \varphi(\overline{x}, \varphi(\_, P))$ positive in P (cf. 8.3.2). Since $F_n^{\psi} = F_{2\cdot n}^{\varphi}$, we have $F_{\infty}^{\psi} = \bigcup_{n \geq 0} F_{2n}^{\varphi} = P_{\text{true}}$. Hence, $(\Pi, P)\bar{t}$ is equivalent to the FO(LFP)-formula

$$[\text{LFP}_{\overline{x},P}\, \psi(\overline{x}, P)]\,\bar{t}.$$

FO(LFP) $\leq$ WF-DATALOG: Let $\varphi(\overline{x})$ be an FO(LFP)-formula. By Theorem 9.5.1(d), it is equivalent to a totally defined FO(PFP)-formula of the form

$$\exists u[\text{PFP}_{\overline{y},X}\, \exists \overline{z} \notin X \psi(\overline{x}, \overline{y}, \overline{z})]\, \tilde{u}$$

where ψ is quantifier-free and does not contain X. Then, $\varphi(\overline{x})$ is equivalent to

$$\exists u[\text{PFP}_{\overline{x}\,\overline{y},Z}\, \exists \overline{z}(\neg Z\overline{x}\,\overline{z} \wedge \psi(\overline{x}, \overline{y}, \overline{z}))]\, \overline{x}\tilde{u}.$$

We can assume that ψ is in disjunctive normal form, e.g., $\psi = \bigvee_{i=1}^{k}(\gamma_{i1} \wedge \ldots \wedge \gamma_{ik_i})$ with atomic or negated atomic γ_{ij}. Then, $\varphi(\overline{x})$ is equivalent to

$$\exists u[\text{PFP}_{\overline{x}\,\overline{y},Z} \bigvee_{i=1}^{k} \exists \overline{z}(\neg Z\overline{x}\,\overline{z} \wedge \gamma_{i1} \wedge \ldots \wedge \gamma_{ik_i})]\, \overline{x}\tilde{u}$$

and hence, equivalent to the totally defined WF-DATALOG formula $(\Pi, P)\overline{x}$, where Π consists of the clauses

$$\begin{array}{rcl} Z\overline{x}\,\overline{y} & \leftarrow & \gamma_{11}, \dots \gamma_{1k_1}, \neg Z\overline{x}\,\overline{z} \\ \vdots & \leftarrow & \vdots \\ Z\overline{x}\,\overline{y} & \leftarrow & \gamma_{s1}, \dots \gamma_{sk_s}, \neg Z\overline{x}\,\overline{z} \\ P\overline{x} & \leftarrow & Z\overline{x}\tilde{u} \end{array}$$

This completes the proof of part (a). At the same time, we have proved part (b), too. □

The following exercise contains two further applications of part (d) of Theorem 9.5.1: In part (a) we give a short proof for the normal form for FO(LFP) contained in Theorem 9.4.7 (thereby restricting ourselves to (2)); in part (b) we show that every FO(LFP) formula is equivalent to a first-order formula containing relations implicitly definable in FO (cf. 8.5.8).

Exercise 9.5.3 Let $\rho(\overline{w})$ be an FO(LFP)-formula. According to 9.5.1(d) choose a totally defined formula equivalent to $\rho(\overline{w})$ of the form

$$(*) \qquad \exists(\forall)u[\mathrm{PFP}_{\overline{x},X}\exists\overline{y} \notin X\chi]\,\tilde{u}$$

where χ is quantifier-free, does not contain X, and $\mathrm{free}(\chi) \subseteq \{\overline{x}, \overline{y}, \overline{w}\}$.

(a) Set $\varphi(\overline{x}, X) := \exists\overline{y} \notin X\chi(\overline{x}, \overline{y})$ and $\psi(\overline{x}, X) := \varphi(\overline{x}, \varphi(\_, X)$. Then, $F^\psi = F^\varphi \circ F^\varphi$. Since the formula in $(*)$ is totally defined, we have $F^\psi_\infty = F^\varphi_\infty$ (cf. 8.1.5 and 8.3.2). Show that $\psi(\overline{x}, X)$ is equivalent to

$$\exists\overline{y}(\chi(\overline{x}, \overline{y}) \wedge \forall\overline{z}(\chi(\overline{y}, \overline{z}) \rightarrow X\overline{z})).$$

Conclude that

$$\models \exists\overline{z}\neg X\overline{z} \rightarrow (\psi(\overline{x}, X) \leftrightarrow \exists\overline{y}\forall\overline{z} \notin X(\chi(\overline{x}, \overline{y}) \wedge \neg\chi(\overline{y}, \overline{z}))$$

and hence, that $\rho(\overline{w})$ is equivalent

$$\exists(\forall)u[\mathrm{LFP}_{\overline{x},X}\exists\overline{y}\forall\overline{z} \notin X(\chi(\overline{x}, \overline{y}) \wedge \neg\chi(\overline{y}, \overline{z}))]\,\tilde{u}.$$

(b) Clearly, $\rho(\overline{w})$ is equivalent to to the formula

$$\exists(\forall)\tilde{u}[\mathrm{PFP}_{\overline{x}\,\overline{w},Y}\exists\overline{y}(\neg Y\overline{y}\,\overline{w} \wedge \chi)]\,\tilde{u}\overline{w},$$

that is totally defined, too. Introduce $\varphi(\overline{x}\,\overline{w}, Y)$ as above. By 8.1.5, F^φ_∞ is the unique fixed-point of F^φ. Let $\psi_0(Y) := \forall\overline{x}\,\overline{w}(Y\overline{x}\,\overline{w} \leftrightarrow \exists\overline{y}(\neg Y\overline{y}\,\overline{w} \wedge \chi))$ and conclude that

$$\models_{\mathrm{fin}} \exists^{=1}Y\psi_0(Y) \quad \text{and} \quad \models_{\mathrm{fin}} \forall Y(\psi_0(Y) \rightarrow (\rho(\overline{w}) \leftrightarrow Y\tilde{w}_1\overline{w})),$$

i.e., that $\rho(\overline{w})$ is equivalent to a section of a relation implicitly definable in FO. □

9.6 Hierarchies of Fixed-Point Logics

We first discuss the problem to what extent the expressive power of fixed-point operators depends on their arities. This problem comes up very naturally with the proofs of the normal form theorems, where the simplifications we have obtained with respect to the number of fixed-point operators, quantifiers, and occurrences of second-order variables had to be paid by enlarging the arity of the fixed-point operators involved.

Denote by FO(LFP^r), FO(IFP^r), and FO(PFP^r) the set of formulas of the corresponding fixed-point logic that contain second-order variables only of arity $= r$.[4] And let FO(DTC^r) be the fragment of FO(DTC) consisting of those formulas, where for any subformula of the form $[\mathrm{DTC}_{\overline{x},\overline{y}}\varphi]\,\overline{s}\overline{t}$ we have length$(\overline{x}) = r$. Define FO(TC^r) similarly. Clearly,

$$(*) \quad \mathrm{FO}(\mathrm{DTC}^r) \leq \mathrm{FO}(\mathrm{TC}^r) \leq \mathrm{FO}(\mathrm{LFP}^r) \leq \mathrm{FO}(\mathrm{IFP}^r) \leq \mathrm{FO}(\mathrm{PFP}^r).$$

Theorem 9.6.1 *Let $\mathcal{L}$ be one of the logics* FO(DTC), FO(TC), FO(LFP), FO(IFP), FO(PFP). *For $r \geq 1$, let $\mathcal{L}^r$ be the corresponding fragment satisfying the arity restriction above. Then*

$$\mathcal{L}^0 < \mathcal{L}^1 < \mathcal{L}^2 < \ldots .$$

More precisely: If τ contains an at least binary relation symbol then

$$\mathcal{L}^0[\tau] < \mathcal{L}^1[\tau] < \ldots .$$

Using dummy variables, it is clear that $\mathcal{L}^0 \leq \mathcal{L}^1 \leq \mathcal{L}^2 \leq \ldots$. In view of $(*)$, the strict inequalities are an immediate consequence of

Theorem 9.6.2 *For any $r \geq 0$ we have*

$$\mathrm{FO}(\mathrm{TC}^{r+1}) \not\leq \mathrm{FO}(\mathrm{PFP}^r) \quad \textit{and} \quad \mathrm{FO}(\mathrm{DTC}^{r+1}) \not\leq \mathrm{FO}(\mathrm{TC}^r)$$

on the class of graphs.

For a proof we refer to [64]. We only sketch the main idea. To obtain $\mathrm{FO}(\mathrm{TC}^{r+1}) \not\leq \mathrm{FO}(\mathrm{PFP}^r)$, one considers graphs definable on the set of $(r+1)$-tuples of the universe of certain structures. These structures are sufficiently homogeneous with respect to r-tuples to cause a collapse of the expressive power of FO(PFP^r) to FO. As a consequence, the query expressing that two $(r+1)$-tuples are connected by a path is not definable in FO(PFP^r) but, of course, it is expressible in FO(TC^{r+1}). The construction is quite intricate. It still has to be refined to obtain structures showing $\mathrm{FO}(\mathrm{DTC}^{r+1}) \not\leq \mathrm{FO}(\mathrm{TC}^r)$. Note that the first claim cannot be strengthened to $\mathrm{FO}(\mathrm{DTC}^{r+1}) \not\leq \mathrm{FO}(\mathrm{PFP}^r)$, as $\mathrm{FO}(\mathrm{DTC}) \leq \mathrm{FO}(\mathrm{PFP}^2)$ by the next proposition. □

[4] Thus FO(LFP^1) is monadic fixed-point logic FO(M-LFP) as introduced in section 8.5.

Throughout this section, given a formula $\varphi(\overline{x}, \overline{y})$, we abbreviate by $\varphi_{\text{det}}(\overline{x}, \overline{y})$ the formula

$$\varphi_{\text{det}}(\overline{x}, \overline{y}) \quad := \quad \varphi(\overline{x}, \overline{y}) \wedge \forall \overline{z}(\varphi(\overline{x}, \overline{z}) \rightarrow \overline{z} = \overline{y})$$

(where $\overline{z}$ are new variables). Then

$$\models_{\text{fin}} [\text{DTC}_{\overline{x},\overline{y}}\, \varphi(\overline{x}, \overline{y})]\, \overline{x}\, \overline{y} \leftrightarrow [\text{TC}_{\overline{x},\overline{y}}\, \varphi_{\text{det}}(\overline{x}, \overline{y})]\, \overline{x}\, \overline{y}.$$

Proposition 9.6.3 FO(DTC) $\leq$ FO(PFP^2).

Proof. For r-tuples $\overline{a}, \overline{e}$ let

$$X_{\overline{e}}(\overline{a}) \quad := \quad \{(a_1, e_1), \ldots, (a_r, e_r)\}.$$

As the main step of an inductive proof we show how to express a DTC operator by a PFP^2 operator: Given an FO(PFP^2)-formula $\varphi(\overline{x}, \overline{y})$, we shall find an FO(PFP^2)-formula $\psi(x, y, X, \overline{x}, \overline{y}, \overline{z})$ with binary X and with parameters $\overline{x}, \overline{y}, \overline{z}$ such that for distinct elements $z_1, \ldots, z_r$,

(1) if there is a deterministic φ-path from $\overline{x}$ to $\overline{y}$, say $\overline{x} = \overline{x}_0, \ldots, \overline{x}_l = \overline{y}$, then $F_0^\psi, F_1^\psi, \ldots$ is the sequence $\emptyset, X_{\overline{z}}(\overline{x}_1), X_{\overline{z}}(\overline{x}_2), \ldots, X_{\overline{z}}(\overline{x}_l), X_{\overline{z}}(\overline{x}_l), \ldots$;
(2) if there is no deterministic φ-path from $\overline{x}$ to $\overline{y}$ then $F_\infty^\psi = \emptyset$.

Then, in structures with at least r elements (the others can be treated separately), $[\text{DTC}_{\overline{x},\overline{y}}\, \varphi]\, \overline{x}\, \overline{y}$ and

$$\exists z_1 \ldots \exists z_r (\bigwedge_{1 \leq i < j \leq r} \neg z_i = z_j \wedge \exists u \exists v [\text{PFP}_{xy,X}\, \psi(x, y, X, \overline{x}, \overline{y}, \overline{z})]\, uv)$$

are equivalent, and we are done.

As $\psi(x, y, X, \overline{x}, \overline{y}, \overline{z})$ we can take the formula (note that the case of a φ-path

$$\overline{x} = \overline{x}_0, \overline{x}_1, \ldots, \overline{x}_m, \overline{x}_m, \ldots$$

not containing $\overline{y}$ needs special care)

$$\begin{aligned}
&(X = \emptyset \wedge \exists \overline{w}(\varphi_{\text{det}}(\overline{x}, \overline{w}) \wedge (\neg \overline{w} = \overline{x} \vee \overline{w} = \overline{y}) \wedge \textstyle\bigvee_{1 \leq i \leq r}(x = w_i \wedge y = z_i))) \\
&\vee \exists \overline{v} \exists \overline{w}(\textstyle\bigwedge_{1 \leq i \leq r} X v_i z_i \wedge \varphi_{\text{det}}(\overline{v}, \overline{w}) \wedge \neg \overline{v} = \overline{y} \wedge \neg \overline{v} = \overline{w} \wedge \\
&\qquad \textstyle\bigvee_{1 \leq i \leq r}(x = w_i \wedge y = z_i)) \\
&\vee (\textstyle\bigwedge_{1 \leq i \leq r} X y_i z_i \wedge \textstyle\bigvee_{1 \leq i \leq r}(x = y_i \wedge y = z_i)).
\end{aligned}$$

□

We now turn to the hierarchy problem for *ordered* structures. We present two propositions which will help us to discuss the general problem.

Proposition 9.6.4 *On ordered structures,* FO(DTC) $\leq$ FO(LFP^3).[5]

Proof. Consider a formula $\varphi(\overline{x}, \overline{y})$, where $\overline{x} = x_1 \dots x_r$ and $\overline{y} = y_1 \dots y_r$. In a structure $\mathcal{A}$ of cardinality n, the (shortest) deterministic φ-path between two tuples must have length $\leq n^r$. Assume $A = \{1, \dots, n\}$. First suppose that there is a deterministic φ-path

$$\text{(1)} \qquad \overline{x} = \overline{x}_0, \overline{x}_1, \dots, \overline{x}_k = \overline{y}$$

from $\overline{x}$ to $\overline{y}$ with $k \leq n$ (where $\overline{x}_i = x_{i1} \dots x_{ir}$). To rewrite φ as an FO(LFP^3)-formula, we shall code the existence of such a φ-path by an LFP operator with stages

$$\begin{aligned} X_1 &= \{(x_{11}, 1, 1), (x_{12}, 1, 2), \dots, (x_{1r}, 1, r)\} \\ X_2 &= X_1 \cup \{(x_{21}, 2, 1), (x_{22}, 2, 2), \dots, (x_{2r}, 2, r)\} \\ &\vdots \\ X_k &= X_{k-1} \cup \{(x_{k1}, k, 1), (x_{k2}, k, 2), \dots, (x_{kr}, k, r)\} \end{aligned}$$

and $X_k = X_{k+1} = \dots$. Using s nested LFP operators appropriately, we can express the existence of a deterministic φ-path of length $\leq n^s$. For $s = r$, we get the desired formula.

To give the details of the elimination of a DTC operator by LFP^3 operators (as the main step of an inductive proof), let $\varphi(\overline{x}, \overline{y})$ be any FO(LFP^3) formula.

Consider the formula

$$\psi(\overline{x}, \overline{y}) \quad := \quad [\mathrm{DTC}_{\overline{x}, \overline{y}}\, \varphi(\overline{x}, \overline{y})]\overline{x}\,\overline{y}.$$

Then, in structures with at least $\max\{3, r\}$ elements, the statement

> "there is a deterministic φ-path of length $\leq$ (cardinality of the universe) from $\overline{x}$ to $\overline{y}$"

can be expressed by the FO(LFP^3)-formula $\varphi^*(\overline{x}, \overline{y})$ below, which uses the above coding of the path (1) prolongated to the path $\overline{x} = \overline{x}_0, \overline{x}_1, \dots, \overline{x}_k = \overline{y}, \overline{y}, \dots, \overline{y}$ of length n: [6]

$$\begin{aligned} \varphi^* \;:=\; & \exists x \exists v [\mathrm{LFP}_{xuv, X} \\ & \exists \overline{y}' (\varphi_{\det}(\overline{x}, \overline{y}') \wedge u = \min + 1 \wedge \textstyle\bigvee_{1 \leq i \leq r} (x = y'_i \wedge v = i)) \\ & \vee \exists u' \exists \overline{x}' \exists \overline{y}' (\varphi_{\det}(\overline{x}', \overline{y}') \wedge \neg \overline{x}' = \overline{y} \wedge \textstyle\bigwedge_{1 \leq i \leq r} X x'_i u' i \wedge u = u' + 1 \\ & \qquad \wedge \textstyle\bigvee_{1 \leq i \leq r} (x = y'_i \wedge v = i) \wedge (u = \max \rightarrow \overline{y}' = \overline{y})) \\ & \vee \exists u' (\textstyle\bigwedge_{1 \leq i \leq r} X y_i u' i \wedge u = u' + 1 \wedge \textstyle\bigvee_{1 \leq i \leq r} (x = y_i \wedge v = i))] x \max v. \end{aligned}$$

---

[5] The next exercise shows that LFP^3 can be replaced by LFP^2.

[6] As usual, for $i \geq 1$ we abbreviate by $z = i$ that z is the i-th element in the ordering.

We now iterate the transition from φ to φ^*, defining

$$\varphi^1 = \varphi^* \quad \text{and} \quad \varphi^{n+1} = (\varphi^n)^*.$$

Then $\varphi^s(\overline{x}, \overline{y})$ expresses

"there is a deterministic φ-path of length $\leq$ (cardinality of the universe)^s from $\overline{x}$ to $\overline{y}$."

Hence, φ^r is an FO(LFP^3)-formula equivalent to $[\mathrm{DTC}_{\overline{x},\overline{y}}\, \varphi(\overline{x}, \overline{y})]\, \overline{x}\, \overline{y}$. □

Exercise 9.6.5 Show that FO(DTC) $\leq$ FO(LFP^2) on ordered structures. Hint: Using arithmetics definable in FO(LFP^2), code the last two components of X in the preceding proof by a single one. □

We know that FO(DTC) $\not\leq$ FO(LFP^1) on ordered structures, since even FO(DTC) $\not\leq$ MSO on ordered structures (cf. 8.6.3). But we have:

Proposition 9.6.6 *On ordered structures,* FO(DTC) $\leq$ FO(PFP^1).

Proof. Consider the formula

$$[\mathrm{DTC}_{\overline{x},\overline{y}}\, \varphi(\overline{x}, \overline{y})]\, \overline{s}\overline{t}$$

with $\overline{x} = x_1 \dots x_r$ and $\overline{y} = y_1 \dots y_r$. We show how the DTC operator can be expressed by PFP^1. For this purpose we consider a deterministic φ-path

$$(*) \qquad \overline{s} = \overline{x}_0, \overline{x}_1, \overline{x}_2, \dots$$

starting from $\overline{s}$. The idea is to code r-tuples $\overline{x}$ by subsets $P_{\overline{x}}$ of the universe and to find a formula of FO(PFP^1) whose PFP^1 operator has the stages $\emptyset, P_{\overline{x}_1}, P_{\overline{x}_2}, \dots, P_{\overline{x}_k}, P_{\overline{t}}, P_{\overline{t}}, \dots$ in case $\overline{x}_k = \overline{t}$ in $(*)$, but does not reach a fixed-point in case the deterministic φ-path started in $\overline{s}$ does not pass $\overline{t}$.

We present the codification of $\overline{x}$ in a set $P_{\overline{x}}$: Let (the interpretations of) $\overline{x}$ be given in an ordered structure. For $1 \leq i \leq r$, let m_i be the position of x_i in the ordering induced on the x_i (for example, if $r = 4$ and $x_3 < x_1 = x_4 < x_2$ then $m_1 = 2, m_2 = 3, m_3 = 1, m_4 = 2$). Note that $1 \leq m_i \leq r$. Order the set $\{(n_1, \dots, n_r) \mid 1 \leq n_1, \dots, n_r \leq r\}$, a set independent of the given structure, lexicographically. If $(m_1, \dots, m_r)$ is the l-th element in this ordering set

$$P_{\overline{x}} := \{\overline{x}\} \cup I$$

where I is the leftmost interval in the given structure having length $r + 1 + l$ and containing no x_i (such an interval exists in structures of cardinality $\geq (r+1)\cdot(r+2+r^r)$). Note that $P_{\overline{x}}$ uniquely codes $\{\overline{x}\}$ and l and hence, $\overline{x}$. Since the coding and decoding of $\overline{x}$ in $P_{\overline{x}}$ is first-order definable, it is easy to write down an FO(PFP^1)-formula equivalent to $[\mathrm{DTC}_{\overline{x},\overline{y}}\, \varphi(\overline{x}, \overline{y})]\, \overline{s}\overline{t}$. □

The hierarchy result for FO(PFP) remains true also on ordered structures.

Theorem 9.6.7 *For all $r \geq 0$, $\mathrm{FO}(\mathrm{PFP}^r) < \mathrm{FO}(\mathrm{PFP}^{r+1})$ on ordered structures.*

Proof. For $e \in \mathbb{R}, e \geq 1$, consider the function $n \mapsto n^e$. We say that a class K is in $\mathrm{SPACE}(n^e)$ if for some $c > 0$ there is an $c \cdot n^e$ space-bounded deterministic Turing machine accepting K. From complexity theory we know that for $r \in \mathbb{N}$,

$$(*) \qquad \mathrm{SPACE}(n^r) \neq \mathrm{SPACE}(n^{r+\frac{1}{2}}).$$

We show that

$$(1) \qquad \text{each class axiomatizable in } \mathrm{FO}(\mathrm{PFP}^r) \text{ is in } \mathrm{SPACE}(n^r)$$

and for $\epsilon \in \mathbb{R}$, $0 \leq \epsilon < 1$ that

$$(2) \qquad \text{each class in } \mathrm{SPACE}(n^{r+\epsilon}) \text{ is axiomatizable in } \mathrm{FO}(\mathrm{PFP}^{r+1}).$$

Then we are finished: If K is a class in $\mathrm{SPACE}(n^{r+\frac{1}{2}}) \setminus \mathrm{SPACE}(n^r)$ then K is axiomatizable in $\mathrm{FO}(\mathrm{PFP}^{r+1})$ (by (2)), but not in $\mathrm{FO}(\mathrm{PFP}^r)$ (by (1)).

Claim (1) is already inherent in the proof of 7.4.2. Claim (2) is obtained by an analysis of the proof showing that classes of ordered structures in PSPACE are axiomatizable in FO(PFP) (cf. 7.3.2). We sketch a proof. To code the configurations of an n^r space-bounded machine, in section 7.3 we used a $(2+r)$-ary relation, where the first two components served to distinguish the distinct ingredients of configurations, their number k being independent of the input structure. Clearly, k distinct numbers of a single component are sufficient for this purpose. The remaining $n-k$ numbers of this component can be used (in sufficiently large structures) to encode the n^ϵ part of an $n^{r+\epsilon}$ space-bounded machine (because of the preceding proposition and 7.3.11 the arithmetics needed for the encoding is available in $\mathrm{FO}(\mathrm{PFP}^{r+1})$). □

Corollary 9.6.8 *On ordered structures,* $\mathrm{FO}(\mathrm{DTC}) < \mathrm{FO}(\mathrm{PFP})$.

Proof. On ordered structures, we have $\mathrm{FO}(\mathrm{DTC}) \leq \mathrm{FO}(\mathrm{PFP}^1)$ (cf. 9.6.6) and $\mathrm{FO}(\mathrm{PFP}^1) < \mathrm{FO}(\mathrm{PFP})$ by the preceding theorem. □

In the following remarks we restrict ourselves to ordered structures. It is open whether the arity hierarchy is strict for FO(DTC), FO(TC), or FO(LFP), that is, whether the analogues of 9.6.7 are true for these logics. The problem for FO(LFP) is related to prominent questions in complexity theory: If $\mathrm{FO}(\mathrm{LFP}) \equiv \mathrm{FO}(\mathrm{LFP}^r)$ for some $r \geq 1$, then by 9.6.7, $\mathrm{FO}(\mathrm{LFP}) \not\equiv \mathrm{FO}(\mathrm{PFP})$ and hence, $\mathrm{PTIME} \neq \mathrm{PSPACE}$ (cf. 7.5.2). On the other hand, in view of $\mathrm{FO}(\mathrm{DTC}) \leq \mathrm{FO}(\mathrm{LFP}^3)$ (cf. 9.6.4), if $\mathrm{FO}(\mathrm{LFP}) \not\equiv \mathrm{FO}(\mathrm{LFP}^3)$ (by 9.6.5, even if $\mathrm{FO}(\mathrm{LFP}) \not\equiv \mathrm{FO}(\mathrm{LFP}^2)$) then $\mathrm{FO}(\mathrm{DTC}) \not\equiv \mathrm{FO}(\mathrm{LFP})$ and hence, $\mathrm{LOGSPACE} \neq \mathrm{PTIME}$ (cf. 7.5.2).

Exercise 9.6.9 Let $\mathcal{L} \in \{\text{FO(DTC)}, \text{FO(TC)}, \text{FO(LFP)}\}$. Show $\mathcal{L}^1 < \mathcal{L}^2$ on ordered structures (cf. 8.6.3). Show that $\text{FO(LFP}^1) < \text{FO(IFP}^1)$. Hint: Use 6.3.3 (and 6.2.3) to show that $\text{FO(IFP}^1) \not\leq \text{MSO}$ on word models. □

Exercise 9.6.10 Show $\text{FO(DTC)} \leq \text{FO(IFP}^2)$. Hint: By 9.6.5, FO(DTC) $\leq \text{FO(LFP}^2)$ on ordered structures. Use deterministic paths to define sufficiently large orderings that allow to carry out the corresponding argument. □

We know that every FO(LFP)-formula is equivalent to an FO(LFP)-formula containing just one fixed-point operation. But, given r, does this result hold for $\text{FO(LFP}^r)$ instead of FO(LFP)? In other words: Do we really have to increase the arity when replacing nested fixed-point operations by single ones? In general, we have. We close by stating a theorem which contains the corresponding result for all fixed-point logics. For a proof we refer to [113].

For an $\text{FO(DTC}^1)$-formula φ let $n(\varphi)$ be the maximal number of nested DTC operations in φ. Define $\rho : \text{FO(PFP)} \to \mathbb{N}$ by

$$\begin{aligned}
\rho(\varphi) &:= 0, \quad \text{if } \varphi \text{ is atomic;} \\
\rho(\neg\varphi) &:= \rho(\varphi); \\
\rho(\varphi \vee \psi) &:= \max\{\rho(\varphi), \rho(\psi)\}; \\
\rho(\exists x\varphi) &:= \rho(\varphi); \\
\rho([\text{PFP}_{\overline{x},X}\varphi]\,\overline{t}) &:= \rho(\varphi) + \text{arity of } X.
\end{aligned}$$

Theorem 9.6.11 *For $k \geq 0$ there is an $\text{FO(DTC}^1)$-sentence φ with $n(\varphi) = k+1$, which is not equivalent to any FO(PFP)-sentence ψ with $\rho(\psi) \leq k$.* □

Corollary 9.6.12 *For every $k \geq 1$ there is an $\text{FO(LFP}^1)$-sentence not equivalent to any $\text{FO(LFP}^k)$-sentence with a single* LFP *operation.* □

Notes 9.6.13 A reference for databases and logic programs is [1]. The main reference for sections 9.1 and 9.2 is [3]. Theorem 9.3.2 goes back to [7]. Most of the results of section 9.4 are due to Grohe and contained in his thesis [62]; cf. also [63]. References for section 9.5 are [112] (for Theorem 9.5.1(d)) and [42] (for Theorem 9.5.2). The main result of section 9.6, Theorem 9.6.2, is due to Grohe [62, 64]. Further references are [67, 88].

10. Optimization Problems

Many of the oldest and more prominent examples of NPTIME-complete decision problems arose from the study of combinatorial optimization problems, the NPTIME-completeness reflecting their apparent intractability. More precisely, the NPTIME-completeness of the decision problem rules out the existence of a polynomial time algorithm for the optimization problem (unless PTIME = NPTIME). Of course, the intractability of an optimization problem does not exclude the existence of efficient algorithms that provide approximative solutions. And in fact, such algorithms may be necessary for practical purposes.

It has turned out that intractable optimization problems may behave quite differently with respect to the existence of approximative algorithms (a phenomenon that led to various notions of approximability).

In this chapter we first show that the class of polynomially bounded optimization problems coincides with the class of optimization problems definable by first-order formulas. This characterization leads to a logical or descriptive classification of those problems. Section 10.2 presents some first results documenting a relationship between this classification and approximability. For further study we refer the reader to the literature.

10.1 Polynomially Bounded Optimization Problems

To prepare our definition of an optimization problem, let us consider two examples, a maximization and a minimization problem.

A *cut* C in an graph $\mathcal{G} = (G, E^G)$ is a subset of G.

MAXCUT: Given a graph $\mathcal{G}$, MAXCUT asks for the maximal number of edges between the two parts of a cut, that is, for the maximum of

$$\|\{(a,b) \mid E^G ab,\ a \in C,\ b \notin C\}\|,$$

where C ranges over all cuts in $\mathcal{G}$.

A *vertex cover* C in a graph $\mathcal{G} = (G, E^G)$ is a subset of G such that

$$\mathcal{G} \models \forall x \forall y (Exy \rightarrow (Cx \vee Cy)).$$

MINVERTEX: Given a graph $\mathcal{G}$, MINVERTEX asks for the minimum of $\|C\|$ where C ranges over all vertex covers in $\mathcal{G}$.

In this chapter we use some sloppy formulations. For example, we say that a class K of (unordered) structures is acceptable in polynomial time, actually meaning that the class $K_<$ of its ordered representations (cf. 7.5.11) is in PTIME. And if we say that a function $f : K \to \mathbb{N}$ is computable in polynomial time, we mean that the function $f_< : K_< \to \mathbb{N}$ with

$$f_<((\mathcal{A}, <)) \quad := \quad f(\mathcal{A})$$

is computable in polynomial time.

Definition 10.1.1 An NPTIME *optimization problem* $\mathcal{Q}$ is given by the data K, F, cost, μ, by short: $\mathcal{Q} = (K, F, \text{ cost}, \mu)$, where

- K is a class of structures (of a fixed vocabulary) acceptable in polynomial time, the class of *input structures* or *input instances.*
- F is a function defined on K; for every instance $\mathcal{A}$, $F(\mathcal{A})$ is the set of *feasible solutions* for $\mathcal{A}$. There is an $s \geq 1$ such that for every $\mathcal{A} \in K$ and $S \in F(\mathcal{A})$,

  $$S \subseteq A^s.$$

  Moreover, the class

  $$\{(\mathcal{A}, S) \mid \mathcal{A} \in K, S \in F(\mathcal{A})\}$$

  is acceptable in polynomial time.
- cost is a polynomial time computable function defined on the class $\{(\mathcal{A}, S) \mid \mathcal{A} \in K, S \in F(\mathcal{A})\}$. The values of cost are natural numbers.
- $\mu \in \{\max, \min\}$.

If $\mu = \max$ ($\mu = \min$) we speak of a *maximization* (*minimization*) problem. □

For $\mathcal{Q}$ we define the function $\text{opt}_\mathcal{Q}$ on the class K of input instances by

$$\text{opt}_\mathcal{Q}(\mathcal{A}) \quad := \quad \mu\{\text{cost}(\mathcal{A}, S) \mid S \in F(\mathcal{A})\}.$$

An NPTIME optimization problem $\mathcal{Q}$ induces the following decision problem: Given an input instance $\mathcal{A}$ and a $k \geq 0$, is there a feasible solution S for $\mathcal{A}$ such that $\text{cost}(\mathcal{A}, S) \geq k$ if $\mu = \max$, and $\text{cost}(\mathcal{A}, S) \leq k$ if $\mu = \min$? The name "NPTIME optimization problem" is justified by the observation that this decision problem is in NPTIME.

Let us see how the above examples, MAXCUT and MINVERTEX, fit into these definitions. In both examples the class of instances is the class of graphs. For MAXCUT the set of feasible solutions $F(\mathcal{G})$ is the power set $\text{Pow}(G)$, and the cost function is given by

$$\text{cost}(\mathcal{G}, C) = \|\{(a,b) \in E^G \mid a \in C \text{ and } b \notin C\}\|$$

and $\mu = \max$. For MINVERTEX, $F(\mathcal{G})$ is the set of vertex covers in $\mathcal{G}$,

$$\text{cost}(\mathcal{G}, C) = \|C\|$$

and $\mu = \min$. In both cases the problem asks for $\text{opt}_Q(\mathcal{G})$.

An NPTIME optimization problem Q is said to be *polynomially bounded* if there is a polynomial $p \in \mathbb{N}[x]$ such that $\text{opt}_Q(\mathcal{A}) \leq p(\|A\|)$ for all input structures $\mathcal{A}$.

For polynomially bounded optimization problems the bridge to logic is given by the following concept.

Definition 10.1.2 An NPTIME optimization problem Q is *first-order definable* if there is a first-order formula $\varphi(\overline{x}, \overline{Y})$ with free variables among $\overline{x} = x_1 \dots x_k$ and $\overline{Y} = Y_1 \dots Y_m$ such that for any input structure $\mathcal{A}$,

$$\text{opt}_Q(\mathcal{A}) = \underset{\overline{R}}{\mu} \; \|\{\overline{a} \in A \mid \mathcal{A} \models \varphi[\overline{a}, \overline{R}]\}\|. \qquad \square$$

Clearly, MAXCUT is first-order definable by the formula

$$\varphi(x, y, Y) \quad := \quad Exy \wedge Yx \wedge \neg Yy$$

and similarly, MINVERTEX by

$$\varphi(x, Y) \quad := \quad \forall x \forall y (Exy \rightarrow (Yx \vee Yy)) \rightarrow Yx.$$

Theorem 10.1.3 *An* NPTIME *optimization problem is polynomially bounded iff it is first-order definable.*

Proof. Clearly, if an optimization problem Q is first-order definable, say by $\varphi(x_1, \dots, x_k, \overline{Y})$, then $\text{opt}_Q(\mathcal{A}) \leq \|A\|^k$ for all instances $\mathcal{A}$; hence, Q is polynomially bounded.

Now suppose that Q is a polynomially bounded optimization problem, say, $\text{opt}_Q(\mathcal{A}) \leq \|A\|^k$ for all instances $\mathcal{A}$. Denote by τ the vocabulary of the input structures of Q and let P be a new k-ary relation symbol. Set

$$K_1 := \{(\mathcal{A}, P^A) \mid \mathcal{A} \in K, P^A \subseteq A^k, \text{ there is a feasible solution } S \in F(\mathcal{A}) \text{ such that } \text{cost}(\mathcal{A}, S) \geq \|P^A\|\}$$

if $\mu = \max$, and

$$K_1 := \{(\mathcal{A}, P^A) \mid \mathcal{A} \in K, P^A \subseteq A^k, \text{ there is a feasible solution } S \in F(\mathcal{A}) \text{ such that } \text{cost}(\mathcal{A}, S) \leq \|P^A\|\}$$

if $\mu = \min$. By the remarks following Definition 10.1.1 we know that the class K_1 is in NPTIME. Therefore, by 7.5.14, there is a Σ_1^1-formula $\exists \overline{X} \psi(\overline{X}, P)$ of vocabulary $\tau \cup \{P\}$ with first-order ψ such that for any structure $(\mathcal{A}, P^A)$,

$$(\mathcal{A}, P^A) \in K_1 \text{ iff } (\mathcal{A}, P^A) \models \exists \overline{X} \psi(\overline{X}, P).$$

Thus,

$$\text{opt}_Q(\mathcal{A}) = \mu\{||P|| \mid \overline{R}, P \text{ relations over } A \text{ such that } \mathcal{A} \models \psi(\overline{R}, P)\}.$$

Therefore, in case opt = max we have

$$\text{(1)} \qquad \text{opt}_Q(\mathcal{A}) = \max_{\overline{R}, P} ||\{\overline{a} \in A \mid \mathcal{A} \models P\overline{a} \wedge \psi(\overline{R}, P)\}||,$$

and in case opt = min,

$$\text{(2)} \qquad \text{opt}_Q(\mathcal{A}) = \min_{\overline{R}, P} ||\{\overline{a} \in A \mid \mathcal{A} \models \psi(\overline{R}, P) \rightarrow P\overline{a}\}||.$$

Hence, Q is first-order definable. □

Denote by MAX PB the class of polynomially bounded maximization problems and by MAX Σ_n the class of first-order definable maximization problems, where the defining formula (cf. 10.1.2) can be chosen Σ_n. Classes such as MAX Π_n, MIN PB, ... are defined similarly (for the definition of Σ_n and Π_n compare 1.B).

Using so-called Skolem relations, one easily shows that every Σ_1^1-formula is logically equivalent to a formula of the form $\exists \overline{X} \psi$ with $\psi \in \Pi_2$. E.g., for $\psi = \forall \overline{y} \exists z \forall \overline{u} \exists v \varphi$ with quantifier-free φ, we show how the formula $\exists \overline{X} \psi$ can be transformed into the desired form, thereby exemplifying the main step of an inductive proof. For a new relation variable Z, the formula $\exists \overline{X} \psi$ is equivalent to $\exists \overline{X} \exists Z \forall \overline{y} (\exists w Z \overline{y} w \wedge \forall z (Z \overline{y} z \rightarrow \forall \overline{u} \exists v \varphi))$ and hence, equivalent to $\exists \overline{X} \exists Z \forall \overline{y} \forall z \forall \overline{u} \exists w \exists v (Z \overline{y} w \wedge (Z \overline{y} z \rightarrow \varphi))$.

Thus the formula ψ used in the proof of the preceding theorem can be assumed to be Π_2. Hence, the formula in (1) is equivalent to a Π_2-formula and that in (2) to a Σ_2-formula. Therefore, we have

Corollary 10.1.4 (a) MAX PB = MAX FO = MAX Π_2.
(b) MIN PB = MIN FO = MIN Σ_2. □

For minimization problems we can do better: We shall see in 10.1.6 that MIN PB = MIN Π_1. In the proof, given a problem in MIN Σ_2, we shall replace the leading existential quantifiers by relations similarly as above. The same idea underlies the following proof.

Proposition 10.1.5 MAX Σ_2 = MAX Π_1 *and hence,* MAX $\Sigma_1 \subseteq$ MAX Π_1.

Proof. Let Q be in MAX Σ_2 and let $\psi(\overline{x}, \overline{y}, \overline{Y})$ be a Π_1-formula such that for any instance $\mathcal{A}$ we have

$$\text{opt}_Q(\mathcal{A}) = \max_{\overline{R}} ||\text{Sat}(\overline{R})||,$$

where

$$\text{Sat}(\overline{R}) \quad := \quad \{\overline{a} \in A \mid \mathcal{A} \models \exists \overline{y} \psi(\overline{a}, \overline{y}, \overline{R})\}.$$

We show

$$(*) \qquad \max_{\overline{R}} \|\text{Sat}(\overline{R})\| = \max_{\overline{R}, P} \|\text{Sat}(\overline{R}, P)\|,$$

where $\text{Sat}(\overline{R}, P)$ is

$$\{\overline{a}\overline{b} \in A \mid \mathcal{A} \models \psi(\overline{a}, \overline{b}, \overline{R}) \wedge P\overline{a}\overline{b} \wedge \forall \overline{y} \forall \overline{z}((P\overline{a}\,\overline{y} \wedge P\overline{a}\,\overline{z}) \rightarrow \overline{y} = \overline{z})\}.$$

This gives the claim, since then $\text{opt}_Q(\mathcal{A}) = \max_{\overline{R},P} \|\text{Sat}(\overline{R}, P)\|$, and the formula in the definition of $\text{Sat}(\overline{R}, P)$ is equivalent to a Π_1-formula.

To prove $(*)$, note that $\overline{a}\overline{b}_1, \overline{a}\overline{b}_2 \in \text{Sat}(\overline{R}, P)$ implies $\overline{b}_1 = \overline{b}_2$ and $\overline{a} \in \text{Sat}(\overline{R})$. Thus

$$(1) \qquad \|\text{Sat}(\overline{R})\| \geq \|\text{Sat}(\overline{R}, P)\|.$$

On the other hand, for $\overline{a} \in \text{Sat}(\overline{R})$ choose a witness $\overline{b}(\overline{a})$ such that $\mathcal{A} \models \psi(\overline{a}, \overline{b}(a), \overline{R})$ and set $P_0 := \{\overline{a}\overline{b}(\overline{a}) \mid \overline{a} \in \text{Sat}(\overline{R})\}$. Then $\text{Sat}(\overline{R}, P_0) = P_0$ and hence,

$$(2) \qquad \|\text{Sat}(\overline{R})\| = \|\text{Sat}(\overline{R}, P_0)\|.$$

From (1) and (2) we conclude $(*)$. □

Proposition 10.1.6 $\text{MIN}\,\Pi_1 = \text{MIN}\,\Sigma_2$ $(= \text{MIN}\,\text{PB})$ *and* $\text{MIN}\,\Sigma_0 = \text{MIN}\,\Sigma_1$.

Proof. For the first assertion, let Q be a problem in $\text{MIN}\,\Sigma_2$ and let $\psi(\overline{x}, \overline{y}, \overline{z}, \overline{Y})$ be a quantifier-free formula such that

$$\text{opt}_Q(\mathcal{A}) = \min_{\overline{R}} \|\text{Sat}(\overline{R})\|$$

where

$$\text{Sat}(\overline{R}) \quad := \quad \{\overline{a} \in A \mid \mathcal{A} \models \exists \overline{y} \forall \overline{z} \psi(\overline{a}, \overline{y}, \overline{z}, \overline{R})\}.$$

We can assume that $\overline{x} = x_1 \ldots x_k$ and $\overline{y} = y_1 \ldots y_m$ are of the same length (if $m < k$ we extend $\overline{y}$ by dummy variables and if $k < m$ we replace ψ by $\psi \wedge x_k = x_{k+1} = \ldots = x_m$). We show

$$(*) \qquad \min_{\overline{R}} \|\text{Sat}(\overline{R})\| = \min_{\overline{R}, P} \|\text{Sat}(\overline{R}, P)\|,$$

where

$$\text{Sat}(\overline{R}, P) \;:=\; \{\overline{a}\overline{b} \in A \mid \mathcal{A} \models (\forall \overline{z}\psi(\overline{a}, \overline{b}, \overline{z}, \overline{R}) \wedge \neg P\overline{a}) \vee (\overline{a} = \overline{b} \wedge P\overline{a})\}.$$

This finishes the proof, since the formula in the definition of $\text{Sat}(\overline{R}, P)$ is equivalent to a Π_1-formula.

To show $(*)$, we verify for every $\overline{R}$ that

$$(+) \qquad \|\text{Sat}(\overline{R})\| = \min_P \|\text{Sat}(\overline{R}, P)\|.$$

Fix $\overline{R}$ and let P be arbitrary. For $\overline{a} \in \text{Sat}(\overline{R})$ choose $\overline{b}$ such that $\mathcal{A} \models \forall \overline{z}\psi(\overline{a}, \overline{b}, \overline{z}, \overline{R})$. If not $P\overline{a}$ then $\overline{a}\overline{b} \in \text{Sat}(\overline{R}, P)$ and if $P\overline{a}$ then $\overline{a}\,\overline{a} \in \text{Sat}(\overline{R}, P)$. Thus, $\|\text{Sat}(\overline{R})\| \leq \|\text{Sat}(\overline{R}, P)\|$. On the other hand, for $P = \text{Sat}(\overline{R})$, we have $\|\text{Sat}(\overline{R})\| = \|\text{Sat}(\overline{R}, P)\|$. Altogether, we see that $(+)$, and thus $(*)$, holds.

To prove MIN Σ_0 = MIN Σ_1, just omit the variables $\overline{z}$ and the quantifiers $\forall \overline{z}$ in the preceding proof. □

By 10.1.4–10.1.6 we have

$$\text{MAX}\,\Sigma_0 \subseteq \text{MAX}\,\Sigma_1 \subseteq \text{MAX}\,\Pi_1 = \text{MAX}\,\Sigma_2 \subseteq \text{MAX}\,\Pi_2 = \text{MAX}\,\text{PB}$$
$$\text{MIN}\,\Sigma_0 = \text{MIN}\,\Sigma_1 \subseteq \text{MIN}\,\Pi_1 = \text{MIN}\,\Sigma_2\ (= \text{MIN}\,\Pi_2) = \text{MIN}\,\text{PB}.$$

All the inclusions are proper (compare [105]). The situation changes for structures with a built-in successor relation (compare [34]).

10.2 Approximable Optimization Problems

For many NPTIME optimization problems $\mathcal{Q} = (K, F, \text{cost}, \mu)$ the corresponding decision problem, in case $\mu = \max$ the set

$$\{(\mathcal{A}, k) \mid \text{ there is a feasible solution } S \in F(\mathcal{A}) \text{ such that } \text{cost}(\mathcal{A}, S) \geq k\},$$

is NPTIME-complete. An example is given by MAXCUT. Unless PTIME = NPTIME, this rules out the existence of a polynomial time algorithm Π giving optimal solutions in the sense that for any instance $\mathcal{A}$,

$$\Pi(\mathcal{A}) \in F(\mathcal{A}) \quad \text{and} \quad \text{opt}_{\mathcal{Q}}(\mathcal{A}) = \text{cost}(\mathcal{A}, \Pi(\mathcal{A})).$$

For practical purposes one is interested in approximation algorithms:

Definition 10.2.1 Let $0 \leq \varepsilon < 1$ and $\mathcal{Q}$ be an NPTIME optimization problem. $\mathcal{Q}$ is *ε-approximable* if there is a polynomial time algorithm Π, which for every instance $\mathcal{A}$ returns a feasible solution $\Pi(\mathcal{A})$ such that

$$|\text{opt}_{\mathcal{Q}}(\mathcal{A}) - \text{cost}(\mathcal{A}, \Pi(\mathcal{A}))| \leq \varepsilon \cdot \text{opt}_{\mathcal{Q}}(\mathcal{A}).$$

That is, $\text{cost}(\mathcal{A}, \Pi(\mathcal{A})) \geq (1 - \varepsilon) \cdot \text{opt}_{\mathcal{Q}}(\mathcal{A})$ if $\mu = \max$, and $\text{cost}(\mathcal{A}, \Pi(\mathcal{A})) \leq (1 + \varepsilon) \cdot \text{opt}_{\mathcal{Q}}(\mathcal{A})$ if $\mu = \min$.

Denote by APX the class of NPTIME optimization problems that are ε-approximable for some ε with $0 \leq \varepsilon < 1$. □

To relate the class MAX Σ_1 to APX, we have to incorporate some effectivity requirements; they are fulfilled in all familiar examples.

Definition 10.2.2 A formula $\varphi(\overline{x}, \overline{Y})$ *effectively represents* the NPTIME optimization problem $\mathcal{Q}$ if, in addition to the conditions stated in Definition 10.1.2, there is a polynomial time algorithm that, given any instance $\mathcal{A}$ and any relations $\overline{R}$ on A, computes a feasible solution $S \in F(\mathcal{A})$ with

$$\text{cost}(\mathcal{A}, S) = \|\{\overline{a} \in A \mid \mathcal{A} \models \varphi(\overline{a}, \overline{R})\}\|. \qquad \Box$$

In the following let MAX$_e$ Σ_1 be the class of maximization problems which can be effectively represented by a Σ_1-formula.

Theorem 10.2.3 MAX$_e$ $\Sigma_1 \subseteq$ APX.

Proof. Consider a problem $\mathcal{Q}$ in MAX$_e$ Σ_1. Then, for a suitable quantifier-free formula $\psi(\overline{x}, \overline{y}, \overline{Y})$, the formula $\exists\overline{y}\psi(\overline{x}, \overline{y}, \overline{Y})$ effectively represents $\mathcal{Q}$; in particular, for any instance $\mathcal{A}$ we have

$$\text{opt}_{\mathcal{Q}}(\mathcal{A}) = \max_{\overline{R}} \|\{\overline{a} \in A \mid \mathcal{A} \models \exists\overline{y}\psi(\overline{a}, \overline{y}, \overline{R})\}\|.$$

Set

$$A_0 \quad := \quad \{\overline{a} \in A \mid \mathcal{A} \models \exists\overline{Y}\exists\overline{y}\psi(\overline{a}, \overline{y}, \overline{Y})\}.$$

Then

$$(*) \qquad \|A_0\| \geq \text{opt}_{\mathcal{Q}}(\mathcal{A}).$$

Assume $\overline{y} = y_1 \ldots y_n$. Since $A_0 = \{\overline{a} \in A \mid \mathcal{A} \models \exists\overline{y}\exists\overline{Y}\psi(\overline{a}, \overline{y}, \overline{Y})\}$, there is a function $f : A_0 \to A^n$ such that

$$\mathcal{A} \models \exists\overline{Y}\psi(\overline{a}, f(\overline{a}), \overline{Y})$$

holds for all $\overline{a} \in A_0$. Consider the probability space Ω of all tuples of relations $\overline{R}$ on A (of arities corresponding to $\overline{Y}$) with the uniform probability distribution and let $\chi_{\mathcal{A}}$ be the random variable with

$$\chi_{\mathcal{A}}(\overline{R}) \quad := \quad \|\{\overline{a} \in A_0 \mid \mathcal{A} \models \psi(\overline{a}, f(\overline{a}), \overline{R})\}\|.$$

Let k be the number of atomic formulas in ψ with a relation variable in $\overline{Y}$. Then for the mean value $E(\chi_{\mathcal{A}})$ of $\chi_{\mathcal{A}}$ we show:

(1) $E(\chi_{\mathcal{A}}) \geq \frac{1}{2^k}\|A_0\|$.
(2) There is a polynomial time algorithm that, given an instance $\mathcal{A}$, yields relations $\overline{R}$ on A such that

$$\chi_{\mathcal{A}}(\overline{R}) \geq E(\chi_{\mathcal{A}}).$$

This proves the claim: Given $\mathcal{A}$, in polynomial time we first pass to relations $\overline{R}$ as in (2) and then, by effective representability, to a feasible solution, say $\Pi(\mathcal{A})$, with

$$\begin{aligned} \text{cost}(\mathcal{A},\Pi(\mathcal{A})) &= \|\{\overline{a} \in A \mid \mathcal{A} \models \exists \overline{y} \psi(\overline{a},\overline{y},\overline{R})\}\| \\ &\geq \|\{\overline{a} \in A_0 \mid \mathcal{A} \models \psi(\overline{a},f(\overline{a}),\overline{R})\}\| \\ &= \chi_{\mathcal{A}}(\overline{R}) \geq \tfrac{1}{2^k}\|A_0\| \quad \text{(by (1) and (2))} \\ &\geq \tfrac{1}{2^k}\text{opt}_Q(\mathcal{A}) \quad \text{(by } (*)). \end{aligned}$$

To show (1), we define the indicator function I by

$$I(\overline{R},\overline{a}) := \begin{cases} 1 & \text{if } \mathcal{A} \models \psi(\overline{a},f(\overline{a}),\overline{R}) \\ 0 & \text{otherwise.} \end{cases}$$

Then $\chi_{\mathcal{A}}(\overline{R}) = \sum_{\overline{a}\in A_0} I(\overline{R},\overline{a})$ and hence,

$$\begin{aligned} E(\chi_{\mathcal{A}}) &= \textstyle\sum_{\overline{R}\in\Omega} \frac{1}{\|\Omega\|} \cdot \chi_{\mathcal{A}}(\overline{R}) = \frac{1}{\|\Omega\|} \sum_{\overline{R}\in\Omega} \sum_{\overline{a}\in A_0} I(\overline{R},\overline{a}) \\ &= \textstyle\sum_{\overline{a}\in A_0} \frac{1}{\|\Omega\|} \sum_{\overline{R}\in\Omega} I(\overline{R},\overline{a}). \end{aligned}$$

Fix $\overline{a} \in A_0$. Let $\alpha_1,\dots,\alpha_k$ be the distinct atomic subformulas of ψ with a relation variable in $\overline{Y}$. Then $\mathcal{A}$ determines the truth values of all atomic subformulas of ψ different from $\alpha_1,\dots,\alpha_k$. By definition of A_0 and f, there are relations $\overline{P}$ such that $\mathcal{A} \models \psi(\overline{a},f(\overline{a}),\overline{P})$. Thus, for the corresponding assignment of truth values to $\alpha_1,\dots,\alpha_k$, the formula ψ gets the truth value TRUE. Hence, for all relations $\overline{R}$ leading to the same truth values of $\alpha_1,\dots,\alpha_k$, and these are at least $\frac{1}{2^k}\|\Omega\|$ many, we have $\mathcal{A} \models \psi(\overline{a},f(\overline{a}),\overline{R})$. Thus, $\sum_{\overline{R}\in\Omega} I(\overline{R},\overline{a}) \geq \frac{1}{2^k}\|\Omega\|$. So the equalities above give $E(\chi_{\mathcal{A}}) \geq \frac{1}{2^k}\|A_0\|$.

The last considerations show (and the same applies to similar mean values below) that $E(\chi_{\mathcal{A}})$ can be evaluated in time polynomial in $\|A\|$: First, we list A_0 and a function f of the kind in question by testing for all $\overline{a} \in A$ whether for some $\overline{b}$ we have $\psi(\overline{a},\overline{b},\overline{R})$ for some $\overline{R}$, putting $\overline{a}$ in A_0, and setting $f(\overline{a}) = \overline{b}$ if $\overline{b}$ is the first tuple with this property (note that we need $\overline{R}$ only on the sets $\{\overline{a},\overline{b}\}$ of limited cardinality). Now, to calculate $E(\chi_{\mathcal{A}})$, for $\sum_{\overline{R}\in\Omega} I(\overline{R},\overline{a})$ we again need consider the behaviour of the relations $\overline{R}$ only on the set $\{\overline{a},f(\overline{a})\}$, the total number of the relations with the same behaviour being easily calculable.

We come to a proof of part (2). Suppose $A = \{1,\dots,n\}$. In polynomial time we construct relations $\overline{R}$ with $\chi_{\mathcal{A}}(\overline{R}) \geq E(\chi_{\mathcal{A}})$. For this purpose we shall stepwise fix the truth value of $R\overline{e}$ for $R \in \overline{R}$ and $\overline{e} \in A$.

Let $\beta_1,\dots,\beta_m$ be an enumeration (in a standard way) without repetitions of the elements of the set

$$\{\alpha_j(\overline{a},f(\overline{a})) \mid 1 \leq j \leq k, \overline{a} \in A_0\}$$

(if, for example, $f(\overline{a}) = \overline{b}$, $f(\overline{b}) = \overline{a}$, $\alpha_i = R\overline{x}\,\overline{y}$, $\alpha_j = R\overline{y}\,\overline{x}$, and $i \neq j$, then $\alpha_i(\overline{a}, f(\overline{a})) = \alpha_j(\overline{b}, f(\overline{b}))$, and this instance occurs only once in $\beta_1, \ldots, \beta_m$).

By induction on s we fix the truth value $t_s \in \{\text{TRUE}, \text{FALSE}\}$ of β_s such that the following inequality holds for the conditional expectations:

$$(*) \quad E(\chi_{\mathcal{A}} | \beta_1^{\mathcal{A}} = t_1, \ldots, \beta_s^{\mathcal{A}} = t_s) \geq E(\chi_{\mathcal{A}} | \beta_1^{\mathcal{A}} = t_1, \ldots, \beta_{s-1}^{\mathcal{A}} = t_{s-1}).$$

Since $E(\chi_{\mathcal{A}} | \beta_1^{\mathcal{A}} = t_1, \ldots, \beta_{s-1}^{\mathcal{A}} = t_{s-1}) =$

$$\begin{aligned} &\tfrac{1}{2} E(\chi_{\mathcal{A}} | \beta_1^{\mathcal{A}} = t_1, \ldots, \beta_{s-1}^{\mathcal{A}} = t_{s-1}, \beta_s^{\mathcal{A}} = \text{TRUE}) \\ +&\tfrac{1}{2} E(\chi_{\mathcal{A}} | \beta_1^{\mathcal{A}} = t_1, \ldots, \beta_{s-1}^{\mathcal{A}} = t_{s-1}, \beta_s^{\mathcal{A}} = \text{FALSE}), \end{aligned}$$

in order to fix t_s we only have to check whether

$$\begin{aligned} & E(\chi_{\mathcal{A}} | \beta_1^{\mathcal{A}} = t_1, \ldots, \beta_{s-1}^{\mathcal{A}} = t_{s-1}, \beta_s^{\mathcal{A}} = \text{TRUE}) \\ \geq \; & E(\chi_{\mathcal{A}} | \beta_1^{\mathcal{A}} = t_1, \ldots, \beta_{s-1}^{\mathcal{A}} = t_{s-1}, \beta_s^{\mathcal{A}} = \text{FALSE}), \end{aligned}$$

and by the remarks above this can be done in polynomial time. For $s = m$ we have by $(*)$

$$\begin{aligned} E(\chi_{\mathcal{A}} | \beta_1^{\mathcal{A}} = t_1, \ldots, \beta_m^{\mathcal{A}} = t_m) &\geq E(\chi_{\mathcal{A}} | \beta_1^{\mathcal{A}} = t_1, \ldots, \beta_{m-1}^{\mathcal{A}} = t_{m-1}) \\ &\geq \ldots \geq E(\chi_{\mathcal{A}} | \beta_1^{\mathcal{A}} = t_1) \geq E(\chi_{\mathcal{A}}). \end{aligned}$$

Since for relations $\overline{R}$ and $\overline{R}'$ that coincide on all atomic subformulas of $\psi(\overline{a}, f(\overline{a}), \overline{Y})$ with $\overline{a} \in A_0$, we have

$$\chi_{\mathcal{A}}(\overline{R}) = \chi_{\mathcal{A}}(\overline{R}'),$$

we see that for any $\overline{R}$ with truth values $t_1, \ldots, t_m$ on $\beta_1, \ldots, \beta_m$, respectively, we have

$$\chi_{\mathcal{A}}(\overline{R}) = E(\chi_{\mathcal{A}} | \beta_1^{\mathcal{A}} = t_1, \ldots, \beta_m^{\mathcal{A}} = t_m) \quad (\geq E(\chi_{\mathcal{A}})).$$

Thus we obtain the relations $\overline{R}$ as claimed in (2) by fixing the values on $\beta_1, \ldots, \beta_m$ according to $t_1, \ldots, t_m$ and arbitrarily (say, as FALSE) on all other tuples. □

In the first part of the preceding proof we have shown

Corollary 10.2.4 *Assume Q is in $\text{MAX}_e\ \Sigma_1$, say, for every instance $\mathcal{A}$ we have*

$$\text{opt}_Q(\mathcal{A}) \quad = \quad \max_{\overline{R}} \|\{\overline{a} \in A \mid \mathcal{A} \models \exists \overline{y} \psi(\overline{a}, \overline{y}, \overline{R})\}\|$$

with quantifier-free $\psi(\overline{x}, \overline{y}, \overline{Y})$. Then there is a polynomial time algorithm that, given an instance $\mathcal{A}$, yields relations $\Pi(\mathcal{A})$ on A such that

$$\text{cost}(\mathcal{A}, \Pi(\mathcal{A})) \quad \geq \quad \frac{1}{2^k} \|A_0\|,$$

where k is the number of atomic formulas in ψ with a relation variable in $\overline{Y}$ and where $A_0 := \{\overline{a} \in A \mid \mathcal{A} \models \exists \overline{Y} \exists \overline{y} \psi(\overline{a}, \overline{y}, \overline{Y})\}$. □

As an application we obtain

Proposition 10.2.5 MAXCUT *is* $\frac{1}{2}$*-approximable.*

Proof. If $\mathcal{Q}$ denotes MAXCUT, we have for any graph $\mathcal{A}$,

$$(+) \qquad \text{opt}_{\mathcal{Q}}(\mathcal{A}) \;=\; \max_{R} \|\{(a,b) \mid \mathcal{A} \models Eab \wedge Ra \wedge \neg Rb\}\|.$$

Thus, the preceding corollary with $\psi = Exy \wedge Yx \wedge \neg Yy$, $\overline{y}$ the empty sequence, $k = 2$, and $A_0 = \{(a,b) \mid \mathcal{A} \models \exists Y(Eab \wedge Ya \wedge \neg Yb)\} = E^A$ yields a polynomial time algorithm Π such that

$$\text{cost}(\mathcal{A}, \Pi(\mathcal{A})) \;\geq\; \frac{1}{4}\|E^A\|.$$

Since $E^A ab$ implies $E^A ba$, we have by (+),

$$\text{opt}_{\mathcal{Q}}(\mathcal{A}) \;\leq\; \frac{1}{2}\|E^A\|.$$

Altogether,

$$\text{cost}(\mathcal{A}, \Pi(\mathcal{A})) \geq \frac{1}{2}\text{opt}_{\mathcal{Q}}(\mathcal{A}).$$ □

The following two observations show that possible extensions of Theorem 10.2.3 fail.

Denote by PTAS the class of NPTIME optimization problems $\mathcal{Q}$ with a polynomial time approximation scheme, i.e., with the property that there is an effective procedure Π that returns, for every rational ϵ with $0 < \epsilon < 1$, a polynomial time algorithm Π_ϵ witnessing that $\mathcal{Q}$ is ϵ-approximable. It has been shown that, unless PTIME = NPTIME, MAXCUT $\notin$ PTAS; in particular, as MAXCUT $\in \text{MAX}_e\Sigma_1$, we have $\text{MAX}_e\ \Sigma_1 \not\subseteq$ PTAS.

We come two the second observation. Let $\mathcal{Q}$ be the problem MAXCLIQUE: Given a graph $\mathcal{G}$ it asks for the maximal size of a clique in $\mathcal{G}$. Hence, MAXCLIQUE $\in \text{MAX}_e\ \Pi_1$, since

$$\text{opt}_{\mathcal{Q}}(\mathcal{G}) \;=\; \max_{R} \|\{a \in G \mid \mathcal{G} \models Ra \wedge \forall x \forall y((Rx \wedge Ry \wedge x \neq y) \rightarrow Exy)\}\|.$$

It is known that, unless PTIME = NPTIME, MAXCLIQUE $\notin$ APX, showing that Theorem 10.2.3 cannot be extended to $\text{MAX}_e\ \Pi_1$.

We close this section with two "positive" remarks on Theorem 10.2.3. The first one gives a generalization, the second one shows how to enlarge its scope of applicability.

We can change the definition of a first-order definable optimization problem by considering in 10.1.2 formulas $\varphi(\overline{x}, \overline{u}, \overline{Y})$ and requiring that

$$\text{opt}_{\mathcal{Q}}(\mathcal{A}) = \mu_{\overline{b}, \overline{R}} \;\|\{\overline{a} \in A \mid \mathcal{A} \models \varphi[\overline{a}, \overline{b}, \overline{R}]\}\|$$

("optimization over elements and relations"). Replacing elements by relations, we see that this does not change the notion of a first-order definable optimization problem; however, the classes MAX Σ_1 and $\text{MAX}_e\ \Sigma_1$ strictly increase, if we allow optimization over elements and relations (note that the replacement of constants by relation symbols uses universal quantifiers). Nevertheless, Theorem 10.2.3 remains true for the larger class $\text{MAX}_e\ \Sigma_1$ (indeed with essentially the same proof).

The following example shows how one can enlarge the scope of applicability of the preceding theorem. Consider the optimization problem known as MAXIMUM CONNECTED COMPONENT (MCC). Given a graph, it asks for the maximum of the cardinalities of connected components. Since the connected components are computable in polynomial time, there is a polynomial time algorithm solving the optimization problem. By 10.1.4(a) we know that MCC $\in$ MAX Π_2. On the other hand, MCC $\notin$ MAX Π_1 (see Exercise 10.2.6 below) and hence, MCC $\notin$ MAX Σ_1 by 10.1.5. Therefore, our logical approach does not yield MCC $\in$ APX, even though there is a polynomial time algorithm giving an optimal solution.

However, since the transitive closure is computable in polynomial time, the approximability of MCC is equivalent to the approximability of MCC$'$, where MCC$'$ is the problem

> Given $\mathcal{G} = (G, E^G, R^G)$ with $R^G = \text{TC}(E^G)$, a "graph with transitive closure", compute the maximal cardinality of its connected components.

Clearly, MCC$'$ is in $\text{MAX}_e\ \Sigma_1$ (with maximazation over elements) since

$$\text{opt}_{\text{MCC}'}(\mathcal{A}) = \max_b \|\{a \mid \mathcal{G} \models Rab\}\|.$$

Now, 10.2.3 (with maximization over elements) yields MCC$'$ $\in$ APX.

Of course, the same idea can be applied to any other optimization problem: The values of polynomial time global relations can be added to the instances by free without changing the behaviour of the optimization problem with respect to the notion of approximability we have considered. In [11] this extra relations have been "added" on the logical side, allowing formulas of FO(LFP) instead of FO.

Exercise 10.2.6 (a) Give a direct proof that MCC $\in$ MAX Π_2. (b) Show that MCC $\notin$ MAX Π_1 (even not with maximization over elements). □

Notes 10.2.7 The main references for this chapter are [104, 105, 129]; they also contain corresponding historical remarks.

11. Logics for PTIME

There are logics (for instance FO(IFP); cf. Chapter 7) that capture PTIME on ordered structures. Instantly, this result evokes the question whether it can be strengthened to the case of not necessarily ordered structures:

Is there a logic strongly capturing PTIME*?*

There are examples of such logics (cf. 11.1.2); however, they are quite artificial, at the same time showing that we should incorporate some obvious requirements of effectivity in our notion of strongly capturing. We thus are led to an effective notion of a logic strongly capturing PTIME and, therefore, may reformulate the question above as follows:

$(*)$ *Is there a logic effectively strongly capturing* PTIME*?*

The question is considered to be the most prominent open problem in finite model theory and, at the same time, a difficult one. For instance, a proof that such a logic does not exist, would yield that PTIME $\neq$ NPTIME (cf. section 11.1) and hence, solve the most prominent open problem in complexity theory. Moreover, in section 11.1 we show that the question $(*)$ is strongly related to the isomorphism problem for finite structures.

The statement "FO(IFP) captures PTIME" can be reformulated as "FO(IFP) effectively strongly captures PTIME on the class of ordered structures". In section 11.2 we are led to investigate specific classes of (not necessarily ordered) structures whether they allow for a similar result, i.e., we are led to search for partial solutions of $(*)$. We present some positive results.

The discussion of specific logics that may be candidates for a positive solution of $(*)$ itself, is transferred to the next chapter. There we shall also see that if there is any logic at all that effectively strongly captures PTIME, then there is such a logic with a familiar syntax.

In the following not only structures, but also formulas and Turing machines will be considered as objects of computations. When doing so, we shall tacitly assume that vocabularies, formulas, and machines are coded by $\{0,1\}$-words in some reasonable way.

11.1 Logics and Invariants

First, we recall the definition of a logic $\mathcal{L}$ capturing a complexity class $\mathcal{C}$ and of $\mathcal{L}$ strongly capturing $\mathcal{C}$.

For a class K of structures we introduced the class $K_<$ of its ordered representations,

$$K_< = \{(\mathcal{A}, <) \mid \mathcal{A} \in K, < \text{an ordering of } A\}.$$

In particular, if $<_1$ and $<_2$ are orderings of A then

$$(*) \qquad (\mathcal{A}, <_1) \in K_< \quad \text{iff} \quad (\mathcal{A}, <_2) \in K_<.$$

The logic $\mathcal{L}$ captures the complexity class $\mathcal{C}$ iff for every class K of *ordered* structures,

$$K \in \mathcal{C} \quad \text{iff} \quad K \text{ is axiomatizable in } \mathcal{L}.$$

And $\mathcal{L}$ strongly captures $\mathcal{C}$ iff for every class K of (not necessarily ordered) structures,

$$K_< \in \mathcal{C} \quad \text{iff} \quad K \text{ is axiomatizable in } \mathcal{L}.$$[1]

Already several times we mentioned the following open question: Is there a logic strongly capturing PTIME? Below we present two examples of such logics which will turn out to be "unsatisfactory", thus leading to a sharper version of the question.

Recall that a sentence φ of vocabulary $\tau_< = \tau\dot{\cup}\{<\}$ (i.e., $<\notin \tau$) is *order-invariant in the finite* if

$$(\mathcal{A}, <_1) \models \varphi \quad \text{iff} \quad (\mathcal{A}, <_2) \models \varphi$$

holds for all (finite) τ-structures $\mathcal{A}$ and orderings $<_1, <_2$ of A.

By $(*)$ above, if $K_< = \text{Mod}(\varphi)$ then φ is order-invariant in the finite.

Exercise 11.1.1 For sufficiently rich τ the set

$$\{\varphi \in \text{FO}[\tau_<] \mid \varphi \text{ a sentence order-invariant in the finite}\}$$

is not decidable and hence, not recursively enumerable. Hint: Let σ be a vocabulary such that

$$\{\psi \in \text{FO}[\sigma] \mid \psi \text{ a sentence unsatisfiable in the finite}\}$$

[1] Note that on the left side of the last two equivalences, K and $K_<$ are classes of ordered structures. This is an accordance with the conventions we agreed upon in Chapter 7, that only classes of ordered structures are considered as members of complexity classes.

is not decidable (cf. Trahtenbrot's Theorem 7.2.1). Let $\tau := \sigma \cup \{P\}$, where P is a unary relation symbol, $P \notin \sigma$. For a sentence $\psi \in \text{FO}[\sigma]$ let φ_ψ be the $\text{FO}[\tau_<]$-sentence

$$\varphi_\psi \quad := \quad (\psi \rightarrow \exists x(\text{“}x \text{ is the } <\text{-minimal element”} \wedge Px)).$$

Then, φ_ψ is order-invariant in the finite iff $\psi \wedge \exists x \exists y\, x \neq y$ is unsatisfiable in the finite. □

We present the promised examples.

Example 11.1.2 Let $\mathcal{L}_1$ be the logic whose sentences of vocabulary τ are the $\text{FO(IFP)}[\tau_<]$-sentences, where the satisfaction relation $\mathcal{A} \models^{\mathcal{L}_1} \varphi$ for a τ-structure $\mathcal{A}$ and an $\mathcal{L}_1[\tau]$-sentence φ is defined by

$$\mathcal{A} \models^{\mathcal{L}_1} \varphi \quad \text{iff} \quad \varphi \text{ is order-invariant in the finite and there is an ordering } <^A \text{ on } A \text{ such that } (\mathcal{A}, <^A) \models^{\text{IFP}} \varphi$$

(on the right side, $(\mathcal{A}, <^A) \models^{\text{IFP}} \varphi$ means that $(\mathcal{A}, <^A)$ is a model of φ in the semantics of FO(IFP)).

Let $\mathcal{L}_2$ be the logic whose sentences of vocabulary τ are the $\text{FO(IFP)}[\tau_<]$-sentences order-invariant in the finite where now the satisfaction relation is defined by

$$\mathcal{A} \models^{\mathcal{L}_2} \varphi \quad \text{iff} \quad \text{there is an ordering } <^A \text{ on } A \text{ such that } (\mathcal{A}, <^A) \models^{\text{IFP}} \varphi.$$

As FO(IFP) captures PTIME, one easily verifies that both $\mathcal{L}_1$ and $\mathcal{L}_2$ strongly capture PTIME. However, both logics have unsatisfactory features: Concerning $\mathcal{L}_1$, in view of 11.1.1 there is no effective procedure assigning to every $\mathcal{L}_1$-sentence φ an algorithm that, given an ordered version of $\mathcal{A}$, evaluates $\mathcal{A} \models \varphi$ in time polynomial in $\|A\|$. On the other hand, given an $\mathcal{L}_2$-sentence φ, there is such an effective procedure. But now, using 11.1.1 again, we see that for sufficiently rich τ the set of $\mathcal{L}_2[\tau]$-sentences is not decidable. □

None of the logics just considered satisfies at the same time that

- the set of sentences is decidable
- there is an effective procedure assigning to every φ a polynomially time-bounded algorithm testing $\mathcal{A} \models \varphi$.

We incorporate both requirements in our notion of capturing given by the following definition.[2] When speaking of complexity classes, we have in mind one of the concrete complexity classes considered so far, and we think of a logic $\mathcal{L}$ as given – for any vocabulary τ – by a set $\mathcal{L}[\tau]_0$, the set of $\mathcal{L}$-*sentences* of vocabulary τ, and by a relation $\models^{\mathcal{L},\tau}$ between τ-structures and $\mathcal{L}$-sentences of vocabulary τ, the $\mathcal{L}$-*satisfaction* relation for τ.

[2] The reader may wonder why, say, in the second requirement we do not ask for a polynomial time procedure instead of an effective procedure. In fact, there is no real difference: By standard techniques one can pass from a logic satisfying the weaker conditions to a logic satisfying the stronger ones.

Definition 11.1.3 Let $\mathcal{L}$ be a logic and $\mathcal{C}$ a complexity class. $\mathcal{L}$ *effectively strongly captures* $\mathcal{C}$, $\mathcal{L} \equiv_{\text{es}} \mathcal{C}$, if

- $\mathcal{L}$ strongly captures $\mathcal{C}$;
- for every vocabulary τ
  (i) $\mathcal{L}[\tau]_0$ is a decidable set;
  (ii) there is an effective procedure that to every sentence $\varphi \in \mathcal{L}[\tau]_0$ assigns a pair (M, f), where M is a Turing machine that accepts $\text{Mod}(\varphi)_<$ and f is (the code of) a function witnessing that M is resource-bounded according to $\mathcal{C}$.[3] □

In all concrete cases considered in Chapter 7 we have implicitly proved that the logics and the corresponding complexity classes satisfy the effectivity conditions (i) and (ii). For example, the proof of 7.4.2(a) yields a procedure that to every $\varphi \in \text{FO(IFP)}[\tau_<]$ assigns a pair (M, d) where M is an x^d time-bounded deterministic Turing machine accepting the class of ordered models of φ.

In the definition above we do not impose limitations on the syntax of a logic (besides (i)). For example, NPTIME is not only effectively strongly captured by the logic of Σ_1^1-sentences, but also by the logic $\mathcal{L}$ with

- $\mathcal{L}[\tau]_0 := \{(M, d) \mid d \geq 1$ and M is a nondeterministic Turing machine for ordered representations of τ-structures$\}$
- for every $\varphi = (M, d) \in \mathcal{L}[\tau]_0$ and every τ-structure $\mathcal{A}$,

$$\mathcal{A} \models \varphi \text{ iff } M \text{ accepts some ordered representation of } \mathcal{A} \text{ in } \leq \|A\|^d \text{ steps.}$$

This logic is very artificial and lacks the logical flavour we are accustomed to. Moreover, it is tied up so closely with the definition of NPTIME that the characterization of NPTIME in terms of this logic does not give us any additional insight. We shall show in 12.3.17 that in case PTIME is effectively strongly captured by any logic at all, it can already be captured by a logic with a familiar syntax.

Altogether, we reformulate our main question as

(*) *Is there a logic effectively strongly capturing* PTIME?

In the remainder of this section we study the relationship of (*) with other problems, thereby presenting methods and results that may be of help in getting an answer.

Proposition 11.1.4 *If the answer to* (*) *is "no" then* PTIME $\neq$ NPTIME.

[3] To be definite, in case $\mathcal{C} \in \{$LOGSPACE, NLOGSPACE, PSPACE$\}$ we code f by a natural number d meaning that M is $d\cdot\log$ space-bounded or (for PSPACE) x^d space-bounded; in case $\mathcal{C} \in \{$PTIME, NPTIME$\}$ we also code f by a number d meaning that M is x^d time-bounded.

Proof. There seems to be a very short proof: Assume PTIME = NPTIME. Since $\Sigma_1^1 \equiv_{\text{es}}$ NPTIME, we have $\Sigma_1^1 \equiv_{\text{es}}$ PTIME. However, this argument has a gap: By $\Sigma_1^1 \equiv_{\text{es}}$ NPTIME we know that, given τ, there is an effective procedure assigning to every $\Sigma_1^1[\tau]$-sentence φ a number $d \geq 1$ and a nondeterministic x^d time-bounded Turing machine M accepting $\text{Mod}(\varphi)_<$. But how do we get a *deterministic* machine effectively?

To close this gap one shows, using an NPTIME-complete problem, that PTIME = NPTIME implies an effective version of PTIME = NPTIME. More precisely: Let SAT be the set of satisfiable propositional formulas in conjunctive normal form. SAT is NPTIME-complete (cf. [87]). Even more: There is a polynomial time algorithm A_0 that assigns to every $d \geq 1$, every nondeterministic x^d time-bounded Turing machine M, and every input string u for M a propositional formula $\alpha(u, (M, d))$ in conjunctive normal form such that

$$(+) \qquad M \text{ accepts } u \quad \text{iff} \quad \alpha(u, (M, d)) \in \text{SAT}.$$

Assume PTIME = NPTIME and let A_1 be a deterministic polynomially time-bounded algorithm for SAT. By (+), using the algorithms A_0 and A_1 and time bounds for them, one can effectively assign to every $d \geq 1$ and nondeterministic x^d time-bounded M a number $d' \geq 1$ and a deterministic $x^{d'}$ time-bounded algorithm accepting the same language as M. □

When defining the logics in Example 11.1.2, we essentially restricted ourselves to sentences order-invariant in the finite to ensure that the model classes $\text{Mod}^{\mathcal{L}_1}(\varphi)$ and $\text{Mod}^{\mathcal{L}_2}(\varphi)$ belong to PTIME. We could have dispensed with this restriction if there would exist a PTIME-algorithm defining a "canonical" ordered version of every structure. Proposition 11.1.6 contains the precise statement and the following Definition 11.1.5 gives the precise notion.

Denote by Str and Str$[\tau]$ the class of all finite structures and of all finite τ-structures, respectively.

Definition 11.1.5 A PTIME-*canonization* **C** consists of functions

$$C_\tau : \text{Str}[\tau] \to \text{Str}[\tau_<]$$

for every vocabulary τ such that:

(1) For all $\mathcal{A} \in \text{Str}[\tau]$, ($\mathcal{A} \cong C_\tau(\mathcal{A})|\tau$ and $<^{C_\tau(\mathcal{A})}$ is an ordering).[4]
(2) For all $\mathcal{A}, \mathcal{B} \in \text{Str}[\tau]$, $\mathcal{A} \cong \mathcal{B}$ implies $C_\tau(\mathcal{A}) \cong C_\tau(\mathcal{B})$.
(3) C_τ is PTIME-computable, more precisely: there is a PTIME-algorithm that, applied to $(\mathcal{A}, <^A) \in \text{Str}[\tau]_<$, gives (the encoding of) the ordered structure $C_\tau(\mathcal{A})$. [5] □

---

[4] Recall that $C_\tau(\mathcal{A})|\tau$ denotes the τ-reduct of $C_\tau(\mathcal{A})$.
[5] In Chapter 7 we agreed upon how to regard ordered structures as inputs to a Turing machine M. By those conventions it is clear that for isomorphic $(\mathcal{A}, <^A)$

Clearly, in view of (1), condition (2) is equivalent to

(2′) For all $\mathcal{A}, \mathcal{B} \in \mathrm{Str}[\tau]$, $\mathcal{A} \cong \mathcal{B}$ iff $C_\tau(\mathcal{A}) \cong C_\tau(\mathcal{B})$.

Proposition 11.1.6 *If there exists a* PTIME-*canonization then there is a logic effectively strongly capturing* PTIME.

Proof. Let **C** be a PTIME-canonization. Consider the logic $\mathcal{L}$ whose sentences of vocabulary τ are the FO(IFP)$[\tau_<]$-sentences. The satisfaction relation is defined by

$$(*) \qquad \mathcal{A} \models^{\mathcal{L},\tau} \varphi \quad \text{iff} \quad C_\tau(\mathcal{A}) \models^{\mathrm{IFP}} \varphi.$$

$\mathcal{L}$ effectively strongly captures PTIME: Clearly, the set of $\mathcal{L}[\tau]$-sentences is decidable. From Chapter 7 we know that there is an effective procedure assigning to every FO(IFP)$[\tau_<]$-sentence φ a pair (M_0, d_0), where M_0 is an x^{d_0} time-bounded Turing machine accepting the class of ordered models of φ. Therefore, using a PTIME algorithm for C_τ, the equivalence $(*)$ shows that one effectively can assign to every $\mathcal{L}[\tau]$-sentence φ a pair (M, d), where M is an x^d time-bounded Turing machine accepting $\mathrm{Mod}^{\mathcal{L}}(\varphi)_<$. In particular, $\mathrm{Mod}^{\mathcal{L}}(\varphi)_<$ is in PTIME.

Conversely, if K is a class of τ-structures with $K_< \in$ PTIME, then $K_< = \mathrm{Mod}^{\mathrm{FO(IFP)}}(\varphi)$ for some FO(IFP)$[\tau_<]$-sentence φ. But then, $K = \mathrm{Mod}^{\mathcal{L}}(\varphi)$. □

For an ordered $\tau_<$-structure $\mathcal{A}$, let $\mathcal{A}^+$ be its numerical representation, i.e., $\mathcal{A}^+$ is the unique $\tau_<$-structure isomorphic to $\mathcal{A}$ with $A^+ = \{0, \ldots, \|A\| - 1\}$ and with $<^{A^+}$ the natural ordering on A^+. Assume that **C** is a PTIME-canonization. By passing to $(C_\tau(\mathcal{A}))^+$, one sees that (2) (compare (2′), too) in the definition of PTIME-canonization can be replaced by

$$\text{For all } \mathcal{A}, \mathcal{B} \in \mathrm{Str}[\tau], \quad \mathcal{A} \cong \mathcal{B} \text{ iff } C_\tau(\mathcal{A}) = C_\tau(\mathcal{B}).$$

The structures $(C_\tau(\mathcal{A}))^+$ can be coded in a canonical way, say, by words over the alphabet $\{0,1\}$. Thus, a canonization leads to an "invariantization", i.e., to a map associating with every structure $\mathcal{A}$ a word only depending on the isomorphism type of $\mathcal{A}$. Again, we give a precise definition and statement.

Definition 11.1.7 A PTIME-*invariantization* **I** consists of functions

$$I_\tau : \mathrm{Str}(\tau) \to \{0,1\}^*$$

such that for every vocabulary τ we have:

---

and $(\mathcal{B}, <^B)$ the result of M applied to $(\mathcal{A}, <^A)$ is the same as that of M applied to $(\mathcal{B}, <^B)$. Therefore, if C_τ is computable then

$$\mathcal{A} \cong \mathcal{B} \quad \text{implies} \quad C_\tau(\mathcal{A}) \cong C_\tau(\mathcal{B}).$$

Thus, in the above definition, property (2) is redundant. Nevertheless, here and in the following, we stay with such redundancies in order to emphasize both the structural and the computational aspects of such functions.

(1) For all $\mathcal{A}, \mathcal{B} \in \mathrm{Str}[\tau]$, $\mathcal{A} \cong \mathcal{B}$ iff $I_\tau(\mathcal{A}) = I_\tau(\mathcal{B})$.
(2) I_τ is PTIME-computable (more precisely: there is a PTIME-algorithm that, applied to $(\mathcal{A}, <^A) \in \mathrm{Str}[\tau]_<$, gives the word $I_\tau(\mathcal{A})$).

Proposition 11.1.8 *If there is a* PTIME-*canonization, then there is a* PTIME-*invariantization.*

Proof. If $\mathbf{C}$ is a PTIME-canonization, then let $I_\tau(\mathcal{A})$ be the $\{0,1\}$-code of $(C_\tau(\mathcal{A}))^+$. By the remarks above, $\mathbf{I}$ is a PTIME-invariantization. □

It may be surprising that the converse of the preceding proposition is also true, that is, an "arbitrary" invariant already leads to a canonical ordering.

Theorem 11.1.9 *If there is a* PTIME-*invariantization, then there is a* PTIME-*canonization.*

Proof. Assume that $\mathbf{I}$ is a PTIME-invariantization. Let τ be an arbitrary vocabulary and let $<$ be a new binary relation symbol. Set $\sigma = \tau_<$. First, using $I_\sigma : \mathrm{Str}(\sigma) \to \{0,1\}^*$, for every τ-structure $\mathcal{A}$ and every ordering $\prec^A$ on A, we stepwise define an ordering $<^A$ on A.

Fix $\mathcal{A}$ (with at least two elements) and an ordering $\prec^A$ on A. If $a_1, \ldots, a_l$ are distinct elements of A, let

$$[a_1, \ldots, a_l] \quad := \quad \{(a_i, a_j) \mid 1 \le i < j \le l\}$$

if $l \ge 2$ and $[a_1] := \{(a_1.a_1)\}$. We start the step by step construction of $<^A$: Let w_1 be the first element in the lexicographic ordering of $\{0,1\}^*$ of the set

$$\{I_\sigma((\mathcal{A}, [a])) \mid a \in A\}.$$

Among the a's with $w_1 = I_\sigma((\mathcal{A}, [a]))$, let a_1 be the $\prec^A$-first one.

In the second step let w_2 be the first element in

$$\{I_\sigma((\mathcal{A}, [a_1, a])) \mid a \in A, a \neq a_1\}.$$

Among the a's with $w_2 = I_\sigma((\mathcal{A}, [a_1, a]))$ and $a \neq a_1$, let a_2 be the $\prec^A$-first one.

After n steps, where $n := \|A\|$, we have obtained $a_1, \ldots, a_n$ with $A = \{a_1, \ldots, a_n\}$. We set

$$<^A \quad := \quad [a_1, \ldots, a_n]$$

and define $\mathbf{C}$ by setting

$$C_\tau(\mathcal{A}) \quad := \quad \text{the numerical representation of } (\mathcal{A}, <^A).$$

To see that $\mathbf{C}$ is a PTIME-canonization, we first observe that the transition from $(\mathcal{A}, \prec^A)$ to $(\mathcal{A}, <^A)$ can effectively be performed in polynomial time.

Trivially, $\mathcal{A} \cong (\mathcal{A}, <^A)|\tau$ and hence, $\mathcal{A} \cong C_\tau(\mathcal{A})|\tau$. Finally, suppose that $\mathcal{B} \cong \mathcal{A}$ and $\prec^B$ is an ordering of B. Let $<^B = [b_1, \ldots, b_n]$ be the ordering given by the stepwise procedure above applied to $(\mathcal{B}, \prec^B)$. Then, for $l = 1, \ldots, n$,

$$(+) \qquad (\mathcal{A}, [a_1, \ldots, a_l]) \cong (\mathcal{B}, [b_1, \ldots, b_l]).$$

For $l = n$ this yields $(\mathcal{A}, <^A) \cong (\mathcal{B}, <^B)$ and hence, $C_\tau(\mathcal{A}) = C_\tau(\mathcal{B})$.

The proof of (+) proceeds by induction on l. We show the case $l = 1$: Suppose $\pi : \mathcal{A} \cong \mathcal{B}$. Then, for all $a \in A$,

$$\pi : (\mathcal{A}, [a]) \cong (\mathcal{B}, [\pi(a)]).$$

Thus, by property (1) of the invariantization $\mathbf{I}$,

$$\{I_\sigma((\mathcal{A}, [a])) \mid a \in A\} \quad = \quad \{I_\sigma((\mathcal{B}, [b])) \mid b \in B\},$$

and hence, $I_\sigma((\mathcal{A}, [a_1])) = I_\sigma((\mathcal{B}, [b_1]))$. Again by property (1), we therefore have $(\mathcal{A}, [a_1]) \cong (\mathcal{B}, [b_1])$. □

Together with 11.1.6 the theorem yields:

Corollary 11.1.10 *If there is a* PTIME-*invariantization, then there is a logic effectively strongly capturing* PTIME. □

Of course, the notion of canonization can be defined for any deterministic complexity class $\mathcal{C}$: A $\mathcal{C}$-canonization $\mathbf{C}$ consists of functions $C_\tau : \mathrm{Str}[\tau] \to \mathrm{Str}[\tau_<]$ satisfying (1), (2) of Definition 11.1.5 and

(3) C_τ is $\mathcal{C}$-computable (more precisely: there is an algorithm according to $\mathcal{C}$ that, applied to $(\mathcal{A}, <^A) \in \mathrm{Str}[\tau]_<$, gives (the encoding of) the ordered structure $C_\tau(\mathcal{A})$). [6]

Exercise 11.1.11 Generalizing the proof of 11.1.6 show: If there are a $\mathcal{C}$-canonization and a logic effectively strongly capturing $\mathcal{C}$ on ordered structures then there is a logic $\mathcal{L}$ with $\mathcal{L} \equiv_{\mathrm{es}} \mathcal{C}$. □

Exercise 11.1.12 Show the existence of a PSPACE-canonization. Hint: Given a τ-structure $\mathcal{A}$, let $C_\tau(\mathcal{A})$ be the numerical representation of $(\mathcal{A}, <)$, where $<$ is a minimal order on $\mathcal{A}$ (compare 8.5.13). □

We conclude with some remarks that shed still further light on the main question.

Suppose there is a logic $\mathcal{L}$ effectively strongly capturing the complexity class $\mathcal{C}$. Then, given τ, we can effectively enumerate the $\mathcal{L}[\tau]$-sentences. Using the effective transition from sentences φ to pairs (M, f), where M is a Turing machine accepting $\mathrm{Mod}(\varphi)_<$ and f a resource bound of M according to $\mathcal{C}$, we get that $\mathcal{C}$ is effectively enumerable in the sense of the following

[6] For $\mathcal{C}$ = LOGSPACE this means that C_τ is computable by a log space-bounded transducer, cf. section 7.5.1.

Definition 11.1.13 Let $\mathcal{C}$ be a complexity class. An *effective enumeration* $\mathbf{F}$ of $\mathcal{C}$ consists of computable functions F_τ, τ a vocabulary, that are defined on the set $\mathbb{N}$ of natural numbers, with the properties

(1) For any $i \in \mathbb{N}$, $F_\tau(i)$ is a pair (M, f) where M is a Turing machine accepting a class $K_<$ with $K \subseteq \mathrm{Str}[\tau]$, and f witnesses that M is according to $\mathcal{C}$.

(2) For any class $K \subseteq \mathrm{Str}[\tau]$ such that $K_<$ is in $\mathcal{C}$, there is an i with $F_\tau(i) = (M, f)$ such that the machine M accepts $K_<$. □

By the remark preceding the definition we have the direction from (i) to (ii) of

Proposition 11.1.14 *For a complexity class $\mathcal{C}$, the following are equivalent:*

(i) *There is a logic effectively strongly capturing $\mathcal{C}$.*

(ii) *There is an effective enumeration of $\mathcal{C}$.*

Proof. We have to show that (ii) implies (i). Let $\mathbf{F}$ be an effective enumeration of $\mathcal{C}$. First assume that, for all τ, the set $\{F_\tau(i) \mid i \in \mathbb{N}\}$ is decidable. Define the logic $\mathcal{L}$ as follows. The set of $\mathcal{L}[\tau]$-sentences is $\{F_\tau(i) \mid i \in \mathbb{N}\}$. And if $F_\tau(i) = (M, f)$ where M accepts the class $K_<$, then

$$\mathcal{A} \models^{\mathcal{L}} F_\tau(i) \quad \text{iff} \quad \mathcal{A} \in K.$$

Hence, $\mathcal{L} \equiv_{\mathrm{es}} \mathcal{C}$.

Otherwise, we replace $\mathbf{F}$ by an enumeration $\mathbf{F}^*$ such that, for all τ, the set $\{F^*_\tau(i) \mid i \in \mathbb{N}\}$ is decidable. We define $F^*_\tau(i)$ as follows. Let $F_\tau(i) = (M, f)$. Set $F^*_\tau(i) = (M^*, f)$, where M^* behaves the same way as M, however, the length of (the coding of) M^* is greater than i; this is achieved, say, by adding to M i-many new instructions that never will be used. To decide whether a pair (M', f') belongs to $\{F^*_\tau(i) \mid i \in \mathbb{N}\}$, we check whether (M', f') coincides with some $F^*_\tau(i)$ for $1 \leq i \leq$ length of M'. □

By the proposition, the complexity classes above PTIME such as NPTIME or PSPACE are effectively enumerable. If we view an effective enumeration as providing a tool for a systematic exploration of the complexity class concerned, NPTIME and PSPACE are explorable in this sense, whereas it is open whether PTIME or its subclasses LOGSPACE and NLOGSPACE are explorable.

11.2 PTIME on Classes of Structures

We have already remarked that the question ("main question") whether there is a logic effectively strongly capturing PTIME, has a positive answer if we restrict ourselves to ordered structures, FO(IFP) being such a logic. We may consider this result as a partial answer to the main question and, thus, are led to a more systematic study of such restrictions. In the following we show, for example, that the main question is equivalent to the question whether there is a logic effectively strongly capturing the PTIME properties on the class GRAPH of graphs. Moreover, we shall see that for certain subclasses of GRAPH such as the class of trees the corresponding question has a positive answer.

If S is a class of structures, $S \subseteq \text{Str}$, and τ a vocabulary, let $S[\tau] := \{\mathcal{A} \in S \mid \mathcal{A} \text{ a } \tau\text{-structure}\}$.

Definition 11.2.1 Let $\mathcal{L}$ be a logic, $\mathcal{C}$ a complexity class, and S a class of structures, $S \subseteq \text{Str}$. *$\mathcal{L}$ effectively strongly captures $\mathcal{C}$ on S*, $\mathcal{L} \equiv_{\text{es}} \mathcal{C}$ on S, if for all τ

(1) $\mathcal{L}[\tau]_0$ is decidable;

(2) $\{K \cap S[\tau] \mid K \subseteq \text{Str}[\tau],\, K_< \in \mathcal{C}\} = \{\text{Mod}^{\mathcal{L}}(\varphi) \cap S[\tau] \mid \varphi \in \mathcal{L}[\tau]_0\}$;[7]

(3) there is an effective procedure that to every $\varphi \in \mathcal{L}[\tau]_0$ assigns a pair (M, f), where f is a function witnessing that M is resource-bounded according to $\mathcal{C}$ and where for every $(\mathcal{A}, <) \in S[\tau]_<$ the machine M accepts $(\mathcal{A}, <)$ iff $\mathcal{A} \models \varphi$. □

Clearly, if S and S' are classes of structures with $S \subseteq S' \subseteq \text{Str}$ and if $\mathcal{L} \equiv_{\text{es}} \mathcal{C}$ on S', then $\mathcal{L} \equiv_{\text{es}} \mathcal{C}$ on S.

In many concrete situations we have $S[\tau]_< \in \mathcal{C}$. Then, condition (2) of the previous definition can be simplified:

Proposition 11.2.2 *Let $\mathcal{L}$ be a logic, $\mathcal{C}$ a complexity class and S a class of structures with $S[\tau]_< \in \mathcal{C}$ for all τ.*

(a) *If $\mathcal{L}$ is a logic that for all τ satisfies* (1), (3) *of Definition 11.2.1 and*

(2*) *for all $K \subseteq S[\tau]$,*

$$K_< \in \mathcal{C} \quad \textit{iff} \quad \textit{there is } \varphi \in \mathcal{L}[\tau]_0 \textit{ such that } K = \text{Mod}^{\mathcal{L}}(\varphi),$$

then $\mathcal{L} \equiv_{\text{es}} \mathcal{C}$ on S.

(b) *Let $\mathcal{L}$ be a logic such that $\mathcal{L} \equiv_{\text{es}} \mathcal{C}$ on S. Define the logic $\mathcal{L}'$ by*

$$\mathcal{L}'[\tau]_0 = \mathcal{L}[\tau]_0 \quad \textit{and} \quad (\mathcal{A} \models^{\mathcal{L}'} \varphi \textit{ iff } (\mathcal{A} \models^{\mathcal{L}} \varphi \textit{ and } \mathcal{A} \in S[\tau])).$$

Then $\mathcal{L}' \equiv_{\text{es}} \mathcal{C}$ on S, too; moreover, $\mathcal{L}'$ satisfies (2*).

The proof is left as an exercise. □

To get a first example we restate a result of Chapter 7:

Proposition 11.2.3 (a) FO(DTC) $\equiv_{\text{es}}$ LOGSPACE *on the class of ordered structures.*

[7] Readers familiar with set theory will apologize our use of classes of classes.

(b) FO(IFP) $\equiv_{\text{es}}$ PTIME *on the class of ordered structures.*

Proof. (a) Let $\prec\in\tau$ and let $\mathcal{O}[\tau]$ be the class of ordered τ-structures ($\prec$ being the ordering). Since $\mathcal{O}[\tau]_< \in$ LOGSPACE and FO(DTC) satisfies the effectivity conditions (1), (3) of Definition 11.2.1, it suffices to show that FO(DTC) satisfies (2*) of part (a) of the preceding proposition. Clearly, for $K \subseteq \mathcal{O}[\tau]$ we have (arguing as for Proposition 7.5.13),

$$K \in \text{LOGSPACE} \quad \text{iff} \quad K_< \in \text{LOGSPACE}$$

and therefore,

$$K_< \in \text{LOGSPACE} \quad \text{iff} \quad \text{there is an FO(DTC)}[\tau]\text{-sentence } \varphi \text{ such that } K = \text{Mod}(\varphi).$$

The proof of (b) is similar. □

We now turn to the class GRAPH of finite graphs and prove that the main question is equivalent to its restriction to GRAPH. The reason is that any structure can feasibly be coded (interpreted) into a graph. We give the exact notion of interpretation.

Fix vocabularies σ and τ where $\sigma = \{R_1, \dots, R_s\}$ with r_i-ary R_i.[8] An $(s+1)$-tuple

$$\Pi := (\pi_{\text{uni}}(\overline{x}), \pi_{R_1}(\overline{x}_1, \dots, \overline{x}_{r_1}), \dots, \pi_{R_s}(\overline{x}_1, \dots, \overline{x}_{r_s}))$$

of $\mathcal{L}[\tau]$-formulas (with free variables among the displayed ones), where for some $k \geq 1$ all tuples $\overline{x}, \overline{x}_1, \dots$ have length k, is called an *$\mathcal{L}$-interpretation of σ in τ of width k.*

Given a τ-structure $\mathcal{A}$, an $\mathcal{L}$-interpretation Π of σ in τ induces a σ-structure $\mathcal{A}^\Pi$ over k-tuples of A, provided $\pi^{\mathcal{A}}_{\text{uni}}(\_)$ is nonempty; the case $\pi^{\mathcal{A}}_{\text{uni}}(\_) = \emptyset$ plays a marginal role for our considerations and will be neglected. The structure $\mathcal{A}^\Pi$ has universe $\pi^{\mathcal{A}}_{\text{uni}}(\_)$, the interpretation of R_i is given by

$$\pi^{\mathcal{A}}_{R_i}(\_,\dots,\_) \cap (\pi^{\mathcal{A}}_{\text{uni}}(\_))^{r_i}.$$

For better reading we often suppress the intersection with the corresponding power of the universe and succinctly write

$$\mathcal{A}^\Pi = (\pi^{\mathcal{A}}_{\text{uni}}(\_), \pi^{\mathcal{A}}_{R_1}(\_,\dots,\_), \dots, \pi^{\mathcal{A}}_{R_s}(\_,\dots,\_)).$$

Exercise 11.2.4 Let $\mathcal{L}$ be one of the logics FO, FO(DTC), FO(TC), FO(IFP), FO(PFP), $L^\omega_{\infty\omega}$, or SO and let Π be an $\mathcal{L}$-interpretation of σ in τ. Show that for every $\mathcal{L}[\sigma]$-sentence ψ there is an $\mathcal{L}[\tau]$-sentence $\psi^{-\Pi}$ such that for all τ-structures $\mathcal{A}$ (with $\pi^{\mathcal{A}}_{\text{uni}}(\_) \neq \emptyset$),

$$\mathcal{A} \models \psi^{-\Pi} \quad \text{iff} \quad \mathcal{A}^\Pi \models \psi.$$

Check whether your proof works for MSO, too. If "yes", it is incorrect. □

---

[8] For simplicity, we restrict ourselves to relational σ. It should be clear how the notion of interpretation can be extended to vocabularies σ containing constants.

With the following proposition we show that "up to FO-interpretations, structures can be viewed as graphs".

Proposition 11.2.5 *Let τ be a vocabulary. There are* FO*-interpretations* Π *of* $\{E\}$ *in* τ *and* Δ *of* τ *in* $\{E\}$ *with the following properties:*

(i) *For all τ-structures $\mathcal{A}$ with at least two elements,* $\mathcal{A}^{\Pi} \in$ GRAPH *and*

$$(\mathcal{A}^{\Pi})^{\Delta} \cong \mathcal{A}.$$

(ii) *For all* $K \subseteq \mathrm{Str}[\tau]$,

$$K_{<} \in \mathrm{PTIME} \quad \textit{iff} \quad (K^{\Pi})_{<} \in \mathrm{PTIME},$$

where $K^{\Pi} := \{\mathcal{B} \mid \mathcal{B} \cong \mathcal{A}^{\Pi}$ *for some* $\mathcal{A} \in K\}$.

Proof. We treat the case $\tau = \{T, R\}$ with unary T and binary R. From the exposition it will be clear how to handle relation symbols of higher arity.

To get an idea of how to define Π, we first code a given τ-structure $\mathcal{A}$ as a graph $\mathcal{G}$ as follows:

- Each $a \in A$ is a vertex in $\mathcal{G}$;
- each $a \in A$ has two new vertices a_1, a_2 linked to it;
- each $a \in T^A$ is linked to each element of a clique of two new elements;
- for each $(a, b) \in R^A$, a is linked to one element and b to the remaining two elements of a clique of three new elements.

The following picture shows $\mathcal{G}$ in case $A = \{a, b\}$, $T^A = \{a\}$, and $R^A = \{(a, b)\}$.

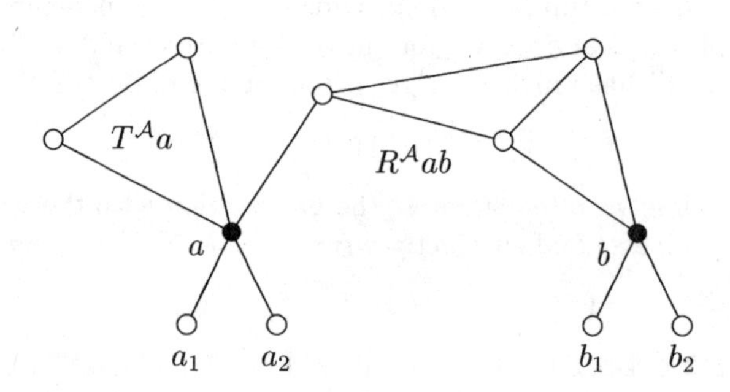

Obviously, $\mathcal{G}$ contains all the data of $\mathcal{A}$. As $\mathcal{G}$ has more elements than $\mathcal{A}$, any interpretation Π such that $\mathcal{A}^{\Pi} \cong \mathcal{G}$ has to have a width $k \geq 2$. Then $(a, \ldots, a) \in A^k$ can be the vertex of $\mathcal{G}$ corresponding to $a \in A$. But, what elements of A^k may serve as "legs" a_1 and a_2 attached to a? In general, for arbitrary $\mathcal{A}$ there is no way to uniquely define two such elements for every $a \in A$. Thus, we replace each leg of a as well as every other auxiliary point attached to a via T or R by all elements of the set $A_a := \{b \in A \mid b \neq a\}$

(as A has at least two elements, A_a is nonempty). Altogether, we are led to the following modification $\mathcal{H}$ of $\mathcal{G}$. We set $k = 9$ and let H contain, for every $a, b \in A$, the following points:

(1) the point a^9 (:=(a,...,a));
(2) the points of $a^2 A_a a^6$ ($:= \{(a, a, c, a, \ldots, a) \mid c \in A, c \neq a\}$) and of $a^3 A_a a^5$;
(3) if $T^A a$, then the points of $a^4 A_a a^4$ and of $a^5 A_a a^3$;
(4) if $R^A ab$, then the points of $aba^4 A_a a^2$, of $aba^5 A_a a$, and of $aba^6 A_a$.

Define the edge relation E^H as follows:

– for $a \in A$, connect the points in (2) with a^9
– for $a \in A$ with $T^A a$ and for $c \in A_a$, link a^4ca^4 and a^5ca^3 to a^9 and to each other
– for $a, b \in A$ with $R^A ab$ and for $c \in A_a$, form a clique out of aba^4ca^2, aba^5ca, and aba^6c, and link aba^4ca^2 to a^9 and aba^5ca and aba^6c to b^9.

Let $\Pi = (\pi_{\mathrm{uni}}(\overline{x}), \pi_E(\overline{x}, \overline{y}))$ with $\overline{x}, \overline{y}$ of length 9 be the interpretation of $\{E\}$ in τ that describes this encoding. For instance, set

$$\begin{aligned} \pi_{\mathrm{uni}}(\overline{x}) \;:=\; & x_1 = x_2 = \ldots = x_9 \\ & \vee (x_1 = x_2 = x_4 = \ldots = x_9 \wedge x_3 \neq x_1) \\ & \vee (x_1 = x_2 = x_3 = x_5 = \ldots = x_9 \wedge x_4 \neq x_1) \\ & \vee (Tx_1 \wedge x_1 = \ldots = x_4 = x_6 = \ldots = x_9 \wedge x_5 \neq x_1) \\ & \vee (Tx_1 \wedge x_1 = \ldots = x_5 = x_7 = x_8 = x_9 \wedge x_6 \neq x_1) \\ & \vee \ldots . \end{aligned}$$

Then we have $\mathcal{A}^{\Pi} \in$ GRAPH.

We now define an FO-interpretation Δ of τ in $\{E\}$ of width 1 that allows to regain the structure $\mathcal{A}$ in $\mathcal{A}^{\Pi}$, more precisely: $(\mathcal{A}^{\Pi})^{\Delta} \cong \mathcal{A}$ for all τ-structures $\mathcal{A}$ with at least two elements.

In view of the definition of Π this is achieved by setting

$\delta_{\mathrm{uni}}(x)$:= "x has (at least) two legs" (y is a leg of x, if y is only linked to x)

$\delta_T(x)$:= "there are two elements which are linked to x and to each other and to no other point"

$\delta_R(x, y)$:= "there are three points u, v, w such that u is linked to x and v, w are linked to y, to u, and to each other, and u, v, w are not linked to other points".

We now come to part (ii). In the argumentation we tacitly assume that the τ-structures in question have at least two elements. (The remaining finitely many cases have to be treated separately.) First, we show that $(\mathrm{Str}[\tau]^{\Pi})_<$ is in PTIME. Namely, given an $\{E\}$-structure $\mathcal{B}$, in polynomial time we can check whether

- $\mathcal{B}$ is a graph
- all points with legs ("domain points") have the same number of legs, this number being equal to twice the number of domain points minus two
- all other points are linked to exactly one domain point and the set of these points correctly codes relations T and R.

Now, let $K \subseteq \text{Str}[\tau]$. Assume that $K_< \in \text{PTIME}$. Given an ordered $\{E\}$-structure $(\mathcal{B}, <)$, check whether $(\mathcal{B}, <) \in (\text{Str}[\tau]^\Pi)_<$. If not, $(\mathcal{B}, <) \notin (K^\Pi)_<$. Otherwise, let $\mathcal{B} \cong \mathcal{A}^\Pi$ for some τ-structure $\mathcal{A}$. Using Δ, construct the τ-structure $\mathcal{B}^\Delta$ in time polynomial in $\|B\|$. By part (i) of our proposition, $\mathcal{B}^\Delta \cong (\mathcal{A}^\Pi)^\Delta \cong \mathcal{A}$. Let $<^\Delta$ be the ordering induced on B^Δ by $<$. Then,

$$(\mathcal{B}, <) \in (K^\Pi)_< \quad \text{iff} \quad (\mathcal{B}^\Delta, <^\Delta) \in K_<.$$

In fact, if $(\mathcal{B}^\Delta, <^\Delta) \in K_<$ then $\mathcal{A} \in K$ and hence, $(\mathcal{A}^\Pi \cong) \mathcal{B} \in K^\Pi$ and $(\mathcal{B}, <) \in (K^\Pi)_<$. Conversely, if $(\mathcal{B}, <) \in (K^\Pi)_<$ then $\mathcal{B} \cong \mathcal{A}_1^\Pi$ for some $\mathcal{A}_1 \in K$ and hence, $\mathcal{B}^\Delta \cong \mathcal{A}_1$ and $(\mathcal{B}^\Delta, <^\Delta) \in K_<$.

By assumption, the right hand side can be checked in time polynomial in $\|B^\Delta\|$ and hence, in time polynomial in $\|B\|$. Altogether, the left hand side can be checked in time polynomial in $\|B\|$.

For the other direction assume that $(K^\Pi)_< \in \text{PTIME}$ and let $(\mathcal{A}, <)$ be an ordered $\tau_<$-structure. We can construct $\mathcal{A}^\Pi$ and, restricting the $<$-lexicographic ordering of tuples to A^Π, an ordering $<^\Pi$ on A^Π in time polynomial in $\|A\|$. Since, by (i),

$$(\mathcal{A}, <) \in K_< \quad \text{iff} \quad (\mathcal{A}^\Pi, <^\Pi) \in (K^\Pi)_<$$

and $(K^\Pi)_< \in \text{PTIME}$ by assumption, we altogether can decide the left hand side in time polynomial in $\|A\|$. □

Now we are able to reduce the problem whether there is a logic $\mathcal{L}$ with $\mathcal{L} \equiv_{\text{es}} \text{PTIME}$ to the class of graphs.

Theorem 11.2.6 *If there is a logic that effectively strongly captures* PTIME *on* GRAPH *then there is a logic effectively strongly capturing* PTIME.

Proof. Choose a logic $\mathcal{L}_G$ with $\mathcal{L}_G \equiv_{\text{es}} \text{PTIME}$ on GRAPH. Define the logic $\mathcal{L}$ as follows. For every vocabulary τ let $\mathcal{L}[\tau] = \mathcal{L}[\tau]_0$ be the set of sentences of $\mathcal{L}_G[\{E\}]$. For a τ-structure $\mathcal{A}$ and $\varphi \in \mathcal{L}[\tau]$ define

$$(*) \qquad \mathcal{A} \models^{\mathcal{L}} \varphi \quad \text{iff} \quad \mathcal{A}^\Pi \models^{\mathcal{L}_G} \varphi$$

(here Π is the interpretation of $\{E\}$ in τ as given by Proposition 11.2.5). We claim that $\mathcal{L} \equiv_{\text{es}} \text{PTIME}$.

Firstly, $\mathcal{L}$ strongly captures PTIME: Assume $K \subseteq \text{Str}[\tau]$. We have to show that

$$K_< \in \text{PTIME} \quad \text{iff} \quad \text{for some } \varphi \in \mathcal{L}[\tau], K = \text{Mod}^{\mathcal{L}}(\varphi)$$

or, by Proposition 11.2.5(ii), that

$$(K^{\Pi})_{<} \in \text{PTIME} \quad \text{iff} \quad \text{for some } \varphi \in \mathcal{L}[\tau], K = \text{Mod}^{\mathcal{L}}(\varphi).$$

If $(K^{\Pi})_{<} \in$ PTIME, then $K^{\Pi} = \text{Mod}^{\mathcal{L}_G}(\varphi)$ for some $\varphi \in \mathcal{L}_G[\{E\}]$, since $\mathcal{L}_G \equiv_{\text{es}}$ PTIME. Therefore, by $(*)$, $K = \text{Mod}^{\mathcal{L}}(\varphi)$. Conversely, assume $K = \text{Mod}^{\mathcal{L}}(\varphi)$ for some $\varphi \in \mathcal{L}[\tau]$. Then, $K^{\Pi} = \text{Mod}^{\mathcal{L}_G}(\varphi) \cap \text{Str}[\tau]^{\Pi}$ and therefore, $(K^{\Pi})_{<} = \text{Mod}^{\mathcal{L}_G}(\varphi)_{<} \cap (\text{Str}[\tau]^{\Pi})_{<}$. Both classes on the right hand side are in PTIME (the first one, as $\mathcal{L}_G \equiv_{\text{es}}$ PTIME, the second one again by 11.2.5(ii)). Thus, $(K^{\Pi})_{<} \in$ PTIME.

Let τ be given. Of course, $\mathcal{L}[\tau]$ is recursive. Moreover, since $\mathcal{L}_G \equiv_{\text{es}}$ PTIME, we effectively can assign to $\varphi \in \mathcal{L}[\tau]$ a pair (M_{φ}, d), where M_{φ} is a deterministic x^d time-bounded machine accepting $\text{Mod}^{\mathcal{L}_G}(\varphi)_{<}$. To check whether a structure $(\mathcal{A}, <) \in \text{Str}(\tau)_{<}$ is in $\text{Mod}^{\mathcal{L}}(\varphi)_{<}$, by a polynomial time algorithm we pass to $(\mathcal{A}^{\Pi}, <^{\Pi})$ (where $<^{\Pi}$ is the lexicographic ordering on A^{Π} induced by $<$) and then use M_{φ} to decide whether $(\mathcal{A}^{\Pi}, <^{\Pi}) \in \text{Mod}^{\mathcal{L}_G}(\varphi)_{<}$, and hence, by $(*)$, whether $(\mathcal{A}, <) \in \text{Mod}^{\mathcal{L}}(\varphi)_{<}$. Clearly, we can effectively calculate a time bound.

Altogether, we have $\mathcal{L} \equiv_{\text{es}}$ PTIME. □

Further notions and results of the preceding section can be restricted to classes of structures, too. E.g., let τ be a vocabulary and S a class of τ-structures and $\mathcal{C}$ a deterministic complexity class. A map $C_{\tau} : \text{Str}[\tau] \to \text{Str}[\tau_{<}]$ is a $\mathcal{C}$-canonization on S, if

(1) For all $\mathcal{A} \in S$, $\mathcal{A} \cong C_{\tau}(\mathcal{A})|\tau$ and $<^{C_{\tau}(\mathcal{A})}$ is an ordering.
(2) For all $\mathcal{A}, \mathcal{B} \in \text{Str}[\tau]$, if $\mathcal{A} \cong \mathcal{B}$ then $C_{\tau}(\mathcal{A}) \cong C_{\tau}(\mathcal{B})$.
(3) C_{τ} is $\mathcal{C}$-computable, more precisely: there is a PTIME-algorithm that, applied to $(\mathcal{A}, <^{A}) \in \text{Str}[\tau]_{<}$, gives (the encoding of an ordered version of) $C_{\tau}(\mathcal{A})$.

Analogously, one defines the notion of $\mathcal{C}$-invariantization. The proof of 11.1.8 showing that PTIME-invariantizations yield PTIME-canonizations, in general, does not work in the restricted case as it uses PTIME-invariantizations for structures with a larger vocabulary. However, in the case of GRAPH one can settle this difficulty by interpreting the expanded structures in graphs:

Exercise 11.2.7 (a) If there is a PTIME-canonization on GRAPH, then there is a PTIME-canonization.

(b) If there is a PTIME-invariantization on GRAPH, then there is a PTIME-invariantization. □

Exercise 11.2.8 Let τ be a vocabulary and S a class of τ-structures. Show: If there is a PTIME-canonization on S then there is a logic $\mathcal{L}$ such that $\mathcal{L} \equiv_{\text{es}}$ PTIME on S. □

We now come to some positive results and start with a simple observation. We saw that the capturing result for PTIME proven in Chapter 7 can be reformulated as

> FO(IFP) effectively strongly captures PTIME on the class of ordered structures.

This can easily be generalized to

Proposition 11.2.9 *Let $S \subseteq Str[\tau]$. Assume that there is an* FO(IFP)$[\tau]$-*formula $\varphi(x, y, \overline{z})$ with free variables among $x, y, \overline{z}$ such that for all $\mathcal{A} \in S$ there are $\overline{b} \in A$ such that $\varphi^{\mathcal{A}}(\_,\_,\overline{b})$ is an ordering of A. Then,* FO(IFP) $\equiv_{\text{es}}$ PTIME *on S.*

Proof. We first treat the case without parameters. The idea is to use the result just quoted, letting $\varphi(x, y)$ play the role of $x < y$. More explicitly: If $K_< \in$ PTIME then $K_< = \text{Mod}(\psi)$ for some sentence $\psi \in$ FO(IFP)$[\tau_<]$. But then, for the FO(IFP)$[\tau]$-sentence $\psi\frac{\varphi(\_\_\_,\dots)}{\_\_\_<\dots}$ we have that $K \cap S = \text{Mod}(\psi\frac{\varphi(\_\_\_,\dots)}{\_\_\_<\dots}) \cap S$. Conversely, if φ is an FO(IFP)$[\tau]$-sentence and $K := \text{Mod}(\varphi)$, then we have $K_< \in$ PTIME and trivially, $K \cap S = \text{Mod}(\varphi) \cap S$.

For the general case one replaces the formula $\psi\frac{\varphi(\_\_\_,\dots)}{\_\_\_<\dots}$ by the formula $\exists\overline{y}(\varphi(\_,\_,\overline{y})$ is an ordering $\wedge\, \psi\frac{\varphi(\_\_\_,\dots,\overline{y})}{\_\_\_<\dots})$. □

The following two exercises contain some applications and generalizations of the preceding proposition.

Exercise 11.2.10 (a) Show that FO(IFP) $\equiv_{\text{es}}$ PTIME on the class of cycles. Remember that a cycle is a graph that is isomorphic to some $\mathcal{G}_l$ (cf. 2.3.8). Hint: Note that there is a formula $\varphi(x, y, u, v)$ in FO(IFP)$[\{E\}]$ such that for any cycle $\mathcal{G}$ and all $a, b \in G$ with $E^{\mathcal{G}}ab$, the relation $\varphi^{\mathcal{G}}(\_,\_,a, b)$ is an ordering of G.

(b) A finite grid is a graph isomorphic to

$$(\{0,\dots,n\}\times\{0,\dots,m\}, \{((i,j),(k,l)) \mid i,k \leq n,\ j,l \leq m,\ |i-k|+|j-l| = 1\})$$

for some n, m. Show that FO(IFP) $\equiv_{\text{es}}$ PTIME on the class of grids. □

Note that in the situation given by the hypotheses of Proposition 11.2.9 we have, using the notations of its proof, that $C_\tau : \text{Str}[\tau] \to \text{Str}[\tau_<]$ with

$$C_\tau(\mathcal{A}) \quad := \quad (\mathcal{A}, \varphi^{\mathcal{A}}(\_,\_))$$

is a PTIME-canonization on S. Moreover, $C_\tau(\mathcal{A}) = \mathcal{A}^{\Pi}$ for the FO(IFP)-interpretation Π of $\tau_<$ in τ given by $\pi_{\text{uni}}(x) := x = x$, $\pi_<(x, y) := \varphi(x, y)$, and $\pi_R(\overline{x}) := R\overline{x}$ for $R \in \tau$. The proof of 11.2.9 contains the main argument needed to show the following exercise:

Exercise 11.2.11 Let $\mathcal{L}$ be one of the logics mentioned in 11.2.4 and $\mathcal{C}$ a deterministic complexity class with $\mathcal{L} \equiv_{\text{es}} \mathcal{C}$ on ordered structures. Assume that S is a class of τ-structures and C_τ a $\mathcal{C}$-canonization on S given by an $\mathcal{L}$-interpretation of $\tau_<$ in τ (in the form $C_\tau : \mathcal{A} \mapsto \mathcal{A}^\Pi$). Then $\mathcal{L} \equiv_{\text{es}} \mathcal{C}$ on S. □

Example 11.2.12 Let K_0 be the class of $\tau := \{E\}$-structures isomorphic to some $\mathcal{H}_l$ with $l \geq 3$, where $\mathcal{H}_l$ is a digraph as pictured by

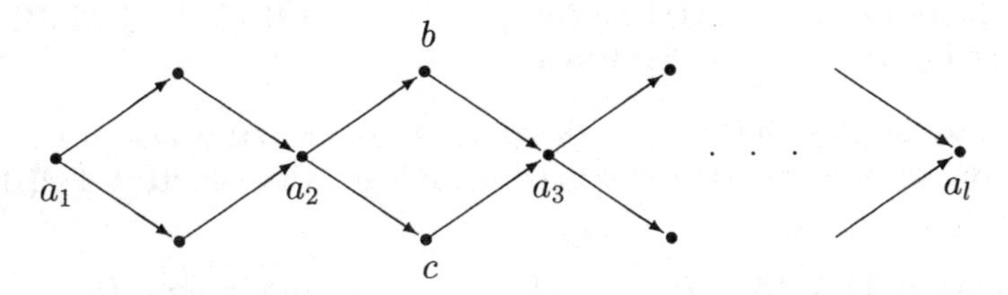

There is no formula that defines an ordering on all the structures in K_0, even if parameters are allowed. In fact, consider a formula $\varphi(x, y, z_1, \ldots, z_n)$ and the digraph $\mathcal{H}_l$ for $l > n+1$. Then, for any choice of parameters $d_1, \ldots, d_n$, there are corresponding elements b, c of in-degree 1 (cf. the figure) not occurring among $d_1, \ldots, d_n$. There is an automorphism of $\mathcal{H}_l$ that fixes $d_1, \ldots, d_n$ and interchanges b and c. Hence,

$$(b, c) \in \varphi^{\mathcal{H}_l}(\_, \_, \overline{d}) \quad \text{iff} \quad (c, b) \in \varphi^{\mathcal{H}_l}(\_, \_, \overline{d});$$

so $\varphi^{\mathcal{H}_l}(\_, \_, \overline{d})$ is not an ordering on $\mathcal{H}_l$.

Thus, we cannot apply 11.2.9 to conclude FO(IFP) $\equiv_{\text{es}}$ PTIME on K_0. We obtain this result from the preceding exercise showing that there is a "definable" LOGSPACE-canonization. In fact, there is an FO(DTC)-interpretation $\Pi := (\pi_{\text{uni}}, \pi_E, \pi_<)$ of $\sigma := \{E, <\}$ in $\tau\ (= \{E\})$ assigning to the digraph $\mathcal{H}_l$ the ordered digraph $\mathcal{H}_l^\Pi$ consisting of the isomorphic copy of $\mathcal{H}_l$

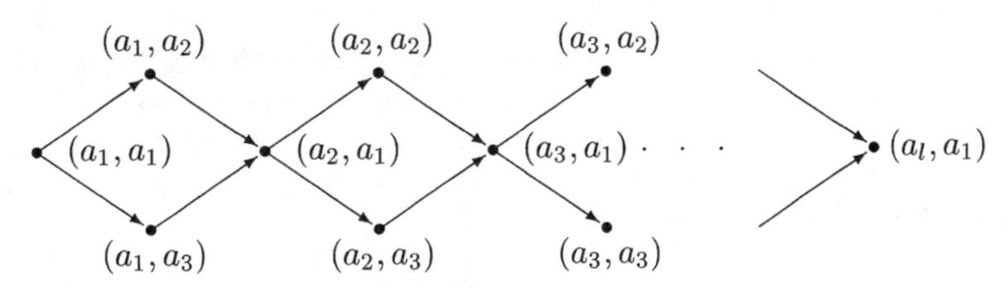

together with the ordering indicated by

$$(a_1, a_1) < (a_1, a_2) < (a_1, a_3) < (a_2, a_1) < (a_2, a_2) < \ldots < (a_l, a_1).$$

To define Π, for $i = 1, 2, 3$ let $\varphi_i(x)$ be a formula defining the i-th "crosspoint" of $\mathcal{H}_l$,

$$\begin{aligned}\varphi_1(x) &:= \forall y \neg Eyx \\ \varphi_2(x) &:= \exists z(\varphi_1(z) \wedge \exists y(Ezy \wedge Eyx)) \\ \varphi_3(x) &:= \exists z(\varphi_2(z) \wedge \exists y(Ezy \wedge Eyx)).\end{aligned}$$

Then, for example, set

$$\pi_{\text{uni}}(x,y) := (\exists^{=2} z\, Exz \wedge (\varphi_1(y) \vee \varphi_2(y) \vee \varphi_3(y)) \vee (\forall z \neg Exz \wedge \varphi_1(y)).$$

Using the preceding exercise we conclude that FO(IFP) $\equiv_{\text{es}}$ PTIME on K_0 and that FO(DTC) $\equiv_{\text{es}}$ LOGSPACE on K_0. □

We turn to the class of trees. For this purpose we first state and prove a variation of the preceding exercise for fixed-point logic with counting, FO(IFP, #) (cf. subsection 8.4.1).

Proposition 11.2.13 *Let* $\tau = \{R_1, \ldots, R_m\}$ *with* r_i*-ary* R_i *and let* S *be a class of* τ*-structures. Assume that there are* FO(IFP, #)$[\tau]$*-formulas* $\varphi_1(\overline{\mu}_1), \ldots, \varphi_m(\overline{\mu}_m)$ *(where* $\overline{\mu}_i$ *is a sequence of number variables of length* r_i*) such that for every* $\mathcal{A} \in S$*,*

$$(*) \qquad \mathcal{A} \cong (\{0, \ldots, \|A\| - 1\}, \varphi_1^{\mathcal{A}}(\_), \ldots, \varphi_m^{\mathcal{A}}(\_))$$ [9]

Then,

$$\text{FO(IFP, \#)} \equiv_{\text{es}} \text{PTIME } \textit{on } S.$$

Proof. First, we observe that the map $C_\tau : \text{Str}[\tau] \to \text{Str}[\tau_<]$ given by

$$C_\tau(\mathcal{A}) := (\{0, \ldots, \|A\| - 1\}, \varphi_1^{\mathcal{A}}(\_), \ldots, \varphi_m^{\mathcal{A}}(\_), <)$$

(with $<$ the natural ordering on $\{0, \ldots, \|A\| - 1\}$) is a PTIME-canonization on S.

Let ψ be an FO(IFP)$[\tau_<]$-sentence. For every point variable x we choose a number variable μ_x (with $\mu_x \neq \mu_y$ for $x \neq y$), and we let ψ^* be the FO(IFP, #)$[\tau]$-sentence that arises from ψ by replacing every subformula $R_i x_1 \ldots x_{r_i}$ by $\varphi_i(\mu_{x_1}, \ldots, \mu_{x_{r_i}})$, by replacing every variable x by μ_x in all other atomic formulas, and by changing every subformula $\forall x \ldots$ to $\forall \mu_x(\mu_x < \max \to \ldots)$ and every subformula $\exists x \ldots$ to $\exists \mu_x(\mu_x < \max \wedge \ldots)$. Then we have for every τ-structure $\mathcal{A}$:

$$\mathcal{A} \models \psi^* \quad \text{iff} \quad C_\tau(\mathcal{A}) \models \psi.$$

It remains to show that

$$\{K \cap S \mid K \subseteq \text{Str}[\tau],\, K_< \in \text{PTIME}\} \;=\; \{\text{Mod}(\varphi) \cap S \mid \varphi \in \text{FO(IFP, \#)}\}.$$

[9] Note that $\{0, \ldots, \|A\|\}$ is the universe of the number part of $\mathcal{A}$; thus, instead of $\varphi_i^{\mathcal{A}}(\_)$ we should write $\varphi_i^{\mathcal{A}}(\_) \cap \{0, \ldots, \|A\| - 1\}^{r_i}$.

If φ is an FO(IFP, #)[τ]-sentence and $K = \text{Mod}(\varphi)$, then $K_< \in$ PTIME and clearly, $K \cap S = \text{Mod}(\varphi) \cap S$. Conversely, assume $K \subseteq \text{Str}[\tau]$ and $K_< \in$ PTIME. Then $K_< = \text{Mod}(\psi)$ for some FO(IFP)[$\tau_<$]-sentence ψ. We claim that $K \cap S = \text{Mod}(\psi^*) \cap S$. In fact, for $\mathcal{A} \in S$,

$$\begin{array}{lll} \mathcal{A} \in K & \text{iff} & C_\tau(\mathcal{A}) \in K_< \quad (\text{as } \mathcal{A} \cong C_\tau(\mathcal{A})|\tau) \\ & \text{iff} & C_\tau(\mathcal{A}) \models \psi \\ & \text{iff} & \mathcal{A} \models \psi^* \quad (\text{see above}). \end{array}$$

□

We apply the preceding result to the class TREE of finite trees (cf. Chapter 1.A1).

Theorem 11.2.14 FO(IFP, #) $\equiv_{\text{es}}$ PTIME *on* TREE.

In view of the preceding proposition it suffices to show:

Lemma 11.2.15 *There is a formula $\varphi(\mu, \nu)$ in* FO(IFP)[$\{E\}$] *such that for all $\mathcal{A} \in$* TREE,

$$\mathcal{A} \cong (\{0, \ldots, \|A\| - 1\}, \varphi^{\mathcal{A}}(\_,\_)).$$

Proof. Let $\mathcal{A} \in$ TREE. By induction going from the leafs to the root, for each $u \in A$ we define a copy of the subtree $\mathcal{A}_u$ of $\mathcal{A}$ with root u on (an initial segment of) the number part of $\mathcal{A}$; the tree relation of the copy of $\mathcal{A}_u$ will be given by $Xu\_\_$ (X is a ternary mixed relation variable with first component for point variables, the last two components for number variables). The case of leaves $u \in A$ being trivial, let $u \in A$ have the E^A-successors $v_1, \ldots, v_l$. By induction hypothesis, the trees $\mathcal{A}_{v_1}, \ldots, \mathcal{A}_{v_l}$ have isomorphic copies on initial segments of the number part of $\mathcal{A}$ given by $Xv_1\_\_, \ldots, Xv_l\_\_$. Using the ordering $<$ of the number part, these copies can be ordered lexicographically in an FO-definable way by

> $Xv_i\_\_ \prec Xv_j\_\_$ (for short : $v_i \prec v_j$) iff "$Xv_i\_\_ \neq Xv_j\_\_$, and the minimal pair (μ, ν), minimal in the lexicographic ordering induced by $<$, where they differ belongs to $Xv_i\_\_$".

We abbreviate $(\neg v_i \prec v_j \wedge \neg v_j \prec v_i)$ by $v_i \equiv v_j$. If, for example, $l = 4$ and $v_1 \prec v_2 \equiv v_3 \prec v_4$, we define the copy of $\mathcal{A}_u$ on the number part of $\mathcal{A}$ by assigning 0 to u and let it be followed by a copy of $Xv_1\_\_$, two copies of $Xv_2\_\_$, and one copy of $Xv_4\_\_$. More precisely, in the general case we define

$$Xu\mu\nu$$

by the disjunction of the following formulas φ_1, φ_2 (φ_1 describes that the number 0 assigned to u is linked to the numbers assigned to the v_i; φ_2 describes the copies associated with the subtrees $\mathcal{A}_{v_i}$).

Using $\delta(u, v)$ for the number term $\#\{w \mid \exists v'(Euv' \wedge v' \prec v \wedge w \in A_{v'})\}$ which counts the number of all elements in trees $\mathcal{A}_{v'}$ of E-successors v' of u that are $\prec v$, we set

$$\begin{aligned}\varphi_1(u,\mu,\nu,X) \;:=\;& \mu=\min \wedge \exists v(Euv \wedge \exists \zeta(0 \le \zeta < \#\{w \mid Euw \wedge w \equiv v\} \\ & \wedge \nu = 1+\delta(u,v)+\zeta\cdot\#\{w \mid w \in \mathcal{A}_v\}));\end{aligned}$$

$$\begin{aligned}\varphi_2(u,\mu,\nu,X) \;:=\;& \exists v(Euv \wedge \exists\zeta(0\le\zeta<\#\{w \mid Euw \wedge w\equiv v\} \wedge \\ & \exists\xi\exists\eta(Xv\xi\eta \\ & \wedge \mu = \xi+1+\delta(u,v)+\zeta\cdot\#\{w \mid w\in\mathcal{A}_v\} \\ & \wedge \nu = \eta+1+\delta(u,v)+\zeta\cdot\#\{w \mid w\in\mathcal{A}_v\}))).\end{aligned}$$

Note that, with u_0 the root of $\mathcal{A}$,

$$Xu_0\,\_\,\_ \text{ defines a copy of } \mathcal{A}.$$

According to Lemma 7.3.11, addition and multiplication on the number part of $\mathcal{A}$ are FO(IFP)-definable as is "$w \in \mathcal{A}_v$". Thus, the desired FO(IFP, #)[$\{E\}$]-formula $\chi(\mu,\nu)$, defining an isomorphic copy of $\mathcal{A}$ on its number part, can be given as

$$\chi(\mu,\nu) \;:=\; \exists u_0(u_0 \text{ is the root } \wedge [\mathrm{IFP}_{u\mu\nu,X}(\varphi_1\vee\varphi_2)]u_0\mu\nu). \qquad \square$$

Exercise 11.2.16 (a) Show that FO(IFP, #) $\equiv_{\mathrm{es}}$ PTIME on the class of acyclic connected graphs. Hint: Each point of an acyclic connected graph can be viewed as the root of a corresponding tree.

(b) Show that FO(IFP, #) $\equiv_{\mathrm{es}}$ PTIME on the class of acyclic graphs. □

In 11.2.6 we have seen that, in order to obtain a logic $\mathcal{L}$ with $\mathcal{L} \equiv_{\mathrm{es}}$ PTIME, it is enough to find a logic $\mathcal{L}'$ such that $\mathcal{L}' \equiv_{\mathrm{es}}$ PTIME on GRAPH. We then have seen that there are subclasses of GRAPH such as the class of acyclic graphs or the class of trees that can effectively strongly be captured by a logic. Further positive results are contained in [65, 68] where it is shown, for example, that FO(IFP, #) $\equiv_{\mathrm{es}}$ PTIME on the class of planar graphs. These positive results may be interpreted as an approximation to a positive solution of the main question, if there should be a positive solution. If not, they might serve to discover the borderline where the possibility for an effective strong capturing gets lost.

Notes 11.2.17 The main question whether there is a logic strongly capturing PTIME was first asked in [19]. The notion of a logic effectively strongly capturing a complexity class originates with Gurevich [71]; Theorem 11.1.9 is from [73]. The capturing result 11.2.14 goes back to [97, 117]. For further capturing results see [82] and the survey paper [66].

12. Quantifiers and Logical Reductions

In Chapter 7 we have seen that the fixed-point logic FO(IFP) captures PTIME. In the last chapter we have remarked that the question

(*) *Is there a logic that effectively strongly captures* PTIME?

is still open. We have discussed strategies for a solution and thereby given some positive results in case we restrict ourselves to certain classes of structures.

The notion of logic underlying the main question is very general. For instance, we have admitted that the sentences of a logic are Turing machines (and the models of a Turing machine are the structures it accepts). Thus, its syntax lacks the characteristic flavour of logics we are accustomed to. A corresponding positive answer to (*) could hardly be considered to be a *logical* characterization of PTIME. However, in the present chapter we shall show that if there is a logic that effectively strongly captures PTIME, then there is a logic with a familiar syntax.

At a first glance the following idea could lead to such a logic: Consider, for example, FO(IFP). We know that there is some class K of structures such that $K_<$ belongs to PTIME, but K is not axiomatizable in FO(IFP). In a first step one could add a "logical construct" to FO(IFP) that allows for an axiomatization of K, and in this way one could continue trying to enlarge the expressive power just up to PTIME. In traditional model theory there is a well-established method to enrich logics by new concepts, namely by adjoining so-called Lindström quantifiers. We have already met such quantifiers, for example, the counting quantifiers (cf. section 3.4).

In the first section we shall introduce the concept of Lindström quantifier and demonstrate its scope by various examples showing that some important logics encountered so far can be obtained as extensions of first-order logic by such quantifiers. In section 2 we prove some limiting results saying us, for example, that no extension of FO(IFP) by a finite set of Lindström quantifiers will strongly capture PTIME. Nevertheless, we shall see there that, if any reasonable logic strongly captures PTIME, then there is already such a logic that can be obtained from first-order logic by adding a simple infinite set of quantifiers. When pursuing this line also for other complexity classes such as NPTIME, a logical analogue of complexity-theoretical reductions turns out to be useful. Section 3 presents results of the corresponding kind.

If we extend a logic $\mathcal{L}$ by the binary Lindström quantifier Q where $Qxy\varphi(x,y)$ says that $\varphi(x,y)$ defines a connected graph on the universe, then in the extended logic $\mathcal{L}(Q)$ we can speak about the connectivity of definable graphs by simply using the quantifier Q. There is a similarity to Turing machines with oracles that have the ability to supply, in one step, correct answers to the question whether a binary relation (encoded on some tape) is a connected graph. We shall see in the last section that this similarity is by no means superficial.

Throughout the chapter all vocabularies are relational and structure means finite structure.

12.1 Lindström Quantifiers

For a binary relation symbol E we know that FO(TC), but not FO, allows to express

(∗) "(x,y) is in the transitive closure of E".

What do we expect from any "natural" logic $\mathcal{L}$ allowing to formalize (∗)? Clearly, for every formula $\varphi(u,v)$ of $\mathcal{L}$ we want to have a formula in $\mathcal{L}$ expressing

"(x,y) is in the transitive closure of $\varphi(\_,\_)$".

Similarly, in a natural logic $\mathcal{L}$ which allows to express that a graph has a Hamiltonian path, we expect to be able to say that an $\mathcal{L}$-definable graph has a Hamiltonian path. If in the following definitions we take as K the class of graphs with a Hamiltonian path, we get a quantifier Q such that FO(Q) will be the smallest natural extension of FO having this ability.

Fix a vocabulary

$$\sigma = \{R_1, \ldots, R_s\}$$

where R_i is r_i-ary, and a class K of σ-structures. Furthermore, let $\mathcal{L}$ be some logic such as FO, FO(IFP), or $\mathrm{L}_{\infty\omega}$. For a new "quantifier symbol" Q_K we enlarge, for any vocabulary τ, the formation rules for $\mathcal{L}[\tau]$-formulas by the following clause:

(+) If for $1 \le i \le s$, $\psi_i(\overline{x}_i)$[1] is a formula and the length of $\overline{x}_i$ is r_i then

$$Q_K \overline{x}_1, \ldots, \overline{x}_s[\psi_1(\overline{x}_1), \ldots, \psi_s(\overline{x}_s)]$$

is a formula, too.

[1] We only display the variables relevant in the given context. The variables in each $\overline{x}_i$ are pairwise distinct, but the same variable may occur in distinct $\overline{x}_i$'s.

In case $s = 1$ we write $Q\overline{x}\psi(\overline{x})$ instead of $Q\overline{x}[\psi(\overline{x})]$. A variable y is free in $Q_K\overline{x}_1, \ldots, \overline{x}_s[\psi_1(\overline{x}_1), \ldots, \psi_s(\overline{x}_s)]$ if, for some i, we have $y \in \text{free}(\psi_i)$ and y not in $\overline{x}_i$.

For any τ-structure $\mathcal{A}$ the meaning of $Q_K\overline{x}_1, \ldots, \overline{x}_s[\psi_1(\overline{x}_1), \ldots, \psi_s(\overline{x}_s)]$ is given by (we suppress the assignment to variables)

$$\mathcal{A} \models Q_K\overline{x}_1, \ldots, \overline{x}_s[\psi_1(\overline{x}_1), \ldots, \psi_s(\overline{x}_s)] \quad \text{iff} \quad (A, \psi_1^{\mathcal{A}}(\_), \ldots, \psi_s^{\mathcal{A}}(\_)) \in K$$

where $\psi_i^{\mathcal{A}}(\_)$ stands for $\{\overline{b} \in A^{r_i} \mid \mathcal{A} \models \psi_i[\overline{b}]\}$.

Q_K is the *Lindström quantifier* (by short: *quantifier*) given by K. Its arity is defined to be $\max\{r_i \mid 1 \leq i \leq s\}$. In case $s = 1$ one speaks of a *simple* quantifier.

The extension of $\mathcal{L}$ by Q_K is denoted by $\mathcal{L}(Q_K)$. If $\mathbf{Q}$ is a class of Lindström quantifiers, the extension $\mathcal{L}(\mathbf{Q})$ of $\mathcal{L}$ is defined by adding clause (+) for all $Q \in \mathbf{Q}$. $\mathcal{L}(\mathbf{Q})$ is called a *Lindström extension* of $\mathcal{L}$.

Obviously, we have for all σ-structures $\mathcal{A}$:

$$\mathcal{A} \models Q_K\overline{x}_1, \ldots, \overline{x}_s[R_1\overline{x}_1, \ldots, R_s\overline{x}_s] \quad \text{iff} \quad \mathcal{A} \in K.$$

Hence, K is axiomatizable in $\mathcal{L}(Q_K)$. Indeed, $\mathcal{L}(Q_K)$ is the least extension of $\mathcal{L}$ that allows to axiomatize K. A proof requires some closure conditions of the logics involved. They are satisfied in the cases covered by the next exercise. As a rule, such closure conditions (cf. 12.1.11) will also be necessary in the following when dealing with statements on arbitrary logics and will be tacitly assumed.

Exercise 12.1.1 Let $\mathcal{L}_1$ be a Lindström extension of FO and $\mathcal{L}_2$ one of the logics FO(TC), FO(DTC), FO(IFP), FO(PFP), or $\text{L}_{\infty\omega}$. Show: If $\mathcal{L}_1 \leq \mathcal{L}_2$ and the class K of structures is axiomatizable in $\mathcal{L}_2$ then $\mathcal{L}_1(Q_K) \leq \mathcal{L}_2$. □

One can show that any logic $\mathcal{L}$ above FO that obeys the closure conditions referred to above, can be represented as a Lindström extension FO($\mathbf{Q}$) of FO, where the quantifiers in $\mathbf{Q}$ correspond to the classes axiomatizable in $\mathcal{L}$, i.e., $\mathcal{L} \equiv \text{FO}(\{Q_K \mid K \text{ axiomatizable in } \mathcal{L}\})$. Of course, one is interested in small sets $\mathbf{Q}$ such that FO($\mathbf{Q}$) and $\mathcal{L}$ are equivalent.

We give some examples of Lindström quantifiers.

Examples 12.1.2 (a) Let $\sigma := \{P\}$ with unary P. For $K := \{(A, B) \mid \emptyset \neq B \subseteq A\}$, the simple unary quantifier Q_K corresponds to the existential quantifier, as $Q_K x\psi(x)$ is equivalent to $\exists x\psi(x)$.

(b) Similarly, for $K := \{(A, A) \mid A \text{ a nonempty set}\}$ we obtain the universal quantifier.

(c) For $n \geq 1$ and $K_n := \{(A, B) \mid B \subseteq A \text{ and } \|B\| \geq n\}$ the quantifier Q_{K_n} corresponds to the quantifier $\exists^{\geq n}$ (cf. section 3.4).

(d) For the class CONN of connected graphs we get a simple binary quantifier Q_{CONN}. So

$$Q_{\text{CONN}} xy\, \psi(x, y)$$

is valid in a structure $\mathcal{A}$ iff $(A, \psi^{\mathcal{A}}(\_,\_))$ is a connected graph.

(e) Let $\sigma := \{E, R\}$ with binary relation symbols E and R. If K is the class of σ-structures

$$\mathcal{A} = \{(A, E^A, R^A) \mid R^A \subseteq \text{TC}(E^A)\},$$

then Q_K can replace the unary transitive closure operator, as

$$[\text{TC}_{x,y}\psi(x, y)]st$$

is equivalent to

$$Q_K xy, uv[\psi(x, y), u = s \wedge v = t],$$

and

$$Q_K xy, uv[\psi_1(x, y), \psi_2(u, v)]$$

is equivalent to

$$\forall u \forall v(\psi_2(u, v) \rightarrow [\text{TC}_{x,y}\psi_1(x, y)]uv).$$

Thus, $\text{FO}(\text{TC}^1) \equiv \text{FO}(Q_K)$.

(f) Let K be a class of σ-structures, $<$ a new binary relation symbol, and $\mathcal{L}$ a logic. Then,

$$\mathcal{L}(Q_{K_<}) \equiv \mathcal{L}(Q_K) \text{ on ordered structures.}$$

In fact, if $\sigma = \{R_1, \ldots, R_s\}$ and $\varphi_1, \ldots, \varphi_s, \psi$ are formulas in some vocabulary τ that contains a binary relation symbol $\prec$, then on ordered τ-structures (that is, structures in which $\prec$ is an ordering) the formulas

$$Q_K \overline{x}_1, \ldots, \overline{x}_s[\varphi_1(\overline{x}_1), \ldots, \varphi_s(\overline{x}_s)]$$

and

$$Q_{K_<} \overline{x}_1, \ldots, \overline{x}_s, xy[\varphi_1(\overline{x}_1), \ldots, \varphi_s(\overline{x}_s), x \prec y]$$

are equivalent, and the formulas

$$Q_{K_<} \overline{x}_1, \ldots, \overline{x}_s, xy[\varphi_1(\overline{x}_1), \ldots, \varphi_s(\overline{x}_s), \psi(x, y)]$$

and

$$\text{“}\psi(x, y) \text{ defines an ordering”} \wedge Q_K \overline{x}_1, \ldots, \overline{x}_s[\varphi_1(\overline{x}_1), \ldots, \varphi_s(\overline{x}_s)]$$

are equivalent. Conclude: If $\mathcal{L}(\mathbf{Q})$ captures, say PTIME, then $K_< \in$ PTIME for all K such that $Q_K \in \mathbf{Q}$. □

Exercise 12.1.3 Let K be a class such that $K_< \in$ PTIME. Show that $\text{Mod}(\psi)_< \in$ PTIME for any sentence $\psi \in \text{FO}(\text{IFP})(Q_K)$. The same is true for sets of such quantifiers. Give the corresponding statements for PSPACE and FO(PFP) and for LOGSPACE and FO(DTC). Hint: For LOGSPACE cf. Appendix 7.5.1. □

Finitely many Lindström quantifiers can be replaced by a single one. We give a precise statement in the next theorem. The following two technical remarks (1) and (2) will be useful for the proof. As above, let $\sigma = \{R_1, \ldots, R_s\}$ with r_i-ary R_i and let K be a class of σ-structures.

(1) For a new r-ary relation symbol R let K_1 be the class of $\sigma \cup \{R\}$-structures

$$K_1 \quad := \quad \{(\mathcal{A}, R^A) \mid \mathcal{A} \in K \text{ and } R^A = A^r\}.$$

Then $\mathcal{L}(Q_K) \equiv \mathcal{L}(Q_{K_1})$, since

$$\models_{\text{fin}} Q_K \overline{x}_1, \ldots, \overline{x}_s[\psi_1, \ldots, \psi_s] \leftrightarrow Q_{K_1} \overline{x}_1, \ldots, \overline{x}_s, \overline{x}\,[\psi_1, \ldots, \psi_s, \overline{x} = \overline{x}]$$

and

$$\models_{\text{fin}} Q_{K_1} \overline{x}_1, \ldots, \overline{x}_s, \overline{x}[\psi_1, \ldots, \psi_s, \psi] \leftrightarrow (Q_K \overline{x}_1, \ldots, \overline{x}_s[\psi_1, \ldots, \psi_s] \wedge \forall \overline{x} \psi(\overline{x})).$$

(2) Let $r \geq \max\{r_i \mid 1 \leq i \leq s\}$ and let σ' arise from σ by replacing each R_i by an r-ary relation symbol R_i'. Finally, let K' be the class of σ'-structures $\mathcal{A}'$ that we get from structures $\mathcal{A}$ in K by replacing each R_i^A by $R_i'^{A'} := R_i^A \times A^{r-r_i}$. Using dummy variables, one easily shows that $\mathcal{L}(Q_K) \equiv \mathcal{L}(Q_{K'})$.

Theorem 12.1.4 *Let $\mathcal{L}$ be a logic. Furthermore, let $\mathbf{Q}$ be a finite set of Lindström quantifiers. Then there is a Lindström quantifier Q_K such that $\mathcal{L}(\mathbf{Q}) \equiv \mathcal{L}(Q_K)$. In particular, if the arities of the quantifiers in $\mathbf{Q}$ are $\leq l$, then Q_K can be chosen of arity $l + 1$.*

Proof. Let $\mathbf{Q}$ be $\{Q_{K_j} \mid 1 \leq j \leq k\}$ where K_j is a class of σ_j-structures. According to the preceding remarks we may assume that the σ_j have the same cardinality and that the Q_{K_j} are of the same arity l as are all relation symbols in the vocabularies σ_j. For simplicity of notation we treat the case of two simple binary quantifiers Q_{K_1} and Q_{K_2}. Let R be a ternary relation symbol. If no structure of the form $(A, \emptyset)$ belongs to K_1 or K_2, we define the class K of $\{R\}$-structures by

$$\begin{aligned} K \quad := \quad & \{(A, B \times \{a\}) \mid (A, B) \in K_1,\ a \in A\} \\ & \cup \{(A, B \times \{a, b\}) \mid (A, B) \in K_2,\ a, b \in A, a \neq b\}. \end{aligned}$$

Then the following formulas are equivalent on structures with at least two elements (structures of cardinality 1 can be taken into consideration explicitly):

(1) $Q_{K_1} xy\varphi(x, y)$ and $\exists v Q_K xyz(\varphi(x, y) \wedge z = v)$
(2) $Q_{K_2} xy\varphi(x, y)$ and $\exists v \exists w(v \neq w \wedge Q_K xyz(\varphi(x, y) \wedge (z = v \vee z = w)))$
(3) $Q_K xyz\psi(x, y, z)$ and $(\exists^{=1} z \exists x \exists y \psi(x, y, z) \wedge Q_{K_1} xy \exists z \psi(x, y, z))$

$$\vee (\exists^{=2} z \exists x \exists y \psi(x, y, z) \wedge \forall xy(\exists z \psi(x, y, z) \rightarrow \exists^{=2} z \psi(x, y, z))$$

$$\wedge\, Q_{K_2} xy \exists z \psi(x, y, z)).$$

If $K_1 \cup K_2$ contains structures of the form $(A, \emptyset)$, we add to K the structure $(A, A \times A \times \{a, b, c\})$ with distinct a, b, c if $(A, \emptyset) \in K_1$, and the structure $(A, A \times A \times \{a, b, c, d\})$ with distinct a, b, c, d if $(A, \emptyset) \in K_2$. Arguing similarly as above one shows that $\mathcal{L}(\{Q_{K_1}, Q_{K_2}\}) \equiv \mathcal{L}(Q_K)$. □

Exercise 12.1.5 Generalize the preceding theorem to a countable set $\mathbf{Q}$ of simple binary quantifiers, thereby weakening the assertion $\mathcal{L}(\mathbf{Q}) \equiv \mathcal{L}(Q_K)$ to $\mathcal{L}(\mathbf{Q}) \leq \mathcal{L}(Q_K)$. □

$\mathcal{A} \models [\mathrm{TC}_{x,y}\varphi(x, y)]st$ means that (s, t) is in the transitive closure of the binary relation on A given by $\varphi^{\mathcal{A}}(\_,\_)$. Similarly, for $k \geq 2$ and $\psi(x_1 \ldots x_k, y_1 \ldots y_k)$ we can read $\mathcal{A} \models [\mathrm{TC}_{\overline{x},\overline{y}}\psi(\overline{x}, \overline{y})]\,\overline{s}\,\overline{t}$ as "$(\overline{s}, \overline{t})$ is in the transitive closure of the *binary* relation on A^k given by $\psi^{\mathcal{A}}(\_,\_)$". Guided by this example we define the k-vectorization of a Lindström quantifier.

Let $\sigma = \{R_1, \ldots, R_s\}$ with r_i-ary R_i and let $k \geq 1$. Let $\sigma(k)$ arise from σ by replacing each relation symbol R in σ, say of arity r, by an $r \cdot k$-ary relation symbol $R(k)$. An r-ary relation on A^k can also be viewed as an $r \cdot k$-ary relation on A just by identifying the r-tuple consisting of the elements $\overline{a}_1, \ldots, \overline{a}_r$ of A^k with the $r \cdot k$-tuple $\overline{a}_1 \ldots \overline{a}_r$ of elements of A. Using this identification on the level of tuples and relations, we define:

Definition 12.1.6 Let K be a class of σ-structures.

(a) K^k, the *k-vectorization* of K, is the class of $\sigma(k)$-structures

$$K^k \quad := \quad \{(A, S_1, \ldots, S_s) \mid (A^k, S_1, \ldots, S_s) \in K\};$$

here, in $(A, S_1, \ldots, S_s)$, the relation S_i is viewed as an $r_i \cdot k$-ary relation over A, and in $(A^k, S_1, \ldots, S_s)$ it is viewed as an r_i-ary relation over A^k; in $(A, S_1, \ldots, S_s)$ it is the interpretation of $R_i(k)$, in $(A^k, S_1, \ldots, S_s)$ the interpretation of R_i.

(b) We set $\mathbf{Q}_K^{\omega} := \{Q_{K^k} \mid k \geq 1\}$. □

Thus, for the class K of Example 12.1.2(a) we have

$$\begin{aligned} \mathcal{A} \models Q_{K^2}xy\varphi(x, y) \quad & \text{iff} \quad (A^2, \varphi^{\mathcal{A}}(\_,\_)) \in K \\ & \text{iff} \quad \mathcal{A} \models \exists x \exists y \varphi(x, y), \end{aligned}$$

and for the class CONN (cf. 12.1.2(d))

$$\mathcal{A} \models Q_{\mathrm{CONN}^2}xx'yy'\varphi(xx', yy') \text{ iff } (A^2, \varphi^{\mathcal{A}}(\_,\_)) \text{ is a connected graph.}$$

Exercise 12.1.7 Let K be as in Example 12.1.2(e) and let $k \geq 1$. Generalize the equivalence $\mathrm{FO}(Q_K) \equiv \mathrm{FO}(\mathrm{TC}^1)$ to $\mathrm{FO}(Q_{K^k}) \equiv \mathrm{FO}(\mathrm{TC}^k)$. Hence, $\mathrm{FO}(\mathbf{Q}_K^{\omega}) \equiv \mathrm{FO}(\mathrm{TC})$. □

In view of 9.6.1 we have for the class K of the preceding exercise:

$$K \text{ but not } K^2 \text{ is axiomatizable in } \mathrm{FO}(\mathrm{TC}^1).$$

The following exercise shows that our standard logics are closed under vectorizations.

Exercise 12.1.8 Let $\mathcal{L}$ be one of the logics FO(TC), FO(DTC), FO(IFP), FO(PFP), or $L_{\infty\omega}$. Show: If K is axiomatizable in $\mathcal{L}$ and $k \geq 1$, then K^k is axiomatizable in $\mathcal{L}$. Conclude: If $\mathcal{L}_1$ with $\mathcal{L}_1 \leq \mathcal{L}$ is a Lindström extension of FO and K is axiomatizable in $\mathcal{L}$, then $\mathcal{L}_1(Q_K^\omega) \leq \mathcal{L}$. □

Two further logics which we have extensively studied can be obtained from FO by adding the vectorizations of a simple quantifier:

Theorem 12.1.9 *There are classes* I *and* P *of structures such that*

$$\text{FO(IFP)} \equiv \text{FO}(\mathbf{Q}_{\mathrm{I}}^\omega) \quad \textit{and} \quad \text{FO(PFP)} \equiv \text{FO}(\mathbf{Q}_{\mathrm{P}}^\omega).$$

Proof. We first consider FO(IFP) and recall (cf. 9.4.3) that any FO(IFP)-formula is equivalent to a formula of the form

$$\forall u[\text{IFP}_{\overline{x},X}(\psi \vee \exists \overline{y} \in X \exists \overline{z} \notin X\chi)]\,\tilde{u}$$

where ψ and χ are quantifier-free and X does not occur in ψ, χ. This normal form motivates the following definition of I.

Let $\sigma = \{U, R, V\}$ with ternary R and unary U, V and let I consist of the σ-structures $\mathcal{A}$ such that there is a sequence $A_0, A_1, A_2, \ldots$ of subsets of A with the following properties:

(1) $A_0 = \emptyset$ and $A_1 = U^{\mathcal{A}}$.
(2) For $i \geq 1$, $A_{i+1} = A_i \cup \{a \in A \mid R^{\mathcal{A}}abc$ for some $b \in A_i$ and $c \notin A_i\}$.
(3) There is an $i \geq 1$ such that $V^{\mathcal{A}} \subseteq A_i$.

To prove $\text{FO}(\mathbf{Q}_{\mathrm{I}}^\omega) \leq \text{FO(IFP)}$, by 12.1.8, it suffices to show that I is axiomatizable in FO(IFP). In fact, I is the class of models of φ_0 where

$$\varphi_0 \quad := \quad \forall u(Vu \rightarrow [\text{IFP}_{x,X}(Ux \vee \exists y \in X \exists z \notin X Rxyz)]\,u).$$

For the other direction we consider an FO(IFP)-sentence φ in the normal form mentioned above, say

$$\varphi = \forall u\,[\text{IFP}_{\overline{x},X}(\psi(\overline{x}) \vee \exists \overline{y} \in X \exists \overline{z} \notin X\chi(\overline{x},\overline{y},\overline{z}))]\,\tilde{u},$$

where X is k-ary. Given a structure $\mathcal{A}$, let $X_0, X_1, X_2, \ldots$ be the stages of the fixed-point operation in φ. Then $X_1 = \psi^{\mathcal{A}}(\_)$, and $\mathcal{A}$ is a model of φ iff $\{\tilde{a} \mid a \in A\} \subseteq X_\infty$. Hence, φ is equivalent to

$$Q_{\mathrm{I}^k}\overline{x}, \overline{x}\,\overline{y}\,\overline{z}, \overline{u}\,[\psi(\overline{x}), \chi(\overline{x},\overline{y},\overline{z}), u_1 = \ldots = u_k].$$

The proof for FO(PFP) is similar; one uses the class P that is defined as I with (2), (3) modified to

(2)′ For $i \geq 1$, $A_{i+1} = \{a \in A \mid R^{\mathcal{A}}abc$ for some $b \in A_i$ and $c \notin A_i\}$.
(3)′ $V^{\mathcal{A}} = \emptyset$ or there is an $i \geq 1$ such that $V^{\mathcal{A}} \subseteq A_i$ and $A_i = A_{i+1}$. □

We finish by representing FO(ATC) as a Lindström extension of FO. Setting $\sigma := \{E, U, R\}$ with binary E, R and unary U, we define A to be the class of σ-structures (A, E^A, U^A, R^A) such that R^A is contained in the alternating transitive closure of (E^A, U^A). Then one easily gets:

Theorem 12.1.10 FO(ATC) $\equiv$ FO($\mathbf{Q}^{\omega}_{\mathrm{A}}$). □

Notes 12.1.11 At many places of this chapter we tacitly assume that $\mathcal{L}$ is a logic extending FO and satisfying certain closure conditions such as closure under negations or conjunctions (cf. (3) in the proof of Theorem 12.1.4). For instance, closure of $\mathcal{L}$ under conjunctions means:

> If φ and ψ are $\mathcal{L}$-formulas then $(\varphi \wedge \psi)$ is an $\mathcal{L}$-formula (or, at least, $\mathcal{L}$ contains a formula equivalent to $(\varphi \wedge \psi)$).

Compare [31] for the precise statement and an analysis of such closure conditions.

12.2 PTIME and Quantifiers

In this section we shall see that no extension of FO(IFP) by means of finitely many Lindström quantifiers strongly captures PTIME. The same is true for all fragments of the extension of $\mathrm{L}^{\omega}_{\infty\omega}$ by all counting quantifiers.

For a set $\mathbf{Q}$ of Lindström quantifiers, $\mathrm{L}^{\omega}_{\infty\omega}(\mathbf{Q})$ consists of those formulas of $\mathrm{L}_{\infty\omega}(\mathbf{Q})$ that contain only finitely many variables. As a first result we have

Theorem 12.2.1 *Let $\mathbf{Q}$ be a finite set of Lindström quantifiers Q_L with $L_<$ in* PTIME. *Then for any relational vocabulary τ there is a class K of τ-structures such that $K_<$ belongs to* PTIME, *but K is not axiomatizable in* $\mathrm{L}^{\omega}_{\infty\omega}(\mathbf{Q})$.

The proof will be given below. As FO(IFP)($\mathbf{Q}$) $\leq \mathrm{L}^{\omega}_{\infty\omega}(\mathbf{Q})$ (mimic the proof of FO(PFP) $\leq \mathrm{L}^{\omega}_{\infty\omega}$, cf. 8.4.2) we have

Corollary 12.2.2 *No extension of* FO(IFP) *by means of finitely many Lindström quantifiers strongly captures* PTIME. □

To give an application of 12.2.1, let E be binary and $\mathbf{Q}$ be the set of all simple binary quantifiers corresponding to classes K of $\{E\}$-structures with $K_<$ in PTIME. According to 12.1.5, $\mathbf{Q}$ may be replaced by one (ternary) quantifier Q_L such that FO($\mathbf{Q}$) $\leq$ FO(Q_L). Therefore, FO(Q_L) and hence $\mathrm{L}^{\omega}_{\infty\omega}(Q_L)$, is able to axiomatize all the classes K of $\{E\}$-structures with $K_<$ in PTIME. By the theorem, $L_<$ cannot be in PTIME, that is, FO(Q_L) is strictly stronger than FO($\mathbf{Q}$) (cf. 12.1.3).

The theorem has a generalization to certain infinite sets of quantifiers. Without proof we mention the following generalization to arity-bounded sets of quantifiers:

Theorem 12.2.3 *Let $n \geq 1$ and let $\mathbf{Q}$ consist of Lindström quantifiers of arity $\leq n$. Then there is a relational vocabulary τ and a class K of τ-structures such that $K_<$ is in* PTIME, *but K is not axiomatizable in $L^\omega_{\infty\omega}(\mathbf{Q})$.*

Corollary 12.2.4 *No extension of* FO(IFP) *by means of Lindström quantifiers of bounded arity strongly captures* PTIME. □

Since the first examples of PTIME classes not axiomatizable in FO(IFP) were related to the class EVEN of structures of even cardinality, the logic FO(IFP, #), fixed-point logic with counting, appeared as a natural candidate for a logic strongly capturing PTIME. However: Let $\mathbf{C}$ be the set of counting quantifiers, i.e., $\mathbf{C} := \{Q_{K_n} \mid n \geq 1\}$, where K_n is as in 12.1.2(c). Then we have

Theorem 12.2.5 *There is a class K of graphs such that $K_<$ is in* PTIME, *but K is not axiomatizable in $L^\omega_{\infty\omega}(\mathbf{C})$.*

Note that $L^\omega_{\infty\omega}(\mathbf{C}) \equiv C^\omega_{\infty\omega}$. By 8.4.18, FO(IFP, #) $\leq C^\omega_{\infty\omega}$. Therefore:

Corollary 12.2.6 FO(IFP, #) *does not strongly capture* PTIME. □

In the following we prove Theorem 12.2.1 and Theorem 12.2.5.

Proof (of 12.2.1). First we show that for $\tau = \emptyset$ there is a class K of structures such that $K_<$ is in PTIME, but K is not axiomatizable in $L^\omega_{\infty\omega}(\mathbf{Q})$.

So let $\tau = \emptyset$ and fix $s \geq 1$. Thus τ-structures are sets and will be denoted by their universes. We start by proving that every $L^\omega_{\infty\omega}(\mathbf{Q})$-sentence (of vocabulary $\tau = \emptyset$ and) with variables among $v_1, \ldots, v_s$ is equivalent to an FO($\mathbf{Q}$)-sentence of a certain simple form.

First we note that there are, up to logical equivalence, only finitely many quantifier-free formulas with variables among $v_1, \ldots, v_s$ and hence, only finitely many formulas $\psi^s_1, \ldots, \psi^s_{l_s}$ that are strings of the form $Q\alpha$ with $Q \in \mathbf{Q} \cup \{\forall, \exists\}$, where α does not contain quantifiers and has variables among $v_1, \ldots, v_s$.

Given an s-tupel $\overline{a} = a_1 \ldots a_s$ of elements of a structure A, the *equality type* of $\overline{a}$ is, by definition, the formula

$$\bigwedge\{v_i = v_j \mid a_i = a_j, 1 \leq i < j \leq s\} \wedge \bigwedge\{\neg v_i = v_j \mid a_i \neq a_j, 1 \leq i < j \leq s\}.$$

For any s-tuples $\overline{a}, \overline{b} \in A$ of the same equality type there is a permutation (i.e., an automorphism) of A that maps $\overline{a}$ onto $\overline{b}$. Hence, for $n \geq 0$, every $L^\omega_{\infty\omega}(\mathbf{Q})$-formula with variables among $v_1, \ldots, v_s$ is equivalent in $A_n := \{0, 1, \ldots, n\}$ to a disjunction of equality types (the disjunction is finite as there are only finitely many equality types of s-tuples). In particular, there are disjunctions $\chi^{(n,s)}_i$ of equality types of s-tuples such that

$$(1) \qquad A_n \models \bigwedge\nolimits_{1 \leq i \leq l_s} \forall v_1 \ldots \forall v_s (\psi^s_i \leftrightarrow \chi^{(n,s)}_i),$$

that is, we have a "local quantifier elimination". Now, a simple induction on formulas shows:

(2) For all $\varphi \in L^{\omega}_{\infty\omega}(\mathbf{Q})$ with variables among $v_1, \dots, v_s$ there is a quantifier-free first-order formula $\chi(v_1, \dots, v_s)$ (and hence, a boolean combination of equality types of s-tuples) such that

$$\bigwedge_{1 \le i \le l_s} \forall v_1 \dots \forall v_s(\psi_i^s \leftrightarrow \chi_i^{(n,s)}) \models \forall v_1 \dots \forall v_s(\varphi \leftrightarrow \chi).$$

We set

$$\delta^{(n,s)} := \begin{cases} \text{“there are exactly } n \text{ elements”}, & \text{if } n < s \\ \text{“there are at least } s \text{ elements”}, & \text{if } n \ge s \end{cases}$$

and

$$\varphi^{(n,s)} \quad := \quad \delta^{(n,s)} \wedge \bigwedge_{1 \le i \le l_s} \forall v_1 \dots \forall v_s(\psi_i^s \leftrightarrow \chi_i^{(n,s)}).$$

Clearly, $A_n \models \varphi^{(n,s)}$ and $\{\varphi^{(n,s)} \mid n \ge 1\}$ is finite. Now we can show:

(3) If $B \models \varphi^{(n,s)}$ then for all sentences φ of $L^{\omega}_{\infty\omega}(\mathbf{Q})$ with variables among $v_1, \dots, v_s$,

$$B \models \varphi \quad \text{iff} \quad A_n \models \varphi.$$

In fact, if $\|B\| < s$ then, by $B \models \delta^{(n,s)}$, we have $n = \|B\|$ and hence, $B \cong A_n$; so (3) is trivially true. If $\|B\| \ge s$ then $n \ge s$. For a sentence φ as in (3) choose a boolean combination $\chi(v_1, \dots, v_s)$ of equality types of s-tuples according to (2). By (1), $A_n \models \bigwedge_{1 \le i \le l_s} \forall v_1 \dots \forall v_s(\psi_i^s \leftrightarrow \chi_i^{(n,s)})$ and by $B \models \varphi^{(n,s)}$, also $B \models \bigwedge_{1 \le i \le l_s} \forall v_1 \dots \forall v_s(\psi_i^s \leftrightarrow \chi_i^{(n,s)})$. Choose $\overline{a} \in A_n^s$ and $\overline{b} \in B^s$ of the same equality type (recall $\|A_n\|, \|B\| \ge s$). Then

$$\begin{aligned} B \models \varphi \quad &\text{iff} \quad B \models \chi[\overline{b}] \\ &\text{iff} \quad A_n \models \chi[\overline{a}] \\ &\text{iff} \quad A_n \models \varphi. \end{aligned}$$

Since $A_n \models \varphi^{(n,s)}$ and every B is isomorphic to some A_n, we see, using (3), that

$$\models_{\text{fin}} \varphi \leftrightarrow \bigvee \{\varphi^{(n,s)} \mid n \ge 1,\, A_n \models \varphi\}$$

holds for any sentence φ of $L^{\omega}_{\infty\omega}(\mathbf{Q})$ with variables among $v_1, \dots, v_s$. As the disjunction on the right side is finite, φ is equivalent to an $\mathrm{FO}(\mathbf{Q})$-sentence.

Altogether, every $L^{\omega}_{\infty\omega}(\mathbf{Q})$-sentence is equivalent to an $\mathrm{FO}(\mathbf{Q})$-sentence ψ of the form

(4) $$\psi \ := \ \bigvee_{j \in J} (\delta_j \wedge \bigwedge_{i \in I_j} \forall v_1 \dots \forall v_s(\psi_i \leftrightarrow \chi_i)),$$

where $s \ge 1$, the χ_i are quantifier-free, and the δ_j and the ψ_i are of the form $\delta^{(n,s)}$ and ψ_i^s, respectively. Let Ψ be the set of these formulas.

Using a diagonalization procedure, we show that there is a class K of sets such that $K_<$ is in PTIME, but K is not axiomatizable by a sentence in Ψ.

For a sentence ψ as in (4) we set

$$\|\psi\| := \text{length of } \psi, \quad s_\psi := s.$$

It is not difficult to define a uniform procedure that, given n and ψ, checks $A_n \models \psi$ in time polynomial in n, $\|\psi\|$, and 2^{s_ψ}. The main point here is that in order to check the validity of $A_n \models \forall v_1 \ldots \forall v_{s_\psi}(\psi_i \leftrightarrow \chi_i)$, it suffices to test the validity of $A_n \models (\psi_i \leftrightarrow \chi_i)[\overline{a}]$ for exactly one s_ψ-tuple $\overline{a}$ of each equality type in A_n. This gives a corresponding procedure, because the number of equality types is $O(2^{s_\psi})$ and the (finitely many) quantifiers in $\mathbf{Q} \cup \{\forall, \exists\}$ belong to classes K with $K_<$ in PTIME.

We now consider a function $f : \mathbb{N} \to \Psi$ computable in time polynomial in n that enumerates Ψ in such a way that

$$\|f(n)\| = O(n) \quad \text{and} \quad s_{f(n)} = O(\log n)$$

(it is not hard to see that such an enumeration exists). We then define K by

(5) $\qquad B \in K \quad \text{iff} \quad \text{for some } n, \ \|B\| = n \text{ and } A_n \not\models f(n).$

By the preceding remarks, $K_<$ is in PTIME; however, by (5), K is not axiomatized by an $f(n)$, that is, K is not axiomatizable by any sentence in Ψ and hence, not by an $L^\omega_{\infty\omega}(\mathbf{Q})$-sentence.

Now, let τ be an arbitrary vocabulary and let C_τ be the class of structures $\mathcal{A}$ such that $R^\mathcal{A}\overline{a}$ for all $R \in \tau$ and $\overline{a} \in A$. As in structures of C_τ any atomic formula $R\overline{t}$ is equivalent to $t_1 = t_1$, the proof above for $\tau = \emptyset$ works for this case too, yielding a class $K \subseteq C_\tau$ such that $K_<$ is in PTIME, but K is not axiomatizable in $L^\omega_{\infty\omega}(\mathbf{Q})$. □

We now prove Theorem 12.2.5. For this purpose we show that for all $s \geq 3$ there are graphs $\mathcal{G}_s$ and $\tilde{\mathcal{G}}_s$ such that the following holds:

(1) $\mathcal{G}_s \equiv_{L^s_{\infty\omega}(\mathbf{C})} \tilde{\mathcal{G}}_s$.

(2) There is a class K of graphs such that $K_< \in$ PTIME and K contains all $\mathcal{G}_s$ but no $\tilde{\mathcal{G}}_s$.

This yields the theorem, because, by (1), no $L^\omega_{\infty\omega}(\mathbf{C})$-sentence φ can axiomatize a class K as in (2).

In section 3.4 we have denoted $L^s_{\infty\omega}(\mathbf{C})$ by $C^s_{\infty\omega}$ and shown that (1) is equivalent to

(1′) $\qquad$ the duplicator wins $\text{C-G}^s_\infty(\mathcal{G}_s, \tilde{\mathcal{G}}_s)$

where $\text{C-G}^s_\infty(\_,\_)$ is the Ehrenfeucht game corresponding to $C^s_{\infty\omega}$.

We define the graphs $\mathcal{G}_s$ and $\tilde{\mathcal{G}}_s$. Let $k := s + 2$. Then $k \geq 5$. Take sets

$$A := \{a_1, \ldots, a_k\}, \quad B := \{b_1, \ldots, b_k\}, \quad C := \{c_1, \ldots, c_k\}$$

of distinct elements, let

$$D := \{S \subseteq \{1, \ldots, k\} \mid \|S\| \text{ is even}\},$$

and define the graph $\mathcal{H} := (H, E)$ by

$$\begin{aligned} H &:= A \cup B \cup C \cup D \\ E &:= \{(a_i, c_i), (c_i, a_i), (b_i, c_i), (c_i, b_i) \mid 1 \le i \le k\} \\ &\quad \cup \{(a_i, S), (S, a_i) \mid S \in D, i \in S\} \\ &\quad \cup \{(b_i, S), (S, b_i) \mid S \in D, i \notin S\} \\ &\quad \cup \{(c_i, c_{i+1}), (c_{i+1}, c_i) \mid 1 \le i < k\} \cup \{(c_k, c_1), (c_1, c_k)\}. \end{aligned}$$

In $\mathcal{H}$, the a_i and b_i have degree $2^{k-2}+1$, being connected with c_i and exactly 2^{k-2} points in D; similarly, the $S \in D$ have degree k and the c_i degree 4. (As $k \ge 5$, these degrees are different.) Therefore, any automorphism π of $\mathcal{H}$ maps the sets $A \cup B$, C, and D onto themselves, and if π is the identity on C then $\pi(a_i), \pi(b_i) \in \{a_i, b_i\}$.

Lemma 12.2.7 *Let π be a permutation of $A \cup B \cup C$ which is the identity on C. Then π can be extended to an automorphism of $\mathcal{H}$ iff π interchanges a_i and b_i for an even number of i's, leaving the other elements in $A \cup B$ fixed.*

Proof. If π interchanges a_i and b_i for $i \in S_0$, then for any automorphism $\tilde{\pi}$ of $\mathcal{H}$ that extends π, and for any $S \in D$ we have

$$\tilde{\pi}(S) = (S \setminus S_0) \cup (S_0 \setminus S) \tag{3}$$

which is in D just in case S_0 is even. If S_0 is even, (3) gives an extension of π which is an automorphism of $\mathcal{H}$. □

Let V be a set of $s+3$ $(= k+1)$ elements. We define $\mathcal{H}(V) = (H(V), E^{H(V)})$ as follows: We replace each $v \in V$ by a copy $\mathcal{H}(v)$ of $\mathcal{H}$, where $a_i(v), b_j(v), \ldots$ correspond to $a_i, b_j, \ldots$. Moreover, if $v, w \in V$ and $v \neq w$, we add edges

$$\begin{aligned} &(a(v,w), a(w,v)),\ (a(w,v), a(v,w)), \\ &(b(v,w), b(w,v)),\ (b(w,v), b(v,w)) \end{aligned} \tag{$*$}$$

where

$$\begin{aligned} &\text{for some } i,\ a(v,w) = a_i(v) \text{ and } b(v,w) = b_i(v), \\ &\text{for some } j,\ a(w,v) = a_j(w) \text{ and } b(w,v) = b_j(w). \end{aligned}$$

We do this in such a way that for every v and i there are uniquely determined w and j such that $a_i(v)$ is linked to $a_j(w)$ (and hence, $b_i(v)$ is linked to $b_j(w)$).

For $v_0, w_0 \in V$, $v_0 \neq w_0$, we define the "twisted" version $\mathcal{H}(V)^{v_0,w_0}$ just as $\mathcal{H}(V)$, but replacing $(*)$ – only for the points v_0, w_0 – by

$$(a(v_0, w_0), b(w_0, v_0)),\ (b(w_0, v_0), a(v_0, w_0)),$$
$$(b(v_0, w_0), a(w_0, v_0)),\ (a(w_0, v_0), b(v_0, w_0)).$$

Fix $v_0, w_0 \in V$, $v_0 \neq w_0$. Set

$$\mathcal{G}_s := \mathcal{H}(V), \quad \tilde{\mathcal{G}}_s := \mathcal{H}(V)^{v_0, w_0}.$$

We show that (1′) and (2) are satisfied. First we prove that the duplicator wins the game C-$\mathrm{G}^s_\infty(\mathcal{G}_s, \tilde{\mathcal{G}}_s)$. Obviously, it is enough to show that he can play in such a way that, after his moves, the following holds:

(+) There is a $v \in V$ such that $\mathcal{H}(v)$ (in $\mathcal{G}_s$) is not pebbled and there is an isomorphism $\pi : \mathcal{G}_s^{v,w_0} \cong \tilde{\mathcal{G}}_s$ which respects the pebbles and maps elements of $H(u)$ to elements of $H(u)$ for all $u \in V$.

In the beginning we take $v := v_0$ and π the identity on $H(V)$. Now assume that, before a move of the spoiler, we have v and π according to (+). Let v' be a further point of V different from w_0 such that $\mathcal{H}(v')$ (in $\mathcal{G}_s$) is not pebbled. Using 12.2.7, choose an automorphism ρ of $\mathcal{H}(v)$ that interchanges $a(v, v')$ and $b(v, v')$ and interchanges $a(v, w_0)$ and $b(v, w_0)$; similarly, choose an automorphism ρ' of $\mathcal{H}(v')$ interchanging $a(v', v)$ and $b(v', v)$, and $a(v', w_0)$ and $b(v', w_0)$. Let σ be the permutation of $H(V)$ that extends $\rho \cup \rho'$ and is the identity outside $H(v) \cup H(v')$. Then it is easy to check that

$$\sigma : \mathcal{G}_s^{v', w_0} \cong \mathcal{G}_s^{v, w_0},$$

i.e., σ shifts the twist from $\{v', w_0\}$ to $\{v, w_0\}$.

Now, if the spoiler chooses a subset X, say, of $\mathcal{G}_s$, then the duplicator answers by the subset Y of $\tilde{\mathcal{G}}_s$ consisting of the π-images of the elements in $X \setminus H(v)$ and of the $\pi \circ \sigma$-images of the elements in $X \cap H(v)$. If then the spoiler pebbles an element of Y, the duplicator answers with the corresponding preimage. If $\mathcal{H}(v)$ remains unpebbled, (+) is again satisfied by v and π; otherwise, it is satisfied by v' and $\pi \circ \sigma : \mathcal{G}_s^{v', w_0} \cong \tilde{\mathcal{G}}_s$.

Before we prove (2), we provide some more information about the graphs $\mathcal{G}_s$ and $\tilde{\mathcal{G}}_s$. $\tilde{\mathcal{G}}_s$ arises from $\mathcal{G}_s$ by a twist corresponding to the pair $\{v_0, w_0\}$. Similarly, we can define twisted versions of $\mathcal{G}_s$ that have twists corresponding to several pairs. We call $\mathcal{G}$ an m-twisted version of $\mathcal{G}_s$, if there are exactly m twists of this kind. So $\mathcal{G}_s$ is a 0-twisted version of $\mathcal{G}_s$ and $\tilde{\mathcal{G}}_s$ a 1-twisted one.

Claim. Let $\mathcal{G}$ and $\mathcal{G}'$ be m-twisted and m'-twisted versions of $\mathcal{G}_s$, respectively. Then

$$\mathcal{G} \cong \mathcal{G}' \text{ iff } m - m' \text{ is even.}$$

Proof (of Claim). Let $\pi : \mathcal{G} \cong \mathcal{G}'$. Then, given $v \in V$, π maps $\mathcal{H}(v)$ isomorphically onto some copy $\mathcal{H}(w)$. By (a simple generalization of) 12.2.7, the number of i's such that $\pi(a_i(v)) \in \{b_1(w), \ldots, b_k(w)\}$ is even. This yields that $m - m'$ is even. Now, let $m - m'$ be even. In the proof of (+) we saw how one can shift twists by suitable isomorphisms, say, a twist corresponding to a pair $\{v', w\}$ to a twist corresponding to a pair $\{v, w\}$. Hence, we may assume that the twists of $\mathcal{G}$ and $\mathcal{G}'$ correspond to pairs that have w_0 as one point. (Note that it may be necessary to shift a twist upon a twist, thus replacing m by $m - 2$ (or, m' by $m' - 2$).) Then by 12.2.7, as $m - m'$ is even, we can find an automorphism of $H(w_0)$ that can be extended to an isomorphism of $\mathcal{G}$ onto $\mathcal{G}'$.

We now come to a proof of (2). For this purpose we describe a deterministic polynomially time-bounded Turing machine M that accepts all $\mathcal{G}_s$ for $s \geq 3$, but no $\tilde{\mathcal{G}}_s$.

Started with (the ordered version of) a graph $\mathcal{G}$, M first checks whether there is a number $s \geq 3$ such that $\mathcal{G}$ contains exactly $s+3$ cycles $C_1, \dots, C_{s+3}$ of length $s + 2$, consisting of points of degree 4 (so-called *c-points*), and whether $\mathcal{G}$ contains no other points of degree 4. If so, M checks

(i) whether each c-point has exactly two neighbours that are no c-points (so-called *ab-points*; different ab-points with the same c-neighbour are called *related*);
(ii) for each pair $\{a, b\}$ of related ab-points, whether a and b are neighbours of exactly one ab-point a' and b', respectively, a' and b' being related via a c-point that belongs to a circle different from that of the c-point that relates a to b;
(iii) for each circle C_i, whether there are exactly 2^{s+1} points (so-called *d-points*) that have exactly $s + 2$ neighbours, these neighbours being ab-points that are related to a c-point in C_i (we let H_i be the set of these d-points, the neighbouring ab-points, and the elements of C_i);
(iv) whether there are no other points or edges.

Finally, M checks for each H_i whether the subgraph $\mathcal{H}_i$ of $\mathcal{G}$ with domain H_i is isomorphic to $\mathcal{H}$, in the positive case also producing a partition of the ab-points into a-points and b-points. (This can be done in a time polynomial in 2^s and hence, polynomial in $||G||$). If all checks have a positive outcome, M counts the number of twists of $\mathcal{G}$. If this number is even, M accepts, otherwise it rejects. By the claim above, M accepts the $\mathcal{G}_s$ for $s \geq 3$, but rejects the $\tilde{\mathcal{G}}_s$. □

12.3 Logical Reductions

In complexity theory there are well-established notions of reducibility, among them the notion of many-one reducibility: Given alphabets $\mathbb{A}_i^+$, languages $L_i \subseteq \mathbb{A}_i^+$ $(i = 1, 2)$, and a deterministic complexity class $\mathcal{C}$, the language L_1 is *$\mathcal{C}$-reducible* (or, *$\mathcal{C}$-many-one reducible*) to L_2 if there is a function $f : \mathbb{A}_1^* \to \mathbb{A}_2^*$ in $\mathcal{C}$ such that

$$w \in L_1 \text{ iff } f(w) \in L_2$$

for all $w \in \mathbb{A}_1^*$. Given a further complexity class $\mathcal{C}'$, a language L is *$\mathcal{C}'$-hard with respect to $\mathcal{C}$-reductions* if any language in $\mathcal{C}'$ is $\mathcal{C}$-reducible to L, and L is *$\mathcal{C}'$-complete with respect to $\mathcal{C}$-reductions* if it is $\mathcal{C}'$-hard and belongs to $\mathcal{C}'$.

We aim at a logical analogue. For this purpose we need the notion of an interpretation as introduced in section 11.2. For convenience, we shortly repeat the definitions and notations.

Fix a logic $\mathcal{L}$ and vocabularies σ and τ where $\sigma = \{R_1, \ldots, R_s\}$ with r_i-ary R_i. An $\mathcal{L}$-interpretation of σ in τ of width k is given by an $(s+1)$-tuple of $\mathcal{L}[\tau]$-formulas

$$\Pi \quad := \quad (\pi_{\text{uni}}(\overline{x}), \pi_{R_1}(\overline{x}_1, \ldots, \overline{x}_{r_1}), \ldots, \pi_{R_s}(\overline{x}_1, \ldots, \overline{x}_{r_s})),$$

where all tuples $\overline{x}, \overline{x}_1, \ldots$ are of length k.

For any τ-structure $\mathcal{A}$, Π induces a σ-structure $\mathcal{A}^{\Pi}$ over k-tuples of A given by

$$\mathcal{A}^{\Pi} \quad = \quad (\pi_{\text{uni}}^{\mathcal{A}}(\_), \pi_{R_1}^{\mathcal{A}}(\_, \ldots, \_), \ldots, \pi_{R_s}^{\mathcal{A}}(\_, \ldots, \_)),$$

provided $\pi_{\text{uni}}^{\mathcal{A}}(\_)$ is nonempty.[2] So any $\mathcal{L}$-interpretation Π of σ in τ induces a map

$$^{\Pi} : \text{Str}[\tau] \to \text{Str}[\sigma].$$

Exercise 12.3.1 Let Π_1 be an $\mathcal{L}$-interpretation of σ in τ and Π_2 be an $\mathcal{L}$-interpretation of τ in ρ. Show that there is an $\mathcal{L}$-interpretation Π of σ in ρ such that for all ρ-structures $\mathcal{A}$,

$$\mathcal{A}^{\Pi} \;=\; (\mathcal{A}^{\Pi_2})^{\Pi_1}. \qquad \square$$

Taking the maps $^{\Pi}$ as candidates for the maps linked to logical reducibility, we are led to the following analogue of the notion of reducibility in complexity theory.

Definition 12.3.2 Let $K \subseteq \text{Str}[\tau]$ and $L \subseteq \text{Str}[\sigma]$. K is *$\mathcal{L}$-reducible* to L, for short: $K \leq_{\mathcal{L}} L$, if there is an $\mathcal{L}$-interpretation Π of σ in τ such that for all τ-structures $\mathcal{A}$,

$$\mathcal{A} \in K \quad \text{iff} \quad \mathcal{A}^{\Pi} \in L.$$

We write $K \leq_{\mathcal{L}}^k L$ to indicate that $K \leq_{\mathcal{L}} L$ can be witnessed by an interpretation of width k. $\square$

Exercise 12.3.3 Show that $\leq_{\mathcal{L}}$ is transitive. $\square$

---

[2] As in section 11.2, we tacitly ignore the case $\pi_{\text{uni}}^{\mathcal{A}}(\_) = \emptyset$ as it will not be of importance; moreover, note that $\pi_{R_i}^{\mathcal{A}}(\_, \ldots, \_)$ stands for $\pi_{R_i}^{\mathcal{A}}(\_, \ldots, \_) \cap (\pi_{\text{uni}}^{\mathcal{A}}(\_))^{r_i}$.

Exercise 12.3.4 Let $K \subseteq \mathrm{Str}[\tau]$ be a class of ordered structures and $L \subseteq \mathrm{Str}[\sigma]$. Show that $K \leq_{\mathcal{L}} L$ implies $K \leq_{\mathcal{L}} L_<$. □

An important example of first-order reductions is given in the following consequence of 11.2.5. In fact, taking as L the class K^{Π} (Π as in 11.2.5) we have

Proposition 12.3.5 *Let τ be an arbitrary vocabulary. Then, for all $K \subseteq$ Str$[\tau]$ there is a class $L \subseteq$ GRAPH such that $K \leq_{\mathrm{FO}} L$. Moreover, L can be chosen such that $K_< \in$ PTIME iff $L_< \in$ PTIME.* □

We study the relationship between reductions and Lindström quantifiers. First, there is a difference: Interpretations Π contain a formula $\pi_{\mathrm{uni}}(\overline{x})$ giving the new universe, whereas Lindström quantifiers do not allow to "relativize", they always refer to the whole universe of the structure in question. To overcome this difference, we proceed as follows:

Let $\sigma = \{R_1, \ldots, R_s\}$ with r_i-ary R_i and let U be a new unary relation symbol. For $L \subseteq \mathrm{Str}[\sigma]$ we introduce the class rel-$L \subseteq \mathrm{Str}[\sigma \cup \{U\}]$ by

$$\text{rel-}L \;:=\; \{(A, U^A, R_1^A, \ldots, R_s^A) \mid (U^A, R_1^A, \ldots, R_s^A) \in L\};$$

more precisely, rel-L is the class

$$\{(A, U^A, R_1^A, \ldots, R_s^A) \mid (U^A, R_1^A \cap (U^A)^{r_1}, \ldots, R_s^A \cap (U^A)^{r_s}) \in L\}.$$

For $k \geq 1$ set $\sigma(k) := \{R_1(k), \ldots, R_s(k)\}$ with $r_i{\cdot}k$-ary $R_i(k)$ and let $U(k)$ be an k-ary relation symbol. Identifying $r{\cdot}k$-tuples over A with r-tuples over A^k as in section 12.1, we "relativize" the definition of vectorization (cf. 12.1.6) as follows:

Definition 12.3.6 Let $L \subseteq \mathrm{Str}[\sigma]$.

(a) rel-k-L, the *relativized k-vectorization of L*, is the following class of $\{U(k)\} \cup \sigma(k)$-structures:

$$\begin{aligned} \text{rel-}k\text{-}L \;&:=\; \{(A, W, S_1, \ldots, S_s) \mid (A^k, W, S_1, \ldots, S_s) \in \text{rel-}L\} \\ (&=\; \{(A, W, S_1, \ldots, S_s) \mid W \subseteq A^k \text{ and } (W, S_1, \ldots, S_s) \in L\}). \end{aligned}$$

Note that rel-L is rel-1-L.

(b) The Lindström quantifier $Q_{\text{rel-}k\text{-}L}$ is called the *relativized k-vectorization* of the quantifier Q_L. We set rel-ω-$\mathbf{Q}_L := \{Q_{\text{rel-}k\text{-}L} \mid k \geq 1\}$. □

Example 12.3.7 For the class CONN of connected graphs we have

$$\mathcal{A} \models Q_{\text{rel-}k\text{-CONN}}\overline{x}, \overline{x}\,\overline{y}[\varphi(\overline{x}), \psi(\overline{x}, \overline{y})] \;\text{ iff }\; (\varphi^{\mathcal{A}}(\_), \psi^{\mathcal{A}}(\_,\_)) \in \text{CONN}. \quad □$$

As a simple observation we have (compare 12.1.6 for the definition of L^k):

Proposition 12.3.8 *Let L be a class of structures and $\mathcal{L}$ a logic. Then for all $k \geq 1$,*

$$\mathcal{L}(Q_{L^k}) \leq \mathcal{L}(Q_{\text{rel-}k\text{-}L}).$$

Proof. If, for instance, $k = 2$ and L is a class of $\{R\}$-structures with r-ary R, the formulas

$$Q_{L^2} x_1 x_1' \ldots x_r x_r' \psi(x_1 x_1', \ldots, x_r x_r'),$$
$$Q_{\text{rel-2-}L} x x'\, x_1 x_1' \ldots x_r x_r' [x = x \wedge x' = x', \psi(x_1 x_1', \ldots, x_r x_r')]$$

are equivalent. □

The direction " $\geq$ " may be wrong:

Exercise 12.3.9 Let P be unary and $L := \{(A, P^A) \mid \|P^A\| \leq \frac{1}{2}\|A\|\}$. Show that

$$\mathcal{A} \models Q_{\text{rel-}L} x, y\, [\varphi(x), \psi(y)] \quad \text{iff} \quad \varphi^{\mathcal{A}}(\_) \neq \emptyset \text{ and } \|\psi^{\mathcal{A}}(\_) \cap \varphi^{\mathcal{A}}(\_)\| \leq \tfrac{1}{2}\|\varphi^{\mathcal{A}}(\_)\|$$

and that $\mathrm{FO}(Q_L) < \mathrm{FO}(Q_{\text{rel-}L})$. □

Exercise 12.3.10 Let I and P be the classes introduced in 12.1.9. Show that $\mathrm{FO}(\text{rel-}\omega\text{-}\mathbf{Q}_{\mathrm{I}}) \equiv \mathrm{FO}(\mathbf{Q}_{\mathrm{I}}^{\omega}) \equiv \mathrm{FO}(\mathrm{IFP})$ and $\mathrm{FO}(\text{rel-}\omega\text{-}\mathbf{Q}_{\mathrm{P}}) \equiv \mathrm{FO}(\mathbf{Q}_{\mathrm{P}}^{\omega}) \equiv \mathrm{FO}(\mathrm{PFP})$.

Exercise 12.3.11 Prove $\mathcal{L}(\text{rel-}\omega\text{-}\mathbf{Q}_L) \equiv \mathcal{L}(\text{rel-}\omega\text{-}\mathbf{Q}_{L_<})$ on ordered structures (compare 12.1.2(f)). □

Exercise 12.3.12 Let K be a class such that $K_< \in \mathrm{PTIME}$. Show that $\mathrm{Mod}(\psi)_< \in \mathrm{PTIME}$ for any sentence $\psi \in \mathrm{FO}(\mathrm{IFP})(\text{rel-}\omega\text{-}\mathbf{Q}_K)$. Moreover show: If $\mathrm{FO}(\mathrm{IFP})(\text{rel-}\omega\text{-}\mathbf{Q}_K)$ or $\mathrm{FO}(\text{rel-}\omega\text{-}\mathbf{Q}_K)$ strongly captures PTIME, then it effectively strongly captures PTIME. □

For $\mathcal{L}$ a logic and $\mathbf{Q}$ a set of Lindström quantifiers, we denote by $\mathcal{L}[\mathbf{Q}]$ the set of formulas of $\mathcal{L}(\mathbf{Q})$ that are of the form

$$Q\overline{x}_1, \ldots, \overline{x}_s[\psi_1(\overline{x}_1), \ldots, \psi_s(\overline{x}_s)]$$

where $Q \in \mathbf{Q}$ and $\psi_1, \ldots, \psi_s$ are $\mathcal{L}$-formulas (thus the ψ_i do not contain the new quantifiers). Then a simple, but crucial link between reducibility and quantifiers can be stated as follows:

Proposition 12.3.13 *Let $\mathcal{L}$ be a logic, $K \subseteq \mathrm{Str}[\tau]$ and $L \subseteq \mathrm{Str}[\sigma]$. Then for $k \geq 1$ the following are equivalent:*

(i) *$K \leq_{\mathcal{L}}^{k} L$.*
(ii) *There is a sentence φ in $\mathcal{L}[Q_{\text{rel-}k\text{-}L}]$ of vocabulary τ such that $K = \mathrm{Mod}(\varphi)$.*

Proof. Assume $\sigma = \{R_1, \ldots, R_s\}$ with r_i-ary R_i. To prove (i) $\Rightarrow$(ii), suppose that $K \leq_{\mathcal{L}}^{k} L$ via

$$\Pi \;:=\; (\pi_{\text{uni}}(\overline{x}), \pi_{R_1}(\overline{x}_1, \ldots, \overline{x}_{r_1}), \ldots, \pi_{R_s}(\overline{x}_1, \ldots, \overline{x}_{r_s})).$$

Then for any τ-structure $\mathcal{A}$ we have:

$$\begin{aligned}\mathcal{A} \in K \ \text{ iff } \ & \mathcal{A}^{\Pi} = (\pi^{\mathcal{A}}_{\text{uni}}(\_), \pi^{\mathcal{A}}_{R_1}(\_,\dots,\_), \dots, \pi^{\mathcal{A}}_{R_s}(\_,\dots,\_)) \in L \\ \text{iff } \ & (A, \pi^{\mathcal{A}}_{\text{uni}}(\_), \pi^{\mathcal{A}}_{R_1}(\_,\dots,\_), \dots, \pi^{\mathcal{A}}_{R_s}(\_,\dots,\_)) \in \text{rel-}k\text{-}L \\ \text{iff } \ & \mathcal{A} \models Q_{\text{rel-}k\text{-}L}\overline{x}, \overline{x}_1, \dots, \overline{x}_{r_s}[\pi_{\text{uni}}, \pi_{R_1}, \dots, \pi_{R_s}].\end{aligned}$$

Concerning the direction from (ii) to (i), assume that K is axiomatized by the sentence

$$Q_{\text{rel-}k\text{-}L}\overline{x}, \overline{x}_1, \dots, \overline{x}_{r_s}[\pi_{\text{uni}}, \pi_{R_1}, \dots, \pi_{R_s}]$$

with $\mathcal{L}$-formulas $\pi_{\text{uni}}, \pi_{R_1}, \dots, \pi_{R_s}$. Then a rearrangement of the equivalences above shows that $K \leq^k_{\mathcal{L}} L$ via $\Pi = (\pi_{\text{uni}}, \pi_{R_1}, \dots, \pi_{R_s})$. □

We now transfer the notions of hardness and completeness from complexity theory to logic.[3]

Definition 12.3.14 Let $\mathcal{C}$ be a complexity class, $L \subseteq \text{Str}[\sigma]$, and $\mathcal{L}$ a logic.

(a) L is *$\mathcal{C}$-hard with respect to $\mathcal{L}$-reductions* if $K \leq_{\mathcal{L}} L$ for all $K \in \mathcal{C}$.

(b) L is *$\mathcal{C}$-complete with respect to $\mathcal{L}$-reductions* if L is $\mathcal{C}$-hard with respect to $\mathcal{L}$-reductions and $L_< \in \mathcal{C}$. □

Corollary 12.3.15 *Let $\mathcal{C}, L$, and $\mathcal{L}$ be as in the preceding definition.*

(a) *The following are equivalent:*
  (i) *L is $\mathcal{C}$-hard with respect to $\mathcal{L}$-reductions.*
  (ii) $\mathcal{C} \subseteq \{\text{Mod}(\varphi) \mid \varphi \in \mathcal{L}[\text{rel-}\omega\text{-}\mathbf{Q}_L]\}$.

(b) *If $\mathcal{C}$ is closed under $\mathcal{L}$-reducibility (i.e., for classes K_1, K_2 of ordered structures, $K_1 \leq_{\mathcal{L}} K_2$ and $K_2 \in \mathcal{C}$ imply $K_1 \in \mathcal{C}$), then the following are equivalent:*
  (i) *L is $\mathcal{C}$-complete with respect to $\mathcal{L}$-reductions.*
  (ii) *$\mathcal{L}[\text{rel-}\omega\text{-}\mathbf{Q}_L]$ captures $\mathcal{C}$.*

Proof. (a) is immediate from the preceding proposition. For the implication (ii) ⇒ (i) in (b) note that $L_<$ (as L) is FO$[Q_{\text{rel-}L}]$-axiomatizable and therefore $\mathcal{L}[\text{rel-}\omega\text{-}\mathbf{Q}_L]$-axiomatizable, hence $L_< \in \mathcal{C}$ by (ii). The rest is clear by (a). For (i) ⇒ (ii) assume that the class K of ordered structures is axiomatizable in $\mathcal{L}[\text{rel-}\omega\text{-}\mathbf{Q}_L]$. Then $K \leq_{\mathcal{L}} L$ by 12.3.13 and hence, $K \leq_{\mathcal{L}} L_<$ (cf. 12.3.4). Since $L_< \in \mathcal{C}$ (by (i)) and $\mathcal{C}$ is closed under $\mathcal{L}$-reducibility, we get $K \in \mathcal{C}$. The rest is clear by (a). □

Corollary 12.3.16 *Let* I *and* P *be the classes introduced in the proof of 12.1.9. Then* I *is* PTIME-*complete with respect to* FO-*reductions and* P *is* PSPACE-*complete with respect to* FO-*reductions.*

[3] Recall our convention that only classes of ordered structures are considered as members of complexity classes.

Proof. By the proof of 12.1.9, FO(IFP) $\leq$ FO[$\mathbf{Q}_{\mathrm{I}}^{\omega}$]. Therefore, FO(IFP) $\leq$ FO[$\mathbf{Q}_{\mathrm{I}}^{\omega}$] $\leq$ FO[rel-ω-$\mathbf{Q}_{\mathrm{I}}$] $\leq$ FO(rel-ω-$\mathbf{Q}_{\mathrm{I}}$) $\leq$ FO(IFP) (by 12.3.10). Since FO(IFP) captures PTIME, so does FO[rel-ω-$\mathbf{Q}_{\mathrm{I}}$]. Thus, I is PTIME-complete with respect to FO-reductions by part (b) of the preceding corollary. The proof for P is similar. □

We are now in a position to show the result announced in section 11.1 that in case PTIME is effectively strongly captured by any logic at all, it can already be captured by a very familiar one, namely by an extension of first-order logic by means of a single vectorized Lindström quantifier.

A class L is *strongly* PTIME-*complete with respect to* FO-*reductions* if $L_<$ is in PTIME and $K \leq_{\mathrm{FO}} L$ for any class K of structures with $K_<$ in PTIME (in Definition 12.3.14 we only considered classes K of ordered structures). The main result is:

Theorem 12.3.17 *The following are equivalent:*

(i) *There is a logic that effectively strongly captures* PTIME.

(ii) *There is a class L of structures such that* FO(rel-ω-$\mathbf{Q}_L$) *effectively strongly captures* PTIME.

(iii) *There is a class L of structures which is strongly* PTIME-*complete with respect to* FO-*reductions.*

Moreover, a class L satisfies (ii) *just in case it satisfies* (iii).

The theorem gets false if "strongly PTIME-complete" is replaced by "PTIME-complete": With respect to FO-reductions the class I is PTIME-complete (cf. 12.3.16) but not strongly PTIME-complete (otherwise, by the last statement of the theorem, FO(rel-ω-$\mathbf{Q}_{\mathrm{I}}$) $\equiv$ FO(IFP) (cf. the proof of 12.3.16) would effectively strongly capture PTIME).

Proof (of 12.3.17). We show (iii) $\Rightarrow$ (ii) and (i) $\Rightarrow$ (iii), the implication (ii) $\Rightarrow$ (i) being trivial.

(iii) $\Rightarrow$ (ii): Let the class L be given by (iii) and let K be any class of structures such that $K_< \in$ PTIME. By (iii), $K \leq_{\mathrm{FO}} L$ and hence, by 12.3.13, K is axiomatizable in FO[rel-ω-$\mathbf{Q}_L$] $\subseteq$ FO(rel-ω-$\mathbf{Q}_L$). Moreover, by induction one can effectively assign to every $\varphi \in$ FO(rel-ω-$\mathbf{Q}_L$) a pair (M, d), where M is a deterministic x^d time-bounded machine accepting $\mathrm{Mod}(\varphi)_<$. Hence, FO(rel-ω-$\mathbf{Q}_L$) effectively strongly captures PTIME (note that $\mathrm{Mod}(\psi)_< \in$ PTIME for $\psi \in$ FO(rel-ω-$\mathbf{Q}_L$), cf. 12.3.12).

(i) $\Rightarrow$ (iii): It suffices to define a class L of structures such that $L_<$ is in PTIME and such that for every class K of $\{E\}$-structures (where E is binary) with $K_<$ in PTIME we have $K \leq_{\mathrm{FO}} L$. Then 12.3.5 together with the transitivity of $\leq_{\mathrm{FO}}$ (cf. 12.3.3) yields the strong PTIME-completeness of L.

Let $\mathcal{L}$ be a logic that effectively strongly captures PTIME. We may assume that the set $\mathcal{L}[\{E\}]_0$ of sentences of $\mathcal{L}$ of vocabulary $\{E\}$ is $\mathbb{N}$. Furthermore, let g be a computable function assigning to $i = \varphi \in \mathcal{L}[\{E\}]_0$ a pair (M, d) such that M is deterministic and x^d time-bounded and accepts $\mathrm{Mod}(\varphi)_<$. Let M_g be a deterministic machine that calculates g.

We set

$$\sigma := \{V, E, P, Q\}$$

with unary V, Q and binary E, P and define L as the class of σ-structures $\mathcal{A}$ such that:

(1) $E^A \subseteq V^A \times V^A$.

(2) P^A is a preordering of A (i.e., it is reflexive and transitive) such that the equivalence classes of the equivalence relation given by

$$a \text{ and } b \text{ are equivalent iff } P^A ab \text{ and } P^A ba$$

are linearly ordered with respect to the ordering $\preceq^A$ induced by P^A.

(3) Q^A is an equivalence class, say, Q^A is the i-th class with respect to $\preceq^A$.

(4) The length of the run that M_g performs on the input $1 \ldots 1$ of length i to calculate $g(i) =: (M, d)$ is $\leq \|A\|$.

(5) $\|A\| \geq \|V^A\|^d$.

(6) M accepts all ordered representations of (V^A, E^A).

$L_<$ is in PTIME. Namely, in order to check in time polynomial in $\|A\|$ whether an ordered $\sigma_<$-structure $(\mathcal{A}, <^A)$ with $A = \{1, \ldots, n\}$ belongs to $L_<$, we

- check (1)
- check whether P^A satisfies (2) and whether Q^A is an equivalence class and determine i
- calculate $g(i) = (M, d)$, thereby checking (4)
- check (5)
- check whether M accepts $(V^A, E^A, <^A \cap (V^A \times V^A))$ (which is equivalent to a check of (6), as M is x^d time-bounded and accepts $\mathrm{Mod}(i)_<$).

Now, let K be a class of $\{E\}$-structures such that $K_<$ is in PTIME. We show that $K \leq_{\mathrm{FO}} L$. For this purpose let $i = \varphi \in \mathcal{L}[\{E\}]_0$ be such that $\mathrm{Mod}(\varphi) = K$. Let $g(i) = (M, d)$. Then the deterministic machine M is x^d time-bounded and accepts $K_<$. Choose l such that there are at least $\max\{i, d\}$ many equality types of l elements (cf. section 12.2) and such that $l^l \geq$ the length of the run that M_g performs on the input $1 \ldots 1$ of length i to calculate $g(i)$. We only take care of $\{E\}$-structures of cardinality $\geq l$.

We show $K \leq_{\mathrm{FO}}^l L$ by defining the following FO-interpretation Π of σ in $\{E\}$ of width l ($\overline{x}$ and $\overline{y}$ are of length l):

$$\begin{array}{lll}
\pi_{\text{uni}}(\overline{x}) & := & x_1 = x_1 \\
\pi_V(\overline{x}) & := & x_1 = x_2 = \ldots = x_l \\
\pi_P(\overline{x}, \overline{y}) & := & \text{"the equality type of } \overline{x} \text{ is smaller than that of } \overline{y}\text{"} \\
& & \text{(in a fixed, first-order definable ordering of the finitely} \\
& & \text{many equality types of } l \text{ elements)} \\
\pi_Q(\overline{x}) & := & \text{"}\overline{x} \text{ satisfies the } i\text{-th equality type"} \\
\pi_E(\overline{x}, \overline{y}) & := & x_1 = \ldots = x_l \land y_1 = \ldots = y_l \land E x_1 y_1.
\end{array}$$

Then for all $\{E\}$-structures $\mathcal{B}$ of cardinality $\geq l$,

$$\mathcal{B} \in K \quad \text{iff} \quad \mathcal{B}^{\Pi} \in L. \qquad \square$$

To conclude this section, we give some further applications of Corollary 12.3.15 for $\mathcal{C} = \text{NPTIME}$.

Let $< \in \sigma$ and let $L \subseteq \mathcal{O}[\sigma]$ be a class of ordered structures that is NPTIME-complete with respect to logspace reductions, i.e., $L \in \text{NPTIME}$ and every NPTIME class K of ordered structures, say of vocabulary τ, is reducible to L in the following sense: There is a log space-bounded[4] (τ, σ)-transducer M (cf. Appendix 7.5.1) such that for all τ-structures $\mathcal{A}$,

$$\mathcal{A} \in K \quad \text{iff} \quad M(\mathcal{A}) \in L$$

where $M(\mathcal{A})$ is the output of M started with $\mathcal{A}$. From complexity theory we know that an example of such an L is given by $\text{HAM}_<$ where

$$\text{HAM} := \{\mathcal{G} \mid \mathcal{G} \text{ is a directed graph that has a Hamiltonian circuit}\}.$$

Proposition 12.3.18 *For classes K and L of ordered structures the following are equivalent:*

(i) *K is logspace reducible to L.*

(ii) *$K \leq_{\text{FO(DTC)}} L$.*

Proof. For the implication (ii) $\Rightarrow$ (i) let Π be an FO(DTC)-interpretation witnessing $K \leq_{\text{FO(DTC)}} L$. Since satisfiability for FO(DTC)-formulas can be checked by log space-bounded machines, there is a (τ, σ)-transducer that, given $\mathcal{A}$, writes an encoding of $\mathcal{A}^{\Pi}$ on the output tapes.

For (i) $\Rightarrow$ (ii) suppose that $L \subseteq \mathcal{O}[\sigma]$ where $\sigma = \{R_1, \ldots, R_s\}$ with r_i-ary R_i. Let $K \subseteq \mathcal{O}[\tau]$ and let M be a log space-bounded (τ, σ)-transducer such that for all $\mathcal{A} \in \mathcal{O}[\tau]$,

$$\text{(1)} \qquad \mathcal{A} \in K \quad \text{iff} \quad M(\mathcal{A}) \in L.$$

By 7.5.23 there are $k \geq 1$ and formulas $\chi_{\text{uni}}(\overline{x})$ and $\chi_{R_i}(\overline{x}_1, \ldots, \overline{x}_{r_i})$ of FO(DTC) with $\overline{x}, \overline{x}_1, \ldots, \overline{x}_{r_s}$ of length k such that for all $\mathcal{A} \in \mathcal{O}[\tau]$,

[4] Recall that log space-bounded means $c \cdot$log space-bounded for a suitable $c > 0$.

(2) $$M(\mathcal{A}) \cong (\chi^{\mathcal{A}}_{\text{uni}}(\_), \chi^{\mathcal{A}}_{R_1}(\_,\dots,\_), \dots, \chi^{\mathcal{A}}_{R_s}(\_,\dots,\_)).$$

Let Π be the FO(DTC)-interpretation $(\chi_{\text{uni}}, \chi_{R_1}, \dots, \chi_{R_s})$. By(1) and (2), we have $\mathcal{A} \in K$ iff $\mathcal{A}^\Pi \in L$; hence, $K \leq^k_{\text{FO(DTC)}} L$. □

Since NPTIME is closed under logspace reductions, we get by 12.3.15(b) and the preceding proposition:

Corollary 12.3.19 *For a class L of ordered structures the following are equivalent:*

(i) *L is* NPTIME-*complete with respect to logspace reductions.*
(ii) *L is* NPTIME-*complete with respect to* FO(DTC)-*reductions.*
(iii) FO(DTC)[rel-ω-$\mathbf{Q}_L$] *captures* NPTIME. □

Corollary 12.3.20 HAM$_<$ *is* NPTIME-*complete with respect to* FO(DTC)-*reductions.* □

Corollary 12.3.21 FO(DTC)[rel-ω-$\mathbf{Q}_{\text{HAM}_<}$] *captures* NPTIME. □

In some cases one can do better. For example, NPTIME is captured already by FO[rel-ω-$\mathbf{Q}_{\text{HAM}_<}$] (in the presence of min and max even quantifier-free formulas suffice); moreover, according to the following exercise we can replace HAM$_<$ by HAM.

Exercise 12.3.22 Show that $\mathcal{L}$[rel-ω-$\mathbf{Q}_L$] $\equiv$ $\mathcal{L}$[rel-ω-$\mathbf{Q}_{L_<}$] on ordered structures (compare 12.1.2(f)). □

In a typical first-order prefix, for example a prefix as in

(1) $$\forall x \exists y \forall u \exists v \varphi(x, y, u, v),$$

the existentially quantified variables depend on all the preceding universally quantified variables. So (1) means (using function symbols f (unary) and g (binary))

(2) $$\exists f \exists g \forall x \forall u \varphi(x, f(x), u, g(x, u)).$$

Due to the linear character of our notation of formulas it is not possible to arrange the quantifiers in the prefix in (1) in such a way that y only depends on x and v only depends on u. Henkin introduced quantifiers which allow such dependencies. Here we only consider the simplest one, the one we just have described. It is written as $\begin{pmatrix} \forall x \exists y \\ \forall u \exists v \end{pmatrix}$, and $\begin{pmatrix} \forall x \exists y \\ \forall u \exists v \end{pmatrix} \varphi(x, y, u, v)$ means $\exists f \exists g \forall x \forall u \varphi(x, f(x), u, g(u))$. We turn $\begin{pmatrix} \forall x \exists y \\ \forall u \exists v \end{pmatrix}$ into a Lindström quantifier by introducing, for $\sigma = \{R\}$ with a 4-ary relation symbol R, the class

$$\text{HENK} \quad := \quad \{(A, R^A) \mid (A, R^A) \models \exists f \exists g \forall x \forall u Rx f(x) u g(u)\}.$$

$\text{FO}(Q_{\text{HENK}})$ is strictly stronger than FO. For example, with $\tau = \{U, V\}$, where U, V are unary, and

$$\begin{aligned} \varphi &:= \text{“}\begin{pmatrix} \forall x \exists y \\ \forall u \exists v \end{pmatrix} ((u = y \rightarrow v = x) \wedge (Ux \leftrightarrow Vy))\text{”} \\ &= \text{“}\exists f \exists g \forall x \forall u ((u = f(x) \rightarrow g(u) = x) \wedge (Ux \leftrightarrow Vf(x)))\text{”} \\ &= Q_{\text{HENK}} xyuv ((u = y \rightarrow v = x) \wedge (Ux \leftrightarrow Vy)) \end{aligned}$$

we have $\text{Mod}(\varphi) = \{(A, U^A, V^A) \mid \|U^A\| \leq \|V^A\|\}$, a class that is not axiomatizable in FO.

Proposition 12.3.23 *For* $\mathbf{Q} = \text{rel-}\omega\text{-}\mathbf{Q}_{\text{HENK}}$ *and for* $\mathbf{Q} = \text{rel-}\omega\text{-}\mathbf{Q}_{\text{HENK}_<}$ *the logic* FO(DTC)[**Q**] *captures* NPTIME.

Proof. By 12.3.22 we may restrict ourselves to $\text{HENK}_<$. Let 3COL be the class of 3-colourable graphs. (A graph is *3-colourable*, if its vertices can be coloured with three colours such that no two adjacent vertices have the same colour.) $3\text{COL}_<$ is NPTIME-complete with respect to logspace reductions and hence, with respect to FO(DTC)-reductions (cf. 12.3.19). We show

$$(*) \qquad 3\text{COL}_< \leq_{\text{FO(DTC)}} \text{HENK}_<.$$

Since $\text{HENK}_<$ is in NPTIME, (*) yields that $\text{HENK}_<$ is NPTIME-complete with respect to FO(DTC)-reductions. By 12.3.19, FO(DTC) $[\text{rel-}\omega\text{-}\mathbf{Q}_{\text{HENK}_<}]$ captures NPTIME.

We show (*), even $3\text{COL} \leq_{\text{FO}} \text{HENK}$: We use a unary function f to represent three colours: a vertex a has the first colour if

$$f(a) = a,$$

the second colour if

$$f(a) \neq a \text{ and } Eaf(a),$$

and the third colour if

$$f(a) \neq a \text{ and not } Eaf(a).$$

Then an easy argumentation shows that a graph $\mathcal{G} = (G, E^G)$ is 3-colourable iff it is of cardinality ≤ 3 or there is a function $f : G \rightarrow G$ such that for all a, b with $E^G ab$ it is not the case that (i), (ii), or (iii) where

(i) $f(a) = a$ and $f(b) = b$;
(ii) $f(a) \neq a, Eaf(a)$ and $f(b) \neq b, Ebf(b)$;
(iii) $f(a) \neq a$, not $Eaf(a)$ and $f(b) \neq b$, not $Ebf(b)$.

Therefore, there is a formula $\varphi_{3\text{COL}}(x, y, u, v) \in \text{FO}[\{E\}]$ such that

$$\text{Mod}(Q_{\text{HENK}} xyuv \varphi_{3\text{COL}}(x, y, u, v)) = 3\text{COL};$$

for instance, we can take

$$\begin{aligned}\varphi_{3\text{COL}}(x,y,u,v) \quad := \quad & \varphi_{\leq 3} \vee ((x = u \rightarrow y = v) \wedge (Exu \rightarrow \\ & \neg((y = x \wedge v = u) \vee (y \neq x \wedge Exy \wedge v \neq u \wedge Euv) \\ & \vee (y \neq x \wedge \neg Exy \wedge v \neq u \wedge \neg Euv)))).\end{aligned}$$

Hence, $3\text{COL} \leq_{\text{FO}}$ HENK via the interpretation given by

$$\begin{aligned}\pi_{\text{uni}}(x) &:= x = x \\ \pi_R(x,y,u,v) &:= \varphi_{3\text{COL}}(x,y,u,v).\end{aligned}$$

□

12.4 Quantifiers and Oracles

In the preceding section we have encountered complexity classes captured by logics of the form $\mathcal{L}(\text{rel-}\omega\text{-}\mathbf{Q}_K)$. In the following we deal with a question in the opposite direction: Let $\mathcal{L}$ be a logic, K a class of structures, and $\mathcal{C}$ a complexity class. Then:

(+) If $\mathcal{L}$ captures $\mathcal{C}$, what complexity class is captured by $\mathcal{L}(\text{rel-}\omega\text{-}\mathbf{Q}_K)$?

Interprete a formula φ, say, of $\mathcal{L}(Q_K)$ as a program to evaluate the meaning of φ, addressing a device which can evaluate instructions of the form

$$Q_K\overline{x}[\varphi_1, \ldots, \varphi_s]$$

in one step, once $\varphi_1, \ldots, \varphi_s$ have been evaluated. There is a concept of such a device in complexity theory that fits into this frame, Turing machines with *oracles*. We let $\mathcal{C}^K$ denote the complexity class defined by Turing machines with resource-bound according to $\mathcal{C}$ that can use an oracle for K to decide membership for K. Then the natural answer to (+) would be

$$\mathcal{L}(\text{rel-}\omega\text{-}\mathbf{Q}_K) \text{ captures } \mathcal{C}^K.$$

As it has turned out, this answer is true in various cases. However, one has to choose the right oracle machine model in each particular case. We only will present one important group of results to give an impression of what is going on. First, we introduce an oracle model (essentially Turing machines with one oracle tape) that meets our purposes.

Let σ and τ be vocabularies with $< \in \sigma$ and $< \in \tau$. A *Turing machine M for τ-structures with oracle tapes for σ-structures*, for short, a *(τ, σ)-oracle machine*, is a (τ,σ)-Turing machine (cf. Appendix 7.5.1), that is, a deterministic Turing machine with input tapes for τ-structures, with work tapes, and with output tapes for σ-structures (now called *oracle tapes*). In addition to the instructions of a (τ, σ)-Turing machine it has so-called *oracle instructions*, that is, instructions of the form

$$(1) \qquad sb_0 \dots b_{k+l}\, c_1 \dots c_m \to s' s''.$$

Given an "oracle class" O of σ-structures, (1) has the meaning

> "If you are in a configuration corresponding to $sb_0 \dots b_{k+l}\, c_1 \dots c_m$, do the following: if the inscription on the oracle tapes encodes a structure in O, go into state s'; otherwise, go into state s''. Delete the inscription on the oracle tapes and move the heads to the initial position."

Configurations which ask for an oracle instruction are called *oracle configurations.* Note that in an oracle configuration the inscriptions on the oracle tapes need not encode a σ-structure.

We write (M, O) to indicate that M works with O as the class fixing the meaning of the oracle instructions.

An oracle instruction is carried out in one step. By definition, space bounds refer only to work tapes; in particular, space used on the oracle tapes does not count. If (M, O) is $s(n)$ space-bounded then – in case (M, O) stops – the number of cells on the oracle tapes that are scanned during the run is $\leq d^{s(n)}$ for a certain constant $d > 0$.

Occasionally we shall also use transducers with oracle tapes, so-called *oracle transducers*; they possess both output tapes and oracle tapes, the latter ones serving the same purpose and working in the same way as with oracle machines. It should not be difficult for the reader to provide an exact definition.

Let $\mathcal{C}$ be a complexity class and $\mathcal{K}$ a class of classes of ordered structures. $\mathcal{C}^{\mathcal{K}}$, the *relativization* of $\mathcal{C}$ to $\mathcal{K}$, consists of those classes of ordered structures that are accepted by some oracle machine (M, O) with $O \in \mathcal{K}$ that is resource-bounded according to $\mathcal{C}$. In particular, LOGSPACE$^{\mathcal{K}}$ consists of those classes of ordered structures that are accepted by some log space-bounded oracle machine (M, O) with $O \in \mathcal{K}$. If L is a class of ordered structures, we write $\mathcal{C}^L$ for $\mathcal{C}^{\{L\}}$.

The following proposition which is stated for the case we are interested in exhibits the relationship between oracles and quantifiers.

Proposition 12.4.1 *Suppose that O is a class of ordered σ-structures and φ an* FO(DTC)[rel-ω-$\mathbf{Q}_O$]*-sentence. Then there is a log space-bounded oracle machine (M, O) that decides* Mod(φ).[5]

Proof. Let

$$\varphi = Q_{\text{rel-}k\text{-}O}\, \overline{x}, \overline{x}_1, \dots, \overline{x}_{r_s} [\varphi_{\text{uni}}(\overline{x}), \varphi_{R_1}(\overline{x}_1, \dots, \overline{x}_{r_1}), \dots, \varphi_{R_s}(\overline{x}_1, \dots, \overline{x}_{r_s})]$$

[5] In particular, the logarithmic space bound applies to rejecting runs, too.

with $\varphi_{\text{uni}}, \ldots, \varphi_{R_s} \in \text{FO(DTC)}$. By 7.4.1 there are log space-bounded Turing machines that evaluate $\varphi_{\text{uni}}, \ldots, \varphi_{R_s}$. The intended machine writes the "structure" induced by $\varphi_{\text{uni}}, \ldots, \varphi_{R_s}$ on the oracle tapes and invokes its oracle to decide whether it belongs to O. □

We prove the following well-known result from complexity theory.

Proposition 12.4.2 *Assume that L is a class of ordered structures that is* NPTIME*-complete with respect to logspace reductions. Then*

$$\text{LOGSPACE}^L = \text{LOGSPACE}^{\text{NPTIME}}.$$

In particular, we have

$$\text{LOGSPACE}^{\text{HAM}_<} = \text{LOGSPACE}^{\text{HENK}_<} = \text{LOGSPACE}^{\text{NPTIME}}.$$

Proof. Obviously, we have $\text{LOGSPACE}^L \subseteq \text{LOGSPACE}^{\text{NPTIME}}$. To prove the converse inclusion, let $K \subseteq \mathcal{O}[\tau]$ and $O \subseteq \mathcal{O}[\sigma]$ with $O \in$ NPTIME, and let (M_K, O) be a log space-bounded (τ, σ)-oracle machine that accepts K. We have to show that $K \in \text{LOGSPACE}^L$.

By 12.3.19 there is a sentence φ_O in $\text{FO(DTC)}[\text{rel-}\omega\text{-}\mathbf{Q}_L][\sigma]$ such that $O = \text{Mod}(\varphi_O)$ and hence, by 12.4.1, there is a log space-bounded oracle machine (M_O, L) that decides O.

We construct a log space-bounded oracle machine (M, L) that accepts K. Roughly, (M, L) runs as follows: Started with an ordered τ-structure $\mathcal{A}$, it imitates (M_K, O). Suppose it reaches an oracle configuration of (M_K, O). (M, L) does not have an oracle to decide whether the inscription J on the oracle tapes of (M_K, O) encodes a structure in O. However, using its oracle for L, it now can imitate a run of (M_O, L) to answer the question whether J encodes a structure in O. For this purpose, (M, L) must use work tapes that carry the inscription J to simulate the input tapes of (M_O, L). But J, in general, is not log space-bounded in $\|A\|$. To overcome this difficulty, one applies the same trick as for 7.5.23, going always back to obtain the information on J needed in a given situation.

To be more precise, consider a configuration c which is the starting configuration of (M_K, O) or the immediate successor of an oracle configuration of (M_K, O). Assume that there is a next oracle configuration c' and that the transition from c to c' leads to the inscription J on the oracle tapes of (M_K, O). Assume that c is stored on an additional tape. In order to decide whether J encodes a structure in O, the information about J that (M, L) really needs, is of the kind

> "What letter stands in a certain cell of a certain oracle tape of (M_K, O)?"

To provide such an information, (M, L) imitates (M_K, O) starting from c, neglecting the oracle tapes, but using counters to mark the positions of the heads on the oracle tapes. Finally, (M, L) replaces the stored c by c'. □

We come back to the question (+) of the introduction. For FO(IFP) and PTIME and for FO(DTC) and LOGSPACE we give positive answers, thereby proceeding in two steps (Proposition 12.4.3 and Proposition 12.4.4).

Proposition 12.4.3 *Let $<\in \sigma$ and let O be a class of ordered σ-structures. Then we have:*

(a) *Every class in* PTIME^O *is* $\text{FO(IFP)}(\text{rel-}\omega\text{-}\mathbf{Q}_O)$*-axiomatizable.*
(b) *Every class in* LOGSPACE^O *is* $\text{FO(DTC)}(\text{rel-}\omega\text{-}\mathbf{Q}_O)$*-axiomatizable.*

In both (a) *and* (b) *the axiomatizing sentence can be chosen without nested quantifiers from* $\text{rel-}\omega\text{-}\mathbf{Q}_O$.

Proof. We first prove (b). Let $<\in \tau$ and $K \subseteq \mathcal{O}[\tau]$ be accepted by some log space-bounded oracle machine (M, O). We have to show that K is axiomatizable in $\text{FO(DTC)}(\text{rel-}\omega\text{-}\mathbf{Q}_O)$.

To give such an axiomatization, we proceed as in 7.3.7, using log space-bounded configurations of (M, O) that only refer to the input tapes and to the work tapes; but now we describe the transition from one oracle configuration to the next one. For this purpose we assume that the *oracle answer states* s', s'' in oracle instructions are distinct from each other and do not occur as successor states of other instructions. The main point here is to show that the predicates $\text{Oracle}^+(\overline{u}, \overline{v})$ and $\text{Oracle}^-(\overline{u}, \overline{v})$ as given below can be expressed in $\text{FO(DTC)}(\text{rel-}\omega\text{-}\mathbf{Q}_O)$. $\text{Oracle}^{+/-}(\overline{u}, \overline{v})$ says:

- $\overline{u}$ is the starting configuration of (M, O) or a configuration with an oracle answer state;
- $\overline{v}$ is an oracle configuration of (M, O);
- $\overline{v}$ is reached from $\overline{u}$ without passing an oracle configuration in between;
- if (M, O) is started in $\overline{u}$ with empty oracle tapes then, when reaching $\overline{v}$, the oracle tapes contain/do not contain the encoding of a structure in O.

The first three points do not cause difficulties; so we restrict ourselves to the last one. We consider the run of (M, O) that leads from $\overline{u}$ to $\overline{v}$ ($\overline{u}$ and $\overline{v}$ obeying the first three points above) as a run of a log space-bounded (τ, σ)-transducer. Then we can apply Theorem 7.5.23 (more precisely: the obvious generalization involving the parameters $\overline{u}, \overline{v}$) to get, with a fixed $e \geq 1$, formulas

$$(+) \qquad \psi_{\text{uni}}(\overline{x}, \overline{u}, \overline{v}) \text{ and } \psi_R(\overline{x}_1, \ldots, \overline{x}_r, \overline{u}, \overline{v}) \text{ for } R \in \sigma,\ R\ r\text{-ary},$$

in $\text{FO(DTC)}[\tau]$ where $\overline{x}$ and the $\overline{x}_i$ have length e, and, with similar methods, a formula $\varphi_{\text{Str}}(\overline{u}, \overline{v})$ in $\text{FO(DTC)}[\tau]$ such that $\varphi_{\text{Str}}(\overline{u}, \overline{v})$ says that

> after the run from $\overline{u}$ to $\overline{v}$ the inscription on the oracle tapes encodes a σ-structure, namely $(\psi_{\text{uni}}(\_, \overline{u}, \overline{v}), (\psi_R(\_, \ldots, \_, \overline{u}, \overline{v}))_{R\in\sigma})$.

Then we see that

$$\psi(\overline{u}, \overline{v}) \quad := \quad Q_{\text{rel-}e\text{-}O}\overline{x}, \ldots (\psi_{\text{uni}}(\overline{x}, \overline{u}, \overline{v}), (\psi_R(\overline{x}_R, \overline{u}, \overline{v}))_{R\in\sigma})$$

expresses that $(\psi_{\text{uni}}(\_,\overline{u},\overline{v}),(\psi_R(\_,\ldots,\_,\overline{u},\overline{v}))_{R\in\sigma})\in O$. Therefore, we can describe the fourth point of Oracle$^+(\overline{u},\overline{v})$ by

$$\varphi_{\text{Str}}(\overline{u},\overline{v})\wedge\psi(\overline{u},\overline{v})$$

and the fourth point of Oracle$^-(\overline{u},\overline{v})$ by $\neg\varphi_{\text{Str}}(\overline{u},\overline{v})\vee\neg\psi(\overline{u},\overline{v})$.

The proof of (a) is similar, referring to a PTIME analogue of Theorem 7.5.23 (see Exercise 7.5.25).

The additional remark of this proposition is obvious. □

The preceding proof cannot be extended to PSPACE and FO(PFP). The reason is that the oracle tapes of a polynomially space-bounded oracle machine may get inscriptions of length exponential in the length of the input structure and hence, may not be definable by means of an interpretation. One way to preserve the validity consists in changing the oracle model, say, by suitably restricting the use of the oracle tapes.

We now turn to model checking:

Proposition 12.4.4 *Let $<\in\sigma$ and let O be a class of ordered σ-structures. Then we have:*

(a) *Every* FO(IFP)(rel-ω-$\mathbf{Q}_O$)*-axiomatizable class of ordered structures is in* PTIME^O *(and hence, is decidable by a machine according to* PTIME$^{\mathcal{O}}$*).*
(b) *Every class of ordered structures axiomatizable by an* FO(DTC)(rel-ω-$\mathbf{Q}_O$)*-sentence without nested quantifiers of* rel-ω-$\mathbf{Q}_O$ *is in* LOGSPACE^O *(and hence, is decidable by a machine according to* LOGSPACE$^{\mathcal{O}}$*).*
(c) *The assertion in* (b) *is true for all* FO(DTC)(rel-ω-$\mathbf{Q}_O$)*-axiomatizable classes of ordered structures if O is* NPTIME*-complete with respect to logspace reductions.*

Summing up:

Corollary 12.4.5 (a) *For any class O of ordered σ-structures,* FO(IFP)(rel-ω-$\mathbf{Q}_O$) *captures* PTIME^O.
(b) *If O is* NPTIME*-complete with respect to logspace reductions, then* FO(DTC)(rel-ω-$\mathbf{Q}_O$) *captures* LOGSPACE^O. □

Note that in (b), by 12.4.2, LOGSPACE^O = LOGSPACE^{NPTIME}, so FO(DTC)(rel-ω-$\mathbf{Q}_O$) captures LOGSPACE^{NPTIME}. For example, we have

$$\text{FO(DTC)(rel-}\omega\text{-}\mathbf{Q}_{\text{HAM}_<}\text{) captures LOGSPACE}^{\text{NPTIME}}.$$

The preceding results yield normal form theorems for both the IFP and the DTC case: Over ordered structures, any FO(IFP)(rel-ω-$\mathbf{Q}_O$)-sentence is equivalent to a sentence without nested quantifiers from rel-ω-$\mathbf{Q}_O$. The same is true for FO(DTC)(rel-ω-$\mathbf{Q}_O$) provided O is NPTIME-complete with respect to logspace reductions.

Proof (of 12.4.4). (a) We proceed by induction on FO(IFP)(rel-ω-$\mathbf{Q}_O$)-formulas and confine ourselves to the rel-ω-$\mathbf{Q}_O$-step. So let

$$\varphi = Q_{\text{rel-}k\text{-}O}\overline{x}, \overline{x}_1, \ldots, \overline{x}_s[\varphi_{\text{uni}}(\overline{x}), \varphi_1(\overline{x}_1), \ldots, \varphi_s(\overline{x}_s)]$$

where $\sigma = \{R_1, \ldots, R_s\}$ and $\varphi_{\text{uni}}, \varphi_1, \ldots, \varphi_s$ are FO(IFP)(rel-ω-$\mathbf{Q}_0$)[τ]-formulas. By induction hypothesis there are polynomially time-bounded oracle machines $(M_{\text{uni}}, O), (M_1, O), \ldots, (M_s, O)$ that decide

$$\{(\mathcal{A}, \overline{a}) \mid \mathcal{A} \in \mathcal{O}[\tau], \mathcal{A} \models \varphi_{\text{uni}}[\overline{a}]\}, \ldots, \{(\mathcal{A}, \overline{a}) \mid (\mathcal{A} \in \mathcal{O}[\tau], \mathcal{A} \models \varphi_s[\overline{a}]\},$$

respectively. Using these machines, we can build a polynomially time-bounded oracle machine (M, O) deciding the class of ordered models of φ: Given an ordered τ-structure $\mathcal{A}$ as input,

(i) (M, O) checks whether $\varphi_{\text{uni}}^{\mathcal{A}}(\_)$ is nonempty and whether (in case $R_1 = <$) $\varphi_1^{\mathcal{A}}(\_)$ is an ordering on $\varphi_{\text{uni}}^{\mathcal{A}}(\_)$; if not, it rejects, otherwise (viewing $(\varphi_{\text{uni}}, \varphi_1, \ldots, \varphi_s)$ as an interpretation Π)

(ii) using the ordering $\varphi_1^{\mathcal{A}}(\_)$, it writes the (encoding of the) structure

$$\mathcal{A}^{\Pi} = (\varphi_{\text{uni}}^{\mathcal{A}}(\_), \varphi_1^{\mathcal{A}}(\_), \ldots, \varphi_s^{\mathcal{A}}(\_))$$

on some work tapes;

(iii) it copies $\mathcal{A}^{\Pi}$ on the oracle tapes;

(iv) it envokes its oracle O to decide whether $\mathcal{A}^{\Pi}$ belongs to O.

Note that in (ii) we cannot write $\mathcal{A}^{\Pi}$ directly on the oracle tapes, because these tapes may be used by the machines $(M_{\text{uni}}, O), (M_1, O), \ldots, (M_s, O)$ when constructing $\mathcal{A}^{\Pi}$.

(b) The proof follows the same pattern as that of (a). But now we write $\mathcal{A}^{\Pi}$ directly on the oracle tapes; we can do so as the formulas $\varphi_{\text{uni}}, \varphi_1, \ldots, \varphi_s$ do not contain quantifiers from rel-ω-$\mathbf{Q}_O$. One only has to check that the arguments remain valid if the polynomial time bounds are replaced by logarithmic space bounds.

(c) The proof is more involved. The reason is that, when following the pattern of the proof of (a), we cannot realize (ii) by a log space-bounded oracle machine as the encoding of $\mathcal{A}^{\Pi}$ may be of polynomial length, thus violating the logspace bound. We therefore proceed as follows: In the rel-ω-$\mathbf{Q}_O$-step of an inductive proof, let

$$\varphi = Q_{\text{rel-}k\text{-}O}\overline{x}, \overline{x}_1, \ldots, \overline{x}_s[\varphi_{\text{uni}}(\overline{x}), \varphi_1(\overline{x}_1), \ldots, \varphi_s(\overline{x}_s)]$$

where $\varphi_{\text{uni}}, \varphi_1, \ldots, \varphi_s$ are FO(DTC)(rel-ω-$\mathbf{Q}_O$)[τ]-formulas. By hypothesis, there are log space-bounded oracle machines $(M_{\text{uni}}, O), \ldots$ deciding $\{(\mathcal{A}, \overline{a}) \mid \mathcal{A} \in \mathcal{O}[\tau], \mathcal{A} \models \varphi_{\text{uni}}[\overline{a}]\}, \ldots$, respectively. Using these machines, we now build a log space-bounded oracle transducer (M, O) that works as follows: Given an ordered τ-structure $\mathcal{A}$ as input, (M, O) checks whether $\varphi_{\mathrm{uni}}^{\mathcal{A}}(\_)$ is nonempty and whether $\varphi_1^{\mathcal{A}}(\_)$ is an ordering on $\varphi_{\mathrm{uni}}^{\mathcal{A}}(\_)$; if not, it gives as output a fixed ordered σ-structure not in O (note that we may assume the existence of such a structure, the other case being trivial), otherwise it produces the output structure $\mathcal{A}^{\Pi}$ (cf. (ii) in the proof of (a)).

By construction of (M, O) we have:

$$\mathcal{A} \models \varphi \quad \text{iff} \quad (M, O), \text{ started with } \mathcal{A}, \text{ gives an output in } O,$$

i.e., the class of ordered models of φ is reducible to O via a log space-bounded oracle transducer. The next lemma tells us that we are done. □

Lemma 12.4.6 *Let O be a class of ordered σ-structures that is* NPTIME-*complete with respect to logspace reductions and let K be a class of ordered τ-structures. Then we have:*

If K is reducible to O via a log space-bounded oracle transducer with oracle class $\mathcal{O}$, then K belongs to LOGSPACE^{O}.

For the proof we need a kind of join of classes: Let σ and τ be disjoint and P a unary relation symbol that does not belong to $\sigma \cup \tau$. For $\mathcal{A} \in \mathrm{Str}[\sigma]$ and $\mathcal{B} \in \mathrm{Str}[\tau]$ let $\mathcal{AB}$ be the $(\sigma \cup \tau \cup \{P\})$-structure $\mathcal{E}$ with (we assume that A and B are disjoint)

- $E := A \dot\cup B$
- $P^{\mathcal{E}} := B$
- $R^{\mathcal{E}} := \begin{cases} R^{\mathcal{A}} & \text{if } R \in \sigma \\ R^{\mathcal{B}} & \text{if } R \in \tau. \end{cases}$

We extend the join to the case that one argument is the "empty structure" by setting

$$\begin{aligned} \mathcal{A}\emptyset \;&:=\; \text{the expansion of } \mathcal{A} \text{ to a } (\sigma \cup \tau \cup \{P\})\text{-structure } \mathcal{E} \\ &\; \text{with } P^{\mathcal{E}} = \emptyset \text{ and } R^{\mathcal{E}} = \emptyset \text{ for all } R \in \tau; \\ \emptyset\mathcal{B} \;&:=\; \text{the expansion of } \mathcal{B} \text{ to a } (\sigma \cup \tau \cup \{P\})\text{-structure } \mathcal{E} \\ &\; \text{with } P^{\mathcal{E}} = B \text{ and } R^{\mathcal{E}} = \emptyset \text{ for all } R \in \sigma. \end{aligned}$$

Note that $\mathcal{AB} \cong \mathcal{A}'\mathcal{B}'$ implies $\mathcal{A} \cong \mathcal{A}'$ and $\mathcal{B} \cong \mathcal{B}'$. For $L \subseteq \mathrm{Str}[\sigma]$ and $K \subseteq \mathrm{Str}[\tau]$ let LK be the closure under isomorphisms of

$$\{\mathcal{A}\emptyset \mid \mathcal{A} \in L\} \cup \{\emptyset\mathcal{B} \mid \mathcal{B} \in K\}.$$

LK is a class of $(\sigma \cup \tau \cup \{P\})$-structures. Up to trivial expansions it consists of the structures in L and K.

The proof of the following fact is routine.

Lemma 12.4.7 *If L and K are in* NPTIME *then $(LK)_<$ is in* NPTIME. □

Proof (of 12.4.6). Let O and K be given accordingly, and let K be reducible to O via the log space-bounded oracle transducer (M, O). By 12.4.2 it suffices to show that $K \in \text{LOGSPACE}^{\text{NPTIME}}$, i.e., that there is a class O^* in NPTIME and a log space-bounded oracle machine (M^*, O^*) that accepts K.

Let (M, O) be started with an input structure of cardinality n. Thus, if n is sufficiently large, the part of the configurations of M comprising

- the current state
- the current inscriptions of the work tapes
- the current position of the heads on both the input and the work tapes

can be encoded by d numbers between 0 and $n-1$ and hence, by d counters of length logarithmic in n. Here, d is a suitable positive number independent of n.

Let C_n be the set of these partial configurations. The cardinality of C_n is $\leq n^d$ and hence, polynomial in n. Given an input structure $\mathcal{A}$ of cardinality n, set

$$\begin{aligned}\text{Or}(\mathcal{A}) := \{(c, \mathcal{B}) \mid\ & c \in C_n \text{ and } (M, O), \text{ started with } \mathcal{A} \text{ and with} \\ & \text{empty oracle tapes in configuration } c, \\ & \text{reaches a next oracle configuration, the oracle tapes then} \\ & \text{having an inscription that encodes } \mathcal{B}, \text{ and } \mathcal{B} \in O\}.\end{aligned}$$

Now, considering sets $\{1, \ldots, r\}$ as structures (of empty vocabulary), we let L consist of the structures $\mathcal{A}\{1, \ldots, r\}$ with $r \in \mathbb{N}$ such that

- $\mathcal{A}$ is an ordered τ-structure;
- there is a set $\{(c_1, \mathcal{B}_1), \ldots, (c_r, \mathcal{B}_r)\} \subseteq \text{Or}(\mathcal{A})$ with distinct $c_1, \ldots, c_r$ such that $(M, \{\mathcal{B}_1, \ldots, \mathcal{B}_r\})$, started with $\mathcal{A}$, gives an output in O.

Then we have:

(1) For $\mathcal{A} \in \mathcal{O}[\tau]$ and $r = \|\text{Or}(\mathcal{A})\|$,

$$\mathcal{A}\{1, \ldots, r\} \in L \quad \text{iff} \quad \mathcal{A} \in K.$$

(2) $L_<$ is in NPTIME.

For (1) we note that $(M, \{\mathcal{B} \mid (c, \mathcal{B}) \in \text{Or}(\mathcal{A}) \text{ for some } c \in C_n\})$ started with $\mathcal{A}$ behaves exactly like (M, O) started with $\mathcal{A}$.

To prove (2), one can define a (nondeterministic) machine that is polynomially time-bounded and, started with an input structure $\mathcal{D}$,

- checks whether $\mathcal{D}$ has the form $(\mathcal{A}\{1, \ldots, r\}, <^D)$
- nondeterministically guesses distinct configurations $c_1, \ldots, c_r \in C_{\|A\|}$
- checks whether for all $i \in \{1, \ldots, r\}$ there are $\mathcal{B}_i$ such that $(c_i, \mathcal{B}_i) \in \text{Or}(\mathcal{A})$ (note that O is in NPTIME)
- runs $(M, \{\mathcal{B}_1, \ldots, \mathcal{B}_r\})$ on $\mathcal{A}$

– accepts if the resulting output codes a structure in O (recall that O is in NPTIME).

Now we set

$$O^* \quad := \quad (OL)_<$$

which, by 12.4.7, belongs to NPTIME. Let σ^* be its vocabulary.

We are finished if we succeed in defining a log space-bounded (τ, σ^*)-oracle machine (M^*, O^*) that accepts K. We construct (M^*, O^*) in such a way that, started with an ordered τ-structure $\mathcal{A}$, it

– counts the number r_0 of configurations $c \in C_{\|A\|}$ such that $(c, \mathcal{B}) \in \text{Or}(\mathcal{A})$ for some $\mathcal{B}$ (to test this condition, (M^*, O^*) imitates (M, O), starting with an ordered τ- structure $\mathcal{A}$ in c; if it reaches a next oracle configuration of (M, O), the oracle tapes of (M^*, O^*), that correspond to symbols of $\sigma^* \setminus \sigma$, are empty (and hence, code the empty relations); therefore, the inscription on the oracle tapes of (M, O) encodes a structure in O iff the inscription on the oracle tapes of (M^*, O^*) encodes a structure in O^*, and (M^*, O^*) can check this by use of its oracle);
– writes $\emptyset(\mathcal{A}\{1, \ldots, r_0\})$ on its oracle tapes (more exactly, an ordered version, using the ordering of $\mathcal{A}$)
– asks the oracle whether $\emptyset(\mathcal{A}\{1, \ldots, r_0\}) \in O^*$ (or, equivalently, whether $(\mathcal{A}\{1, \ldots, r_0\}) \in L$)
– accepts just in case the answer is positive.

By (1) we see that (M^*, O^*) accepts K. □

In 7.5.18 we have introduced the polynomial hierarchy PH. The following exercise contains an alternative definition. The complexity classes $\text{NPTIME}^{\mathcal{C}}$ appearing there are defined in a similar way as the classes $\text{LOGSPACE}^{\mathcal{C}}$.

Exercise 12.4.8 Set $\text{PH}'_0 := \text{PTIME}$ and $\text{PH}'_k := \text{NPTIME}^{\text{PH}'_{k-1}}$ for $k \geq 1$. Show that $\text{PH}'_k = \text{PH}_k$ for all $k \geq 0$.

Notes 12.4.9 The notion of Lindström quantifier goes back to Lindström [118]; for further information see [31]. Various notions and results that we have treated in the book can be extended to logics with Lindström quantifiers; see, for example, [106, 27]. [89] discusses problems concerning the interplay between quantifiers and logical operators. The results of section 11.2 go back to [28] (Theorem 11.2.1), [80] (Theorem 11.2.3), and [16] (Theorem 11.2.5). For section 11.3, in particular Theorem 11.3.17, we refer to Dawar [24, 25]. Early references for reductions by interpretations are [120] and [22]. The results in section 11.4 are based on [51]; see also [53, 26]. Corollary 11.4.5(b) for $\text{HAM}_<$ goes back to Stewart [134].

References

1. S. Abiteboul, R. Hull, and V. Vianu. *Foundations of Data Bases.* Addison-Wesley, 1995.
2. S. Abiteboul and V. Vianu. Fixpoint extensions of first-order logic and Datalog-like languages. In *Proc. 4th IEEE Symp. on Logic in Computer Science*, pages 71–79, 1989.
3. S. Abiteboul and V. Vianu. Datalog extensions for database queries and updates. *Journal of Computer and System Sciences* **43**, 62–124, 1991.
4. S. Abiteboul and V. Vianu. Generic computation and its complexity. In *Proc. 23rd ACM Symp. on Theory of Computing*, pages 209–219, 1991.
5. M. Ajtai and R. Fagin. Reachability is harder for directed than for undirected finite graphs. *Journal of Symbolic Logic* **55**, 113–150, 1990.
6. M. Ajtai and Y. Gurevich. Monotone versus positive. *Journal of the ACM* **34**, 1004–1015, 1987.
7. M. Ajtai and Y. Gurevich. DATALOG vs. first-order logic. In *Proc. 30th IEEE Symp. on Foundations of Computer Science*, pages 142–146, 1989.
8. J. Barwise. *Admissible Sets and Structures.* Springer, 1975.
9. J. Barwise. An introduction to first-order logic. In J. Barwise (ed.), *Handbook of Mathematical Logic*, pages 5–46. North Holland, 1977.
10. J. Barwise. On Moschovakis closure ordinals. *Journal of Symbolic Logic* **42**, 292–296, 1977.
11. Th. Behrendt, K. Compton, and E. Grädel. Optimization problems: Expressibility, approximation properties, and expected asymptotic growth of optimal solutions. In E. Börger et al. (eds.), *CSL '92, Lecture Notes in Computer Science* 702:43–60. Springer, 1993.
12. A. Blass, Y. Gurevich, and D. Kozen. A zero-one law for logic with a fixed-point operator. *Information and Control* **67**, 70–90, 1985.
13. E. Börger, E. Grädel, and Y. Gurevich. *The Classical Decision Problem.* Springer, 1997.
14. U. Bosse. An "Ehrenfeucht-Fraïssé Game" for fixpoint logic and stratified fixpoint logic. In E. Börger et al. (eds.), *CSL '92, Lecture Notes in Computer Science* 702:100–114. Springer, 1993.
15. J. R. Büchi. Weak second-order arithmetic and finite automata. *Zeitschrift für mathematische Logik und Grundlagen der Mathematik* **6**, 66–92, 1960.

16. J. Cai, M. Fürer, and N. Immerman. An optimal lower bound on the number of variables for graph identification. *Combinatorica* **12**, 389–410, 1992. Extended abstract in *Proc. 30th Symp. on Foundations of Computer Science*, pages 612–617, 1989.

17. A. Calo and J.A. Makowsky. The Ehrenfeucht-Fraïssé game for transitive closure logic. In A. Nerode and M. Taitsin (eds.), *Logical Foundation of Computer Science - Tver '92, Lecture Notes in Computer Science* 620:57–68. Springer, 1992.

18. A. Chandra and D. Harel. Structure and complexity of relational queries. *Journal of Computer and System Sciences* **25**, 99–128, 1982. Extended abstract in *Proc. 21st Symp. on Foundations of Computer Science*, pages 333–347, 1980.

19. A. Chandra and D. Harel. Horn clause queries and generalizations. *Journal of Logic Programming* **1**, 1–15, 1985.

20. C. C. Chang and H. J. Keisler: *Model Theory*. North Holland, 3rd edition, 1990.

21. K. J. Compton. 0–1 laws in logic and combinatorics. In I. Rival (ed.), *NATO Advanced Study Institute on Algorithms and Order*, pages 353–383. D. Reidel, 1989.

22. E. Dahlhaus. Reductions to NP-complete problems by interpretations. In E. Börger et al. (eds.), *Logic and Machines: Decision Problems and Complexity, Lecture Notes in Computer Science* 171:357–365. Springer, 1983.

23. E. Dahlhaus. Skolem normal forms concerning the least fixpoint. In E. Börger (ed.), *Computation theory and Logic, Lecture Notes in Computer Science* 270:101–106. Springer, 1987.

24. A. Dawar. *Feasible Computation through Model Theory*. PhD thesis, University of Pennsylvania, 1993.

25. A. Dawar. Generalized quantifiers and logical reducibilities. *Journal of Logic and Computation* **5**, 213–226, 1995.

26. A. Dawar, G. Gottlob, and L. Hella. Capturing relativized complexity classes without order. *Mathematical Logic Quarterly* **44**, 109–122, 1998.

27. A. Dawar and E. Grädel. Generalized quantifiers and 0–1 laws. In *Proc. 10th IEEE Symp. on Logic in Computer Science*, pages 54–64, 1995.

28. A. Dawar and L. Hella. The expressive power of finitely many generalized quantifiers. In *Proc. 9th Symp. on Logic in Computer Science*, pages 20–29, 1994.

29. A. Dawar, L. Hella, and Ph. G. Kolaitis. Implicit definability and infinitary logic in finite model theory. In *Proceedings of the 22nd International Colloquium on Automata, Languages, and Programming (ICALP 95), Lecture Notes in Computer Science* 944:624–635. Springer, 1995.

30. A. Dawar, S. Lindell, and S. Weinstein. Infinitary logic and inductive definability over finite structures. *Information and Computation* **119**, 160–175, 1995.

31. H.-D. Ebbinghaus. Extended logics: The general framework. In J. Barwise and S. Feferman (eds.), *Model-Theoretic Logics*, pages 25–76. Springer, 1985.

32. H.-D. Ebbinghaus, J. Flum, and W. Thomas: *Mathematical Logic.* Springer, 2nd edition, 1994.

33. A. Ehrenfeucht. An application of games to the completeness problem for formalized theories. *Fundamenta Mathematicae* **49**, 129–141, 1961.

34. T. Eiter, G. Gottlob, and Y. Gurevich. Normal forms for second-order logic over finite structures, and classification of NP optimization problems. *Annals of Pure and Applied Logic* **78**, 111–125, 1996.

35. H. B. Enderton: *A Mathematical Introduction to Logic.* Academic Press, New York, 1972.

36. R. Fagin. Generalized first-order spectra and polynomial-time recognizable sets. In R. M. Karp (ed.), *Complexity of Computation, SIAM-AMS Proceedings* 7:43–73, 1974.

37. R. Fagin. Probabilities on finite models. *Journal of Symbolic Logic* **41**, 50–58, 1976.

38. R. Fagin. The number of finite relational structures. *Discrete Mathematics* **19**, 17–21, 1977.

39. R. Fagin, L. J. Stockmeyer, and M. Y. Vardi. A simple proof that connectivity separates existential and universal monadic second-order logics over finite structures. Research Report RJ 8647, IBM, 1992.

40. J. Flum. On bounded theories. In E. Börger et al. (eds.), *CSL '91, Lecture Notes in Computer Science* 626:111–118. Springer, 1992.

41. J. Flum. On the (infinite) model theory of fixed point logics. In X. Caicedo et al. (eds.), *Models, Algebras, and Proofs, Lecture Notes in Pure and Applied Mathematics* 203:67–75. Dekker, 1999.

42. J. Flum, M. Kubierschky, and B. Ludäscher. Total and partial well-founded datalog coincide. In E. Afrati and Ph. G. Kolaitis (eds), *Proc. 6th International Conference on Database Theory (ICDT), Lecture Notes in Computer Science* 1186:113-124. Springer, 1997.

43. R. Fraïssé. Sur quelques classifications des systèmes de relations (English summary). *Université d'Alger, Publications Scientifiques, Série A*, **1**, 35–182, 1954.

44. M. Fürer. The computational complexity of the unconstrained limited domino problem (with implications for logical decision problems). In *Logical Machines: Decision Problems and Complexity, Lecture Notes in Computer Science* 171:312–319. Springer, 1981.

45. H. Gaifman. On local and nonlocal properties. In J. Stern (ed.), *Logic Colloquium '81*, pages 105–135. North Holland, 1982.

46. M. R. Garey and D. S. Johnson: *Computers and Intractability - A Guide to the Theory of NP-Completeness.* W. H. Freeman and Co., 1979.

47. Y. V. Glebskij, D. I. Kogan, M. I. Liogon'kij, and V. A. Talanov. Range and degree of realizability of formulas in the restricted predicate calculus. *Cybernetics* **5**, 142–154, 1969.

48. K. Gödel. Ein Spezialfall des Entscheidungsproblems der theoretischen Logik. *Ergebnisse eines mathematischen Kolloquiums* **2**, 27–28, 1929.

49. W. Goldfarb. The unsolvability of the Gödel class with identity. *Journal of Symbolic Logic* **49**, 1237–1252, 1984.

50. W. D. Goldfarb, Y. Gurevich, and S. Shelah. A decidable subclass of the minimal Gödel class with identity. *Journal of Symbolic Logic* **49**, 1253–1261, 1984.

51. G. Gottlob. Relativized logspace and generalized quantifiers over finite structures. *Journal of Symbolic Logic* **62**, 545–574, 1997.

52. G. Gottlob, N. Leone and H. Veith. Succinctness as a source of complexity in logical formalisms. *Annals of Pure and Applied Logic* **97**, 231–260, 1999.

53. E. Grädel. On logical descriptions of some concepts in structural complexity theory. In E. Börger et al. (eds.), *CSL '89, Lecture Notes in Computer Science* 440:163–175. Springer, 1990.

54. E. Grädel. On transitive closure logic. In E. Börger et al. (eds.), *CSL '91, Lecture Notes in Computer Science* 626:149–163. Springer, 1992.

55. E. Grädel, Ph. G. Kolaitis, and M. Y. Vardi. On the decision problem for two-variable first-order logic. *Bulletin of Symbolic Logic* **3**, 53–69, 1997.

56. E. Grädel and G. McColm. Deterministic versus nondeterministic transitive closure logic. In *Proc. 7th IEEE Symp. on Logic in Computer Science*, pages 58–63, 1992.

57. E. Grädel and M. Otto. Inductive definability with counting on finite structures. In E. Börger et al. (eds.), *CSL '92, Lecture Notes in Computer Science* 702:231–247. Springer, 1993.

58. E. Grädel and M. Otto. On logic with two variables. *Theoretical Computer Science*, to appear.

59. E. Grädel, M. Otto and E. Rosen. Two-variable logic with counting is decidable. In *Proc. 12th IEEE Symp. on Logic in Computer Science*, pages 306–317, 1997.

60. E. Grädel, M. Otto and E. Rosen. Undecidability results on two variable logics. *Archive for Mathematical Logic*, to appear. Extended abstract in *Proc. 14th Symp. on Theoretical Aspects of Computer Science STACS '97. Lecture Notes in Computer Science* 1200:249–260. Springer, 1997.

61. M. Grohe. *Fixpunktlogiken in der endlichen Modelltheorie.* Master's thesis, University of Freiburg i.Br., 1992.

62. M. Grohe. *The structure of fixed-point logics.* PhD thesis, University of Freiburg i.Br., 1994.

63. M. Grohe. Complete problems for fixed-point logics. *Journal of Symbolic Logic* **60**, 517–527, 1995.

64. M. Grohe. Arity hierarchies. *Annals of Pure and Applied Logic* **82**, 103–163, 1996.

65. M. Grohe. Fixed-point logics on planar graphs. In *Proceedings 13th IEEE Symp. on Logic in Computer Science*, pages 6–15, 1998.

66. M. Grohe. Finite variable logic in descriptive complexity. *Bulletin of Symbolic Logic* **4**, 345–398, 1998.

67. M. Grohe and L. Hella. A double arity hierarchy theorem for transitive closure logic. *Archive for Mathematical Logic* **35**, 157–171, 1996.

68. M. Grohe and J. Mariño. Definability and descriptive complexity on databases of bounded tree width. In *Proc. 7th Intern. Conference on Database Theory*, Jerusalem 1998.

69. M. Grohe and T. Schwentick. Locality of order-invariant first-order formulas. In *Proc. 23rd Intern. Symp. on Math. Foundations in Computer Science, Lecture Notes in Computer Science* 1450:75–95. Springer, 1998.

70. Y. Gurevich. Toward logic tailored for computational complexity. In M. M. Richter et al. (eds.), *Computation and Proof Theory, Lecture Notes in Mathematics* 1104:175–216. Springer, 1984.

71. Y. Gurevich. Logic and the challenge of computer science. In E. Börger (ed.), *Current Trends in Theoretical Computer Science*, pages 1–57. Computer Science Press, 1988.

72. Y. Gurevich. Zero-one laws. *Bulletin of the European Association for Theoretical Computer Science* **46**, 90–106, 1992.

73. Y. Gurevich. From invariants to canonization. *Bulletin of the European Association for Theoretical Computer Science* **63**, 115–119, 1997.

74. Y. Gurevich, N. Immerman, and S. Shelah. McColm's conjecture. In *Proc. 9th IEEE Symp. on Logic in Computer Science*, pages 10–19, 1994.

75. Y. Gurevich and S. Shelah. Random models and the Gödel case of the decision problem. *Journal of Symbolic Logic* **48**, 1120–1124, 1983.

76. Y. Gurevich and S. Shelah. Fixed-point extensions of first-order logic. *Annals of Pure and Applied Logic* **32**, 265–280, 1986.

77. Y. Gurevich and S. Shelah. On finite rigid structures. *Journal of Symbolic Logic* **61**, 549–562, 1996.

78. P. Hájek. On logics of discovery. In G. Goos and J. Hartmanis (eds.), *Proc. 4th Conference on Mathematical Foundations of Computer Science, Lecture Notes in Computer Science* 32:100–114. Springer, 1977.

79. W. Hanf. Model-theoretic methods in the study of elementary logic. In J. Addison, L. Henkin, and A. Tarski (eds.), *The Theory of Models*, pages 132–145. North Holland, 1965.

80. L. Hella. Logical hierarchies in PTIME. In *Proc. 7th IEEE Symp. on Logic in Computer Science*, pages 360–368, 1992.

81. L. Hella, Ph. G. Kolaitis, and K. Luosto. How to define a linear order on finite models. In *Proc. 9th IEEE Symp. on Logic in Computer Science*, pages 40–49, 1994.

82. L. Hella, Ph. G. Kolaitis, and K. Luosto. Almost everywhere equivalence of logics in finite model theory. *Bulletin of Symbolic Logic* **2**, 422–443, 1996.

83. L. Hella, L. Libkin, and Y. Nurmonen. Notions of locality and their logical characterizations over finite models. *Journal of Symbolic Logic*, to appear.

84. L. Henkin. Logical systems containing only a finite number of symbols. Les Presses de l'Université de Montreal. 48 pages, 1967.

85. J. Hintikka. Distributive normal forms in the calculus of predicates. *Acta Philosophica Fennica* **6**, 1–71, 1953.

86. I. Hodkinson. Finite variable logics. *Bulletin of the European Association for Theoretical Computer Science* **51**, 111–140, 1993.

87. J. Hopcroft and J. Ullman: *Introduction to Automata Theory, Languages and Computation.* Addison Wesley, 1979.

88. H. Imhof. Computational aspects of arity hierarchies. In: D. van Dalen and M. Bezem (eds.), *CSL '96, Lecture Notes in Computer Science* 1258:211–225. Springer, 1997.

89. H. Imhof. Fixed-point logic, generalized quantifiers, and oracles. *Journal of Logic and Computation* **7**, 405–425, 1997.

90. N. Immerman. Upper and lower bounds for first-order expressibility. *Journal of Computer and System Sciences* **25**, 76–98, 1982.

91. N. Immerman. Relational queries computable in polynomial time. *Information and Control* **68**, 86–104, 1986. Extended abstract in *Proc. 14th Annual ACM Symp. on Theory of Computing*, pages 147–152, 1982.

92. N. Immerman. Expressibility as a complexity measure: results and directions. In *Second Structure in Complexity Conference*, pages 194–202. IEEE Computer Society Press, 1987.

93. N. Immerman. Languages that capture complexity classes. *SIAM Journal on Computing* **16**, 760–778, 1987. Extended abstract in *Proc. 15th Annual ACM Symp. on Theory of Computing*, pages 347–354, 1983.

94. N. Immerman. Nondeterministic space is closed under complement. *SIAM Journal on Computing* **17**, 935–938, 1988.

95. N. Immerman. *Descriptive Complexity.* Springer, 1998.

96. N. Immerman and D. Kozen. Definability with a bounded number of bound variables. *Information and Computation* **83**, 121–139, 1989.

97. N. Immerman and E. S. Lander. Describing graphs: a first-order approach to graph canonization. In A. Selman (ed.), *Complexity Theory Retrospective*, pages 59–81. Springer, 1990.

98. M. Kaufmann. A counterexample to the 0–1 law for existential monadic second-order logic. Computational Logic Inc., Internal Note 32, 1987.

99. M. Kaufman and S. Shelah. On random models of finite power and monadic logic. *Discrete mathematics* **54**, 285–293, 1985.

100. H. J. Keisler: *Model Theory for Infinitary Logic.* North Holland, 1971.

101. Ph. G. Kolaitis. Implicit definability on finite structures and unambiguous computations. In *Proc. 5th IEEE Symp. on Logic in Computer Science*, pages 168–180, 1990.

102. Ph. G. Kolaitis. The expressive power of stratified logic programs. *Information and Computation* **90**, 50–66, 1991.

103. Ph. G. Kolaitis and M. N. Thakur. Polynomial-time optimization, parallel approximation, and fixedpoint logic. In *Proc. 8th Annual Structure in Complexity Conference*, pages 31–41, 1993.

104. Ph. G. Kolaitis and M. N. Thakur. Logical definability of NP-optimization problems. *Information and Computation* **115**, 321–353, 1994.

105. Ph. G. Kolaitis and M. N. Thakur. Approximation properties of NP minimization classes. *Journal of Computing and System Sciences* **50**, 391–411, 1995. Extended abstract in *Proc. 6th IEEE Symp. on Structure in Complexity Theory*, pages 353–366, 1991.

106. Ph. G. Kolaitis and J. Väänänen. Generalized quantifiers and pebble games on finite structures. *Annals of Pure and Applied Logic* **74**, 23–75, 1995.

107. Ph. G. Kolaitis and M. Y. Vardi. The decision problem for the probabilities of higher order logics. In *Proc. 19th ACM Symp. on Theory of Computing*, pages 425–435, 1987.

108. Ph. G. Kolaitis and M. Y. Vardi. 0-1 laws and decision problems for fragments of second-order logic. *Information and Computation* **87**, 302–338, 1990. Extended abstract in *Proc. 3rd IEEE Symp. on Logic in Computer Science*, pages 2–11, 1988.

109. Ph. G. Kolaitis and M. Y. Vardi. Infinitary logic and 0–1 laws. *Information and Computation* **98**, 258–294, 1992. Extended abstract in *Proc. 5th IEEE Symp. on Logic in Computer Science*, pages 156–167, 1990.

110. Ph. G. Kolaitis and M. Y. Vardi. Fixpoint logic vs. infinitary logic in finite model theory. In *Proc. 7th IEEE Symp. on Logic in Computer Science*, pages 46–57, 1992.

111. Ph. G. Kolaitis and M. Y. Vardi. 0–1 laws for fragments of second-order logic: an overview. In Y. N. Moschovakis (ed.), *Logic from Computer Science*, pages 265–286. Springer, 1992.

112. M. Kubierschky. *Remisfreie Spiele, Fixpunktlogiken und Normalformen*. Master's thesis, University of Freiburg i.Br., 1995.

113. M. Kubierschky. Yet another hierarchy theorem. *Journal of Symbolic Logic*, to appear.

114. J.-M. Le Bars. Fragments of existential second-order logic without 0–1 laws. In *Proc. 13th IEEE Symp. on Logic in Computer Science* 1998, to appear.

115. L. Libkin. On the forms of locality over finite models. In *Proc. 12th IEEE Symp. on Logic in Computer Science*, pages 204–215, 1997.

116. L. Libkin. On counting and local properties. In *Proc. 13th IEEE Symp. on Logic in Computer Science* 1998, to appear.

117. S. Lindell. An analysis of fixed point queries on binary trees. *Theoretical Computer Science* **85**, 75–95, 1991.

118. P. Lindström. First order predicate logic with generalized quantifiers. *Theoria* **32**, 186–195, 1966.

119. A. Livchak. The relational model for process controll. (Russian) *Automated Documentation and Mathematical Linguistics* **4**, 27–29, 1983.

120. L. Lovász and P. Gács. Some remarks on generalized spectra. *Zeitschrift für Mathematische Logik und Grundlagen der Mathematik* **23**, 547–554, 1977.

121. M. Mortimer. On languages with two variables. *Zeitschrift für Mathematische Logik und Grundlagen der Mathematik* **21**, 135–140, 1975.

122. Y. N. Moschovakis: *Elementary Induction on Abstract Structures.* North Holland, 1974.

123. W. Oberschelp. Asymptotic 0–1 laws in combinatorics. In D. Jungnickel (ed.), *Combinatorial theory, Lecture Notes in Mathematics* 969:276–292. Springer, 1982.

124. M. Otto. The expressive power of fixed-point logic with counting. *Journal of Symbolic Logic* **61**, 147–176, 1996.

125. M. Otto. *Bounded Variable Logics and Counting. A Study in Finite Models.* Lecture Notes in Logic 9. Springer, 1997.

126. L. Pacholski and W. Szwast. The 0–1 law fails for the class of existential second-order Gödel sentences with equality. In *Proceedings 30th IEEE Symp. on Foundations of Computer Science*, pages 160-163, 1989.

127. L. Pacholski, W. Szwast, and L. Tendera. Complexity of two-variable logic with counting. In *Proc. 12th IEEE Symp. on Logic in Computer Science*, pages 318–327, 1997.

128. C. H. Papadimitriou. *Computational Complexity.* Addison-Wesley, 1994.

129. C. H. Papadimitriou and M. Yannakakis. Optimization, approximation, and complexity. In *Proc. 20th ACM Symp. on Theory of Computing*, pages 229–234, 1988.

130. B. Poizat. Deux ou trois choses que je sais de L_n. *Journal of Symbolic Logic* **47**, 641–658, 1982.

131. D. Scott. A decision method for validity of sentences in two variables. *Journal of Symbolic Logic* **27**, 477, 1962.

132. S. Shelah and J. Spencer. Zero-one laws for sparse random graphs. *Journal of the AMS* **1**, 97–115, 1988.

133. J. R. Shoenfield: *Mathematical Logic.* Addison-Wesley, Reading, MA, 1967.

134. I. A. Stewart. Using the Hamiltonian path operator to capture NP. *Journal of Computer and System Sciences* **45**, 127–151, 1992.

135. A. Stolboushkin. Finitely monotone properties. In *Proc. 10th IEEE Symp. on Logic in Computer Science*, pages 324–330, 1995.

136. H. Straubing: *Finite Automata, Formal Logic, and Circuit Complexity.* Birkhäuser, 1994.

137. R. Szelepcsényi. The method of forced enumeration for nondeterministic automata. *Acta Informatica* **26**, 279–284, 1988.

138. W. Thomas. *Languages, Automata, and Logic.* Report 9607. Institute for Computer Science and Applied Mathematics. University of Kiel, 1996.

139. B. Trahtenbrot. The impossibility of an algorithm for the decision problem for finite domains. (Russian) *Doklady Academii Nauk SSSR* **70**, 569–572, 1950. English translation: *American Mathematical Society, Translations, Series 2* **23**, 1–5, 1963.

140. B. Trahtenbrot. Finite automata and the logic of monadic predicates. (Russian). *Sibirskij Mat. Zhurnal* **3**, 103-131, 1962. Extended abstract in *Doklady Academii Nauk SSSR* **140**, 326–329, 1961. English translation: *Soviet Physics. Doklady* **6**, 753–755, 1961.

141. J. Tyszkiewicz. On asymptotic probabilities of monadic second order properties. In E. Börger et al. (eds.), *CSL '91, Lecture Notes in Computer Science* 626:425–439. Springer, 1992.

142. M. Y. Vardi. The complexity of relational query languages. In *Proc. 14th ACM Symp. on Theory of Computing*, pages 137–146, 1982.

143. M. Y. Vardi. On the complexity of bounded-variable queries. In *Proc. 14th ACM Symp. on Principles of Data Base Systems*, pages 266–276, 1995.

144. M. Y. Vardi. What makes a modal logic so robustly decidable? In *Descriptive Complexity and Finite Models*. American Mathematical Society, 1997.

145. S. Vinner. A generalization of Ehrenfeucht's game and some applications. *Israel Journal of Mathematics* **12**, 279–298, 1972.

146. P. Winkler. Random structures and zero-one laws. In N. W. Sauer et al. (eds.), *Finite and infinite combinatorics in sets and logic, NATO Advanced Science Institutes Series*, pages 399–420. Kluver, 1993.

Index

τ, σ, 1

$R^{\mathcal{A}}$, $c^{\mathcal{A}}$, R^A, c^A, 1

$R^A a_1 \dots a_n$, 1

$\{<, S, \min, \max\}$, 3

τ_0, 3

$\mathcal{O}[\sigma]$, 3

$\mathcal{A}|\tau$, 3

$\mathcal{A} \cong \mathcal{B}$, 3

$\mathcal{A} \times \mathcal{B}$, 3

$\mathcal{A} \dot{\cup} \mathcal{B}$, $\mathcal{A} \lhd \mathcal{B}$, 4

$\mathcal{A}^I$, $\dot{\bigcup}_I \mathcal{A}$, $\lhd_I \mathcal{A}$, 4

$\|I\|$, 4

$\neg$, $\vee$, $\exists$, 4

$\wedge$, $\to$, $\leftrightarrow$, $\forall$, 5

$\varphi(x_1, \dots, x_n)$, 6

$\overline{x}$, $\varphi(\overline{x})$, 6

$\mathcal{A} \models \varphi[a_1, \dots, a_n]$, $\mathcal{A} \models \varphi$, 6

$\Phi \models \psi$, $\models \psi$, 6

$\Phi \models_{\text{fin}} \psi$, $\models_{\text{fin}} \psi$, 7

$\bigwedge \Phi$, $\bigvee \Phi$, 7

qr(φ), 7

Σ_n, Π_n, Δ_n, 7

$\varphi\frac{t_1 \dots t_n}{x_1 \dots x_n}$, $\varphi(t_1, \dots, t_n)$, 7

$\exists^{\geq n}$, 7

$\exists^{=n}$, $\exists^{\leq n}$, 8

$\varphi_{\geq n}$, $\varphi_{=n}$, $\varphi_{\leq n}$, 8

$\mathcal{A} \equiv \mathcal{B}$, 9

$\varphi^{\mathcal{A}}(\_)$, 10

$\mathcal{A} \equiv_m \mathcal{B}$, 14

$p \subseteq q$, $\overline{a} \mapsto \overline{b}$, 15

$\mathrm{G}_m(\mathcal{A}, \overline{a}, \mathcal{B}, \overline{b})$, 16

$\mathrm{G}_m(\mathcal{A}, \mathcal{B})$, 17

$\varphi^m_{\mathcal{A},\overline{a}}$, $\varphi^m_{\mathcal{A}}$, $\varphi^m_{\overline{a}}(\overline{v})$, 18

$W_m(\mathcal{A}, \mathcal{B})$, 20

$\mathcal{A} \cong_m \mathcal{B}$, $(I_j)_{j \leq m} : \mathcal{A} \cong_m \mathcal{B}$, 20

$\mathcal{G}_l$, 23

$\mathcal{B}_l$, $\mathcal{D}_l$, 24

$\mathrm{IT}^m(\mathcal{A}, \overline{a})$, 26

$S(r, a)$, 26

$\mathcal{A} \equiv^{\mathrm{MSO}}_m \mathcal{B}$, 38

MSO-$\mathrm{G}_m(\mathcal{A}, \mathcal{B})$, 38

$\psi^j_{\overline{a},\overline{P}}$, 38

(M)Σ^1_n, (M)Π^1_n, 39

Δ^1_n, 40

$\mathrm{L}_{\infty\omega}$, $\mathrm{L}_{\omega_1\omega}$, 40

$\bigvee \Psi$, 40

$\bigwedge \Psi$, 41

$\mathcal{A} \equiv^{\mathrm{L}_{\infty\omega}} \mathcal{B}$, 42

$\mathrm{G}_\infty(\mathcal{A}, \overline{a}, \mathcal{B}, \overline{b})$, 42

$\mathcal{A} \cong_{\text{part}} \mathcal{B}$, $I : \mathcal{A} \cong_{\text{part}} \mathcal{B}$, 43

$W_\infty(\mathcal{A}, \mathcal{B})$, 43

T_{rand}, 45

$\mathrm{L}^s_{\infty\omega}$, $\mathrm{L}^\omega_{\infty\omega}$, 47

$\mathcal{A} \equiv^s \mathcal{B}$, $\mathcal{A} \equiv^{\mathrm{L}^s_{\infty\omega}} \mathcal{B}$, $\mathcal{A} \equiv^s_m \mathcal{B}$, 48

$\overline{a}\frac{a}{i}$, 50

$\mathrm{G}^s_m(\mathcal{A}, \overline{a}, \mathcal{B}, \overline{b})$, $\mathrm{G}^s_\infty(\mathcal{A}, \overline{a}, \mathcal{B}, \overline{b})$, 50

$G_m^s(\mathcal{A},\mathcal{B})$, $G_\infty^s(\mathcal{A},\mathcal{B})$, 50
$\mathcal{A} \cong_m^s \mathcal{B}$, $(I_j)_{j\le m} : \mathcal{A} \cong_m^s \mathcal{B}$, 51
$\mathcal{A} \cong_{\text{part}}^s \mathcal{B}$, $I : \mathcal{A} \cong_{\text{part}}^s \mathcal{B}$, 51
$\psi_{\overline{a}}^m$, ${}^s\psi_{\mathcal{A},\overline{a}}^m$, $\psi_{\mathcal{A}}^m$, 52
$W_m^s(\mathcal{A},\mathcal{B})$, $W_\infty^s(\mathcal{A},\mathcal{B})$, 52
$\mathcal{A}/s$, 54
$r(\mathcal{A})$, 57
FO(C), $L_{\infty\omega}(C)$, $C_{\infty\omega}$, 59
$\exists^{\geq l}$, 59
$FO(C)^s$, $C_{\infty\omega}^s$, $C_{\infty\omega}^\omega$, 59
C-$G_m^s(\mathcal{A},\overline{a},\mathcal{B},\overline{b})$, C-$G_\infty^s(\mathcal{A},\overline{a},\mathcal{B},\overline{b})$, 60
$K_<$, 64
$L_n(K)$, $L_n(\varphi)$, $L_n(\tau)$, $l_n(K)$, 71
$l(K)$, $l_n(\varphi)$, $l(\varphi)$, 72
T_{rand}, 73
$l_n(K|H)$, $l(K|H)$, 74
$T_{\text{rand}}(\varphi_0)$, 76
$U_n(K)$, $u_n(K)$, $u(K)$, 77
$u_n(K|H)$, 77
$u(K|H)$, 78
$\Sigma_1^1(\forall^*\exists^*)$, $\Sigma_1^1(\exists^*\forall^*)$, 86
$\mathbb{A}$, $\mathbb{A}^*$, $\mathbb{A}^+$, 105
λ, 105
NDA, $\tilde{\delta}$, $L(M)$, 106
$|u|$, 107
L^+, 108
$\mathcal{B}_u$, 109
F_n, F_∞, F^φ, 120
$\mathcal{L}_1 \le \mathcal{L}_2$, $\mathcal{L}_1 \equiv \mathcal{L}_2$, $\mathcal{L}_1 < \mathcal{L}_2$, 122
$\mathcal{L}[\tau]$, 122
TC, DTC, 123
s_0, s_+, s_-, 125, 133
$|j|_r$, $|j|_r^n$, 131
co-$\mathcal{C}$, 155
$\tau_<$, $K_<$, 156
Str$[\tau]$, 158
PH, 159
$[\text{IFP}_{\overline{x},X}\varphi]$, 169
$[\text{GFP}_{\overline{x},X}\,\psi(\overline{x},X)]\,\overline{x}$, 172
$F_{(\infty)}^i$, 174
$\exists(\forall)x\psi(x)$, 179
$\delta_i^l(x_1,\ldots,x_l,v,w)$, 179
$\varphi(Y_-\overline{y})$, 179
$\#_x\varphi$, 208
$\mathcal{A} \equiv_m^{\text{M}} \mathcal{B}$, 213
MLFP-$G_m(\mathcal{A},\mathcal{B})$, 213
$\mathcal{A} \cong_m^{\text{M}} \mathcal{B}$, 217
TC-$G_m^r(\mathcal{A},\mathcal{B})$, 229
$\gamma \leftarrow \gamma_1,\ldots,\gamma_l$, 239
$(\tau,\Pi)_{\text{int}}$, $(\tau,\Pi)_{\text{ext}}$, 239
τ_{int}, τ_{ext}, 239
$P_{(n)}^i$, $P_{(\infty)}^i$, 240
$\mathcal{A}[\Pi]$, $(\Pi,P)\overline{t}$, 240
$\mathcal{A}[\Sigma]$, $(\Sigma,P)\overline{t}$, 241
$(\Pi,P)\overline{t}$, 246
$(\Pi,Q)\overline{t}$, 249
$\mathcal{L}[\tau]_0$, $\models^{\mathcal{L},\tau}$, 289
$\mathcal{L} \equiv_{\text{es}} \mathcal{C}$, 290
(M,f), 290
$\mathcal{L} \equiv_{\text{es}} \mathcal{C}$ on S, 296
$\mathcal{A}^\Pi$, $\psi^{-\Pi}$, 297
$Q_K\overline{x}_1,\ldots,\overline{x}_s[\psi_1(\overline{x}_1),\ldots,\psi_s(\overline{x}_s)]$, 308
$Q\overline{x}\psi(\overline{x})$, 309
$\mathcal{L}(Q_K)$, $\mathcal{L}(\mathbf{Q})$, 309
K^k, 312
$\mathbf{Q}_K^\omega$, 312
$L_{\infty\omega}^\omega(\mathbf{Q})$, 314
$K \le_{\mathcal{L}} L$, $K \le_{\mathcal{L}}^k L$, 321
rel-L, rel-k-L, 322
$Q_{\text{rel-}k\text{-}L}$, rel-$\omega$-$\mathbf{Q}_L$, 322
$\mathcal{L}[\mathbf{Q}]$, 323
(M,O), 331
$\mathcal{C}^{\mathcal{K}}$, $\mathcal{C}^L$, 331
$\mathcal{AB}$, 336

accept, 126, 132
acceptable, 126
acceptor, 162
almost all, 72
alphabet, 105
antitone, 167, 172, 192
approximable, 280
APX, 281
arity, 1
– of a quantifier, 309
arity hierarchy, 268, 272, 273
assignment, 6
ATC, 261
automaton, 106
– deterministic, 106
– nondeterministic, 106
axiomatizable
– in a logic, 124
– in FO, 14, 20, 150, 252
– in FO(DTC)(rel-ω-$\mathbf{Q}_O$), 333, 334
– in FO(IFP)(rel-ω-$\mathbf{Q}_O$), 333, 334
– in FO(M-LFP), 217
– in FO^s, 54
– in $L_{\omega_1\omega}$, 41
– in $L^s_{\infty\omega}$, 53
– in $L^\omega_{\infty\omega}$, 53
– in SO, 150

back property, 20
– s-, 51
ball, 26
– r-, 26
base of an instruction, 132
basic term, 248
Beth property, 63
Beth's Theorem, 63
blank letter, 125, 130
body of a clause, 239
bounded, 57, 58
– s-, 57
– fixed-point, 204
breadth
– of a DATALOG program, 242
– of an S-DATALOG program, 242

canonization
– $\mathcal{C}$-, 294
– $\mathcal{C}$- on S, 301
– LOGSPACE-, 294
– PSPACE-, 294
– PTIME-, 291–293
capture, 157
– a complexity class, 151, 288
– effectively strongly, 290, 295, 325
– effectively strongly on S, 296
– PTIME effectively strongly, 290, 292, 294, 300
– PTIME effectively strongly on GRAPH, 300
– strongly, 157, 288
cardinality, 4
cell, 125
– virtual, 125, 130
circuit
– Hamiltonian, 2, 113, 327
class
– bounded, 57
– elementary, 14
– elementary relative to, 14
– finitely axiomatizable, 14
– fixed-point bounded, 204
– free parametric, 80
– indiscernible, 222
– nontrivial free parametric, 80
– nontrivial parametric, 77
– of finite structures, 14
– of structures, 10
– parametric, 75
– s-bounded, 57
– s-rigid, 204
clause, 239
clique, 2, 113
closed subset, 222
closed under isomorphisms, 10
closure
– alternating transitive, 261
– deterministic transitive, 123, 220
– plus, 108
– positive, 108
– transitive, 123, 220
colour type, 61
Compactness Theorem, 8
compatible, 76
complement of a complexity class, 155
complete
– NPTIME- with respect to FO(DTC) reductions, 328
– NPTIME- with respect to logspace reductions, 327, 328
– strongly PTIME- with respect to FO-reductions, 325
– with respect to $\mathcal{C}$-reductions, 321
– with respect to $\mathcal{L}$-reductions, 324
Completeness Theorem, 8
complexity class
– as a class of ordered structures, 153

– in complexity theory, 153
component
– connected, 2, 30
– of a graph, 2
computable
– PTIME-, 204, 291
concatenation of languages, 107
configuration, 128, 134
– accepting, 134
– oracle, 331
– starting, 136
CONN, 2, 23, 122
connected component, 30
connective, 4
co-NPTIME, 157
consequence, 6
constant, 1
constant symbol, 1
cost, 276
counting quantifier, 59, 309
Craig property, 64
Craig's Theorem, 64
cut, 275
cycle, 2, 302

database, 119
– relational, 12
DATALOG, 239
– I-, 245
– P-, 248
– positive, 245, 250
– S-, 241
– stratified, 241
– totally defined WF-, 265
– well-founded, 264
– WF-, 264
– with counting, 248
DATALOG $\equiv$ E(LFP), 243, 245
DATALOG program, 239
DATALOG$_r$
– S-, 242
DATALOG(#), 248
decidability
– of validity for $\forall^2\exists^*$, 100
– of validity for FO^2, 98
decidable, 126
definable
– explicitly, 63
– implicitly, 63, 217
– in first-order logic, 114
degree of a point in a graph, 2
depth, 170
deterministic transitive closure, 123, 220

diameter of a graph, 85
digraph, 1
– acyclic, 24, 30, 253
distance function
– in a graph, 2, 23
– in an ordering, 22
distinct, 75
do(p), 15
domain, 1
DTC, 123, 220
duplicator, 16

E(LFP), 243
edge, 1
Ehrenfeucht's Theorem, 18
Ehrenfeucht-Fraïssé *see* game
element
– existential, 261
– universal, 261
enumerable, 126
enumeration
– effective of a complexity class, 295
equality symbol, 4
equivalence relation
– invariant, 107, 111
equivalent, 7
– elementarily, 9
– logically, 6
– $\mathrm{L}_{\infty\omega}$-, 42
– m-, 14
– m- in MSO, 38, 39
– on ordered structures, 152
EVEN, 199, 209
EVEN[τ], 21, 37, 41, 53, 59, 72, 156
existence of a fixed-point, 166
expression
– plus free regular, 114
– regular, 108
expressive power of a logic, 122
extension axiom, 45, 53, 72
extensional, 239

F (truth value), 7
FALSE, 7, 10, 11, 241
feasible solution, 276
finite model property, 95
– of $\forall^2\exists^*$, 100
– of $\exists^*\forall^2\exists^*$, 103
– of FO^2, 96, 99
– of modal logic, 99
fixed-point, 120, 166
– greatest, 167, 172
– least, 167

– simultaneous, 174
– simultaneous least, 175
fixed-point logic *see* logic, fixed-point
FO, 4
FO $\equiv$ S-DATALOG$_0$, 243
FO^s, 47
FO(ATC), 261, 314
FO(BFP), 235
FO(BFP) $<$ FO(LFP), 237
FO(BFP) $\equiv$ S-DATALOG, 242
FO(BFP$_r$), 236
FO(C), 59
FO(C)^s, 59
FO(DTC), 123, 327
FO(DTC) and LOGSPACE, 142, 148, 151, 296, 304
FO(DTC) $\leq$ FO(IFP^2), 273
FO(DTC) $\leq$ FO(LFP^2), 271
FO(DTC) $\leq$ FO(LFP^3), 270
FO(DTC) $<$ FO(PFP), 272
FO(DTC) $\leq$ FO(PFP^1), 271
FO(DTC) $\leq$ FO(PFP^2), 269
FO(DTC) $\not\equiv$ FO(TC), 223
FO(DTC)(rel-ω-$\mathbf{Q}_{\mathrm{HAM}_<}$), 334
FO(DTC)[rel-ω-$\mathbf{Q}_{\mathrm{HAM}_<}$], 328
FO(DTC^r), 268
FO(IFP), 121, 169
FO(IFP) and FO(PFP), 154
FO(IFP) and PTIME, 138, 149, 151, 297, 302
FO(IFP) $\equiv$ FO(S-IFP), 179
FO(IFP) $\equiv$ I-DATALOG, 246
FO(IFP) $\equiv \Sigma_1^1$, 155
FO(IFP) $\equiv$ SO, 155
FO(IFP^r), 268
FO(IFP, #), 208
FO(IFP, #) and PTIME, 305, 306, 315
FO(IFP, #) $\equiv$ DATALOG(#), 248
FO(IFP, #) $\equiv$ FO(PFP$_{\mathrm{PTIME}}$, #), 210
FO(LFP), 170
FO(LFP) $\leq \Delta_1^1$, 212
FO(LFP) $\equiv$ E(LFP), 244
FO(LFP) $\equiv$ FO(ATC), 261
FO(LFP) $\equiv$ FO(IFP), 173
FO(LFP) $\equiv$ FO(PFP), 203, 204
FO(LFP) $\equiv$ FO(PFP$_{\mathrm{PTIME}}$), 205, 206
FO(LFP) $\equiv$ FO(S-LFP), 179
FO(LFP) $\leq$ IMP(FO), 219
FO(LFP) $<$ $\mathrm{L}^\omega_{\infty\omega} \cap$ PTIME , 206
FO(LFP) $\equiv$ WF-DATALOG, 265
FO(LFP^r), 268
FO(LFP^1) $<$ FO(IFP^1), 273
FO(M-LFP), 212
FO(PFP), 121, 191
FO(PFP) and PSPACE, 137, 149, 151
FO(PFP) $\equiv$ FO(S-PFP), 193
FO(PFP) $\equiv$ IMP(FO(PFP)), 219
FO(PFP) $\leq \mathrm{L}^\omega_{\infty\omega}$, 199
FO(PFP) $\equiv$ P-DATALOG, 249
FO(PFP^r), 268
FO(PFP$_{\mathrm{PTIME}}$), 205
FO(PFP$_{\mathrm{PTIME}}$, #), 210
FO(PFP, #), 209
FO(PFP, #) $\leq \mathrm{C}^\omega_{\infty\omega}$, 209
FO(3-PFP), 198
FO(posDTC), 224
FO(posDTC) $\equiv$ FO(DTC), 224
FO(posTC), 143, 159, 224
FO(posTC) and NLOGSPACE, 143, 148
FO(posTC) $\equiv$ FO(TC), 226
FO(S-IFP), 175
FO(S-LFP), 176
FO(S-PFP), 192
FO(TC), 123, 159, 220
FO(TC) and NLOGSPACE, 143, 151
FO(TC) $<$ FO(BFP), 237
FO(TC) $<$ FO(LFP), 231
FO(TC) $\equiv$ S-DATALOG$_1$, 243
FO(TC^r), 221, 268
FO(TC^1) $\equiv$ MSO, 221
forest, 3
– <-, 59, 60
formula
– atomic, 5
– bounded DATALOG, 250
– bounded positive DATALOG, 250
– DATALOG, 240
– existential, 65
– existential FO(posTC), 225
– first-order, 4, 169
– I-DATALOG, 246
– $\mathrm{L}_{\infty\omega}$, 40, 42
– m-Hintikka, 18
– monotone in a relation, 67
– negative in X, 169
– nontrivial, 254
– normal, 95, 169
– P-DATALOG, 249
– positive DATALOG, 245
– positive in a relation, 67
– positive in X, 169
– S-DATALOG, 241
– s-Scott, 57
– totally defined, 194
– totally defined WF-DATALOG, 265

- universal, 65
- WF-DATALOG, 265
forth property, 20
- s-, 51
FO[τ], 5
frame, 99
Fraïssé's Theorem, 21
free parametric, 80
free(φ), 5
function
- antitone, 167, 172, 192
- inductive, 166, 175
- inflationary, 120, 166, 175
- monotone, 166, 175
- polynomial time computable, 190
function symbol, 11
function system
- inductive, 175
- inflationary, 175
- monotone, 175

Gaifman graph, 26
Gaifman's Theorem, 31
game
- associated with a digraph, 168, 172, 198
- Ehrenfeucht-Fraïssé for counting quantifiers, 60
- Ehrenfeucht-Fraïssé for FO, 16, 21
- Ehrenfeucht-Fraïssé for FO(M-LFP), 213
- Ehrenfeucht-Fraïssé for FO^s, 50
- Ehrenfeucht-Fraïssé for FO(TC), 229
- Ehrenfeucht-Fraïssé for $\mathrm{L}_{\infty\omega}$, 42
- Ehrenfeucht-Fraïssé for $\mathrm{L}^s_{\infty\omega}$, 50
- Ehrenfeucht-Fraïssé for MSO, 38
- Ehrenfeucht-Fraïssé with infinitely many moves, 42
- of life, 197
- pebble, 50, 60
global relation, 10, 12
- $\mathrm{L}^s_{\infty\omega}$-definable, 54
graph, 1
GRAPH, 1, 298, 322
graph
- acyclic, 2, 306
- acyclic connected, 306
- balanced, 112
- bipartite, 112, 113, 221, 223
- coloured, 61
- connected, 2, 23, 29, 40, 41, 85, 86, 122, 123, 169, 221, 309
- connected ordered, 23
- directed, 1
- Gaifman, 26
- of a structure, 26
- planar, 85
- random, 85
- rigid, 86
- stable coloured, 61
- undirected, 1
- with a Hamiltonian circuit, 113
grid, 302

Halting Problem, 127
HAM, 113, 327
Hamiltonian circuit, 2, 113, 327
Hamiltonian path, 2
Hanf's Theorem, 27
hard
- with respect to $\mathcal{C}$-reductions, 321
- with respect to $\mathcal{L}$-reductions, 324
head
- of a clause, 239
- of a Turing machine, 125
- read-and-write, 125, 130
- read-only, 130
HENK, 328
hierarchy
- arity, 268, 273
- arity for FO(PFP), 272
- polynomial, 159, 338
Hintikka formula
- m-, 18
hold
- almost surely, 72
- in almost all finite structures, 72
homomorphism, 34
- strict, 34
Horn sentence, 251

$\mathrm{IMP}(\mathrm{FO}) \leq \Delta^1_1$, 219
$\mathrm{IMP}(\mathrm{FO}(\mathrm{PFP})) \not\leq \mathrm{L}^\omega_{\infty\omega}$, 219
IMP($\mathcal{L}$), 218
in-degree, 2
index of an equivalence relation, 107
induction
- on the length of formulas, 5
- simultaneous, 178
- simultaneous for LFP and IFP, 179
- simultaneous for PFP, 193
inductive, 166, 175
inflationary, 120, 166, 175
input inscription, 131
input instance, 276
input tape, 127, 130

Instr(M), 125, 131
instruction, 125, 131
– oracle, 331
intentional, 239
interpolant, 64
interpolation property, 64
interpretation, 297, 321
– of σ in τ, 297, 321
invariant
– s- of a structure, 55, 200
invariantization
– $\mathcal{C}$- on S, 301
– PTIME-, 292–294, 301
– PTIME- on GRAPH, 301
isomorphic, 3
– m-, 20
– partially, 43
– s-m-, 51
– s-partially, 51
isomorphism, 3
– generalized partial, 213
– partial, 15
– s-partial, 50
isomorphism type, 18, 77
– m-, 18, 25, 26
– s-m-, 52

join
– simultaneous, 178

king, 97

language, 106, 126
– acceptable, 126
– accepted by a Turing machine, 126
– accepted by an NDA, 106
– decidable, 126
– decided by a Turing machine, 126
– definable by a Σ_1^1-sentence, 111
– definable in FO, 114, 116
– definable in monadic second-order logic, 109, 111
– enumerable, 126
– noncounting, 116
– plus free regular, 114
– recognized by an automaton, 111
– recognized by an NDA, 106, 111
– regular, 108, 110, 111
– ultimately periodic, 112
law
– 0–1 for FO, 77
– 0–1 for $L^\omega_{\infty\omega}$, 77
– labeled 0–1, 72
– labeled 0–1 for FO, 73
– labeled 0–1 for FO(PFP), 200
– labeled 0–1 for $L^\omega_{\infty\omega}$, 73
– labeled 0–1 for parametric classes, 77
– labeled 0–1 for Ψ, 77
– labeled 0–1 for $\Sigma_1^1(\exists^*\forall^*)$, 88
– unlabeled 0–1, 80
– unlabeled 0–1, failure for Σ_1^1-FO^2, 98
– unlabeled 0–1 for FO, 80
– unlabeled 0–1 for FO(PFP), 200
– unlabeled 0–1 for $L^\omega_{\infty\omega}$, 80
– unlabeled 0–1 for parametric classes, 80
– unlabeled 0–1 for $\Sigma_1^1(\exists^*\forall^*)$, 88
leaf, 3
Lemma
– Pumping, 107, 111
– Simultaneous Induction, 178
– Simultaneous Induction for LFP and IFP, 179
– Simultaneous Induction for PFP, 193
– Transitivity, 182
– Transitivity for LFP, 182
– Transitivity for PFP, 195
length of a path, 2
Lindström extension, 309
Lindström quantifier, 309
– vectorization of, 312
log, 132
log space-bounded, 160
logarithmic space
– deterministic, 132
– nondeterministic, 132
logic, 159, 289, 314
– alternating transitive closure, 261, 314
– bounded fixed-point, 235, 263
– closed under order-invariant sentences, 64
– deterministic transitive closure, 123
– existential fixed-point, 243
– first-order, 4, 13
– first-order with counting quantifiers, 59
– fixed-point with counting, 208, 248
– inductive fixed-point, 169
– infinitary, 40
– inflationary fixed-point, 121, 169, 313
– least fixed-point, 170, 263
– monadic least fixed-point, 212
– monadic second-order, 38, 110
– partial fixed-point, 121, 191, 263, 313
– partial fixed-point with counting, 209
– positive existential fixed-point, 245

– second-order, 37, 210
– simultaneous inflationary fixed-point, 175
– simultaneous least fixed-point, 176
– simultaneous partial fixed-point, 192
– stratified fixed-point, 235
– transitive closure, 123, 220, 263
logic program
– general, 239
logically equivalent, 6
logically valid, 6
LOGSPACE, 132
LOGSPACE and FO(DTC), 142, 148, 151, 296, 304
$\text{LOGSPACE}^{\mathcal{K}}$, 331
$\text{LOGSPACE}^{\text{NPTIME}}$, 332, 334
$\text{LOGSPACE}^{\text{NPTIME}} = \text{LOGSPACE}^{\text{HAM}_<}$, 332
$\text{LOGSPACE}^{\text{NPTIME}} = \text{LOGSPACE}^{\text{HENK}_<}$, 332
LOGSPACE^{O}, 334
Löwenheim-Skolem Theorem, 8

MAXCLIQUE, 284
MAXCUT, 275, 276, 284
MAX FO, 278
maximization, 276
MAXIMUM CONNECTED COMPONENT, 285
MAX PB, 278
MAX Π_n, 278
MAX Σ_n, 278
MAX_e Σ_1, 281
MAX_e $\Sigma_1 \subseteq$ APX, 281
method
– Ehrenfeucht-Fraïssé, 21
MIN FO, 278
MIN PB, 278
minimization, 276
MINVERTEX, 276, 277
Mod(Φ), 14
Mod(φ), 10
modal logic, 99
model
– minimal, 34, 35
– of a formula, 6
– random, 84
model checker, 147
model class, 10
model theory, 26
monotone, 166, 175
move, 16, 168, 213, 229
– $\forall$-, 25
– $\exists$-, 25
– in $\text{G}^s_m(\mathcal{A}, \bar{a}, \mathcal{B}, \bar{b})$, 50
– LFP, 213
– negative LFP, 213
– point, 38, 213, 229
– positive LFP, 213
– set, 38
– TC, 229
MSO, 38, 212
MSO-$\text{G}_m(\mathcal{A}, \mathcal{B})$, 38

NDA, 106
NLOGSPACE, 132
NLOGSPACE and FO(posTC), 143, 148
NLOGSPACE and FO(TC), 143, 151
NLOGSPACE=co-NLOGSPACE?, 161
noncounting, 116
nontrivial parametric, 75
normal form
– for FO(IFP), 138, 152, 189, 253, 255
– for FO(LFP), 174, 188, 254, 259
– for FO(PFP), 138, 152, 192, 194, 196, 197, 255
– for FO(posTC), 226
– for FO(TC), 228
– for FO^2-sentences, 96
– prenex, 7, 39
not, 4
NPSPACE, 126
NPTIME, 126, 132, 328, 329
NPTIME and Σ^1_1, 139, 150, 151, 157
NPTIME optimization problem, 276, 280
– first-order definable, 277
– polynomially bounded, 277
NPTIME=co-NPTIME?, 158
number part, 208
number sort, 208
number variable, 208
numerical invariant, 204

occurrence
– bound, 5
– free, 5, 6, 37, 41, 172
operation
– boolean, 111
opt_Q, 276
optimization problem
– first-order definable, 277
– NPTIME, 276, 280
– polynomially bounded, 277
or, 4

oracle answer state, 333
oracle configuration, 331
oracle instruction, 331
oracle machine
– (τ, σ), 330
oracle tape, 330
oracle transducer, 331
ORD, 88
order-invariant in the finite, 64, 156, 288
ordered representation, 156
ordering, 3, 88
– as $\{<, S, \min, \max\}$-structure, 3
– lexicographic, 131
– minimal, 219
– of even cardinality, 22
ordMod, 130
out-degree, 2
output tape, 162

parametric, 75
Part($\mathcal{A}$,$\mathcal{B}$), 15
partial isomorphism, 15
partially isomorphic, 43
path, 2
pebble, 50, 60
pebble game, 50, 60
PFP(k, l), 263
plus closure, 108
plus free regular, 114
point
– drawn, 168, 172, 192, 198
– of a graph, 1
– won, 168, 172, 192, 198
point sort, 208
point variable, 208
polynomial hierarchy, 159, 338
polynomial space, 126, 132
– nondeterministic, 126
polynomial time, 126, 132
– nondeterministic, 126, 132
positive closure, 108
Pow, 106
predecessor of a vertex, 2
prenex normal form, 7, 39
preservation
– under extensions, 66
– under homomorphisms, 34
– under strict homomorphisms, 34, 35
– under substructures, 65
preservation theorem, 65
probability
– labeled, 74
– labeled asymptotic, 72, 74, 79, 80, 89
– unlabeled, 77
– unlabeled asymptotic, 77–80
problem
– maximization, 276
– minimization, 276
product of structures, 3, 24
program
– DATALOG, 239
– general logic, 239
– I-DATALOG, 245
– P-DATALOG, 248
– positive DATALOG, 250
– S-DATALOG, 241
– stratified DATALOG, 241
– totally defined WF-DATALOG, 265
programming language, 152
projective Horn, 251–253
proof
– formal, 8
PSPACE, 126, 132, 150
PSPACE and FO(PFP), 137, 149, 151
PTAS, 284
PTIME, 126, 132, 261, 314, 315, 325
PTIME and FO(IFP), 138, 149, 151, 297, 302
PTIME and FO(IFP, #), 305, 306, 315
PTIME^{O}, 334
PTIME=NPTIME?, 155, 290
PTIME=PSPACE?, 153, 154, 204, 206
Pumping Lemma, 107, 111

quantifier, 309
– counting, 59, 309
– existential, 4, 309
– Lindström, 309
– simple, 309
– universal, 309
quantifier rank, 7
– for FO(M-LFP), 213
query
– PTIME-computable, 204
query language, 12

random structure
– infinite, 45
random structure theory, 45
rank
– s- of a structure, 57
rank function, 185
reducibility
– many-one, 320
reducible

– $\mathcal{C}$-, 320
– $\mathcal{C}$-many-one, 320
– $\mathcal{L}$-, 321
– logspace, 327
reduct, 3
refinement
– stable coloured, 62
regular expression, 108
regular language, 108
reject, 126
relation, 1
– defined in a structure, 10
– global, 10, 12
– global on a class, 10
– nontrivial, 254
– transitive closure, 11, 24
relation symbol, 1
– 0-ary, 241
– extensional, 239
– intentional, 239
relativization
– of a complexity class, 331
represent effectively, 281
rg(p), 15
RIG, 78, 83
rigid, 78
root, 3
round, 168
run, 125
– accepting, 126
– rejecting, 126

satisfaction relation, 6
satisfiable, 6
satisfy in a structure, 6
scattered
– l-, 30
S-DATALOG $\equiv$ S-DATALOG$_2$, 243
second-order logic, 37
– monadic, 38
section, 177
sentence, 5
– $\forall^2\exists^*$-, 100
– basic local, 31
– $\exists^*\forall^2\exists^*$-, 103
– existential positive, 34
– first-order, 5
– $L_{\infty\omega}$, 41
– $L_{\omega_1\omega}$-, positive in a relation, 67
– local, 31
– monadic Σ^1_1-, 89
– monotone in a relation, 67
– nontrivial free parametric, 89
– nontrivial parametric, 75, 86, 87
– order-invariant in the finite, 64, 156, 288
– parametric, 75
– positive in a relation, 67
– Σ^1_1-, 110
– $\Sigma^1_1(\forall^*\exists^*)$-, 86
– $\Sigma^1_1(\exists^*\forall^*)$-, 86, 87
– universal $L_{\omega_1\omega}$-, 67
– universal Horn, 251
Σ^1_1 and NPTIME, 139, 150, 151, 157
simple, 309
simultaneous join, 178
SO, 37
SO(PFP), 159
SO $\leq$ FO(PFP), 211
space bound for oracle machines, 331
space-bounded, 126
SPACE(n^e), 272
spectrum of a sentence, 46
spoiler, 16
square, 125
stage, 166, 199
stage comparison, 185
start of a Turing machine, 125, 131
starting configuration, 136
state, 131
– accepting, 106, 125, 131
– final, 106
– initial, 106, 125, 131
– of an automaton, 106
– of a Turing machine, 125
– oracle answer, 333
– rejecting, 125, 131
– starting, 131
State(M), 125, 131
Str, 291
Str[τ], 158
string, 105
strongly PTIME-complete, 325
structure, 1
– finite, 13
– indiscernible, 222
– infinite random, 45, 74
– input, 276
– labeled, 71
– numerical, 161
– of vocabulary τ, 1
– ordered, 3, 130
– random, 72
– rigid, 78
– s-rigid, 204
– τ-, 1

structures
– $C^s_{\infty\omega}$-equivalent, 61
– elementarily equivalent, 9
– equivalent in FO(M-LFP), 214
– FO^s-equivalent, 48, 56
– FO(TC)-equivalent, 230
– $L_{\infty\omega}$-equivalent, 42, 44
– $L^s_{\infty\omega}$-equivalent, 48, 52, 55, 57
– m-equivalent, 14, 19, 21
– m-equivalent in FO^s, 50, 52
– m-equivalent in $L^s_{\infty\omega}$, 50
– m-equivalent in MSO, 38
– m-isomorphic, 21
– s-m-isomorphic, 51, 52
– s-partially isomorphic, 51, 52
– partially isomorphic, 43, 44
successor
– of a configuration, 134
– of a vertex, 2
successor relation
– k-adic, 172
sum
– ordered, of m-equivalent structures, 24, 25
– ordered, of structures, 4, 24, 39
supp($\overline{a}$), 50
support, 50, 83

T (truth value), 7
tape, 125
– input, 127, 130
– oracle, 330
– output, 162
– work, 127, 130
τ-structure, 1
TC, 11, 24, 123, 220
term, 4
Theorem
– Beth's, 63
– compactness, 8
– completeness, 8
– Craig's, 64
– Ehrenfeucht's, 18
– Fraïssé's, 21
– Gaifman's, 31
– Hanf's, 27
– Löwenheim-Skolem, 8
– Trahtenbrot's, 8, 127
time-bounded, 126
tournament, 75
Trahtenbrot's Theorem, 8, 127
$T_{\text{rand}}(\varphi_0)$, 76, 84
transducer
– log space-bounded, 163
– oracle, 331
– polynomially time-bounded, 164
– space-bounded, 162
– (τ, σ)-, 162
transitive closure logic *see* logic, transitive closure
transition relation, 106
transitive closure, 24, 123, 220
– alternating, 261
transitive closure relation, 11
Transitivity Lemma, 182
– for LFP, 182
– for PFP, 195
tree, 3, 170, 305
– labeled binary, 206
TREE, 3, 305
tree-width, 171
true, 6
TRUE, 7, 10, 11, 241
Turing machine, 125
– deterministic, 125, 132
– for τ-structures, 130
– f space-bounded, 126, 132
– f time-bounded, 126, 132
– nondeterministic, 125, 132
– (τ, σ)-, 162
– with oracle tapes, 330
type
– m-isomorphism, 18, 25, 26
– r-ball, 27
– s-m-isomorphism, 52

ultimately periodic, 112
undecidability
– of finite satisfiability for FO, 127
– of order-invariance, 288
– of satisfiability in finite graphs, 129
– of the Halting Problem, 127
union
– disjoint, of structures, 4, 24, 39
– of m-equivalent structures, 24
– of structures, 4
universal algebra, 26
universe, 1

valid
– logically, 6
variable, 4
– free, 5, 6, 37, 41, 172
– predicate, 37
– relation, 37
vectorization

- of a class, 312
- of a Lindström quantifier, 312
- relativized of a class, 322
- relativized of a Lindström quantifier, 322

vertex, 1
vertex cover, 275
virtual letter, 125
vocabulary, 1
- containing function symbols, 11
- function-free, 11
- relational, 1

width of an interpretation, 297, 321
win
- a pebble game, 50
- $\mathrm{G}_m(\mathcal{A},\mathcal{B})$, 19, 21
- $\mathrm{G}_m(\mathcal{A},\mathcal{B})$, 17
- $\mathrm{G}_m(\mathcal{A},\overline{a},\mathcal{B},\overline{b})$, 18, 21
- $\mathrm{G}_\infty(\mathcal{A},\mathcal{B})$, 44
- $\mathrm{G}_\infty(\mathcal{A},\overline{a},\mathcal{B},\overline{b})$, 42, 43
- $\mathrm{G}^s_m(\mathcal{A},\mathcal{B})$, 52
- $\mathrm{G}^s_m(\mathcal{A},\overline{a},\mathcal{B},\overline{b})$, 52
- $\mathrm{G}^s_\infty(\mathcal{A},\mathcal{B})$, 52
- $\mathrm{G}^s_\infty(\mathcal{A},\overline{a},\mathcal{B},\overline{b})$, 52
- MSO-$\mathrm{G}_m((\mathcal{A},\overline{P},\overline{a}),(\mathcal{B},\overline{Q},\overline{b}))$, 38

winning position, 20, 43, 52
winning strategy, 17
witness
- negative, 263
- positive, 263
- strongly, 148, 153

word, 105
- empty, 105

word model, 109, 221
work tape, 127, 130

Springer **M***onographs in* **M***athematics*

This series publishes advanced monographs giving well-written presentations of the "state-of-the-art" in fields of mathematical research that have acquired the maturity needed for such a treatment. They are sufficiently self-contained to be accessible to more than just the intimate specialists of the subject, and sufficiently comprehensive to remain valuable references for many years. Besides the current state of knowledge in its field, an SMM volume should also describe its relevance to and interaction with neighbouring fields of mathematics, and give pointers to future directions of research.

Abhyankar, S.S. **Resolution of Singularities of Embedded Algebraic Surfaces** 2nd enlarged ed. 1998
Alexandrov, A.D. **Convex Polyhedra** 2005
Andrievskii, V.V.; Blatt, H.-P. **Discrepancy of Signed Measures and Polynomial Approximation** 2002
Angell, T. S.; Kirsch, A. **Optimization Methods in Electromagnetic Radiation** 2004
Ara, P.; Mathieu, M. **Local Multipliers of C*-Algebras** 2003
Armitage, D.H.; Gardiner, S.J. **Classical Potential Theory** 2001
Arnold, L. **Random Dynamical Systems** corr. 2nd printing 2003 (1st ed. 1998)
Arveson, W. **Noncommutative Dynamics and E-Semigroups** 2003
Aubin, T. **Some Nonlinear Problems in Riemannian Geometry** 1998
Auslender, A.; Teboulle M. **Asymptotic Cones and Functions in Optimization and Variational Inequalities** 2003
Bang-Jensen, J.; Gutin, G. **Digraphs** 2001
Baues, H.-J. **Combinatorial Foundation of Homology and Homotopy** 1999
Brown, K.S. **Buildings** 3rd printing 2000 (1st ed. 1998)
Cherry, W.; Ye, Z. **Nevanlinna's Theory of Value Distribution** 2001
Ching, W.K. **Iterative Methods for Queuing and Manufacturing Systems** 2001
Crabb, M.C.; James, I.M. **Fibrewise Homotopy Theory** 1998
Chudinovich, I. **Variational and Potential Methods for a Class of Linear Hyperbolic Evolutionary Processes** 2005
Dineen, S. **Complex Analysis on Infinite Dimensional Spaces** 1999
Dugundji, J.; Granas, A. **Fixed Point Theory** 2003
Ebbinghaus, H.-D.; Flum J. **Finite Model Theory** 2006
Elstrodt, J.; Grunewald, F. Mennicke, J. **Groups Acting on Hyperbolic Space** 1998
Edmunds, D.E.; Evans, W.D. **Hardy Operators, Function Spaces and Embeddings** 2004
Fadell, E.R.; Husseini, S.Y. **Geometry and Topology of Configuration Spaces** 2001
Fedorov, Y.N.; Kozlov, V.V. **A Memoir on Integrable Systems** 2001
Flenner, H.; O'Carroll, L. Vogel, W. **Joins and Intersections** 1999
Gelfand, S.I.; Manin, Y.I. **Methods of Homological Algebra** 2nd ed. 2003
Griess, R.L. Jr. **Twelve Sporadic Groups** 1998
Gras, G. **Class Field Theory** corr. 2nd printing 2005
Hida, H. ***p*-Adic Automorphic Forms on Shimura Varieties** 2004
Ischebeck, F.; Rao, R.A. **Ideals and Reality** 2005
Ivrii, V. **Microlocal Analysis and Precise Spectral Asymptotics** 1998
Jech, T. **Set Theory** (3rd revised edition 2002)
Jorgenson, J.; Lang, S. **Spherical Inversion on SLn (R)** 2001
Kanamori, A. **The Higher Infinite** corr. 2nd printing 2005 (2nd ed. 2003)
Kanovei, V. **Nonstandard Analysis, Axiomatically** 2005
Khoshnevisan, D. **Multiparameter Processes** 2002
Koch, H. **Galois Theory of *p*-Extensions** 2002
Komornik, V. **Fourier Series in Control Theory** 2005
Kozlov, V.; Maz'ya, V. **Differential Equations with Operator Coefficients** 1999
Landsman, N.P. **Mathematical Topics between Classical & Quantum Mechanics** 1998
Leach, J.A.; Needham, D.J. **Matched Asymptotic Expansions in Reaction-Diffusion Theory** 2004
Lebedev, L.P.; Vorovich, I.I. **Functional Analysis in Mechanics** 2002
Lemmermeyer, F. **Reciprocity Laws: From Euler to Eisenstein** 2000

Malle, G.; Matzat, B.H. **Inverse Galois Theory** 1999
Mardesic, S. **Strong Shape and Homology** 2000
Margulis, G.A. **On Some Aspects of the Theory of Anosov Systems** 2004
Murdock, J. **Normal Forms and Unfoldings for Local Dynamical Systems** 2002
Narkiewicz, W. **Elementary and Analytic Theory of Algebraic Numbers** 3rd ed. 2004
Narkiewicz, W. **The Development of Prime Number Theory** 2000
Parker, C.; Rowley, P. **Symplectic Amalgams** 2002
Peller, V. (Ed.) **Hankel Operators and Their Applications** 2003
Prestel, A.; Delzell, C.N. **Positive Polynomials** 2001
Puig, L. **Blocks of Finite Groups** 2002
Ranicki, A. **High-dimensional Knot Theory** 1998
Ribenboim, P. **The Theory of Classical Valuations** 1999
Rowe, E.G.P. **Geometrical Physics in Minkowski Spacetime** 2001
Rudyak, Y.B. **On Thom Spectra, Orientability and Cobordism** 1998
Ryan, R.A. **Introduction to Tensor Products of Banach Spaces** 2002
Saranen, J.; Vainikko, G. **Periodic Integral and Pseudodifferential Equations with Numerical Approximation** 2002
Schneider, P. **Nonarchimedean Functional Analysis** 2002
Serre, J-P. **Complex Semisimple Lie Algebras** 2001 (reprint of first ed. 1987)
Serre, J-P. **Galois Cohomology** corr. 2nd printing 2002 (1st ed. 1997)
Serre, J-P. **Local Algebra** 2000
Serre, J-P. **Trees** corr. 2nd printing 2003 (1st ed. 1980)
Smirnov, E. **Hausdorff Spectra in Functional Analysis** 2002
Springer, T.A. Veldkamp, F.D. **Octonions, Jordan Algebras, and Exceptional Groups** 2000
Sznitman, A.-S. **Brownian Motion, Obstacles and Random Media** 1998
Taira, K. **Semigroups, Boundary Value Problems and Markov Processes** 2003
Talagrand, M. **The Generic Chaining** 2005
Tauvel, P.; Yu, R.W.T. **Lie Algebras and Algebraic Groups** 2005
Tits, J.; Weiss, R.M. **Moufang Polygons** 2002
Uchiyama, A. **Hardy Spaces on the Euclidean Space** 2001
Üstünel, A.-S.; Zakai, M. **Transformation of Measure on Wiener Space** 2000
Vasconcelos, W. **Integral Closure. Rees Algebras, Multiplicities, Algorithms** 2005
Yang, Y. **Solitons in Field Theory and Nonlinear Analysis** 2001